Recent Trends
in Regeneration
Research

NATO ASI Series

Advanced Science Institutes Series

A series presenting the results of activities sponsored by the NATO Science Committee, which aims at the dissemination of advanced scientific and technological knowledge, with a view to strengthening links between scientific communities.

The series is published by an international board of publishers in conjunction with the NATO Scientific Affairs Division

A	**Life Sciences**	Plenum Publishing Corporation
B	**Physics**	New York and London
C	**Mathematical and Physical Sciences**	Kluwer Academic Publishers Dordrecht, Boston, and London
D	**Behavioral and Social Sciences**	
E	**Applied Sciences**	
F	**Computer and Systems Sciences**	Springer-Verlag
G	**Ecological Sciences**	Berlin, Heidelberg, New York, London,
H	**Cell Biology**	Paris, and Tokyo

Recent Volumes in this Series

Volume 166—Vascular Dynamics: Physiological Perspectives
  edited by N. Westerhof and D. R. Gross

Volume 167—Human Apolipoprotein Mutants 2: From Gene Structure to
  Phenotypic Expression
  edited by C. R. Sirtori, G. Franceschini,
  H. B. Brewer, Jr., and G. Assmann

Volume 168—Techniques and New Developments in Photosynthesis Research
  edited by J. Barber and R. Malkin

Volume 169—Evolutionary Tinkering in Gene Expression
  edited by Marianne Grunberg-Manago, Brian F. C. Clark,
  and Hans G. Zachau

Volume 170—*ras* Oncogenes
  edited by Demetrios Spandidos

Volume 171—Dietary $\omega 3$ and $\omega 6$ Fatty Acids: Biological Effects
  and Nutritional Essentiality
  edited by Claudio Galli and Artemis P. Simopoulos

Volume 172—Recent Trends in Regeneration Research
  edited by V. Kiortsis, S. Koussoulakos, and H. Wallace

Series A: Life Sciences

Recent Trends in Regeneration Research

Edited by
V. Kiortsis and S. Koussoulakos
University of Athens
Athens, Greece

and
H. Wallace
University of Birmingham
Birmingham, United Kingdom

Plenum Press
New York and London
Published in cooperation with NATO Scientific Affairs Division

Proceedings of a NATO Advanced Research Workshop on
Recent Trends in Regeneration Research,
held September 19–23, 1988,
in Saronis, Athens, Greece

Library of Congress Cataloging in Publication Data

NATO Advanced Research Workshop on Recent Trends in Regeneration
Research (1988: Athens, Greece)
Recent trends in regeneration research / edited by V. Kiortsis and S.
Koussoulakos and H. Wallace.
        p.   cm. — (NATO ASI series. Series A, Life sciences: vol. 172.)
    "Proceedings of a NATO Advanced Research Workshop on Recent Trends in
Regeneration Research, held September 19–23, 1988, in Saronis, Athens,
Greece"—T.p. verso.
    "Published in cooperation with NATO Scientific Affairs Division."
    Bibliography: p.
    Includes indexes.

    1. Regeneration (Biology)—Congresses. I. Kiortsis, V. (Vassili), 1925-   II.
Koussoulakos, S. III. Wallace, H. (Hugh). IV North Atlantic Treaty Organization.
Scientific Affairs Division. V. Title. VI. Series: NATO ASI series. Series A, Life
sciences; v. 172.
QH499.N32   1988   89-8444
591.3′1—dc20   CIP

ISBN 978-1-4684-9059-6   ISBN 978-1-4684-9057-2 (eBook)
DOI 10.1007/978-1-4684-9057-2
© 1989 Plenum Press, New York
A Division of Plenum Publishing Corporation
233 Spring Street, New York, N.Y. 10013
Softcover reprint of the hardcover 1st edition
All rights reserved

PREFACE

Regeneration, i.e. the replacement of lost body parts by new outgrowths or by remodelling existing tissues, has been studied for centuries. However, in recent years important developments took place in this field too, owing to new sophisticated techniques and to novel theoretical concepts. Advances in Molecular Genetics, Biochemistry, Cell - and Neurobiology, Immunology, to mention a few of them, are the main causes of this resurgence of interest in regeneration. As a consequence, more and more meetings and publications are devoted, either exclusively or for a large part, to basic and applied research of regenerative processes. "Regeneration-ists" scattered in laboratories all over the world and accustomed to know each other through exchange of reprints - occasionally an encounter in a large conference - tend now to form small groups, even societies and to institutionalize their meetings.

Although the critical mass of scientists involved in regeneration research does not seem yet to be reached, for an autonomous development of this sector, regular and frequent meetings of experts appear useful, even necessary.

Such a meeting was convened in Saronis, near Athens, Greece, from 19 to 23 September 1988 and sponsored by the NATO Science Committee and the University of Athens. The present volume contains the contributions to this Advanced Research Workshop on "Recent Trends in Regeneration Research".

About 50 biologists from different countries, either members of the Alliance or outside it (U.R.S.S., India, Egypt, Switzerland, Sweden) took part, mostly as invited speakers. They presented the latest results of their investigations, critically reviewed previous works, exchanged ideas with their colleagues and discussed controversial points of view. Other participants contributed additional papers, mainly in a poster form. Those are marked with an asterisk in the text.

Each platform presentation was followed by formal and informal discussions, to which many participants took an active part. The meeting was crowned by a General Discussion which included recommendations and proposals for future research.

It is a real challenge to deal judiciously with all the various facets of the subject, during the few days of the workshop. Out of necessity, we had to limit ourselves to crucial issues, leaving out peripheral yet important aspects of regeneration. The same was true for topics treated recently in meetings and books, such as neural regeneration. Another limiting factor was the freedom granted to all authors to deal only with topics belonging to their special competence. The consequence of compromising between our choices and limitations is reflected in the six major sections of the book: Gene expression; Cellular and immunological aspects; Skeletal and cardiac muscle; Regeneration and the nervous system; Retinoids; Pattern formation and morphogenesis. It should be noticed also that vertebrates (mostly amphibians) were favoured against invertebrates and that many authors focused on the urodele limb as the system of choice for experimental work. No doubt that other aspects of animal regeneration and repair merit equal attention; on the other hand no one can deny that the topics chosen are in the frontline of modern research.

Convening a group of experts, across national and continental borders, and editing a book to become more than a mere transcript of the proceedings is not an easy task. To meet this challenge the arrangement of the text was planned as follows: All papers belonging to a main subject were grouped under that heading; they were followed by short versions of related posters. At the end of each chapter a commentary, written by an authority, presents a record of the discussions and expresses some personal views on the actual state, the controversies, the problems and the prospects of research in that particular area. A record of the general discussion, rendering also the atmosphere of the meeting, is given in a final commentary written by one of us (H.W.).

As organisers of the ARW and editors of this volume we wish to thank the NATO Scientific Affairs Division and its Programme Director Dr. C. Sinclair for sponsoring the meeting and the publication of the proceedings. The support of the University of Athens is kindly acknowledged. Thanks are due to all participants, especially the chairpersons of the sessions and authors of the commentaries for their stimulating contributions. As one of us (S.K.), assisted by his wife, carried out most of the organizational work, the other two express their gratitude to him and her. The help of Dr. C. Zafeiratos, of Betty Michael, who typed the manuscript and of the staff of "Plenum Press" is acknowledged.

V. Kiortsis
S. Koussoulakos
H. Wallace

CONTENTS

MUSCLE

NEURAL ASPECTS

RETINOIDS

PATTERN FORMATION

Contributions labeled by an (*) were presented at the Meeting
mainly in poster form.

AN OVERVIEW OF THE HISTORICAL ORIGIN OF THE NERVE

INFLUENCE ON LIMB REGENERATION

Marcus Singer and Jacqueline Géraudie

Department of Anatomy, School of Medicine
Case Western Reserve University
Cleveland, Ohio 44106, U.S.A.

SUMMARY

The interest in the nervous control of limb regeneration in the Amphibia stems from the original finding of Todd in 1823. This research done at Naples in 1823 was largely unsupported for 100 years. The present paper speculates on the reason for Todd's experiment against the background of arguments of the day on the neurotrophic activity of the nervous system, derangements of which were believed to cause all the diseases of the body (Cullen, 1776; see Rath, 1959). This paper also touches on modern developments in the study of nervous control of regeneration, all of which stem from Todd's original findings.

In 1823, an English medical doctor, Tweedy John Todd, went to Naples, ostensibly on a holiday, where he repeated the experiments on limb regeneration of Spallanzani. But, in addition to affirming Spallanzani's observations that the limb of the newt (salamander) completely regenerates, he cut the brachial nerves of the limb at the time of amputation or shortly after regeneration was initiated. He observed that in consequence of denervation, the stump did not regenerate and if regeneration had been initiated, it stopped and the young bud was resorbed. An important question not discussed in his brief report is: What induced Todd to denervate the stump? There is no absolute answer to this question. But we will speculate upon it, since it was a landmark experiment; the results of which are pursued to this day.

If one looks at the experiment against the background of scientific thinking of that day, it is possible to understand the reason for it. Johannes Mueller (1843), the father of neurophysiology, speculated upon the nutritive power of nerves and considered it the first solid demonstration of the so-called "neurotrophic" phenomenon. At that time, there was a controversy between two schools on the importance of nerves in diseases of organs of the body. There was the Edinburgh school of William Cullen, the famous anatomist and medical doctor, who asserted that all bodily diseases were due to derangements of the nervous system. Although somewhat akin to modern psychosomatic medicine, this theory stemmed from the ancient belief that vital spirits or nutritive substances were carried in the nerve. It had many advocates for many years who supported Cullen's views. This theory was disputed by Virchow (1855), the powerful father of modern pathology, who extended his cellular theory to the pathology of cells and organs. He asserted, instead, that each organ of the body, including the nervous system, has its own peculiar diseases. So powerful was Virchow's voice denying a neurotrophic control that it won out, a circumstance reviewed well in the works by Rath (1959), Riese (1958) and Mueller, except for Todd's experiment on the newt; which Mueller, in his textbook of 1843, remarked that if the experiments on limb regeneration of the newt were true they were a real demonstration of a neurotrophic phenomenon, although he did not then know who had done the experiment: "The reproduction of an extremity in the salamander is prevented by its nerve being divided at a second point above the surface of the stump (?)".

Elsewhere in his book, Mueller was careful to cite the authors of their research. Obviously Mueller had heard about it second-hand from someone, perhaps Todd himself, who may have visited his laboratory, as did many scientists travelling to Berlin; or possibly from someone else in his many sojourns to other laboratories in Germany and abroad. The term neurotrophic, then used to mean the effect of nerves on other tissues, has been preempted in recent years to mean the effect of NGF on nerve growth. This term is implied but not used in the work by Mueller. T.H. Morgan was unaware of Todd's experiment in his monumental book on regeneration (1901) and later when he attempted to induce regeneration in non-regenerating forms [the adult frog] (1906-1907).

It took about 100 years for Todd's findings to be generally accepted, but then it initiated considerable experimentation; and the name of the late Dr. Schotté, working then in the laboratory of Guyénot in Geneva, must be remembered in this respect (Schotté, 1926; Guyénot et Schotté, 1926; Schotté and Butler, 1941). The experiments were largely morphological and continue to this day. These various works are very well reviewed and assessed in the excellent book on regeneration by Wallace (1981), a book which also reviews some of the modern developments in this field (see also Sicard, 1985).

Successful induction of regeneration in the postmetamorphic (Singer, 1954) frog occurred as a direct line of research stemming from Todd's findings and the subsequent analysis that axon number, particularly the sensory components, was essential to induce regeneration. Such an experiment was also possible in the lizard limb which was then induced to regenerate by increasing the nerve supply at the amputation surface (Simpson, 1961; Singer, 1961).

Rzehak and Singer (1956) demonstrated in Anurans that it is not only the number of nerve fibers which is important in inducing regeneration, but rather the amount of axoplasm. Recently, Sisken and colleagues (Sisken et al., 1988; Fowler and Sisken, 1982) have extended these inductions to the chick.

2

The mammal is capable of partial regrowth of the fingers as has been shown for the human by the remarkable studies of Illingworth (1974). Our recent studies on the monkey (Singer et al., 1987), those of Borgens (1982) on the mouse, and Mizell and Isaacs (1970) in the opossum supports this view, although the role of nerves in the regrowth has yet to be established. Moreover, it appears that such regrowth, at least in the monkey, is not epimorphic, thus differing from amphibian regeneration (Singer et al., 1987).

It is interesting to remember that Thomas Hunt Morgan spent his early years in research on regeneration embodied in his book "Regeneration", (1901) and embryonic development. He left these pursuits because he felt that classical experimental methods could not answer the fundamental questions of the how and why of the developmental and regeneration processes. It is more interesting to note that his book, "Theory of the Gene", (1929) has eventually led back to applications in development and regeneration.

Modern methods in biology are now providing some tools for fundamental analysis of regeneration and especially the control by the nervous system. Such new work is embodied in various cytogenetic and molecular work of studies, such as the ones by Brockes (1984) and his colleagues (Fekete and Brockes, 1987, 1988, Kintner and Brockes, 1984, 1985), Globus (1988), Vethamany - Globus (1988), Tassava (1988), Mescher and Munaim (1986), Maden (1983, 1985), Stocum and colleagues (Stocum and Thoms, 1984; Stocum and Crawford, 1987) among many others (see also some articles in this book).

It is not chimerical to predict that in a few years the neurotrophic substance(s) will be isolated and provide important studies for diseases stemming from nervous derangements; including muscular dystrophy, amyotrophic lateral sclerosis and multiple sclerosis since it seems likely at the moment that demyelination disorders also are caused by a lack of trophic control by the axon.

ACKNOWLEDGEMENTS

The authors are grateful for the generous support by the Monsanto Corporation, St. Louis, Mo.; and to Gwyn Jackson, Department of Anatomy, for her help with typing and editing, and Thomas Tumbry, Jr., Office of Medical Education, for his expertise in the use of the computer.

REFERENCES

Borgens, R.B., 1982, Mice regrow the tips of their foretoes, Science, 217:747-750.

Brockes, J.P., 1984, Mitogenic growth factors and nerve dependence of limb regeneration, Science, 225:1280-1287.

Cullen, W., 1776, "Practice of Physic", Elliot and Cardell, Publ., Edinborough.

Fekete, D.M. and Brockes, J.P., 1987, A monoclonal antibody detects a difference in the cellular composition of developing and regenerating limb of newts, Development, 99:589-602.

Fekete, D.M. and Brockes, J.P., 1988, Evidence that the nerve controls molecular identity of progenitor cells for limb regeneration, Development, 103:567-673.

Fowler, I. and Sisken, B.F.J., 1982, Effect of augmentation of nerve supply upon limb regeneration in the chick embryo, J. Exp. Zool., 221:49-59.

Globus, M., 1988, A neuromitogenic role for substance p in urodele limb regeneragion, In: "Regeneration and Development", S. Inoue, ed., Okada Printing and Publishing Company, Maebashi, Japan.

Guyénot, E. and Schotté, O.E., 1926, Le rôle du systeme nerveux dans l'édification des régénérats de pattes chez les urodèles, C.R. Soc. Phys. Hist. Nat. Genève, 43:32-36.

Kintner, C.R. and Brockes, J.P., 1984, Monoclonal antibodies identify blastemal cells derived from dedifferentiating muscle in newt limb regeneration, Nature, 308:67-69.

Kintner, C.R. and Brockes, J.P., 1985, Monoclonal antibodies cells of a regenerating limb, J. Embryol. Exp. Morph., 89:37-55.

Illingworth, C.M., 1974, Trapped fingers and amputated finger tips in children, J. Pediatr. Surg., 9:853-858.

Maden, M., 1983, The effect of vitamin A on limb regeneration in Rana temporaria, Dev. Biol., 98:409-416.

Maden, M., 1985, Retinoids and the control of pattern in regenerating limbs, Ciba Found. Symp., 113:132-155.

Mescher, A.L. and Munaim, S.I., 1986, Changes in the extracellular matrix and glycosaminoglycan synthesis in adult newt forelimbs, Anat. Rec., 214:424-431.

Mizell, M. and Isaacs, J.J., 1970, Induced regeneration of hindlimbs in the newborn opossum, Am. Zool., 10:141-153.

Morgan, T.H., 1901, "Regeneration", Macmillan, New York.

Morgan, T.H., 1906-1907, The extent and limitations of the power to regenerate in man and other vertebrates, The Harvey Lectures (1905-1906), Lipincott, Philadelphia.

Morgan, T.H., 1929, "The Theory of the Gene", 2nd printing, 1964, Hafner Publishing Company, New York.

Mueller, J., 1843, "Elements of Physiology", Arranged for the second London edition by John Belt, Translated from the German by William Maly, Lea and Rather Standard, Stanford University Press.

Rath, G., 1959, A pathogenetic concept of the 18th and 19th centuries, Bull. Hist. Med., 33:526-541.

Riese, W., 1958, Descartes idea of brain function, In: "The Brain and Its Functions:, an Anglo-American symposium, July 1957, C.C. Thomas and Co., Springfield, Illinois.

Rzehak, K. and Singer, M., 1956, Limb regeneration and nerve fiber numbers in Rana sylvatica and Xenopus laevis, J. Exp. Zool., 162:15-22.

Schotté, O.E., 1926, Système nerveux et régénération chez le triton, Action globale des nerfs, Rév. Suisse Zool., 33:1-211.

Schotté, O.E. and Butler, E.G., 1941, Morphological effects of denervation and amputation of limbs in urodele larvae, J. Exp. Zool., 87:279-322.

Sicard, R.E., 1985, "Regulation of Vertebrate Limb Regeneration", R.E. Sicard, ed., Oxford University Press, New York, Oxford.

Simpson, S.B., 1961, Induction of limb regeneration in the lizard, *Lygosoma laterale*, by augmentation of the nerve supply, *Proc. Soc. Exp. Biol. Med.*, 107:108-111.

Singer, M., 1952, The influence of the nerve in regeneration of the amphibian extremity. *Quart. Rev. Biol.*, 27:169-200.

Singer, M., 1954, Induction of regeneration of the forelimb of the postmetamorphic frog by augmentation of the nerve supply, *J. Exp. Zool.*, 126:419-471.

Singer, M., 1961, Induction of regeneration of body parts in the lizard, *Anolis*, *Proc. Soc. Exp. Biol. Med.*, 107:106-108.

Singer, M., Weckesser, E.C., Geraudie, J., Maier, C.E. and Singer, J., 1987, Open finger tip healing and replacement after distal amputation in Rhesus monkey with comparison to limb regeneration in lower vertebrates, *Anat. Embryol.*, 177:29-36.

Sisken, B.F., Fowler, I. and Maley, B.E., 1988, Ultrastructural characteristics of neural tube implants in regenerating and non-regenerating chick limbs, *In*: "Regeneration and Development", S. Inoue, ed., Okada Printing and Publishing Company, Maebashi, Japan.

Spallanzani, L., 1768, "Prodromo da un opera da imprimersi sopra le riproduzioni animali", Modena.

Stocum, D.L., Crawford, K., 1987, Use of retinoids to analyze the cellular basis of positional memory in regenerating amphibian limbs, *Biochem. Cell Biol.*, 65:750-761.

Stocum, D.L., Thoms, S.D., 1984, Retinoic-acid-induced pattern completion in regenerating double anterior limbs of urodeles, *J. Exp. Zool.*, 232:207-215.

Tassava, R.A., 1988, Limb regeneration in newts: an immunological search for developmentally significant antigens, *In*: "Regeneration and Development", S. Inoue, ed., Okada Printing and Publishing Company, Maebashi, Japan.

Todd, J.T., 1823, On the process of reproduction of the members of the aquatic salamander, *Quart. J. Sci. Lit. Arts*, 16:84-96.

Vethamany - Globus, S., 1988, A critical analysis of hormone action in newt limb regeneration: a possible opiate (B-endorphin) connection, *In*: "Regeneration and development", S. Inoue, ed., Okada Printing and Publishing Company, Maebashi, Japan.

Virchow, R., 1855, Diseases, life and man, *In*: "Cellular Pathology", translated by Lelard J. Rather, Stanford University Press.

Wallace, H., 1981, "Vertebrate limb regeneration", J. Wiley and Sons, New York.

GENE EXPRESSION

IN

REGENERATION

IDENTIFICATION OF A HOMEOBOX GENE AND A KERATIN PAIR
EXPRESSED IN AMPHIBIAN LIMB REGENERATION

J.B. Brockes, P. Ferretti and P. Savard

Ludwig Institute for Cancer Research
91 Riding House Street
London W1P 8BT, England

SUMMARY

We have identified a newt homeobox gene called NvHbox1 which is the homologue of Xenopus Hbox1 and human HHO.C8. The gene is expressed in the limb and limb blastema of the newt Notophthalmus viridescens, as a transcript of 1.8 kb. It is of interest that the transcript is expressed in a related but distinct form in tail tissues, that it is expressed in adult limb of the newt but not of Xenopus, and that its expression is higher in a proximal than a distal blastema.

By using specific monoclonal antibodies, we have detected transient expression of the keratin pair 8 and 18 in mesenchymal cells of the newt blastema. While the majority of blastemal cells express these antigens, they are not detectable in sections of the developing limb bud of newt embryos. Keratin 8 and 18, which are usually expressed in simple epithelia, represent a new marker for regeneration versus development.

INTRODUCTION

Recent work in this laboratory has identified two quite different molecules in the limb blastema which promise to shed some light on the problems of regeneration. In one set of experiments, we screened blastemal cDNA libraries with probes for homeobox containing cDNAs. The homeobox is a rather conserved motif of approximately 180 base pairs that specifies a protein domain believed to mediate DNA binding. This motif is found in several Drosophila genes specifying pattern in early development (McGinnus et al., 1984; Scott and Weiner, 1984), and in vertebrate genes whose spatial and temporal patterns of expression are broadly consistent with a similar function (Hogan and Holland, 1988). We identified and sequenced a newt blastemal cDNA containing a homeobox. The expression of this transcript shows several features of interest in relation to limb regeneration.

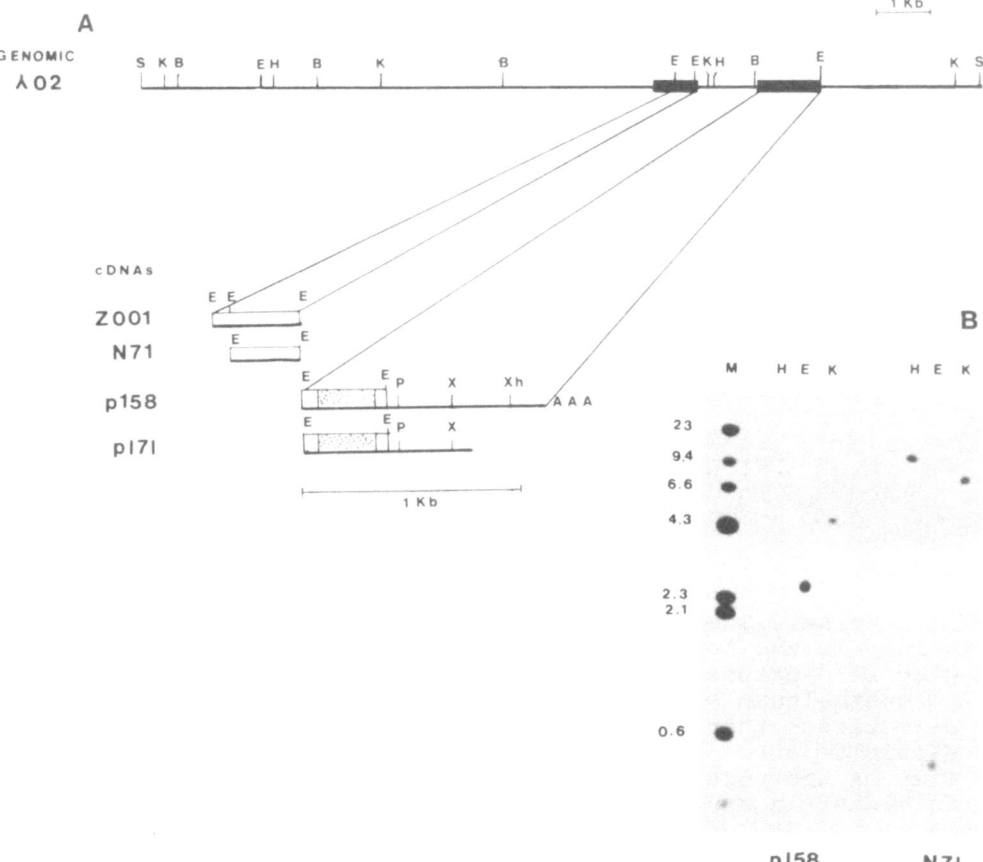

Fig. 1. Structure of the NvHbox1 locus and summary of
the cDNA cloned. (A) The solid rectangles are
the two exons identified in the genomic clone.
The open rectangles represent the predicted
coding region in the cDNA clones. Included in
this region is the homeobox sequence (hatched
area). E=EcoR1; H=Hind III; K=Kpn1;X, Xbal;
B=Bgl1; Xh=Xhol. (B) Analysis of newt genomic
DNA (25 µg per lane) digested with H, E and K.
The fragment sizes of lambda DNA digested with
H are indicated on the left in kilobase pairs.
The Southern blots were hybridised to the
probes indicated below the lanes. These
probes are diagrammed in A.

In a second set of experiments, we screened sections of
the newt limb blastema with a large panel of monoclonal antibodies directed against various members of the keratin family
of intermediate filaments (Ferretti et al.,1988). The keratin
family is made of type I genes (acidic proteins) and type II
genes (neutral to basic) that have been highly conserved in
evolution (Fuchs and Marchuk, 1983; Hoffman and Franz, 1984;

10

Franz and Franke, 1986). The expression of at least one keratin of each type is necessary for filament assembly, and particular pairs are characteristically associated with different
epithelia (Moll et al., 1982; Sun et al., 1984; Lane et al.,
1985; Osborn and Weber, 1986). A restricted subset of these
antibodies reacted with the blastemal cells both in vivo and
in vitro, and this subset identifies a particular pair of
keratins called k8 (type II) and k18 (type I), that are usually expressed in simple epithelia. The expression of keratins in the mesenchymal progenitor cells of the blastema
raises several questions for future investigation.

MATERIALS AND METHODS

    Details of the materials and methods used in these
studies are given by Savard et al.(1988), Ferretti and Brockes
(1988) and Ferretti et al.(1988).

RESULTS

Identification of a homeobox gene transcript

    The homeobox probes have identified a transcript of a
newt gene that we refer to as NvHbox1 (Savard et al.,1988).
Its sequence shows sufficiently high identity in both translated and 5'-untranslated regions with the Xenopus XLHbox1
(Choo et al.,1988) and human HHO.c8 (Simeone et al.,1987) that
there is little doubt that it is the newt homologue of these
genes. Figure 1 shows the cDNAs cloned. In the limb and tail
tissues of the newt, the gene is expressed as a single transcript of 1.8 kb derived from two exons which we have identified and sequenced in a genomic EMBL3 clone (see Savard et
al., 1988). There are certain features of the expressions of
the transcript that are of interest in relation to limb regeneration.

    1) In the limb and the limb blastema (mid-bud) the
transcript shown in Fig. 1 is detectably by Northern analysis
of poly A^+RNA, but in the tail and the tail blastema there is
a related but distinct molecule. Whereas probes for the second of the two exons detect a 1.8 kb transcript in both limb
and tail, probes for the first exon do not react with tail
poly A^+RNA, suggesting that an alternative first exon is
inserted in the tail tissues.

    2) Although the transcript is expressed in the adult
limb, its homologue is not detected in adult Xenopus limb
(Savard et al., 1988). It is expressed, however, during limb
development in Xenopus and human (Simeone et al., 1987).

    3) The transcript is expressed at higher levels in a
proximal (mid-humerus) blastema as compared to a distal one
(mid-radius and ulna). When distal blastemata were "proxima-
lised" by injection of retinoic acid, there was no detectable
change in the level of expression at a single time point after
injection (Savard et al., 1988).

Table 1. Apparent Molecular Weight ($\times 10^{-3}$) of Newt Cytoskeletal Proteins Detected with Anti-Keratin Monoclonal
Antibodies by Immunoblotting.

mAb	LP1K	LE41	RGE53	CK18.2
Blastema	52-57	52-59	43	43
Cultured cells	52	52	43	43

Expression of keratin proteins 8 and 18 in blastemal cells

We have used immunofluorescence and immunoblotting to
study the reactivity of monoclonal antibodies with blastemal
cells _in_ _vivo_ and _in_ _vitro_. Of the antibodies that show
restricted specificity to particular members of the keratin

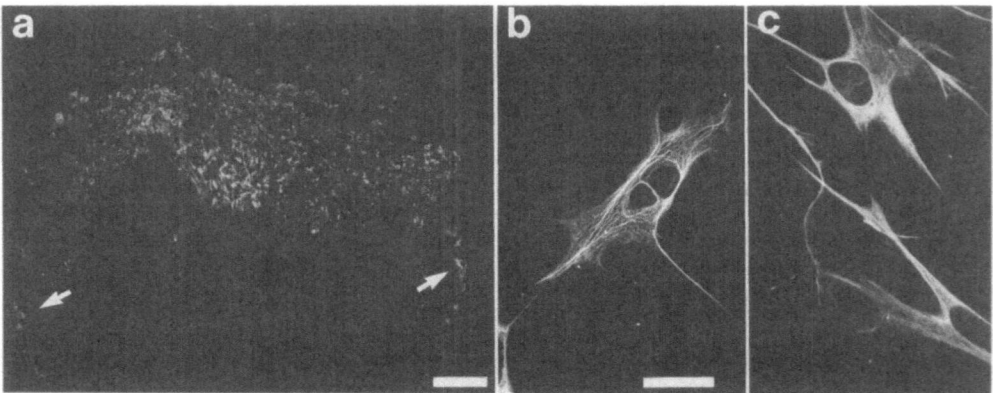

Fig. 2. Reactivity of the anti-keratin 18 monoclonal
antibodies RGE53 and CK18.2 and anti-keratin
8 monoclonal antibody LP1K was assayed by
immunofluorescence on regenerating blastema
15 days after forelimb amputation and on
myogenic newt cells in culture (Ferretti and
Brockes, 1988). RGE53 (a) staining of
blastema cryostat section: the majority of
blastemal cells are highly stained, whereas
the mesenchymal tissues of the stump and the
epidermis are completely negative. Subepidermal glands, as observed in normal tissue,
are also stained (arrows, scale bar 200 µm).
LP1K (b) and CK18.2 (c) reactivity with cultured cells: a bright filamentous staining
pattern is observed with both antibodies
(scale bar is 50 µm).

family, only those reactive with k8 and k18 stain the blastema (Ferretti et al., 1988). A summary of the results with the four key monoclonals is shown in Table 1. Antibodies LP1K and LE41 react with keratin 8, while RGE53 and CK18.2 react with keratin 18 (Lane et al., 1985, Broers et al., 1986). An example of anti-keratin 18 reactivity with sections of the blastema is shown in Figure 2a. The majority of blastemal cells are strongly labeled by two weeks after amputation, although reactivity is lost progressively as differentiation ensued in the blastema. Cultured cells from the blastema also react with the antibodies to k8 and k18 and the filamentous nature of the staining is particularly clear in this case (see Fig. 2b, 2c). As detailed in Table 1, the molecular weights of the newt antigens in the blastema are consistent with those of human keratin 8 and 18 (Moll et al., 1982).

In view of the reactivity of several antibodies in recognising the two members of a characteristic pair, there can be no doubt that the mesenchymal progenitor population in the blastema transiently express the newt homologues of these keratins.

DISCUSSION

The expression of the homeobox gene transcript has several features that are circumstantially consistent with interesting functional roles for the protein product. The difference between limb and tail suggests that it could be involved in the different specification of limb and tail blastemal cells. The persistent expression in the adult newt, as compared to adult Xenopus, suggests that the loss or regenerative ability in anurans and other vertebrates might be related in part to the absence of expression in the adult of genes that are involved in specification of the developing limb bud. Finally, the variation in expression along the proximal-distal axis might indicate a role in limb patterning. In future work it will be important to derive antibodies that allow identification of the protein and permit analysis of its cellular distribution (see Choo et al., 1988). We also wish to complete the identification of the exon that is inserted in the tail. Ultimately it will be necessary to manipulate the expression of the gene in cultured cells or in the animal, and assay the consequences for limb regeneration and morphogenesis.

The keratin family is generally thought to be characteristic of epithelial cells, and hence it is interesting that the keratin pair 8 and 18 are expressed in the blastema. This particular pair are expressed by all cells of Xenopus and mouse early embryos (Franz and Franke, 1986; Chisholm and Houliston 1987; Lehtonen et al., 1988), and also in endothelial cells and a subset of smooth muscle cells in Xenopus (Jahn et al., 1987). While the function of keratins is not known, this pair is characteristic of simple epithelia, including the mesothelium. In at least some circumstances expression of keratin 8 and 18 is dependent on retinoic acid (for example see Kim et al.,1987), a molecule that has aroused considerable attention in limb development and regeneration because of its possible role as a morphogen. Retinoic acid can respecify morphogenetic axes in the regenerating but not the developing limb (Maden, 1985), and the differential expression of the keratins in the two contexts is consistent with this. It will be interesting to determine if 8 and 18 are a marker for the action of retinoids on the blastema _in vivo_. Finally they are a well defined example of genes activated after amputation, and should allow the identification of regeneration promoters and transactivators involved in specific gene expression in the blastema.

REFERENCES

Broers, J.L., Carney, D.N., Klein Rot, M., Schaart, G., Lane, E.B., Vooijs, G.P. and Ramaekers, F.,1986, Intermediate filament proteins in classic and variant types of small cell lung carcinoma cell lines: a biochemical and immunological analysis using a panel of monoclonal and polyclonal antibodies, <u>J.Cell Sci.</u>, 83:37-60.

Chisholm, J.C. and Houliston, 1987, Cytokeratin filament assembly in the preimplantation mouse embryo, <u>Development</u>, 101:565-582.

Choo, K.W.Y., Goetz, J., Wright, C.V.E., Foitz, A., Hardwicke, J. and De Robertis, E.M., 1988, Differential utilization of the same reading frame in a <u>Xenopus</u> homeobox gene encodes two related proteins sharing the same DNA-binding specificity, <u>EMBO J.</u>, 7:2139-2149.

Fekete, D.M. and Brockes, J.P., 1987, A monoclonal antibody detects a difference in the cellular composition of developing and regenerating limbs of newts, <u>Development</u>, 99:589-602.

Fekete, D.M. and Brockes, J.P., 1988, Evidence that the nerve controls molecular identity of progenitor cells fore-limb regeneration, <u>Development</u>, 99:589-602.

Ferretti, P. and Brockes, J.P., 1988, Culture of newt cells from different tissues and their expression of a regeneration-associated antigen, <u>J.Exp.Zool.</u>, (in press).

Ferretti, P., Fekete, D.M., Patterson, M. and Lane, E.B., 1988, Transient expression of simple epithelial keratins by mesenchymal cells of regenerating limb, (submitted for publication).

Franz, J.K. and Franke, W.W., 1986, Cytokeratin filament assembly in the preimplantation mouse embryo, <u>Proc. Natl. Acad. Sci. USA</u>, 83:6475-6479.

Fuchs, E. and Marchuk, D., 1983, Type I and II keratins have evolved from lower eukaryotes to form the epidermal intermediate filaments in mammalian skin, <u>Proc.Natl.Acad.Sci. USA</u>, 80:5857-5861.

Hoffmann, W. and Franz, J.K., 1984, Amino acid sequence of the carboxy-terminal part of an acidic type I cytokeratin of molecular weight 51000 from <u>Xenopus laevis</u> epidermis as predicted from the cDNA sequence, <u>EMBO J.</u>, 6:1301-1306.

Holland, P.W.H., and Hogan, B.L.M., 1988, Expression of homeobox genes that control development: a review, <u>Genes and Development</u>, 2:773-782.

Jahn, L., Fouquet, B., Rohe, K. and Franke, W.W., 1987, Cytokeratins in certain endothelial and smooth muscle cells of two taxonomically distant vertebrate species, <u>Xenopus laevis</u> and man, <u>Differentiation</u>, 36:234-254.

Kim, H.K., Stellmach, V., Javors, J. and Fuchs, E., 1987, Regulation of human mesothelial cell differentiation: opposing roles of retinoids and epidermal growth factor in the expression of intermediate filament protein, <u>J. Cell Biol.</u>, 105:3039-3051.

Lane, E.B., Bartek, J., Purkis, P.E., Leigh, I.M., 1985, Keratin antigens in differentiating skin, <u>Ann. NY. Acad. Sci.</u>, 445:241-258.

Lehtonen, E., Ordanez, G. and Reima, I., 1988, Cytoskeleton in preimplantation mouse development, <u>Cell Differ.</u>, 24:165-178.

Maden, M., 1985, Retinoids and the control of pattern in limb development and regeneration, <u>Trends Genet.</u>, 4:103-107.

Moll, R., Franke, W., Schiller, D.L., Geiger, B. and Krepler, R., 1982, The catalog of human cytokeratins: pattern of expression in normal epithelia, tumours and cultured cells, <u>Cell</u>, 31:11-24.

Osborn, M. and Weber, K., 1986, Intermediate filament proteins: a multigene family distinguishing major cell lineages, <u>Trends Biochem. Sci.</u>, 11:469-472.

Savard, P., Gates, P.B. and Brockes, J.P., 1988, Position dependent expression of a homeobox gene transcript in relation to amphibian limb regeneration, <u>EMBO J.</u>, (in press).

Scott, M.P. and Weiner, A.J., 1984, Structural relationship among genes that control development: sequence homology between antennapedia, ultrabithorax and fushi tarazu, <u>Proc. Natl. Acad. Sci. USA</u>, 81:4115-4125.

Simeone, A., Mavilio, F., Acampora, D., Giampaolo, A., Faiella, A., Zappavigna V., D'Esposito, M., Pannese, M., Russo, G., Boncinelli, E., and Peschle, C., 1987, Two human homeobox genes, c1 and c8: Structure analysis and expression in embryonic development. <u>Proc. Natl. Acad. Sci.</u>, 84:4914-4918.

Sun, T.T., Eichner, R., Schermor, A., Cooper, D., Nelson, W.G., Weiss, R.A., 1984, Classification, expression and possible mechanism of evolution of mammalian epithelial keratins: an unifying model, <u>in</u>: "Cancer Cells I, The transformed phenotype" Levine, A.J., Vandewoude, G.F., Topp, W.C. and Watson, J.D., eds, Cold Spring Harbor, 169-176.

AN EXAMINATION OF HEAT SHOCK AND TRAUMA-INDUCED PROTEINS
IN THE REGENERATING FORELIMB OF THE NEWT, <u>NOTOPHTHALMUS</u>
<u>VIRIDESCENS</u>

Robert L. Carlone and Gordon A.D. Fraser

Department of Biological Sciences
Brock University, St. Catharines, Ontario
Canada L2S 3A1

SUMMARY

Heat shock protein (hsp) synthesis and accumulation was
studied in the limb tissues of the newt, <u>Notophthalmus viri-
descens</u>, in response to the stresses of hyperthermia and amputation. We also examined the developmental regulation and
heat inducible expression of these proteins during subsequent
stages of regeneration. Both heat shock and amputation cause
a decrease in total protein synthesis and the selective synthesis of 70 kD proteins. Two-dimensional gel electrophoretic
analysis reveals differences in these two sets of 70 kD proteins, the amputation induced form (amp 70) being more acidic
than hsp 70. These stresses, in addition, have quite different
effects on the accumulation of hsp 70 in limb stump tissues.
Heat shock only slightly elevates the levels of what appears
to be the constitutive form of hsp 70 (hsc 70) in the limb,
whereas amputation leads to a large decrease in the levels of
this protein within one hour. The titre of the putative constitutive hsc 70 rises gradually after the wound healing stage
and eventually increases to a level greater than controls during the late bud stage of regeneration. The results are discussed in light of a possible role for hsps and trauma induced
proteins in the epimorphic regeneration of the amphibian limb.

INTRODUCTION

Prokaryotic and eukaryotic cells respond to environmental
stresses such as heat by selectively synthesizing heat shock
proteins (hsps) whose function, although unclear, appears to
relate to their ability to both protect cellular structures
from further damage and to assist in their reassembly (Ashburner and Bonner, 1979; Burdon, 1986; Pelham, 1986; Subjeck
and Shyy, 1986; Welch and Suhan, 1986). Three major families
of hsps with molecular weight ranges of 80-90.000, 68-74.000,
and 15-30.000 are synthesized in most organisms, with the two
higher molecular weight families being particularly highly
conserved (Hunt and Morimoto, 1985; Burdon, 1986).

In addition to their induction by heat and other stresses, recent evidence suggests that the expression of at least some of the members of the major hsp families is developmentally regulated. For example, there are developmental stages, in a number of organisms, in which certain hsp genes are active under normal physiological conditions (Bensaude et al., 1983; Bensaude and Morange, 1983; Heikkila et al., 1986). Alternatively, there are specific cell types and developmental stages during which the expression of certain hsp70 genes cannot be induced, even by heat (Zimmerman et al., 1983; Morange et al., 1984).

The developmental regulation of hsps, the demonstration of their induction in response to wounding (Currie and White, 1981; Heikkila and Schultz, 1984) and the recent interest in their possible roles in mitogen stimulation (Wu and Morimoto, 1985; Ferris et al., 1988) and retinoic acid-induced differentiation (Bensaude and Morange, 1983; Imperiale et al., 1984) of cultured cells, has led us to examine their expression in the regenerating forelimbs of the newt. Specifically, we have analyzed the synthesis and accumulation of hsps of the 70.000 class by SDS-PAGE, 2-D PAGE and Western blotting in stump tissues of the regenerating forelimb of the newt in response to hyperthermia and to the trauma of amputation. We have also looked at the heat inducibility and normal expression of these proteins at specific regeneration stages corresponding to periods of dedifferentiation, proliferation, and redifferentiation of the blastema cells.

MATERIALS AND METHODS

Animals

All experiments were performed on adult newts Notophthalmus viridescens obtained from Charles Sullivan, Nashville, Tennessee. Animals were housed in distilled water in vented Tupperware^{TM} containers and fed live Tubifex or frozen brine shrimp twice weekly. Water, maintained at 20±2°C was changed every other day. Bilateral amputations through the distal third of the radius and ulna were performed after benzocaine anaesthesia. For heat shock experiments, animals were maintained in bowls with water at 34±0.5°C for one hour and left to recover for at least one hour prior to removal of tissue.

(^{35}S-methionine labelling and tissue preparation

At the appropriate times prior to or after heat shock or amputation (see results), animals were injected with approximately 100 µCi ^{35}S-methionine (Tran^{35}S-label^{TM}, Sp. Act. 1160 Ci/mmole, ICN). After four hours of incorporation, the distal two to five millimeters of limb tissue were removed and either frozen in liquid N$_2$ in sample buffer for eventual use in one-dimensional SDS PAGE or immediately homogenized in IEF sample buffer and submitted to isoelectric focusing. In some experiments, tail tissue was removed to monitor the stress of limb amputation systemically.

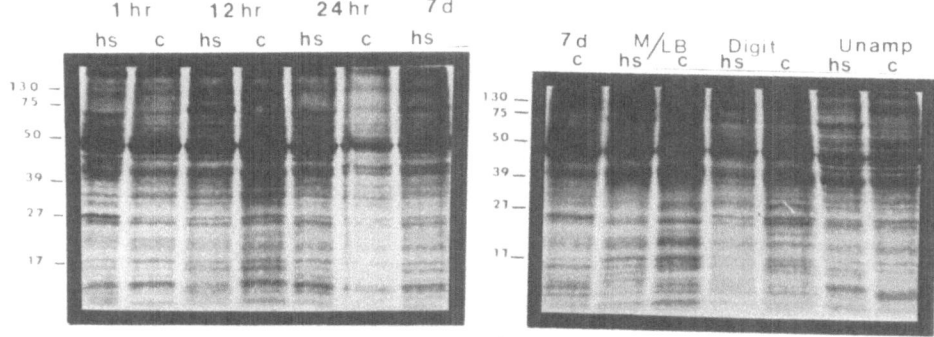

Fig. 1. Fluorographs of polyacrylamide gels showing
the effects of heat shock on protein synthesis
at various stages of limb regeneration.   (hs)
heat shock, 34.5°C, (c) control, 22°C.   Proteins were isolated from stump tissues at 1
hour, 12 hours, 24 hours and 7 days postamputation and at the mid to late bud stage, the
digit stage and from unamputated limbs. Lanes
were loaded with equivalent TCA-precipitable
counts per minute (cpm).

Electrophoresis and immunoblotting

Equal amounts of protein or equal amounts of TCA-
precipitable counts were separated on SDS-PAGE (4% acrylamide
stacking and 12% acrylamide separation gels) after the method
of Laemmli (1970). Slab gels were stained with Coomassie Blue
R-250, destained with 40% methanol: 10% acetic acid, and saturated with En^{3}Hance^{TM} (Amersham) prior to fluorography. X-ray
film was exposed to dried gels for periods up to three weeks.
Two dimensional electrophoresis of blastema proteins was run
according to the method of O'Farrell (1975) with electrofocus-
ing between pH 3.5-9.5 using a mixture of ampholytes (3.5-9.5
and 6-8; LKB).

Electrophoretic transfer of proteins from 12% polyacrylamide gels to nitrocellulose was accomplished overnight (30V,
0.1A) in transfer buffer (25 mM Tris, 192 mM glycine, 20%
methanol). After blocking with gelatin, blots were probed with
mouse monoclonal antibody N-27 (1:200), (generously provided
by W. Welch, Cold Spring Harbor) specific to both the highly
inducible and constitutive forms of human hsp70 (hsps72 and
73 respectively). The membranes were then washed and incubated
with a secondary antibody, alkaline phosphatase conjugated
goat anti-mouse IgG (1:3000). Colour development was with a
mixture of p-nitro blue tetrazolium chloride (NBT) and 5-bro-
mo-4-chloro-3-indolyl phosphate (BCIP) (Biorad).

RESULTS

Stage-specific inducibility of hsp70 synthesis

Thermal induction of hsp70 synthesis occurs in tissues of the unamputated limb as well as at all subsequent post amputation stages examined (Fig. 1). Relatively little hsp70 synthesis occurred in the stump tissues in response to heat during the first hour post-amputation. By twelve hours post amputation however, hsp70 synthesis is highly inducible.

Beginning in the mid-late bud stages, and continuing on into the mid-digit stage, we also see the heat-induced synthesis of a higher molecular weight hsp, possibly hsp90, that is absent at all other stages. The resolution of the bands in the low molecular weight range of our fluorographs was not sufficient to assess the stage-specific heat-inducibility of the hsps of the 15-30.000 family.

The synthesis of hsps in response to the trauma of amputation

Figure 2 demonstrates that the mechanical injury of forelimb amputation is sufficient to cause an apparent heat shock-like response in the distal limb stump tissues. That is to say, the total protein synthesis is greatly diminished when compared to unstressed tissues (Fig. 2B), while the synthesis of a select set of proteins in the 70.000 molecular weight range is greatly increased. The limb amputation stress response is local and is not reflected in tissues removed from the tail at the same time.

More detailed analysis by two dimensional gel electrophoresis however, reveals significant differences between the array of proteins synthesized in the limb in response to heat and injury. Heat shock causes the characteristic synthesis of a set of 70kD proteins not induced by the trauma of amputation (Fig. 3 A,B, and D). These hsps represent the highly heat inducible forms (analogous to hsp72 in man and hsp68 in the mouse). A very complex array of lower molecular weight hsps are induced when compared to the control limb (Fig. 3A,B).

At one hour post amputation, on the other hand, very little synthesis of hsp70 is found (Fig. 3D). In contrast, there is a dramatic increase in the synthesis of 70kD protein with a more acidic pI. The identity of this protein we have termed amp 70, remains to be determined.

The stage-specific accumulation of hsp70 protein as determined by immunoblotting

The great increase in the synthesis of hsp70 detected in limb tissues in response to heat shock is not strictly paralleled by an increased accumulation of hsp70 protein measured by immunoblotting with monoclonal antibody N-27. Only a marginal increase in the amount of hsp70 (hsc70?) protein was observed when compared to unstressed limb tissues. Amputation, however, results in an almost total loss of hsp70 (hsc70) protein in the distal stump tissues within one hour, with levels remaining low for at least the first 24 hours (Fig. 4A).

20

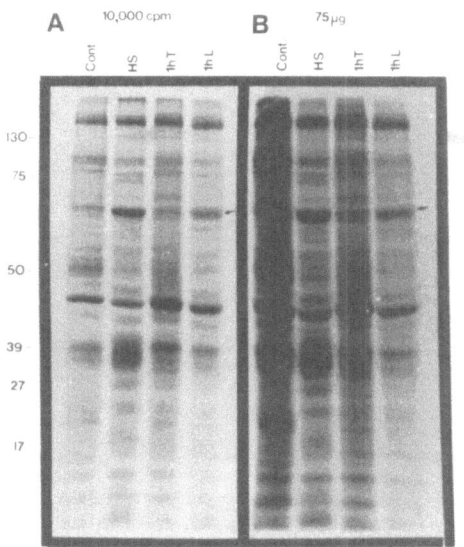

Fig. 2. Fluorographs of polyacrylamide gels showing
the effects of a one hour heat shock at 34.5°C
(HS) and amputation through the radius and
ulna on protein synthesis in the forelimbs of
the adult newt. Cont.=unamputated forelimb;
1hT=tail proteins at one hour post-amputation
of the forelimb; 1hL=distal forelimb stump
proteins, one hr p.a. (A) 10,000 cpm of ^{35}S-
methionine labelled protein loaded per
lane,(B) 75 µg protein loaded per lane.

By the early bud stage, hsp70 (hsc70) levels in the
blastema are higher than in unamputated/unstressed limbs and
considerably more abundant than during the early wound healing
stages (Fig. 4B). Levels decline slightly after the late bud
stage, ultimately reaching control values by the late digit
stage (data not shown).

DISCUSSION

We have documented the effects of hyperthermia and
mechanical injury (limb amputation) on the synthesis and
accumulation in distal limb tissues of hsps of the 70.000 family. Upon initial examination, it appeared that each stress
resulted in a similar response in terms of the array of newly
synthesized proteins detectable on fluorographs of SDS-
polyacrylamide slab gels. Both heat shock and amputation
caused a general decrease in total synthesis of new proteins
with a concomitant increase in the synthesis of proteins with
molecular weights in the 70.000 range. Heikkila and Schultz
(1984) had previously demonstrated that mechanical injury of
rabbit blastocysts caused increased synthesis of a 70 kD protein that they suggest is identical to the heat inducible
form. Currie and White (1981), have provided evidence from
two-dimensional gels that both heat and mechanical trauma

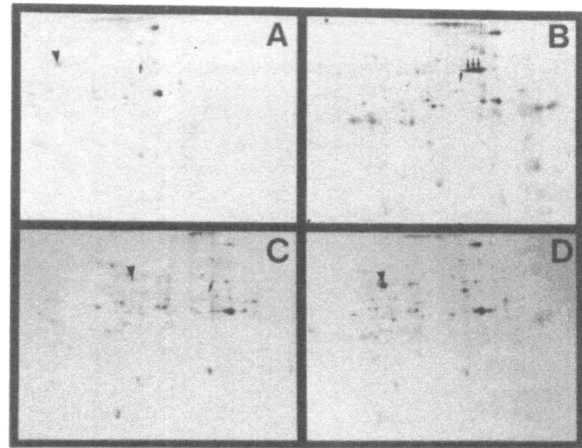

Fig. 3. Fluorographs of two-dimensional polyacrylamide gels showing the effects of the treatments outlined in Fig. 1 on protein synthesis
in the forelimbs of the newt. 5,000 cpm were
loaded per IEF gel. (A) unamputated limb, (B)
heat shock, (C) tail proteins, one hour pa.
of limb, (D) limb stump proteins on hour pa.
The downward arrows in (B) point to the inducible family of hsp70, the upward arrow
points to the constitutive hsp70. The large
arrowheads in (A),(C) and (D) point to the
70kD protein induced by amputation.

induce the synthesis of the same 71 kD "heat shock" protein in
a number of rat tissues.

When our tissue samples were analyzed by two dimensional
gel fluorography, however, we observed significant differences
between the major 70 kD proteins induced by heat and amputation. A one hour heat shock at 34°C was sufficient to cause a
large increase in the synthesis of at least three 70 kD hsps
with pIs in the range from 6.6 to 6.8. In addition, heat
shock caused the selective synthesis of a complex array of
other lower molecular weight proteins, particularly in the
40-50.000 range.

The 2-D pattern of proteins synthesized in response to
amputation was quite different. Very little hsp70 was synthesized, whereas a 70 kD protein (amp 70) with a considerably
more acidic pI was the predominant species (along with actin)
of newly synthesized protein. The identity of this amputa-
tion-induced protein remains to be determined. The possibility exists that it belongs to the 70 kD family of stress proteins and may represent an isoform of hsp70 which, under these
conditions, is covalently modified by, for example, phosphorylation (Burdon, 1986). Alternatively, this protein may prove
to be unique and unrelated to the hsp70 family. Peptide mapping and immunological studies may shed light on the degree of
similarity between these two stress inducible proteins.

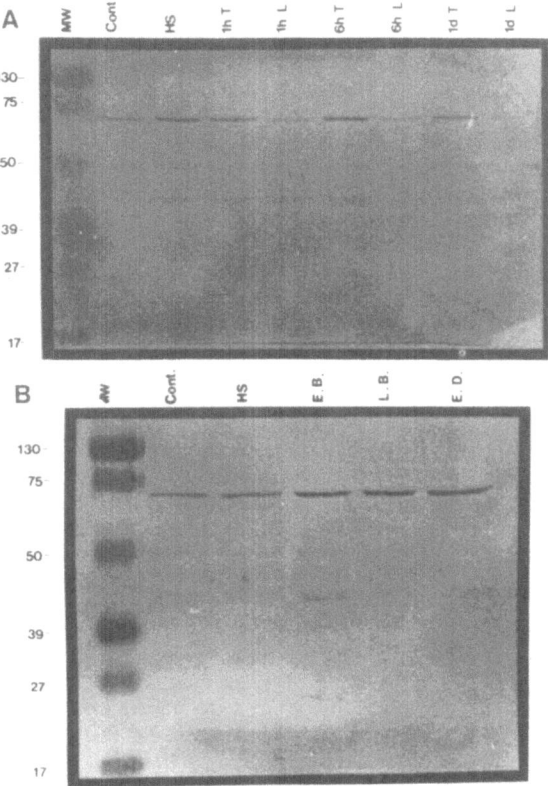

Fig. 4.   (A) Western immunoblot of polyacrylamide gel
          showing the effects of heat shock and amputa-
          tion on the levels of hsp70. The blot was
          probed with a monoclonal antibody (N27) which
          recognizes both the inducible and constitut-
          ive forms of human hsp70.   (HS) heat shock;
          (1h,6h, and 1d T)=Tail proteins from one hour
          six hours and one day pa of the limb. (1h,6h,
          and 1d L)=limb stump proteins at one hour,six
          hours and one day pa of the limb. (B) Stage-
          specific levels of hsp70 in the regenerating
          forelimb of the newt quantitated by Western
          immunoblot as described in (A) above.

     Results from our immunoblotting studies on the develop-
mental regulation of hsp70 synthesis are confounded by the
apparent lack of cross reactivity of antibody N-27 with the
heat inducible hsp70 of the newt.   Although it cross-reacts
with both forms of human hsp70 (W. Welch, personal communication) and strongly reacts with antigen in unstressed cells of
the newt, heat shock only marginally increased the titre of
the protein recognized by this particular antibody.   We have
attempted to address this question of specificity by using
another monoclonal antibody (C92) reputedly specific to the
inducible form of human hsp70 alone (W. Welch, personal communication). Unfortunately, in preliminary experiments, this

antibody showed an unacceptable degree of non-specific cross-reactivity with a number of newt antigens.

Despite the above caveat, it appears that amputation causes a dramatic decline in the accumulation of hsp70 (hsc 70) in the blastema within one hour. This decline is maintained for at least the first 24 hours after which the levels gradually increase, peaking in the late bud stage. The titre then declines to the preamputation control levels by the late digit stage. These data are in general agreement with the results of Ferris et al. (1988) relating increases in constitutive hsp70 levels with increased mitotic activity in human lymphocytes. The decline in hsp70 (hsc70) seen after the early bud stage may reflect a shift in the state of differentiation of blastema cells, and is reminiscent of that induced by retinoic acid in F9 teratocarcinoma cells of the mouse (Imperiale et al., 1984). Data from preliminary experiments in our laboratory on the effects of intraperitoneal injection of retinoic acid on hsp70 synthesis in the blastema support this contention. Retinoic acid induces the transient and selective synthesis of a set of 70 kD proteins. Based on peptide mapping results, these RA inducible proteins are distinct from hsp70. Indeed, as determined by immunoblotting with N-27, retinoic acid appears to cause a decrease in the levels of constitutive hsp70 (hsc70) in these tissues, analogous to the response of limb tissues to amputation (Carlone et al., manuscript in preparation). In studies already underway, we are trying to define more clearly the possible roles played by hsp70, hsc70, amp 70, and other hsps in the molecular events underlying the genesis of pattern in the regenerating limb.

ACKNOWLEDGEMENTS

This work was supported by a grant from the Natural Sciences and Engineering Research Council of Canada to R.L.C. We would like to thank Mrs. Marilyn Ferracuti for typing the manuscript and Mrs. Marion Vijh for technical assistance and for help in preparation of the figures. We also thank Dr. William Welch (Cold Spring Harbor) for his generous gift of the anti hsp70 monoclonal antibodies.

REFERENCES

Ashburner, M., and Bonner, J.J., 1979, The induction of gene activity in _Drosophila_ by heat shock, _Cell_, 17:241-254.
Bensaude, O., and Morange, M., 1983, Spontaneous high expression of heat-shock proteins in mouse embryonal cells and ectoderm from day 8 mouse embryo, _EMBO J._, 2:173-177.
Bensaude, O., Babinet, C., Morange, M., and Jacob, F., 1983, Heat-shock proteins, first major product of zygotic gene activity in mouse embryo, _Nature_, 305:331-333.
Burdon, R.H., 1986, Heat shock and the heat shock proteins, _Biochem. J._, 240:313-324.
Currie, R.W., and White, F.P., 1981, Trauma induced protein in rat tissues: A physiological role for a "heat shock" protein? _Science_, 214:72-73.

Ferris, D.K., Harel-Belan, A., Morimoto, R.I., Welch, W.J., and Farrar, W.L., 1988, Mitogen and lymphokin stimulation of heat shock proteins in T-lymphocytes, Proc. Natl. Acad. Sci. USA, 85:3850-3854.

Heikkila, J.J., and Schultz, G.A., 1984, Different environmental stresses can activate the expression of a heat shock gene in rabbit blastocyst, Gamete Res., 10:45-56

Heikkila, J.J., Browder, L.W., Gedamu, L., Nickells, R.W., and Schultz, G.A., 1986, Heat-schock gene expression in animal embryonic systems, Cab. J. Genet. Cytol., 28:1093-1105.

Hunt, C., and Morimoto, R.I., 1985, Conserved features of eukaryotic hsp70 genes revealed by comparison with the nucleotide sequence of human hsp70, Proc. Natl. Acad. Sci. USA, 82:6455-6459.

Imperiale, M.J., Kao, H.T., Feldman, L.T., Nevins, J.R., and Strickland, S., 1984, Common control of the heat shock gene and early adenovirus genes: evidence for a cellular E1A-like activity, Mol. Cell. Biol., 4:867-874.

Laemmli, U.K., 1970, Cleavage of structural proteins during the assembly of the head of bacteriophage T4, Nature, 277:680-685.

Morange, Babinet, C., 1984, Altered expression of heat shock proteins in embryonal carcinoma and mouse early embryonic cells, Mol. Cell. Biol. 4:730-735.

O'Farrell, P.H., 1975, High resolution two-dimensional electrophoresis of proteins, J. Biol. Chem., 250:4007-4021.

Pelham, H.R.B., 1986, Speculations on the functions of the major heat shock and glucose-regulated proteins, Cell, 46:959-961.

Subjeck, J.R., and Shyy, T.T., 1986, Stress protein systems of mammalian cells, Am. J. Physiol., 250:C1-C17.

Welch, W.J., and Suhan, J.P., 1986, Cellular and biochemical events in mammalian cells during and after recovery from physiological stress, J. Cell. Biol., 103: 2035-2052.

Wu, B.J., and Morimoto, R.I., 1985, Transcription of the human hsp70 gene is induced by serum stimulation, Proc. Natl. Acad. Sci. USA, 82:6070-6074.

Zimmerman, J.L., Petri, W., and Meselson, M., 1983, Accumulation of a specific subset of D. melanogaster heat shock mRNAs in normal development without heat shock, Cell, 32:1161-1170.

C-MYC PROTO-ONCOGENE EXPRESSION DURING NEWT LIMB

REGENERATION

Jacqueline Géraudie (1), Jacques Hourdry (2),
Keith Boehm (3) Marcus Singer (3) and Marcel
Mechali (2)

(1) Laboratoire d'Anatomie Comparee, Université
Paris VII, and (2) Institut Jacques Monod, 2
Place Jussieu, 75271 Paris Cedex 05, France
(3) Department of Anatomy, School of Medicine
Case Western Reserve University, Cleveland
Ohio 44106, USA

SUMMARY

Utilizing the method of in situ hybridization, myc-homologous RNA transcripts were detected in histological sections of the newt limb regenerate. A radiolabeled Xenopus myc-cDNA was used as probe. Autoradiographic grains representing myc-homologous RNAs were observed in cells of the thickened wound epidermis, irrespective of the layer examined. Further, mesenchymatous-like cells which form the blastema also contained multiple grains, indicating that these cells possess myc-like RNA molecules. The presence of RNAs, homologous to the myc proto-oncogene in cells of the regenerating newt limb indicate that a myc-related gene is expressed in these cells during the regenerative process and, therefore, may play an important regulatory role, as myc gene expression has been implicated in the control of cellular proliferation in other systems. Our results are discussed against the background of literature on c-myc proto-oncogene expression.

INTRODUCTION

Most of the urodelan Amphibia possess the remarkable ability of regenerating limbs after injury in the wild or experimental amputation in the laboratory (reviewed in Wallace, 1981). Once wound healing is achieved by epidermal spreading from the stump level, a period of proliferation of progenitor cells follows the cell "dedifferentiation" step. It allows the progressive construction of a blastema, the forerunner of the new limb, through cell differentiation and morphogenesis (reviewed in Singer, 1952, Schmidt, 1968; Wallace, 1981).

Whatever the origin of the progenitor blastemal cells

capping the stump tissues, their active proliferation is the key event that assures an adequate number of cells for limb regenerate growth. Recently, we became interested in biochemical factors that may participate in the regulation of that aspect of limb regeneration. Our initial studies have been focused on examining the expression of the c-myc proto-oncogene within the blastema during: 1) normal conditions of growth and 2) after stump denervation.

Proto-oncogenes [cellular oncogenes or c-oncogenes (c-oncs)] are believed to be genes important in the regulation of cell proliferation. The term "oncogenes" relates to the initial detection of viral gene sequences [viral oncogenes or v-oncogenes (v-oncs)] in various cancerous cells. Oncogene protein products have been shown, in some instances, to be important in carcinogenesis, and are believed to be involved in the maintenance and progression of various cancers (Varmus, 1984; Bishop, 1985). The oncogenes associated with cancers, though, have been shown in most cases, to be altered, in some fashion, from their normal cellular homologues. Therefore, the cellular proto-oncogenes (c-oncs) most likely participate in the regulation of normal cell growth mechanisms.

The c-myc proto-oncogene was first identified as the transforming sequence (v-myc oncogene) in the avian retrovirus MC29 which is responsible for myelocytomatosis in the chicken. It has been found in a number of organisms including the amphibian Xenopus laevis (Taylor et al., 1985; King et al., 1986). The c-myc gene encodes for a phosphorylated 62-64Kd protein which is located in the nucleus (Cole, 1986, for a review), but is not permanently associated with DNA as it dissociates from metaphasic chromosomes. Its role is still not really known, but the protein may be involved as a mediator of the response to the influences of growth factors.

One rationale for study of c-myc proto-oncogene expression in the blastema deals with the relation between c-myc mRNA levels, mitogenic signals and cell proliferation. Indeed, previous studies have shown in mammals or avian cells, that the level of c-myc mRNA increases when cells are stimulated to divide by mitogens or growth factors (Kelly et al., 1983; Heldin and Westermark, 1984). C-myc levels are low in non-dividing fibroblasts or lymphocytes and are subject to increase after exogenous administration of growth factors such as PDGF, EGF, and FGF, among others (reviewed in Cole, 1986; and Kaczmarek, 1986). C-myc mRNA has been detected in developing murine embryos (Zimmerman et al., 1984), tumor cells (Cole, 1986), the mammal liver regenerate (Makino et al., 1984; Thompson et al., 1986; Biesada and Chorazy, 1988), and in Xenopus oocyte (Taylor et al., 1986; King et al., 1986). Furthermore, c-myc expression seems to be related to the entry of cells into the cell cycle (Kelly et al., 1983), and more specifically, in the transition from the G_0 state into the G_1 phase, although, the level of expression of c-myc mRNA and protein has been found to be steady during the whole cell cycle of continuously proliferating cells (Thompson et al., 1985; Hann et al., 1985).

In the limb regenerate, numerous studies have already been devoted to understanding the cycle of blastema cells (reviewed in Wallace, 1981 and this book). They were intended to examine either the pattern of proliferation of the progenitor blastema cells or the disturbances of the cell physiology after nerve withdrawal, which stops the growth of the early regenerate (reviewed in Singer, 1952 and see also Maden, 1978; Olsen et al., 1984; Barger and Tassava, 1985; Boilly et al., 1985; among others).

Consequently, it was of interest to attempt to identify c-myc or myc-homologous gene expression in cells of the newt limb blastema. We are presenting here the first results of an ongoing research program. These have been the subject of a poster presentation at a previous meeting (Hourdry et al., 1988).

MATERIALS AND METHODS

This work has been carried out on limb regenerates of the american newt, Notophthalmus viridescens, taken at the medium bud stage of their development (Singer, 1952). In situ hybridization (ISH) permits visual detection of specific RNA transcripts in the cells of fixed tissues.

Tissue treatment: Fixation with 4% paraformaldehyde in 0.5x phosphate-buffered saline (PBS) for 2-3 hours at room temperature yielded good retention of cellular RNAs in isolated blastemata. Fixed specimens were then enclosed in agargel, before infiltration with ester wax, according to the method of Petavy (1985). Seven (7) µm longitudinal sections were deposited on slides coated with poly-L-lysine. Pretreatment of sections with proteinase K (1 µg/ml) (Brahic and Haase, 1978; also reviewed in Bentley-Lawrence and Singer, 1985), was then performed, as was acetylation, to degrade proteins and reduce background, respectively.

In situ hybridization: ISH sensitivity is related to the radioprobe, itself, as well as to the abundance and retention of target RNA in blastemal cells.

The expression of c-myc or homologous RNAs was detected using a radiolabeled Xenopus c-myc cDNA probe whose sequence has been shown to be highly conserved in various vertebrates (Taylor et al., 1986). Single-stranded DNA probes or randomly-primed DNA probes [25-250 base pairs (bp) in length], labeled with ^{3}H or ^{35}S, were added to hybridization medium (prepared, essentially, according to Brahic and Haase, 1978). Briefly, the hybridization solution was composed of: 50% deionized formamide, 4X STE [1X STE = 10 mM Tris HCl (pH 8.0), 100 mM NaCl, 1 mM EDTA (pH 8.0)], 1X Denhardt's solution (0.02% bovine serum albumin - 0.02% Ficoll -0.02% polyvinylpyrrolidone), 10% dextran sulfate, 300 µg/ml t-RNA and Xenopus myc cDNA probe (5 x 10^5 - 10^6 CPM/20 µl). Twenty (20) µl aliquots were applied to the sections for 16-20 hours at 30°C, in a humidified container. Post-hybridization treatments included washes in 50% deionized formamide/4X STE and 50% deionized formamide/2X STE, at room temperature.

Autoradiography was carried out with a Kodak NTB-2 emulsion and exposure was for 4 weeks (^{3}H-labeled probes) or 4-7 days (^{35}S-labeled probes) at 20°C. Sections were then developed (Kodak D-19), fixed, and counterstained with Giemsa for

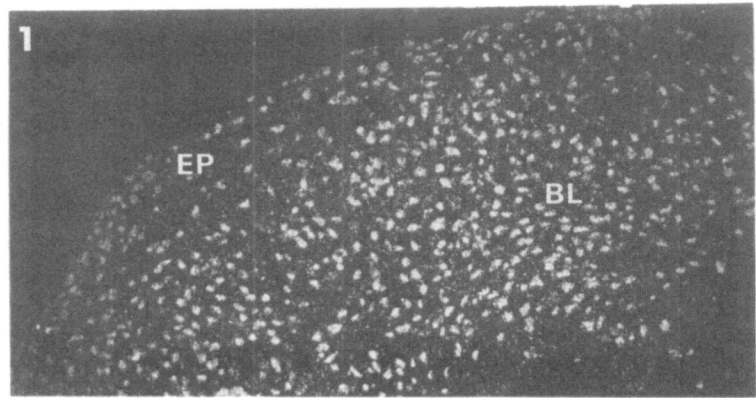

Fig. 1. ISH on the isolated limb blastema of the newt, taken at the medium bud stage. Grain densities are obvious in the cells of the wound epidermis (EP) and blastema (BL); Darkfield illumination (80x mag.).

viewing and photography under light microscopy (Zeiss Universal Microscope) with either darkfield or brightfield illumination.

Controls: In all experiments, similarly radiolabeled pUC18 bacterial plasmid insert vector DNA or M13 phage insert vector DNA (both lacking Xenopus myc cDNA inserts) was used on a few sections as a control for background, non-specific interaction(s). Sections of Xenopus oocytes were hybridized in parallel, as positive controls (Hourdry et al., 1988).

RESULTS

Myc-homologous RNA transcripts were shown to be present in cells of the thickened wound epidermis and in mesenchymatous cells which form the newt limb blastema (Fig. 1), using

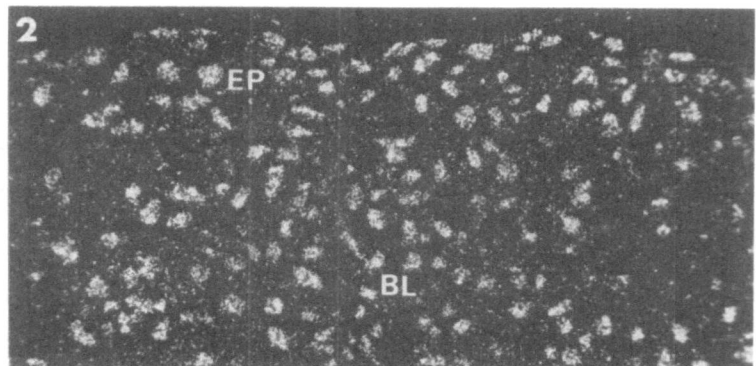

Fig. 2. Slightly higher magnification. Grains are present in most of the cells of the regenerate, indicating the presence of myc-homologous mRNA (165x mag.).

the method of _in situ_ hybridization with a radiolabeled Xenopus myc cDNA, as probe.

Grain densities, representing myc-related RNAs, were
observed in all layers of the wound epidermis. Similarly,
most, if not all, of the mesenchymatous cells gave a positive
signal, indicating the presence of myc-like RNA transcripts in
these cells, as well (Figure 2).

DISCUSSION

The positive results obtained with the _in situ_ hybridization technique using Xenopus (Anura) c-myc probe applied to
Notophthalmus (Urodele) tissue sections show that homologous
myc sequences exist within these two orders of Amphibia. It
demonstrates for the first time the presence of c-myc related
proto-oncogene RNA in urodele Amphibia. This finding extends
previously reported data demonstrating the conservation of
c-myc nucleic acid sequences in various vertebrates, such as
Xenopus, chicken and human (Cole, 1986; Taylor et al.,
1986).

In the growing limb regenerate, the presence of c-myc-
like mRNA in the cells is most likely related to their ability
to be stimulated to divide. Indeed, c-myc transcripts are
also known to be induced during mammalian liver regeneration
(Makino et al., 1984), although in this model it would be more
appropriate to describe a compensatory growth from the intact
liver lobes than a true regeneration (Goyette et al., 1984),
Lombard, 1988). Nevertheless, labeling indicates that cells
are induced to enter or stay in the cell cycle in order to
insure a rapid and steady regrowth of the organ.

We intended to interpret our data in light of our knowledge about cell proliferation in the newt limb regenerate. It
is accepted that the blastema is composed of a rather homogeneous population of cells which could all divide according to
a rather asynchronous pattern (Wallace, 1981 and this book).
This interpretation has been challenged by recent works using
pulse (^{3}H) thymidine labeling and continuous labeling combined
with determination of mitotic index (Tomlinson et al., 1985;
Goldhamer and Tassava, 1987). The new data confirm that most
of the blastemal cells divide during epimorphic limb regeneration - leaving aside a very small population of cells which
seem unable to divide. Consequently, most of the cells
(90-100%), theoretically, belong to the proliferation fraction
(Pf) of the blastemal cell population according to Goldhamer
and Tassava (1987). Nevertheless it emerges from their
studies that two subclasses of cells exist according to their
temporal capacity to incorporate thymidine. In the Pf population which is in a dynamic state of cell cycle evolution -only
about 1/4 of the cells would be actively cycling cells (AC).
These cells would be randomly distributed within the blastema.
The remaining, quite large, cell population would be transiently non-dividing, so called transiently quiescent (TQ)
cells which are in the G_1 phase of the cell cycle. Cells of
the large Pf population would move from the AC phase to the TQ
phase and back during blastemal growth.

The presence of myc-related transcripts in most of the

cells of the blastema could reflect the proliferative fraction of the blastema - without discrimination between the two sub-populations (AC and TQ) and, thus, may suggest that most of the cells are able to divide, although, at a different place.

Consequently, our first results are not surprising when compared with the results of others. Indeed, at the state examined, blastema cell proliferation and myc-like RNA expression are coincident, leading to the suggestion that the production of the myc-homologous RNAs is related to the observed cellular reproduction. Thus, it could be postulated that newt myc-related protein, translated from the mRNA detected, may be involved in maintaining blastemal cells in the proliferating state. Such a continuous mode of cellular replication is incompatible with cell differentiation, as demonstrated by experiments performed in vitro (see Muller, 1986). Therefore, it is possible that myc-homologous RNA translation products, produced in cells of the whole limb regenerate, during the moderate bud stage, may prevent these cells from entering the terminal differentiation phase. Our observations of high levels of myc expression on the lens placode and optic cup of developing Xenopus embryos could be in agreement with this result (Hourdry et al., 1988).

From this hypothesis, we can speculate that when the regenerate enters later stages of growth, i.e., when cartilage and muscle cells differentiate in continuity with the stump tissues, in a proximal-distal sweep, myc expression may be confined only to the growing distal regions. This hypothesis is currently under examination in our laboratory.

Our results demonstrate that a c-myc homologous gene is expressed in all the cells of the regenerating newt limb blastema. This raises the question of how this activity can be interpreted in terms of the phases of their cell cycles. There are still various views about the identity of the cells forming the blastema as far as their mitotic activity is concerned. If the myc-related gene is expressed only in G_1, we may deduce that cells are synchronous in the blastema. But, our results do not really allow us to propose arguments in favor of this hypothesis for the following reason: some studies have pointed out that the c-myc gene is not only expressed when a cell enters the cell cycle (G_0/G_1), but that the c-myc gene is expressed throughout the whole cell cycle, when cells are stimulated to proliferate (Thompson et al., 1985; Hann et al., 1985).

The growth of the limb regenerate is known to be controlled by numerous factors such as hormones (reviewed by Sicard, 1985) and the peripheral nervous system (Singer, 1978; and also reviewed in Sicard, 1985). Growth factors such as EDGF I and II have been shown to induce, in vitro, blastemal cell division (P. Albert, personnal communication). Therefore, blastemal cells are subject to various mitogens and any of them, including nerve products, could be involved in the myc expression observed.

The presence of myc-related transcripts in the whole thickness of the epidermis favors the interpretation that in the wound epidermis, all epidermal cells may divide, not only the cells located in the germinative layer of the epidermis, as has been observed in the normal newt epidermis (Hoffman and Dent, 1977). This hypothesis is supported by studies tracing (^{3}H) thymidine incorporation in the regenerating limb (Riddiford, 1969, Hay and Fischman, 1961; O'Steen and Walker, 1961; Schmidt, 1968). Thymidine labeling is randomly observed within a few cells of the wound-thickened epidermis showing that cells are in the S phase of their cell cycle at the time of labeling, and can possibly undergo mitosis no matter what their position is in the regenerated epidermis. Thus, in the wound epidermis, all cells can divide or are stimulated to do so.

In conclusion, these initial experiments demonstrate that c-myc related RNA transcripts are expressed in most, if not all, of the cells of the newt limb regenerate. The expression of a myc-homologous gene in these blastemal cells is most likely important for the maintenance of the proliferative state in this dynamic cell population.

ACKNOWLEDGEMENTS

The authors thank Martine Robin, IJM, and Gwyn Jackson, Case Western Reserve University for their help in typing and editing this work. This work was supported by grants from Association pour la Recherche sur le Cancer and INSERM.

REFERENCES

Barger, P.M. and Tassava, R.A., 1985, Kinetics of cell cycle entry in innervated and denervated forelimb stumps of larval Ambystoma, J.Exp.Zool., 233:151-154.

Bentley-Lawrence, J. and Singer, R.H., 1985, Quantitative analysis of in situ hybridization methods for the detection of actin gene expression, Nucleic Acids Res., 13:1777-1799.

Biesiada, E. and Chorazy, 1988, Expression of "cell-cycle-dependent" genes in regenerating rat liver, Cell Biol. Int. Rep., 12:483-492.

Bishop, J.M., 1985, Trends in oncogenes, TIGS., 1:245-249.

Boilly, B., Oudkhir,M. and Lassalle, B., 1985, Control of the blastemal cell cycle by peripheral nervous system during newt limb regeneration; continuous labelling analysis, Biol. Cell., 55:107-112.

Brahic, C. and Haase, A.T., 1978, Detection of viral sequences of low reiteration frequency by in situ hybridization, Proc. Natl. Acad. Sci. USA., 75:6125-6129.

Cole, M.D., 1986, The myc oncogene: its role in transformation and differentiation, Ann. Rev. Genet., 20:361-384.

Goldhamer, D.J. and Tassava, R.A., 1987, An analysis of proliferative activity in innervated and denervated forelimb regenerates of the newt, Notophthalmus viridescens, Development, 100:619-628.

Goyette, M., Petropoulos, C.J., Shank, P.R. and Fausto, N., 1984, Regulated transcription of c-ki-ras and c-myc during compensatory growth of rat liver, Mol. Cell. Biol., 4:1493-1498.

Hann, S.R., Thompson, C.B. and Eisenman, R.N., 1985, c-myc oncogene protein synthesis is independent of the cell cycle in human and avian cells, Nature, 314:366-369.

Hay, E.D. and Fischman, D.A., 1961, Origin of the blastema in regenerating limb of the newt Triturus viridescens. An autoradiographic study using tritiated thymidine to follow cell proliferation and migration, Dev. Biol., 3:20-59.

Heldin, CH., and Westermark, B., 1984, Growth factors: mechanism of action and relation to oncogenes, Cell, 37:9-20.

Hoffman, C.W. and Dent, J.N., 1977, Hormonal effects on mitotic rhythm in the epidermis of the red-spotted newt, Gen. Comp. Endocrinol., 32:512-521.

Hourdry, J., Brulfert, A., Gusse, M., Schoevaert, D., Taylor, M. and Mechali, M., Localization of c-myc expression during oogenesis and embryonic development in Xenopus laevis, Development (in press).

Hourdry, J., Géraudie, J., Singer, M. and Mechali, M., 1988, Expression of the c-myc proto-oncogene in the forelimb regenerate of the newt Notophthalmus viridescens, visualized by in situ hybridization. in: "Regeneration and Development", S. Inoue, ed., Okada Printing and Publishing Company, Maebashi, Japan.

Kaczmarek, L., 1986, Proto-oncogene expression during the cell cycle, Lab. Invest., 543:365-376.

Kelly, K., Cochran, B.H., Stiles, C.D. and Leder, P., 1983, Cell-specific regulation of the c-myc gene by lymphocyte mitogens and platelet-derived growth factor, Cell, 35:603-610.

King, M.W., Roberto, J.M. and Eisenman, R.A., 1986, Expression of the c-myc proto-oncogene during development of Xenopus laevis, Mol. Cell. Biol., 6:4499-4508.

Lombard-Des Gouttes, M.N., 1988, Rat liver regeneration and plasma factors: an overview of some early events in the compensatory growth process. in: "Regeneration and Development", S. Inoue Ed., Okada Printing and Publishing Company, Maebashi, Japan.

Maden, M., 1978, Neurotrophic control of the cell cycle during amphibian limb regeneration, J. Embryol. Exp. Morph., 48:169-175.

Makino, R., Hasaki, K. and Sugimura, T., 1984, c-myc transcript is induced in rat liver at a very early stage of regeneration by cycloheximide treatment, Nature, 310:697-698.

Muller, R., 1986, Proto-oncogenes and differentiation, TIBS, 11:129-132.

Olsen, C.L. and Tassava, R.A., 1984, Cell cycle and histological effects of reinnervation in denervated forelimb stumps of larval Ambystoma, J. Exp. Zool., 229:247-258.

O'Steen, W.K. and Walker, B.E., 1961, Radioautographic studies of regeneration in the common newt: II. Regeneration of the forelimb, Anat. Rec., 139:547-555.

Petavy, G., 1985, Reliable preparative procedures for the cytological study of yolk-laden insect eggs and embryos, Stain Tech., 60:321-330.

Pfeiffer-Ohlsson, S., Goustin, A.S., Rydnert, J., Wahlstrom, T., Bjersing, L., Stehelin, D. and Ohlsson, R., 1984, Spatial and temporal pattern of cellular myc oncogene expression in developing human placenta: implications for embryonic cell proliferation, <u>Cell</u>, 38:585-596.

Riddiford, L.M., 1960, Autoradiographic studies of tritiated thymidine infused into the blastema of the early regenerate in the adult newt <u>Triturus</u>, <u>J. Exp. Zool.</u>, 144:25-31.

Schmidt, A.J., 1968, "Cellular biology of vertebrate regeneration and repair", The University of Chicago Press, Chicago and London.

Sicard, R.E., 1985, "Regulation of vertebrate limb regeneration", R.E. Sicard, ed., Oxford University Press, New York, Oxford.

Singer, M., 1952, The influence of the nerve in regeneration of the amphibian extremity, <u>Quart. Rev. Biol.</u>, 27:169-200.

Singer, M., 1978, On the nature of the neurotrophic phenomenon in urodele limb regeneration, <u>Am. Zool.</u>, 18:829-841.

Taylor, M.V., Gusse, M., Evan, G.I., Dathan, N. and Mechali, M., 1986, <u>Xenopus</u> c-<u>myc</u> protooncogene during development: Expression as a stable maternal mRNA uncoupled from cell division, <u>EMBO J.</u>, 5:3563-3570.

Thompson, C.B., Challoner,P.B., Neiman, P.E. and Groudine, M., 1985, Levels of c-<u>myc</u> oncogene mRNA are invariant throughout the cell cycle, <u>Nature</u>, 314:363-366.

Thompson, N.L., Mead, J.E., Braun, L., Goyette, M., Shank, P.R. and Fausto, N., 1986, Sequential protooncogene expression during rat liver regeneration, <u>Cancer Research</u>, 46:3111-3117.

Tomlinson, B.L., Goldhamer, D.J., Barger, P.M. and Tassava, R.A., 1985, Punctuated cell cycling in the regeneration blastema of urodele amphibians: A hypothesis, <u>Differentiation</u>, 28:195-199.

Varmus, H.E., 1984, The molecular genetics of cellular oncogenes, <u>Ann. Rev. Genet.</u>, 8:553-612.

Wallace, H., 1981, "Vertebrate limb regeneration", John Wiley and Sons, New York.

Zimmerman, K.A., Yancopoulos, G.D., Collum, R.G., Smith, R.K., Kohl, N.E., 1986, Differential expression of <u>myc</u> family genes during murine development, <u>Nature</u> 319:783-790.

EXPRESSION OF THE WE3 ANTIGEN IN THE NEWT WOUND EPITHELIUM

Roy A. Tassava, Bruce L. Tomlinson
David J. Goldhamer, and Noorullah Akhtar

Department of Molecular Genetics
The Ohio State University
Columbus, Ohio 43210, USA

SUMMARY

mAb WE3 identifies an antigen abundant in wound epithelium but virtually absent from skin epidermis. The antigen is
developmentally expressed, being absent from the initial wound
epithelium but appearing by the 2nd week after amputation. The
present study was designed to investigate the origin of WE3
reactive wound epithelial cells. The results are consistent
with the view that skin epidermal cells migrate over the amputation surface and subsequently express the WE3 antigen. WE3
reactive wound epithelial cells do not appear to originate
from integumentary glands or "ovoid" cells of skin epidermis.

INTRODUCTION

After amputation of a newt limb, cells from the adjacent
epidermis migrate over the cut surface and form the wound
epithelium; this epithelium subsequently thickens but remains
dermis-free through blastema stages of regeneration (Hay and
Fischman, 1961; Singer and Salpeter, 1961). Results from several experiments have shown that a wound epithelium must be
present over the amputation surface for regeneration to proceed. Regeneration does not occur if whole skin is grafted
over the amputation surface (Mescher, 1976) nor if the distal
end of a freshly amputated limb stump is inserted into the
body cavity (Loyd and Tassava, 1980). It seems likely that
the wound epithelium is functionally and biochemically different from skin epidermis (Singer and Salpeter, 1961). Identifying these differences might reveal clues to the role of the
wound epithelium during regeneration and help to clarify
which, if any, of the various heretofore postulated roles for
the wound epithelium is correct (Singer and Salpeter, 1961;
Tassava et al., 1987).

Evidence has been presented from studies with monoclonal
antibody WE3 (mAb WE3) that the wound epithelium in fact does

differ from skin epidermis in at least one antigen (Tassava et al., 1986). WE3 reactivity is abundant in wound epithelium but virtually absent in skin epidermis. In addition, WE3 is not present in the epithelium that initially closes the wound but, during the 2nd week after amputation, WE3 positive cells begin to appear, increase in number thereafter, and remain abundant throughout the growth phase of regeneration (Tassava et al., 1986). Thus, WE3 positive cells are present when the wound epithelium is likely functioning in the promotion of cell cycle activity of the blastema mesenchyme (Mescher, 1976; Globus et al., 1980). At the four digit stage, reactive cells are restricted to the wound epithelium associated with the digits (Tassava et al., 1986). As WE3 represents the only known antigenic difference between skin epidermis and wound epithelium (Goldhamer et al., submitted), studies concerning the developmental expression and function of the WE3 antigen seem worthy of investigation.

One possible explanation for the change in antigenicity of the wound epithelium, as revealed by mAb WE3 reactivity, is that cells which are already WE3 positive migrate into the wound epithelium and subsequently become the dominant cell type. It has been suggested (Tassava et al., 1986; 1987) that a source of WE3 positive wound epithelial cells might be integumentary glands at the level of amputation. This glandular origin view is favored by several pieces of circumstantial evidence: Glands throughout the body react to WE3 at all times, WE3 positive wound epithelial cells have been observed in close proximity to glands at the level of amputation and, these distal-most glands exhibit extensive cellular proliferation (Tassava et al., 1986; 1987). A second possible source of WE3 positive wound epithelial cells is the small, ovoid, WE3 positive cell present in low frequency in skin epidermis (Tassava et al., 1986; Goldhamer et al., submitted); this cell could preferentially proliferate and populate the wound epithelium (Tassava et al., 1986).

A second possible explanation for the change in antigenicity of the wound epithelium is that the early WE3 negative wound epithelial cells, derived from skin epidermis, subsequently express or "turn on" the WE3 antigen (Tassava et al.,1986). If this latter view is correct, then it will be of interest to determine what factors cause this "turn on". Perhaps the wound epithelium begins to express WE3 in response to wound factors or a mesenchymal interaction.

In the present study, various possible origins of WE3 positive cells of the wound epithelium were investigated (Tassava et al., 1986;1987). We tested whether the small, ovoid, WE3 positive cells of skin epidermis exhibit proliferative activity before and after amputation. A timing/ontogeny study of WE3 reactivity in normal regenerates was conducted to determine if WE3 positive cells in the wound epithelium are always associated with glands. A skin cuffing strategy was used to increase the distance between the wound epithelium and distal-most glands to test whether the appearance of WE3 positive cells might be delayed. Finally, we established a grafting system to test whether a 5 day, WE3 negative wound epithelium, separated from glands, could "turn on" the WE3 antigen without a contribution from glands. The results favor the "turn on" model.

MATERIALS AND METHODS

Animal Care and Operative Procedures

Adult newts, <u>Notophthalmus viridescens</u>, collected from ponds in Southern Ohio, were kept in aerated tap water and fed a diet of lean beef. For all operations, newts were anaesthetized in neutralized MS-222 (Tricaine methane sulfonate, Sigma). Forelimbs were bilaterally amputated through the mid zeugopodium or distal third of the stylopodium; in any given experiment the level of amputation was constant.

Monoclonal Antibody Procedures

Monoclonal antibody WE3 was obtained by ammonium sulfate precipitation from hybridoma culture medium as previously described (Tassava et al., 1986). Precipitated antibody was resuspended in phosphate buffered saline (PBS), dialyzed extensively with PBS containing 0.1% sodium azide, and stored at -20°C. Ab was diluted between 1:50 to 1:100 prior to use.

Immunohistochemistry and Autoradiography

For immunofluorescence, tissues were fast frozen in Tissue Tek (Miles) on an isopropanol/dry ice slurry and cryostat sections were cut at a thickness of 10 microns. For immunoperoxidase staining, limbs were quick-frozen in liquid nitrogen-cooled isopentane and 8 micron cryostat sections were collected on acid-cleaned microscope slides and dried overnight at room temperature. Indirect immunofluorescence and immunoperoxidase staining has been described (Tassava et al., 1986; Goldhamer et al., submitted).

For autoradiography, immunoperoxidase-stained sections were coated with NTB-2 emulsion (Kodak; diluted 1:2 in distilled water), emulsion was allowed to harden at 24°C for 2 hours in a high humidity atmosphere, and slides were exposed for 2 weeks in the dark at 4°C. Autoradiograms were developed in half strength D-19 (Kodak) at 15°C for 4 minutes, fixed in Kodak fixer, and washed for 30 minutes in tap water.

Experimental Approaches:

Proliferative Activity of WE3 Positive Ovoid Cells. A unique oval-shaped WE3 positive cell found scattered throughout the skin epidermis is one possible progenitor of WE3 positive wound epithelial cells. To determine if these cells are active in proliferation, and therefore a potential source of the WE3 positive wound epithelium, their proliferative activity in skin epidermis and in the early wound epithelium was examined using a technique combining immunoperoxidase staining and autoradiography. Prior to amputation, newts were injected intraperitoneally with 5 μCi (60 Ci/mmole, ICN Pharmaceuticals) ^{3}H-thymidine every 12 hours for 96 hours. Some limbs were quick-frozen 12 hours after the last injection (Day 0), while limbs of other labeled newts were bilaterally amputated through the mid radius/ulna, and allowed to regenerate for 7 days before freezing. Cryostat sections were immunoperoxidase stained, washed in three changes of Tris-buffered saline (pH 7.4), rinsed three times in distilled water (5 minutes each), air dried, and immediately processed for autoradiography. Subsequently, sections were lightly (2 to 3 seconds) counterstained with Delafields's hematoxylin to aid in identifying nuclei, dehydrated, cleared in toluene, and coverslipped. Representative sections through each limb were observed, and every WE3 positive ovoid cell in skin epidermis and wound epithelium was scored for the presence of silver grains. A total of 381 ovoid cells in skin epidermis and 37 in wound epithelia were scored. The overall labeling indices (LIs; percentage of labeled cells) of skin epidermis and wound epithelium were estimated by counting between 200 and 400 cells (chosen by random grid placement) from each epithelium per limb.

Temporal and Spatial Appearance of WE3 Positive Cells. Wound epithelial reactivity patterns to WE3 were examined by immunofluorescence at regular intervals after amputation. The forelimbs of 22 adult newts were bilaterally amputated through the mid-zeugopodium. At 3 day intervals, through 33 days post amputation, four representative limbs were frozen for cryostat sectioning. The experiment was repeated once with an equal number of newts.

Skin Cuff Experiments. Skin glands were eliminated from a portion of the limb in the region near the future amputation surface by removing a complete circumference of skin from a 2 mm segment of the limb at the level of the zeugopodium: care was taken to avoid unnecessary injury to the underlying soft tissues. Gland-derived cells would thus have to migrate through a considerable length of cuff wound epithelium in order to reach the distal wound epithelium. Unoperated limbs and limbs in which the skin cuff was loosened but not removed (i.e., a sham) served as controls. Two weeks after removal of the skin-cuff or sham-operation, the limbs were amputated through the distal third of the operated region. Unoperated limbs were amputated at the same level. At 5 day intervals, three representative limbs from each group were frozen for cryostat sectioning and immunofluorescence; the experiment was repeated once.

Flank Wound Epithelium Grafts to the Jaw. To determine if the wound epithelium could express the WE3 antigen in the absence of glands, early, non-reactive wound epithelium was grafted into a jaw site, either alone or combined with muscle or blastema mesenchyme. In preliminary experiments limb wound epithelium was utilized in recombinants. Subsequently we found that WE3 reactivity appeared in wound epithelia of full-thickness skin wounds on the flank, although somewhat delayed and in lesser abundance than in regenerate epithelia. Flank wound epithelium was chosen as a test of the "turn on" view since large areas of wound epithelium could be obtained thus making the recombinants easier to handle in the grafting procedure.

At 5 days after removal of a 5x5 mm patch of full thickness skin from the lateral flank region posterior to the forelimb, a portion of this WE3 negative flank wound epithelium was removed and wrapped around blastema mesenchyme (from which its own WE3 positive wound epithelium had been removed); the combined tissues were then inserted into a site in the lower jaw. The host jaw site was prepared by peeling back three sides of a rectangle of skin. After positioning the graft tissue under the skin flap, the skin was sutured to hold the graft in place. Control grafts included flank wound epithelium with an insert of freshly dissected limb muscle, flank wound epithelium graftet without a mesenchymal insert flank wound epithelium that remained in situ, and flank epithelium that was loosened but immediately repositioned on the flank (sham operation). At 6 days after grafting or sham operation (11 days total time after making the flank wound), from five to seven wound epithelia were examined for WE3 reactivity in each experimental or control situation.

RESULTS

Ovoid Cells are not Active in Proliferation

Analysis of skin epidermis in unamputated limbs revealed that the ovoid cells are withdrawn from the cell cycle. Thus, despite the fact that 15-20% of all epidermal cells were labeled, none of over 200 ovoid cells scored were labeled (Table 1). When newts were labeled prior to amputation, between 70 and 90% of skin epidermal cells were labeled by 7 days post-amputation (Table 1). This increased labeling compared to that in unamputated limbs is due mostly to cell cycling after amputation in the presence of residual, unchasable isotope, which can persist in unknown stores in the newt body for weeks (Reyer, 1983; Goldhamer, 1988). Here, residual isotope served as a continuous supply of the DNA precursor and was used as a convenient means of continuous labeling. Despite 7 days of continuous labeling after amputation, none of the ovoid cells in skin epidermis became labeled.

These infrequent, round to oval-shaped WE3 positive cells are also found in the early wound epithelium that covers the amputation surface. Despite the prolonged labeling period, the ovoid cells of the wound epithelium also failed to incorporate ^{3}H-thymidine (Table 1; Fig. 1). The heavily labeled cells in Fig. 1 likely incorporated the isotope during the injection period, whereas the lighter labeling over the majority of wound epithelial cells was due to incorporation of residual isotope after amputation and, to a lesser extent, to label dilution by cell division (Goldhamer, 1988). These results show that ovoid cells of skin epidermis and wound epithelium do not participate in cell cycle activity, and therefore do not represent a WE3 positive progenitor cell type.

Temporal and Spatial Ontogeny of WE3 Positive Cells

Neither the skin epidermis nor the wound epithelium that initially formed over the amputation surface reacted to mAb WE3, and the time at which WE3 positive cells first appeared in the wound epithelium varied somewhat from limb to limb. Reactivity consistently appeared during the 2nd week, never later than 10 days after amputation, and infrequently during the 1st week. The earliest appearing WE3 reactive cells exhibited a diffuse or granular type of reactivity (Fig. 2A). As the intensity of WE3 reactivity increased, the fluorescence pattern became fibrous with the highest intensity near the margin of the cell (Fig. 2B). In many cases the first cells to react to mAb WE3 lined the basal layer of the wound epithe-

Table 1.   Lack of Proliferative Activity of WE3 Positive Ovoid
           Cells in Newt Skin Epidermis and Wound Epithelium.

	n	#^{3}H-T labeled ovoid cells in:[a]		Approx. LI of:[b]	
		skin epid.	wound epith.	skin epid.	wound epith.
Day 0[c]	4	0/207	–	15–20%	–
Day 7[d]	5	0/174	0/37	70–90%	30–80%

[a]For each limb, every oval cell in 4-7 representative sections
  was evaluated for labeling.
[b]LI=labeling index.   Ranges represent approximate LIs from
  lowest to highest limbs.
[c]On day 0 a wound epithelium had not yet formed.
[d]This group in effect represents limbs continuously labeled
  from 4 days pre-amputation through 7 days post-amputation
  due to the continual presence of residual isotope after the
  end of the injection period.

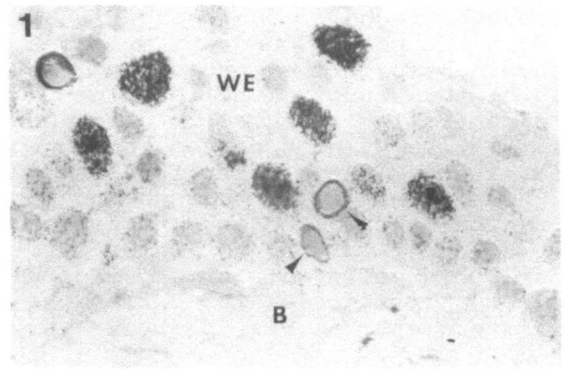

Fig. 1.   Ovoid cells of the wound epithelium are not
          active in proliferation.   This newt was
          labeled with ^{3}H-thymidine for 4 days before
          amputation and the forelimbs were frozen 7
          days after amputation. Unfixed cryostat sec-
          tions were stained by indirect immunoperoxi-
          dase, followed by autoradiography. Most of the
          wound epithelial cells are labeled but the
          ovoid cells are unlabeled (arrowheads). Ovoid
          cells of skin epidermis also remained unla-
          beled.   WE = wound epithelium.   B = blastema.
          x400. Light hematoxylin counterstaining.

lium but in other instances early reacting cells were scattered throughout all but the outermost layer of the wound epithelium. In general, as regeneration progressed, WE3 reactivity became more widely distributed and eventually nearly all wound epithelial cells reacted to mAb WE3, as described previously (Tassava et al., 1986).

In most cases, WE3 first appeared in the distal-most wound epithelium and persisted there, not obviously associated with glands, often to the early bud stage of regeneration (Fig. 3). Subsequently, reactive cells were seen more laterally and, as regeneration progressed, appeared to be more closely associated with glands. In only a few cases were early reacting cells closely associated with glands. At the beginning of regeneration the glands were of normal size and apices were perpendicular to the skin. By the early-bud stage of regeneration, the number of cells per gland appeared to have increased and the apices of the distal-most glands were redirected toward the distal tip of the limb, as described previously (Tassava et al., 1986; 1987).

Skin Cuff Experiments

Removing a complete circumference of skin from the middle third of the zeugopodium initiated the formation of a dermis-free wound epithelium around the full circumference of these unamputated limbs. These wound epithelia were similar in appearance to the epithelia which cover amputated limbs. The appearance of WE3 in skin cuff wound epithelia was delayed somewhat compared to wound epithelia over the amputation surface of regenerating limbs. During the first 14 days after removal of the skin cuff few, if any, WE3 positive cells were seen in the wound epithelium.

Limbs were amputated through the distal portion of the cuffed region at 14 days after cuff removal. At 10 days after amputation (24 days after the skin cuff operation), the distal wound epithelia covering the amputation surface exhibited the typical WE3 reactivity of early regenerate epithelia, which subsequently increased in intensity and in number of reactive cells as regeneration progressed. Serial sectioning revealed that the WE3 positive wound epithelial cells covering the amputation surface were separated from the distal-most glands by a considerable length of nonreactive wound epithelial cells along the cuffed region (Fig. 4). Thus, the wound epithelium of skin cuff limbs, despite being some distance from skin glands, attained WE3 reactivity in a manner similar to that described above in the temporal and spatial ontogeny study.

Flank Wound Epithelia Grafted to the Jaw Express WE3

When a patch of full-thickness flank skin was removed, a wound epithelium formed from cells that migrated from the cut edges of the skin epidermis. In histological sections, this wound epithelium resembled the epithelium that covers an amputated limb, but was much larger in surface area. In situ controls at 5 days were totally non-reactive to WE3; at 11 days, low intensity WE3 reactivity was present in a few cells bordering the mesenchyme. Reactivity increased during the next 2 weeks in intensity and number of cells. As was the case with limb wound epithelia, the first cells to react exhibited a diffuse or granular type of intracellular reactivity while later reactive cells exhibited more intense, fibrous reactivity (see Figs. 2A and B).

In contrast, wound epithelia combined with blastema mesenchyme and grafted to the jaw for a period of 6 days, exhibited abundant mAb WE3 reactivity (Fig. 5A; Table 2). Results from controls showed that the appearance of the WE3 antigen in these grafts was not related to the blastema mesenchyme grafted along with the flank wound epithelium (Table 2). WE3 reactivity was also observed in flank wound epithelium/limb muscle combinations. Furthermore, flank epithelium grafted alone also became WE3 positive (Fig. 5B) indicating that the expression of the WE3 antigen was independent of accompanying tissues. This response was not simply related to injury associated with manipulation of the tissue, since flank wound epithelium lifted and replaced back on the flank did not exhibit an enhanced appearance of WE3 reactivity (Table 2).

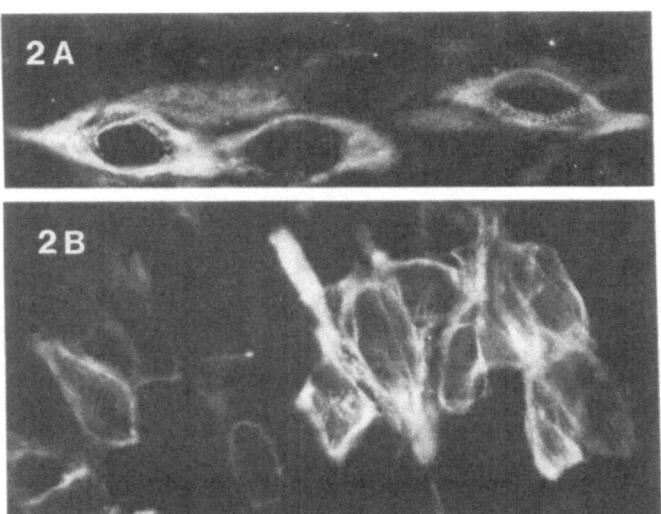

Fig. 2.   The initial WE3 reactivity in the wou...
epithelium, is a diffuse or granular form (A).
Later in the 2nd week post-amputation, cells
with a fibrous type of reactivity are seen
(B). Occasionally, a single cell can exhibit
diffuse, granular and fibrous reactivity. More
frequently, different cells of the same wound
epithelium exhibit either diffuse and granular
or fibrous reactivity. The wound epithelial
cells at bud stages of regeneration predominantly exhibit only fibrous reactivity. Different intensities of reactivity between different cells of the wound epithelium can be
seen at all stages of regeneration but particularly at early stages. Immunofluoresence.
x1200.

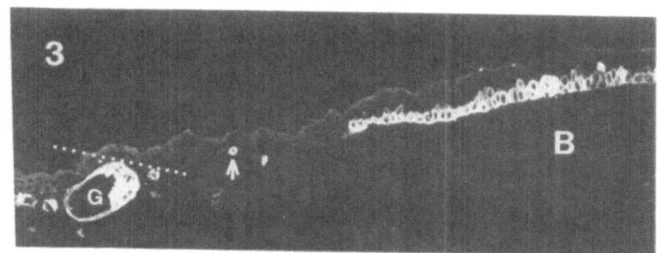

Fig. 3.   A portion of a cryostat cut section of an ear-
           ly-bud stage regenerate showing abundant WE3
           positive cells in the wound epithelium. Note
           that most of the WE3 reactive cells border the
           blastema mesenchyme (B). A gap of non-
           reactive wound epithelium can be seen between
           the distal-most gland (G) and the WE3 reactive
           cells of the wound epithelium. A WE3 reactive
           ovoid cell (arrow) is present in the gap
           region. The level and plane of amputation is
           indicated by the dotted line. Immunofluores-
           cence. x 120.

DISCUSSION

       Taken together, the results argue in favor of a "turn on"
origin   of WE3 positive wound epithelial cells.   The most con-

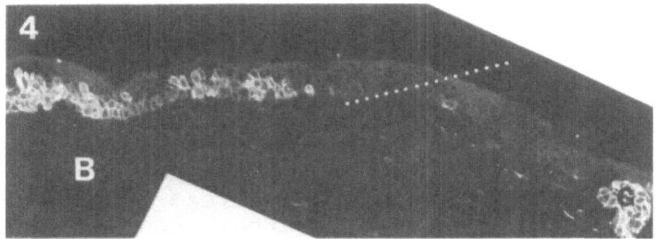

Fig. 4.   A portion of a cryostat cut section of an ear-
           ly-bud stage regenerate developing on a limb
           from which a cuff of skin had been removed
           before amputation. Note that most of the WE3
           reactive cells are near to or border the
           mesenchyme (B). The level and plane of ampu-
           tation (though the cuffed area) is indicated
           by the dotted line. Note that WE3 reactive
           wound epithelial cells of the regenerate (dis-
           tal to the dotted line) are continuous with
           WE3 reactive cells of the cuff wound epithe-
           lium (proximal to the dotted line). A region
           of WE3 negative wound epithelium separates
           the most distal gland(G, lower right corner of
           micrograph) from the most proximal WE3 reac-
           tive cells of the cuffed region. Immunofluor-
           escence. x 130.

clusive data are from the grafting experiment in which case
wound epithelia became WE3 reactive in the absence of a glandular association. It is of interest that blastema mesenchyme
was not essential for WE3 expression; epithelia grafted with
muscle or alone also became reactive. Thus the causative factors in WE3 expression are likely related to the grafting
procedure or the graft environment. Whatever the nature of
these factors, an enhancement in the appearance of WE3 in the
grafted epithelia occurred, compared to flank wound epithelia
in situ, a result that argues against a pre-programming of WE3
"turn-on" within the epithelium prior to grafting. Injury to
the grafted wound epithelium was not by itself sufficient to
cause earlier WE3 appearance since "sham injured" flank
epithelia left in situ did not exhibit enhanced expression
compared with flank wound epithelium that was not injured. It
is reasonable to suggest that the factors responsible for
"turning on" WE3 in grafted flank wound epithelia are similar
or related to those which cause WE3 expression in the limb
wound epithelium during regeneration. This experimental system has the potential to reveal insights into the nature of
these factors/interactions.

The data from the temporal and spatial ontogeny and the
cuffing experiments are also consistent with the "turn-on"
origin of WE3 positive cells. Serial sectioning revealed that
the first cells to show any WE3 reactivity were usually not
contiguous with the distal-most skin glands. The initial
reactivity was therefore separated from direct association
with the distal-most glands. If the reacting cells were coming from glands, one would expect to observe WE3 positive
epithelial cells contiguous with the distal-most skin glands.
Furthermore, the removal of a cuff of skin prior to amputation
should have delayed the appearance of WE3 positive cells in
the distal wound epithelium until cells migrated from the
glands through the cuffed region of wound epithelium and into
the wound epithelium, some 500 μm from the distal-most glands;
such delays were not observed. Finally, in the wound epithelium of both normally regenerating and cuffed limbs and in the
epithelium of flank wounds in situ, a step wise change in the
type of reactivity to WE3 was seen. The earliest reactivity
was weak, diffuse or granular, and spread throughout the cytoplasm but, with increased time after amputation, the reactivity pattern became stronger, appeared fibrous, and was usually concentrated immediately inside the cell membrane. The
diffuse and granular type of reactivity likely represents the
early expression of the WE3 antigen.
WE3 positive cells of the wound epithelium clearly do not
originate from the distinctive infrequently seen WE3 positive
ovoid cells present in skin epidermis, since this cell type
did not incorporate labeled thymidine and thus rarely, if
ever, participates in cell cycle activity. The few ovoid
cells seen in the wound epithelium are probably carried in by
migrating skin epidermis. The function of ovoid cells in the
newt integument remains a mystery.

Experiments by Hay and Fischman (1961) with adult newt
demonstrated that the initial wound epithelium originates from
migrating skin epidermal cells adjacent to the amputation surface. From the present results, we can now suggest that this
epidermally derived wound epithelium, due to some unknown
influences, begins to express the WE3 antigen at the beginning

46

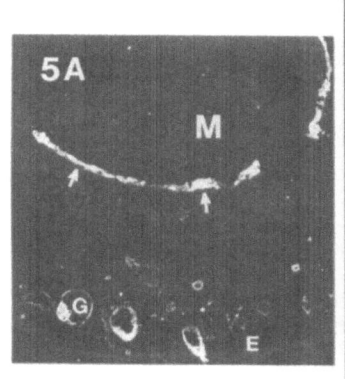
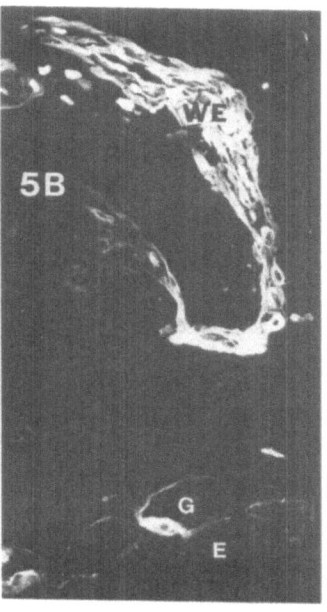

Fig. 5. WE3 negative flank wound epithelium becomes
WE3 positive after grafting to the lower jaw
in the absence of glandular association. (A)
WE3 reactivity in cells of 5 day WE3 negative
flank wound epithelium (arrows) after a 6 day
graft period in the lower jaw. This wound
epithelium was combined with mesenchyme (M)
of a mid-bud stage blastema. Glands (G) and
epidermis (E) of the lower jaw are some distance away. Immunofluorescence. X 30. (B)
WE3 reactivity in cells of a 5 day WE3 negative flank wound epithelium (WE) after a 6
day graft period in the lower jaw. This
wound epithelium was grafted alone. Immunofluorescence. x 130.

of the 2nd week after amputation. The antigen is first dif-
fusely distributed throughout the cytoplasm, then becomes
granular, and finally fibrous. Increasing numbers of WE3 positive cells during early stages of regeneration likely originate from increased numbers of cells "turning on" the antigen,
as opposed to proliferation of a few "turned on" cells. Later,
during blastema stages of regeneration, WE3 positive cells in
the wound epithelium clearly become active in proliferation,
as seen by their frequent incorporation of ^{3}H-thymidine after
a pulse labeling period (Goldhamer, unpublished observations).

That a functional wound epithelium might originate from
skin glands is a novel and interesting hypothesis (Tassava et
al., 1986; 1987) even though not supported by the present
results. In those cases where a close association was

Table 2. Reactivity of WE3 in Flank Wound Epithelia Grafted to the Jaw Either Alone or With Accompanying Mesenchyme of Muscle.

Experimental or Control[a]	WE3 Reactivity
5 day flank wound epithelium _in situ_	-
11 day flank wound epithelium _in situ_	+
11 day flank wound epithelium , sham-operated on day 5 and examined 6 days later	+
5 day flank wound epithelium in combination with blastema mesenchyme for 6 days	+++
5 day flank wound epithelium in combination with limb muscle for 6 days	+++
5 day flank wound epithelium after a 6 day graft period without accompanying tissues	+++

[a]In all cases, at least five tissue samples were examined

observed between WE3 positive cells of the wound epithelium and distal glands (Tassava et al., 1986; 1987) it is likely due to the expression of WE3 in epidermally-derived epithelial cells near glands, rather than reflecting a relationship between glands and wound epithelium. It remains to be explained why there is increased proliferation in these distal skin glands and why glands reorient to "point" toward the wound epithelium (Tassava et al., 1986; 1987). Gland cell proliferation, which occurs concomitant with the appearance of WE3 in the wound epithelium and with the onset of blastema cell proliferation (Goldhamer, 1988), may be stimulated by injury factors present in the vicinity of the amputation surface. Reorientation, on the other hand, appears to occur during the phase of regeneration when the blastema is undergoing the greatest elongation and may be a consequence of the gland necks being pulled distally by epithelial cells migrating to cover the evermore distal end of the limb (Tomlinson and Goldhamer, unpublished observations).

Monoclonal antibody WE3 recognizes the only developmentally regulated wound epithelial antigen thus far described. The change in antigenicity of wound epithelial cells from WE3 negative to WE3 positive suggests that the wound epithelium is performing a function not undertaken by the skin epidermis. Other research with mAb WE3 suggests that the antigen marks cells that may be involved in ion transport and/or secretion (Goldhamer et al., submitted). Because the blastema is initially an avascular environment, metabolic wastes and various ions may need to be eliminated from the blastema environment. Perhaps it is the role of WE3 positive cells to maintain a balanced ionic environment in the blastema and/or to rid the blastema of metabolic wastes (Golhamer et al., submitted).

Further work with mAb WE3 is designed to reveal the nature of the wound epithelium.

ACKNOWLEDGMENTS

Supported by NIH Grant # HD 22024 to R.A. Tassava. We thank Donna E. Tomlinson for technical assistance.

LITERATURE CITED

Globus, M., Vethamany-Globus, S., and Lee, Y.C.I., 1980, Effect of apical epidermal cap on mitotic cycle and cartilage differentiation in regeneration blastemata in the newt, Notophthalmus viridescens. Dev. Biol., 75:358-372.

Goldhamer, D.J., 1988, Analysis of cell proliferation and the roles of nerves and wound epithelium during forelimb regeneration in the adult newt, Notophthalmus viridescens, Ph. D. Dissertation, The Ohio State Univ.

Goldhamer, D.J., Tomlinson, B.L., and Tassava, R.A., 1988, A developmentally regulated wound epithelial antigen is also present in a variety of secretory/transport cell types. [submitted for publication].

Hay, E.D., and Fischman, D.E., 1961, Origin of the blastema in regenerating limbs of the newt, Triturus viridescens. An autoradiographic study using tritiated thymidine to follow cell proliferation and migration, Dev. Biol., 3:26-59.

Loyd, R.M., and Tassava, R.A., 1980, DNA synthesis and mitosis in adult newt limbs following amputation and insertion into the body cavity, J. Exp. Zool., 214:61-69.

Mescher, A.L., 1976, Effects on adult newt limb regeneration of partial and complete skin flaps over the amputation surface, J.Exp.Zool., 195:117-128.

Reyer, R.W., 1983, Availability time of tritium-labeled DNA precursors in newt eyes following intraperitoneal injection of ^{3}H-thymidine, J.Exp.Zool., 226:101-121.

Singer, M., and Salpeter, M.M., 1961, Regeneration in Vertebrates: the role of the wound epithelium, in: "Growth in Living Systems", M.X. Zarrow, ed., Basic Books, Inc., New York.

Tassava, R.A., Goldhamer, D.J., and Tomlinson, B.L., 1987, Cell cycle controls and the role of nerves and the regenerate epithelium in urodele forelimb regeneration; possible modification of basic concepts, Biochem. Cell Biol., 65:739-749.

Tassava, R.A., Johnson-Wint, B. and Gross, J., 1986, Regenerate epithelium and skin glands of the adult newt react to the same monoclonal antibody, J. Exp. Zool., 239:229-240.

by Bruce M. Carlson, on

GENE EXPRESSION IN REGENERATION

During the course of the meeting, Professor Singer told the story of Thomas Hunt Morgan, one of the founding fathers of the field of regeneration. Morgan switched from regeneration to the newly established field of genetics because he felt that with the tools available at that time the major problems of regeneration would never be solved in his lifetime. History has proved his prediction to be correct.

Before leaving regeneration, Morgan made a major contribution to the field by articulating some of the fundamental problems facing researchers on regeneration. These problems have been investigated throughout this century, primarily with the aid of descriptive or classical experimental embryological techniques, and most of them have still not yet been resolved to the satisfaction of all.

Studies of genetics have made several quantum leaps from the seemingly simple time of Morgan, during which the many mutations of _Drosophila_ provided the material basis for validating the fundamental hypothesis of Mendel and solidifying the concept of the gene. Now, many decades later, the molecular nature of the gene is understood in remarkably great detail, and many molecular biologists, geneticists and developmental biologists have moved on to study gene expression. Although a disarmingly simple term, gene expression is used to cover an enormous spectrum of topics, ranging from the identification of specific gene products, to studying the molecular mechanisms underlying the individual or coordinate activation or repression of genes, to investigating the effects of specific gene activity on cellular or supercellular phenomena. All of these studies are made possible by an array of techniques that now permit exquisitely sensitive analysis of minute amounts of material. It is ironic that in this era of technological sophistication some fields of biology have rapidly advanced to a stage where complex processes are rather well understood at a molecular level, whereas others are seemingly refractory to the influence of the new biology. For years, the study of regeneration has certainly been a prime example of the latter category. With investigators just making the first reports of molecular investigations in some areas of regeneration, it is an appropriate time to look at some major problems in regeneration in relation to the techniques currently available.

One of the earliest types of investigation involved the identification of specific gene products in regenerating cells and tissues and the comparison of the types and sequence of appearance of these gene products with that of normal ontogenesis. Studies on both regenerating lens and muscle involving electrophoretic and immunocytochemical analysis have shown that fully regenerated tissues produce basically the same sets of marker proteins as those that characterized the original undamaged tissues. Studies on the process of regeneration are now showing the transitions of isoforms of these molecules. A key question is whether or not the sequences of isoform transitions in these tissues are the same during normal ontogenesis and regeneration. Blotting techniques, which are useful on a mixed tissue if they are used to detect a cell-specific molecule, are now being used to look at the temporal sequence of mRNA formation in relation to the appearance of the corresponding proteins or key cellular events in the regeneration of these tissues. With the powerful techniques of *in situ* hybridization, it should soon be possible to piece together spatiotemporal relationships between mRNA formation and the formation of specific proteins in the regenerating cells. Systems such as muscle have been relatively easy to analyze because to date most investigations have been directed toward only a single cell type and a single phase (cytodifferentiation) of the regenerative process. However, as was shown by Yamada in the regenerating amphibian lens, a real understanding of a regenerative process involves an integrated knowledge of both the cellular and molecular events that extend in time from the initial tissue lesion to the activation of a population of precursor cells, and then the differentiation of these cells.

The regenerating limb poses a problem of far greater complexity. First of all, the limb consists of many tissue types, which are arranged into extremely complex patterns. Although there are good molecular markers that will allow one to distinguish between cartilage and bone, how does one distinguish between a humerus and a phalanx or an anconeus and an interosseous muscle other than by morphological analysis? At this point it is hard to say whether more detailed analysis of the mature structure or investigating earlier stages of normal development will provide a clue. Gene expression is not limited to simple proteins and RNA's, and it is highly likely that questions like the above can be profitably approached by other techniques, such as the monoclonal antibody analysis used by Tassava on the wound epidermis or the cell behavior analysis reported by Stocum.

The crux of the problem of limb regeneration is understanding how the tissues of a mature limb give rise to a population of blastemal cells that have built into them the blueprint for the reconstruction of a limb that is virtually identical to the original one. One cannot forget that superimposed upon what might seem to be a simple recapitulation of an embryological process is a complex overlay of neural, hormonal and humoral controls that must act in concert to permit the regenerative process to occur.

Where does one begin? One of the long unanswered questions is the origin of cells that make up the regeneration blastema. Such an understanding is essential to analysis of many properties of the regeneration blastema. Over the years, many attempts have been made to trace cells into and out of the blastema, and for various reasons, all have failed. Any good cell tracing study must meet the following criteria:
1) The origin of the cells to be traced must be known exactly. This can now be dealt with by techniques of cloning cells. 2) A label of some sort must be attached to these cells only. This requirement has not been met by many labels used in the past. 3) The label must remain in these cells throughout the

regenerative process and not be diluted or transferred to other cells. 4) The label should be present and/or expressed in all cells descended from the originally labelled cells. Such a requirement is important in attempts to label cells with retroviruses. 5) One should be able to identify exactly the phenotypes of the cells that contain the label in the mature regenerate. Satisfying all these criteria is not a trivial task, and all labels used to date have failed to meet one or more of the above criteria. At this symposium, Brockes presented a novel approach to tracing cells using the hypo-methylation of heavy chain myosin DNA as a tool for tracing the fate of cells desired from the muscle line. His finding of the labelled DNA in regenerated cartilage is a fascinating example of the use of a purely molecular tool to attack a cel-lular problem. Only time and additional experimental verifi-cation will tell whether the conclusions drawn from this study will stand up, but such innovative applications of new molecu-lar approaches are vital for the further understanding of problems such as this.

Any expression of pattern or even increase in mass is not possible without cell proliferation, and understanding how external influences, such as neurotrophic factors, growth fac-tors, etc. are translated into blastemal cell division is a large, but potentially attainable goal at the present time. However, it involves the painstaking identification of a long cascade of cellular factors starting with receptors and extending through a complex sequence of membrane-related mole-cules, such as protein kinase C, to cytoplasmic binding pro-teins, oncogenes and transcription factors before there can be an effect on genes that are involved in regeneration.

Studies on oncogenes, such as that reported by Geraudie and coworkers, could be very informative, but it will be important to determine whether c-myc, for instance, behaves in the regeneration blastema as it does in many other populations of proliferating cells or if, as in skeletal muscle, it plays a more paradoxical role. A similar comment applies to the study on heat shock and heat shock-like proteins reported by Carlone. The ability to detect these proteins at such a fine level of discrimination carries with it the challenge of determining, by experimentally manipulating the system, their role in the regenerative process.

In dealing with pattern formation, I frequently remember Professor Charles Thornton's admonition to me many years ago, "Stay away from morphogenesis. It is a bag of worms". Yet, it is one of the most fascinating and compelling aspects of regeneration. Although our traditional approaches to the problem have resulted in several good conceptual frameworks for investigating pattern formation and morphogenesis, we are still almost completely ignorant of morphogenetic mechanisms operating at or below the level of the cell. We can take some comfort from the spectacular success of the Drosophilists in relating pattern formation in their system to the coordinate expression of several families of genes, but our hope must be tempered with the realization that the vast store of mutants available to the Drosophilists is almost totally lacking for those who work on amphibians. We are at this point reduced to poor cousins who must beg for crumbs from the banquet in the forms of <u>Drosophila</u> probes that might cross-react with amphibian tissue. However, screening gene libraries is a must, and there are already some fascinating findings, such as these reported by Brockes on the difference between limb and tail in the expression of certain of the homeobox transcripts. This kind of information could be very helpful in explaining why the regenerating limb and tail seem to follow completely different sets of rules.

One of our basic deficiencies is that despite the extensive use of the concepts of positional information and positional memory, we have no concrete idea of what they are and how they become established. Retinoic acid is proving to be a powerful tool for investigating some aspect of the positional properties of cells. Since positional properties, according to the current concepts, represent an important element of gene expression, it is vital to find cellular or molecular markers of position and to determine which cells are the carriers of positional memory. Investigations at the level of genes that are known to be of morphogenetic importance in some systems represent one fruitful approach. The use of _in situ_ hybridization or monoclonal antibodies to localize morphogenetically important gene products in normal or experimentally altered regenerating limbs should provide a valuable link between purely molecular and morphological studies. At this point it is too early to tell from what source or direction the first major breakthrough in the molecular understanding of morphogenesis in regeneration will occur.

Regeneration is a field in which is all too easy for the uninitiated to become overconfident and the old-timers to become overwhelmed by the total complexity and many ramifications of the problem. But it is the very complexity of the problem that makes it so attractive. There are many valid and effective styles of conducting research on regeneration, and if its long history can teach us any lessons, it is that by remembering the old while embracing the new that many of the greatest advances have taken place.

CELLULAR AND IMMUNOLOGICAL
ASPECTS OF REGENERATION

HOW MANY POPULATIONS OF CELLS

OCCUR IN THE MESENCHYMAL BLASTEMA?

H. Wallace

Department of Genetics, University of Birmingham
Birmingham B15 2TT, U.K.

SUMMARY

The article contains a study designed to test the punctuated cell cycle hypothesis of limb regeneration and its postulate of temporarily quiescent cells. The labelling index of 6-13 day axolotl limb regenerates was assessed after incorporating tracer for 3-24 hours. The results indicate that the blastemal mesenchyme is never an entirely homogeneous population of proliferating cells, but that a persistent dedifferentiation of stump tissues and the later condensations of skeletal primordia entirely account for the non-cycling fraction found at any one time. Although large axolotls regenerate more slowly than younger and smaller ones according to morphological stages, there is no reliable evidence of a corresponding change in the regenerate growth rate. A literature survey finds no distinct subpopulations have been identified in the blastema by other criteria. Consequently, blastemal mesenchyme cells can still be validly considered as a single population.

INTRODUCTION

The formation and growth of the mesenchymal blastema is the basic process of limb regeneration. Chalkley (1954) provided the most useful description of the cellular events underlying this process in adult newts (Notophthalmus viridescens). All mesodermal tissues in the distal millimetre or so of the limb stump contribute dedifferetiated cells to the mesenchyme, which then increase mainly by cell division but remains an apparently uniform population until local condensations indicate the formation of skeletal primordia. Cell division occurs in both the stump and the mesenchyme during the second and third weeks after amputation, but cell counts indicated that blastemal growth could mainly be accounted for by the repeated division of the initial few thousand mesenchymal cells. Two autoradiographic studies have provided some confirmation of this description by showing the general distribution of cells which incorporate radioactive thymidine and by recording a persistent recruitment of dedifferentiating

Table 1. Duration in Hours of Cell Cycle Phases in Blastemal Mesenchyme.

	Gl	S	G2	M	Total	
Adult N. viridescens	4.6	32	5.1	1	42.7	Tassava et al.
at 24°C	6	32.6	3.3	1	42.9	(1985)
Young A. mexicanum	3.2	40	4.8	1	49	Wallace and
at 20°C	9.5	38	4.5	1	53	Maden (1976)
	11	30	6	1	48	Maden (1976)
	12.5	28	6.5	1	48	

stump cells to the growing blastema (Hay and Fischmann, 1961; O'Steen and Walker, 1961). The fact that tritiated thymidine labelled only a proportion of mesenchymal cells aroused no interest at the time and, of course, could not be interpreted until the rate of clearance of the tracer and the duration of cell cycle phases were established.

Several estimates of the cell cycle in blastemal mesenchyme have now been obtained. The more reliable ones used a chase to ensure genuine pulse-labelling and are shown in Table 1; others have been summarised previously (Wallace, 1981). Even the best technique only provides median estimates, with errors and an inherent variability amounting to ±10% for the whole cycle and more for individual phases. Results reported here provide minor refinements for two of these estimates. A consistent feature of Table 1 is that both the whole cell cycle and the S phase are unusually protracted, so that S phase occupies 70±10% of the cycle in repeatedly dividing cells.

If all mesenchyme cells are dividing repeatedly and asynchronously, this calculation predicts that about 70% of them should become labelled soon after exposure to the tracer. Furthermore, this labelling index (LI) should steadily increase as long as the tracer persists, to approach 100% after 10-20 hours. The variability mentioned above prompts caution about the time required to reach a uniform labelled population. Tomlinson et al. (1985) explained these predictions in detail and reported most observed LIs were grossly below the predicted values under both pulse and continuous labelling conditions. Consequently, an appreciable proportion of the mesenchyme cells cannot be cycling at any one time. In addition to considering the possibility that a blastema might contain a non-proliferative fraction of cells which never divide, they suggested that proliferative population should be subdivided into actively cycling (AC) and transiently quiescent (TQ) subpopulations. TQ cells are supposedly held in an extended G1 phase but will become labelled after prolonged exposure to tracer, when they eventually enter S phase. Transient quiescence also implies that they have divided previously during regeneration. Otherwise, TQ would merely be a pretentious term for those cells which dedifferentiate later in regeneration and the entire 'Punctuated Cell Cycle' hypothesis would be misnamed.

According to this hypothesis, the relative size of the AC and TQ subpopulation could be affected by innervation, hormones etc., and thus modulate the rate of regeneration in urodeles or restrict its occurrence in other vertebrates. Considering these potential implications, the hypothesis has not attracted much attention outside its originating group, which perhaps reflects doubts about the methods and interpretation of labelling studies. My own reservations centred on the small amounts of isotope used and the dangers of scoring lightly labelled cells in sections, which should theoretically lead to underestimating the true labelling index. Three years ago, I fancied I could avoid these potential defects to achieve more consistent results, or at least provide an independent confirmation of such studies, by labelling 6,7 and 8 day axolotl blastemata for periods of up to 24 hours. This study was extended in the following two years to 9, 10 and 11 day cone stages and finally to 12 and 13 day palette stages.

MATERIALS AND METHODS

The study was conducted on 7-8 cm long juvenile axolotls (Ambystoma mexicanum) maintained at 20°C, to make use of prior experience. For instance, a completely labelled blastema could only be expected when the mesenchymal cell cycle duration is equal to the population doubling time, about 8-9 days after amputation in these conditions (Maden, 1976). Limited incubator space compelled the study to be split over three spawnings, with some attendant size discrepancy. After rearing about 100 larvae for 4-5 months at room temperature, three groups of 16 larvae were subjected to bilateral forearm amputation when 6 cm long on three consecutive days, before placing at 20°C for the specified period of regeneration (6-13 days). Each specimen was then injected with methyl-^{3}H-thymidine to exceed 1 μCi/g body weight. Both arms from duplicate specimens were preserved at intervals of 3-24 hours after this injection. Some injections were delayed up to 6 hours to permit fixation at a more convenient time, but each labelling period was accurate to within 10 minutes. All operations were performed under anaesthesia (0.05% MS222, pH 7) and the specimens were finally killed without recovering consciousness.

The arms were preserved in ethanol-acetic (3:1) for 1-2 days at 4°C before storage at -20°C. Subsequent processing by Feulgen staining, the isolation, dissociation and squashing of mesenchymal cells and the details of autoradiography have been described before (Wallace and Maden, 1976). This technique is designed to achieve the efficient detection of labelled cells. Unincorporated tracer is removed by the acid hydrolysis and washes involved in the staining. Squashing disperses the cytoplasm to leave flattened nuclei in direct contact with the film. An autoradiographic exposure of 25 days gives satisfactory labelling, with a background of 5-10 grains per nucleus. Only nuclei with twice that grain count were scored as labelled, but most of them could be easily identified as having 50-200 grains which darken the Feulgen stain appreciably. All mitotic figures are labelled in those squash preparations made 12-72 hours after a single injection of the tracer, so all actively cycling cells should become labelled.

Autoradiographs of sections cannot match this efficiency, as they will only register electrons emanating from portions of nucleus within 1 μm of the emulsion. Nevertheless, sections provide spatial information which is lost in squashed preparations. Consequently, four limbs from each groups were fixed in Bouin and sectioned longitudinally at 8 μm for autoradiography. These limbs were all fixed 24 hours after the injection of tracer and thus constitute a control series of 7-14 day regenerates used for histology rather than an assessment of the labelling index. A small extra series of squashes were obtained after 24,48 and 72 hours of labelling, obtained by injecting duplicate specimens at daily intervals prior to fixation of 11 day regenerates. This supplements the earlier series of squashes made up to 72 hours after a single injection of tracer (Wallace and Maden, 1976).

RESULTS

The observations cover three stages of regeneration: the growing blastema at 6-8 days, cone at 9-11 days and flattened palette at 12-14 days after amputation. I shall first describe briefly my observations on sectioned material and then consider the labelling and mitotic indices recorded from squashed preparations.

Histology

Longitudinal sections of 7 day regenerates showed an accumulated mass of undifferentiated mesenchyme at the apex of the arm stump. The epithelium has already thickened as an apical cap 6-8 cells thick. The apical cap thickens further during the next two days and persists up to 12 days. The mesenchyme mass increases steadily until the first procartilage condensations form against the ends of the stump skeleton at 10 days. Subsequent growth is probably slower but is also obscurred by the changing shape of the regenerate. Dedifferentiation continues at least 14 days after amputation, as indicated by the eroded skeleton and frayed muscles at the end of the arm stump in all the material examined. The coordination of external appearance and histological change is precisely that recorded by Maden (1976), although the sequence is appreciably slower in these larger specimens.

Autoradiographs of these sections, after exposure to tracer for much of the previous 24 hours, show that labelled nuclei are scattered throughout the epithelium of the stump but virtually absent from the distal half of the apical cap. Labelled nuclei occur throughout the mesenchymal mass with an irregular or patchy distribution. Labelled nuclei are also present in all tissues at the end of the arm stump, some in chondrocytes and myocytes but mostly in elongated fibroblasts or unidentified cells of the connective tissue. The presence of labelled nuclei in the eroded skeleton of 10-14 day regenerates, in particular, shows that dedifferentiation of cartilage continues even while the blastemal mesenchyme is condensing to produce extensions of the skeleton. In the elbow region, well above the site of amputation, occasional labelled nuclei are present in all tissues except cartilage. This constitutes the only notable discrepancy between the present observations and those of Hay and Fischman (1961) on the limb regenerates of adult newts. Such labelled cells are presumably associated with the normal growth of the arm rather than with its regeneration.

The zone of dedifferentiating stump tissue contains labelled and unlabelled nuclei in both overtly differentiated and nondescript cells. Some cells evidently enter S phase before they are liberated into the blastemal mesenchyme, while other cells dedifferentiate completely before entering the mitotic cycle. The presence of labelled and unlabelled cells throughout the blastemal mesenchyme tends to support that conclusion, as quantified in the next section. The condensed skeletal primordia of palette stages have fewer labelled nuclei centrally than at their sides or in the intervening mesenchyme, suggesting that much of their growth is by accretion rather than by cell division.

Squashed Preparations

The mesenchyme is converted to a layer of dispersed nuclei in these preparations, allowing direct contact with the emulsion. Some clusters of nuclei in most preparations include unusually few labelled interphases and no division figures, indicating that the nuclei are not completely randomised and that a series of sample counts are required to estimate the labelling index in each regenerate. In practice, 8 counts of 100 nuclei from each preparation achieved a reasonable consistency.

The average percentage of nuclei which were labelled (labellilng index, LI) is fairly constant for replicate regenerates and does not vary appreciably with stage of regeneration between 6 and 11 days after amputation, but increases steadily with time of exposure to the tracer up to 24 hours (Fig. 1). The linear regression from the data gives LI=35+0.8t (where t is the number of hours between the injection and fixation). That is slightly less than half prediction from cell cycle analysis (70+2t), implying that the proliferating population only constitutes about half the mesenchyme cells in each regenerate at these blastema to cone stages. The LIs from 12 and 13 day palettes are significantly lower than in the earlier stages (Fig. 1), as expected from the additional unlabelled procartilage cells recognised in the corresponding sections. Regenerates fixed at 11 days after 2 or 3 days labelling show an increased LI, 62% and 71% respectively. At that rate of increase, a labelling period of 6 days would be required to approach 100% LI.

Mitotic figures were scored either by scanning the entire preparation to accumulate 20-100 cases from the younger regenerates or from a sample of 100 at palette stages. The percentage of labelled mitoses is constant for these different stages of regeneration but rises abruptly between 3 and 9 hours after injection (Fig. 1). A few prophases are lightly labelled at 3 hours, the majority of division figures are labelled at 6 hours, 98% by 9 hours, 99.6% by 12 hours and 100% at 18 and 24 hours. The 50% intercept of this curve yields a median value for the duration of G2+1/2M of slightly

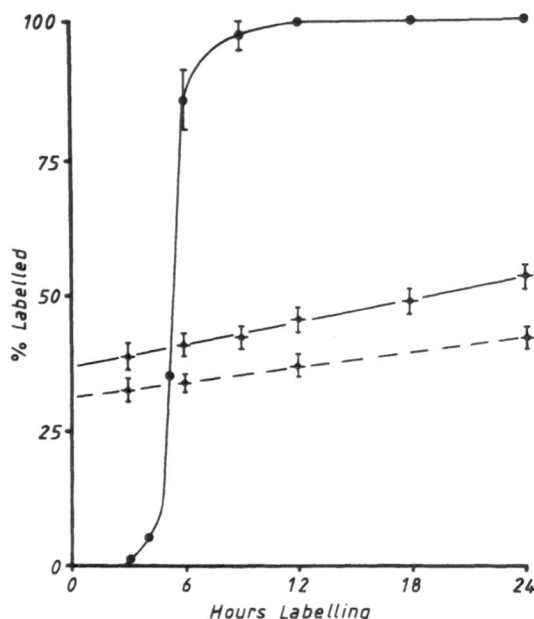

Fig. 1. Mesenchymal labelling indices (small symbols with standard deviations) for 6-11 day regenerates and 12-13 day palettes (broken line) both shown a linear increase with labelling time. The percentage of mitotic figures which are labelled (large symbols, continuous curve) includes some previously published data (Wallace and Maden, 1976).

under 6 hours, but a few mesenchyme cells take only half that time while some others take twice as long. The mitotic index was only roughly estimated in most preparations as being in the range of 1-2%. More accurate counts were only made after 24 hours exposure to the tracer, when all mitotic figures are labelled and most of the proliferating population is also expected to be labelled. Samples of 1000 labelled nuclei were scored then to yield a labelled mitotic index of 3.44±0.55%. That is probably an underestimate, as early prophase and late anaphase stages are easily obscured by the dense cover of silver grains. The duration of mitosis derived from this estimate, Tm=3.44% of 50 hour cycle, is 1.72 hours. Consequently, G2 has a median duration of approximately 5 hours.

DISCUSSION

The most striking observation recorded here is that dedifferentiation, which is one of the earliest events of regeneration, continues throughout blastemal growth up to the late palette stage. The final period of dedifferentiation may be more concerned with effecting a continuity between stump and regenerate tissues than with contributing to the mass of blastemal mesenchyme, but a persistent contribution at least equal to the number of proliferating cells would largely explain the labelling index data. Consider the early palette stage when the total number of mesenchyme cells doubles in 2 days, i.e. the time of one cell cycle, according to Maden (1976). We cannot now assume that all the cells divide, for only 50-60% of them become labelled. Probably half of the mesenchyme must be composed of newly liberated dedifferentiated cells, some of which are also labelled. A clear majority of cells must be the undivided products of dedifferentiation at earlier stages, when the doubling time is less than 2 days. The doubling time in adult newts is considerably longer than 2 days throughout regenerate growth (Chalkley, 1954) and only a quarter of the cells are actively cycling at any one time (Goldhammer and Tassava, 1987). The predominant category of mesenchyme here seems to be cells which have been liberated for several days but are still not ready to enter the cell cycle. The rate at which they are supplied to the mesenchyme and their transit time before entering S phase are unknown. A transit time of 8 days, for instance, could easily offset a low rate of dedifferentiation to produce the observed labelling index and growth rate.

My main objective of trying to improve upon the techniques employed in earlier studies has been partly overtaken by the protagonists of the punctuated cell cycle hypothesis. Tomlinson and Barger (1987) have demonstrated that thymidine injections of 1 μCi/g are required to give adequate labelling in axolotl regenerates, and have used that dosage to record labelling indices somewhat higher than found here. Their 95% LI after 3 days of continuous labelling actually exceeds the theoretical expectation for scoring sections of a completely labelled blastema. They reasonably conclude that virtually all the mesenchymal cells are in the proliferative fraction, that 61% are actively cycling at the early accumulation phase (when dedifferentiation surely predominates) and that 88% are cycling at the cone stage. These data apply to 4-5 cm axolotls, while corresponding tests on cones of 15-16 cm axolotls show actively cycling cells to be 57% of the proliferating population - much closer to the results recorded here. My reservations about discriminating between AC and TQ cells are explained in the introduction. Goldhammer and Tassava (1987) have conducted an equivalent study on adult <u>N. viridescens</u> regenerates. Adequate doses of tracer here only produced an LI of 20% initially in early blastema, rising to 90% after 8 days of continuous labelling. Thus virtually all mesenchyme cells divide at some time during regeneration but only about 26% are actively cycling at any one time, so the remainder are considered to be transiently quiescent.

Both of these studies clearly include newly dedifferentiated cells among the TQ subpopulation and lack the means of identifying any other category of TQ cell. That point is reinforced by the evidence from denervated limbs shown in both studies: the continuous presence of mesenchyme cells with a reduced labelling index and mitotic index is certainly the consequence of continued dedifferentiation in regressing denervated limbs of larvae. Conceivably, that might also apply to the apparently static denervated blastemata of adults, where a steady replacement offers an explanation of the residual labelling index in predominantly G1 or G0 phase cells found in these species and in <u>Pleurodeles</u> (Boilly et al., 1985; Oudkhir et al.,1985). Boilly et al. (1986) have also estimated the frequency of mesenchyme cells in different phases by image analysis. This not only confirms their previous conclusions but should indicate relative phase durations. I should predict from their data that <u>Pleurodeles</u> blastemal cells have a distinctly shorter S phase and probably a shorter cycle time than the corresponding figures shown in Table 1.

RATE OF REGENERATION

The rate of limb regeneration gradually declines with age in growing axolotls and is reduced at or after metamorphosis in other urodeles, according to widely known but poorly documented observations (Wallace, 1981). Tomlinson et al. (1985) suggested that a controlled partition of mesenchyme cells into AC or TQ fractions, according to their punctuated cell cycle hypothesis, could account for such changes. Tomlinson and Barger (1987) correlated the partition they estimated to the size of axolotls, where 5 cm larvae regenerated at about twice the speed of 15 cm juvenile specimens. The curious feature of this comparison is that the rate of regeneration is always measured in terms of morphological stages which are independent of regenerate size, whereas mesenchymal proliferation is essentially concerned with growth in volume. I have checked the retardation of regeneration as axolotls of a single spawning grew from 3-10 cm, to complement the data mentioned above and that from adult axolotls (Tank et al., 1976). There are several defects in the data obtained (Table 2). Firstly a mistaken incubator setting resulted in a water temperature of 19°C. Secondly, running several samples at the same time allowed relatively slow growing specimens to predominate in some size groups. Thirdly, almost 20% of regenerates from 6-10 cm specimens showed the totally unanticipated propensity of regenerating hands with five fingers. Table 2 therefore only provides a rough indication of the regeneration rate, especially as individuals could reach a particular stage one day ahead or behind the median time. The rate of regeneration measured in this way certainly slows down steadily during larval growth. Figure 2 shows typical examples of 14 days regenerates taken from this study revealing that those from larger specimens are at least equal in size to the faster developing regenerates of smaller specimens. At any particular morphological stage, of course, larger specimens have larger regenerates. There is no evidence that they grow more slowly.

The punctuated cell cycle hypothesis has little merit then, being based on a rather subjective partition of mesenchymal cells and leading to quite ambiguous predictions. I still prefer the classical concept of a single proliferating population of mesenchyme cells which is continually augmented by the dedifferentiation of stump tissues and eventually depleted as cells aggregate into tissue primordia and redifferentiate.

FUTURE PROSPECTS

The search for discrete subpopulations will continue, no doubt, for there is a strong tendency to believe that dedif-

Table 2.   Median Time to Reach Arbitrary Stages of Limb
            Regeneration, Given in Days After Amputation at
            19°C.

Stage*	Body length at time of amputation							mean±3 mm
	30	40	50	60	70	80	90	100
Blastema	6.5	7	8	8.5	9	9.5	10	10.5
ConBe	8.5	9	9.5	10	10.5	11	11.5	12
Palette	10.5	11.5	12	12	13	13	14	15
Notch	11.5	13	14	15	16	17	18	19
3 digits	16	17	18	20	21	22	24	26
4 digits	18	19.5	21	22.5	24	25	26	29
Body growth (mm/week)	7	7.5	7	3.5	3.8	4.1	4.4	3.6

*Based on samples of 20 regenerates for 30-80mm specimens or
samples of 10 for larger ones. Notch is a slight indentation
between the first two digits; digits 3 and 4 are only scored
when they project enough to form acute angled tips.

ferentiated cells are more likely to revert to their former
state than to transform into another cell type. Brockes
(1984; Kintner and Brockes, 1984,1985) has described what
seems to be the most sensible way of identifying such subpopulations, by screening antibodies and immunostaining. The
specific antigen chosen for further study was produced temporarily in most mesenchyme cells of the early blastemata of
some species but not others, whether the cells were derived
from Schwann cells or muscle fibres or connective tissue, and
seemed to be nerve dependent (Fekete and Brockes, 1987, 1988).
More appropriate antigens may await discovery, of course, but
neither this example nor any other known to me contradicts the
general conclusion from experimental evidence that tissue-
specific characteristics are effaced during dedifferentiation
and positional characteristics can be modified. Tracing particular groups of cells through regeneration will also be difficult, considering the indirect evidence that they may become
scrambled in the blastema (Faber, 1960) and have the ability
to migrate to a preferred position later (Nardi and Stocum,
1983).

    The identification of extrinsic molecules which control
blastemal growth may prove more rewarding, if equally difficult to achieve. This survey can only touch on the kinds of
controlling factors, selectively citing recent articles.   It
is generally accepted, I believe, that the wound epithelium
encourages dedifferentiation or inhibits redifferentiation of
internal tissues (Stocum, 1985). That might happen simply by
the local leakage of electrolytes, as measured by ionic currents (Borgens, 1985), or it might involve an epidermal secretion of collagenase (Vanrapenbusch and Lassalle, this volume).
The pervasive influence of nerves maintains mesenchymal prolif-

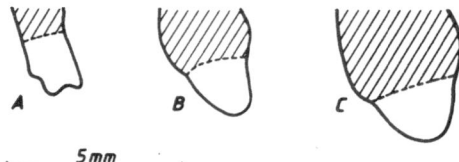

Fig. 2. 14 day regenerates illustrating data in
Table 2.
A = late notch from amputation at 40 mm
B = palette from amputation at 70 mm
C = cone from amputation at 100 mm

eration, apparently by stimulating postmitotic G1 cells to
remain in the cycling population. Extracts of nervous tissue
cause a similar stimulation, implying the presence of one or
more neurotrophic chemicals, as do a variety of growth factors
(Gospodarowicz and Mescher, 1980; Brockes and Kintner, 1986).
Opinion is more divided about the effects of the better known
hormones. Several of them stimulate proliferation at physiological concentrations, but regeneration still occurs without
them in larvae or well-nourished adults (Wallace, 1981). In
the absence of fluctuations correlated to the progress of
regeneration, these hormones may be considered as merely permissive factors, part of the general background like oxygen or
nutrients which we must take for granted (the "conditions
banales" of Guyenot, 1927). The concept of a hormonal milieu
(Liversage et al., 1985), which strikes me as an elaborate and
closely argued version of this view, has already provoked
testy dissent concerning especially the influence of insulin
and opiates (Globus, 1988). The concept exposes a genuine
problem, however, even if it seems defeatist. Altering the
availability of one hormone is liable to affect the levels of
several others. Any consequent perturbation of regeneration
cannot be attributed with certainty to the manipulated hormone, while the absence of any perturbation could mean either
that the hormone has no influence over regeneration or that
other hormones have compensated for the alteration.

The obvious means of escaping from this dilemma is to
test the effect of hormones or other substances on cultured
blastemata. It is still hazardous to distinguish between controlling and permissive factors, however, as an explanted
blastema achieves little beyond survival and some differentiation in a simple culture medium. Additives to the medium
which enhance synthetic activity and cell division include
growth hormone, prolactin, hydrocortisone and insulin; a
source of neurotrophic factors such as co-cultured ganglia or
substance P (Globus and Globus, 1985); fetal serum and transferrin (which may also be a neurotrophic factor according to
Mescher and Munaim, 1984); an ionophore to mediate calcium
entry (Globus et al., 1987) and probably others which I have
missed. I suspect the activity stimulated by any particular
combination of additives is still subnormal in comparison to
in vivo activity, and wonder if all of them together would
enhance the metabolism of blastemal cells still further. Such
a complete mixture might approximate to the hormonal milieu
but could be tested more precisely, by eliminating particular

ingredients to see which (if any) of them have a major effect on mesenchymal proliferation.

REFERENCES

Boilly, B., Moustafa, Y., and Oudkhir, M., 1986, Image analysis of blastema cell proliferation in denervated limb regenerates of the newt, _Pleurodeles waltlii_ M., _Differentiation_, 32:208.

Boilly, B., Oudkhir, M., and Lassalle, B., 1985, Control of the blastemal cell cycle by the peripheral nervous system during newt limb regeneration; continuous labelling analysis, _Biol. Cell_, 55:107.

Borgens, R.B., 1985, Natural voltage gradients and the generation and regeneration of limbs, _in_: "Regulation of Vertegrate Limb Regeneration", R.E. Sicard, ed., University Press, Oxford.

Brockes, J.P., 1984, Mitogenic growth factors and nerve dependence of limb regeneration, _Science_, 225:1280.

Brockes, J.P., and Kintner, C.R., 1986, Glial growth factor and nerve dependent proliferation of the regeneration blastema of urodele amphibians, _Cell_, 45:301.

Chalkley, D.T., 1954, A quantitative histological analysis of forelimb regeneration in _Triturus viridescens_, _J. Morph._, 94:21.

Faber, J., 1960, An experimental analysis of regional organization in the regenerating fore limb of the axolotl (_Ambystoma mexicanum_, _Arch. Biol._, 71:1.

Fekete, D.M., and Brockes, J.P., 1987, A monoclonal antibody detects a difference in the cellular composition of developing and regenerating limbs of newts. _Development_, 99:589.

Fekete, D.M., and Brockes, J.P., 1988, Evidence that the nerve controls molecular identity of progenitor cells for limb regeneration, _Development_, 103:567.

Globus, M., and Globus, S.V., 1985, In vitro studies of controlling factors in newt limb regeneration, _in_: "Regulation of Vertebrate Limb Regeneration", R.E. Sicard, ed., University Press, Oxford.

Globus, M., Globus, S.V., and Kesik, A., 1987, Control of blastemal cell proliferation by possible interplay of calcium and cyclic nucleotides during newt limb regeneration, _Differentiation_, 35:94.

Globus, S.V., 1988, A critical analysis of hormone action in newt limb regeneration: a possible opiate (endorphin) connection, _in_: "Regeneration and Development", S. Inoue, ed., Okada, Maebashi.

Goldhammer, D.J., and Tassava, R.A., 1987, An analysis of proliferative activity in innervated and denervated forelimb regenerates of the newt, _Notophthalmus viridescens_, _Development_, 100:619.

Gospodarowicz, D., and Mescher, A.L., 1980, Fibroblast growth factor and the control of vertebrate regeneration and repair, _Ann. N.Y. Acad. Sci._, 339:151.

Guyenot, E., 1927, Le probleme morphogenetique dans la regeneration des urodeles: determination et potentialites des regenerat,. _Rev.Suisse Zool._, 34:127.

Hay, F.D., and Fischmann, D.A., 1961. Origin of the blastema in regenerating limbs of the newt _Triturus viridescens_, _Dev. Biol._, 3:26.

Kintner, C.R., and Brockes, J.P., 1984, Monoclonal antibodies
    identify blastemal cells derived from the dedifferen-
    tiating muscle in newt limb regeneration, Nature,
    308:67.
Kintner, C., and Brockes, J.P., 1985, Monoclonal antibodies to
    the cells of a regenerating limb, J. Embryol. exp.
    Morph., 89:37.
Liversage, R.A., McLaughlin, D.S., and MacLaughlin, H.M.G.,
    1985, The hormonal milieu in amphibian appendage regen-
    eration, in: "Regulation of Vertebrate Limb Regener-
    ation", R.E. Sicard, ed., University Press, Oxford.
Maden, M., 1976, Blastemal kinetics and pattern formation dur-
    ing amphibian limb regeneration, J. Embryol. exp.
    Morph., 36:561.
Mescher, A.L., and Munaim, S.I., 1984, Trophic effect of
    transferrin on amphibian limb regeneration blastemas,
    J. Exp. Zool., 230:485.
Nardi, J.B., and Stocum, D.L., 1983, Surface properties of
    regenerating limb cells: evidence for gradation along
    the proximodistal axis, Differentiation, 25:27.
O'Steen, W.K., and Walker, B.E., 1961, Radioautographic
    studies of regeneration in the common newt. II.
    Regeneration of the forelimb, Anat. Rec., 139:547.
Oudkhir, M., Boilly, B., Lheureux, E., and Lassalle, B., 1985,
    Influence of denervation on the regeneration of Pleuro-
    dele limbs, Differentiation, 29:116.
Stocum, D.L., 1985, The role of the skin in urodele limb
    regeneration, in: "Regulation of Vertebrate Limb Regen-
    eration", R.E. Sicard, ed., University Press, Oxford.
Tank, P.W., Carlson, B.M., and Connelly, T.G., 1976, A staging
    system for forelimb regeneration in the axolotl,
    Ambystoma mexicanum, J. Morph., 150:117.
Tassava, R.A., Laux, D.L., and Treece, D.P., 1985, The effects
    of partial and complete denervation on adult newt fore-
    limb blastema cell-cycle parameters, Differentiation,
    29:121.
Tomlinson, B.L., and Barger, P.M., 1987, A test of the punctu-
    ated-cycling hypothesis in Ambystoma forelimb regener-
    ates: the role of animal size, limb innervation, and
    the aneurogenic condition, Differentiation, 35:6.
Tomlinson, B.L., Goldhammer, D.J., Barger, P.M., and Tassava,
    R.A., 1985, Punctuated cell cycling in the regeneration
    blastema of urodele amphibians, Differentiation,
    28:195.
Wallace, H., 1981, "Vertebrate Limb Regeneration", Wiley,
    Chichester.
Wallace, H., and Maden, M., 1976, "The cell cycle during
    amphibian limb regeneration", J. Cell Sci., 20:539.

PROTEIN KINASE C ACTIVITY DURING LIMB REGENERATION OF AMPHIBIANS

M. Oudkhir[1,2], I. Martelly[3],
M. Castagna[4], J. Moraczewski[5] and
B. Boilly[1]

[1]Laboratoire de Biologie des Facteurs de
 Croissance, Université de Lille, 59655
 Villeneuve d'Ascq Cedex, France
[2]Laboratoire de Biochimie, Université de
 Marrakech, Maroc
[3]MYREM, Université de Paris XII, France
[4]Laboratoire sur le Cancer, Villejuif, France
[5]Laboratoire de Zoologie, Université de Varsovie
 Pologne

SUMMARY

PKC activity was measured during limb regeneration of
Pleurodeles both in blastemata and in the central nervous
system (CNS). We showed that specific PKC activity increased
post-amputation until 14 days both in membrane and cytosolic
compartments. A partial translocation from cytosol to membrane occurred at 14 days postamputation; denervation prevented this translocation only in blastema mesenchyme but
co-culture of spinal ganglion and blastemata stimulated it.
Limb regeneration, or denervation of non amputated limb had
few effects on specific PKC activity in CNS but increased
significantly translocation from cytosol to membrane both in
spinal cord and brain. The results are discussed with regard
to the control of blastema cell proliferation by nerves and
the influence of nerve regeneration on translocation of PKC in
CNS.

INTRODUCTION

Limb regeneration in Amphibians is a complex phenomenon

Abbreviations

BNTF = blastema neurotrophic factor; CNS = central nervous system; EGTA = ethyleneglycol-bis (b amino ethyl ether)
N',-tetraacetic acid; NdBGF = nerve derived blastema growth
factor; NTF = neurotrophic factor; pa = post amputation;
PKC = protein kinase C; PMSF = phenylmethysulfonyl fluoride;

which results from specific cellular interactions. Among them, nerve-blastema interactions have received particular attention.

It is well established that limb regeneration is controlled by nerves which stimulate blastema cell proliferation. The nerve influence on the blastema is considered to be mediated by growth factor(s)-like molecule(s), the so called NTF (Singer, 1978) or NdBGF (Boilly and Albert, 1988), the nature of which is poorly understood. Recently it was shown that regenerating nerves produce one of the mitogen(s) required for blastema cell proliferation (Boilly and Bauduin, 1988). Moreover, proliferating blastemata stimulate the growth of brachial nerves into blastemata by releasing a neurotrophic factor, the BNTF different from NGF (Bauduin et al., in press). So, in this system nerves and blastemata have a stimulatory effect on each other.

A known action of NdBGF is the regulation of the cell cycle of blastema cells (Boilly et al., 1986; Tassava et al., 1987) and protein synthesis (Singer, 1978). In regenerating neurons, protein synthesis is under the control of growth factor (Bauduin et al., in press). There is also an elevation of protein synthesis in spinal cord (Boilly and Scaps, 1988) and spinal ganglia (Bao et al., 1986) during limb regeneration. Although we know some of the biochemical effects of growth factors related to nerve-blastema interactions during limb regeneration, we are unaware of the mechanism of transduction of these growth signals. In order to approach this problem we studied the activity of PKC during limb regeneration of Pleurodeles waltlii M. in buds and in the CNS.

MATERIALS AND METHODS

Forelimbs of Pleurodeles waltlii Michah from our laboratory colony, aged 6 months to 1 year, were amputated through the distal stylopod under anaesthesia (MS 222, Sandoz). Limbs were denervated by cutting spinal nerves 3,4 and 5 at the shoulder level. Samples consisting either of blastemata, spinal cords, or brains were harvested by dissection of animals on a cooled glass plate, then quickly frozen in liquid nitrogen and stored at -70°C before use.

PKC activity was determined by measuring the incorporation of ^{32}P from gamma-^{32}P ATP into histone III S in the absence or the presence of PKC activators as described in Couturier et al., (1984). Briefly, the samples (30-40 blastemata, 2 brains or 2 spinal cords) were homogenized at 4°C in 1 ml 20 mM Tris HCl pH 7.5 with 10 mM EGTA, 2 mM EDTA, 2mM PMSF, 0.01% leupeptin and 250 mM sucrose (i.e homogenization buffer). Homogenates were centrifuged 1 h (105.000g, 4°C) and supernatant (i.e. cytosolic fraction) was stored at -70°C. The pellet was mixed in 0.5 ml of homogenization buffer with 0.5% Triton X100 (1 h, 0°C), then centrifuged (105.000 g, 1 h, 4°C); the supernatant (i.e. membrane fraction) was stored at -70°C and the pellet discarded. PKC activity was measured on 4-6 ug protein (Bradford's method, 1976) in 20 ul for each fraction. In each assay 25 ug histone III S was used with or without 5 μg phospholipid and 100 ng phorbol ester in the presence of 2 mM EGTA. The reaction was carried out at 25°C

☑ Membrane
☐ Cytosol

Fig. 1. Percentage of total PKC activity in membrane
and cytosol fractions of limb blastemata of
<u>Pleurodeles</u> at 3 times after amputation: 7,
14,24 days. 0 day = non amputated newt.

for 5 min. PKC activity was quantified as the difference
between histone phosphorylation in the presence of phospholipid and phorbol ester and in the absence of these activators
of PKC. The results are expressed as PKC specific activity
(pmol ^{32}P/mg protein/min) and distribution (in %) of total PKC
activity between cytosolic and membrane compartments.

The assays were performed on whole buds, epidermal caps,
blastemata, whole spinal cords and whole brains at 3 times
after amputation, 7 days (i.e. early-bud stage), 14 days (i.e.
mid-bud stage) and 24 days (i.e. 2 digits stage) after amputation. Unamputated animals served as controls. The effect of
denervation on PKC activity was studied in mid-bud stage
blastemata and PKC activity was measured 48 or 96 hours later.

RESULTS

A) <u>PKC Activity in Blastemata During Limb Regeneration of
Pleurodeles</u>

1) Specific PKC activity increased post amputation and
plateaued at 14 days pa (about 40 pmol/mg/min, i.e. 4X the
control). This increase in activity occured both in the membrane and in the cytosolic compartments until 14 days pa when
the percentage of total PKC activity reached its highest value
in the membrane compartment (46% versus 28.7% for the control,
Fig. 1).

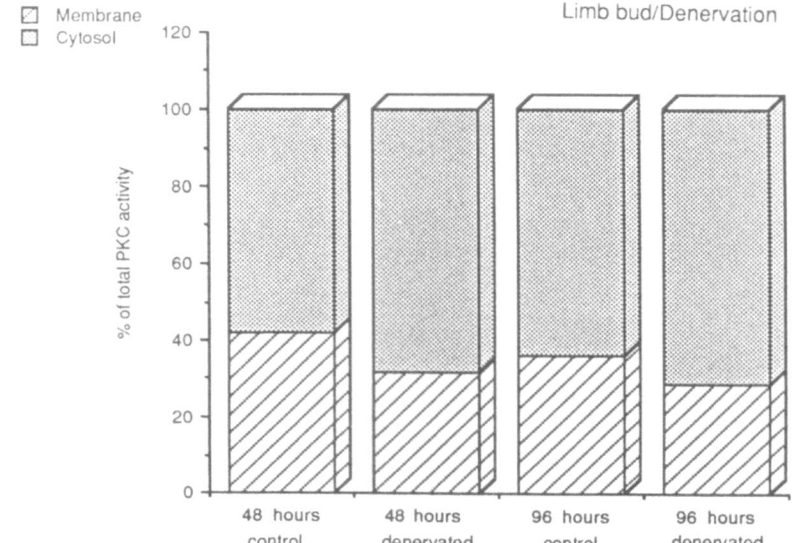

Fig. 2. Percentage of total PKC activity in membrane
and cytosol fractions of mid bud stage
blastemata of <u>Pleurodeles</u> 48 and 96 h after
denervation.

2) Specific PKC activity did not change 48 hours after
denervation of mid-bud stage blastemata. 96 hours after
denervation specific PKC activity increased 1.5 fold in the
cytosolic compartment while there was no significant change in
the membrane compartment. Total PKC activity declined in
denervated animals, as expected based on the fact that denervation negatively affects protein synthesis. So in the membrane compartment the percentage of total PKC activity
decreased from 42% in control to 32% in experimental, 48 h
after denervation and from 36% in control to 29% in experimental, 96 h after denervation. In the same period, the percentage of total PKC activity in the cytosolic compartment
increased from 58% to 68%, 48 hrs after denervation and from
64% to 71%, 96 hrs after denervation (Fig. 2).

3) 96 hours after denervation of mid-bud stage blastemata, PKC activity changed in the bud mesenchyme but not in
the epidermal cap. In denervated blastemata, specific PKC
activity of mesenchyme increased (2.56X) in the cytosolic
compartment and decreased (0.16X) in the membrane. So after
96 hours denervation most of the total PKC activity of
blastema mesenchyme is localized in the cytosolic compartment
(88.5% versus 45% for the control) leaving only 11.5% of the
total PKC activity in the membrane compartment (versus 55% for
the control, Fig. 3).

4) Placing spinal ganglion close to mid-bud stage
blastema (14 days pa) in culture increased PKC activity in the
membrane fraction from 29% to 42% (Fig. 4).

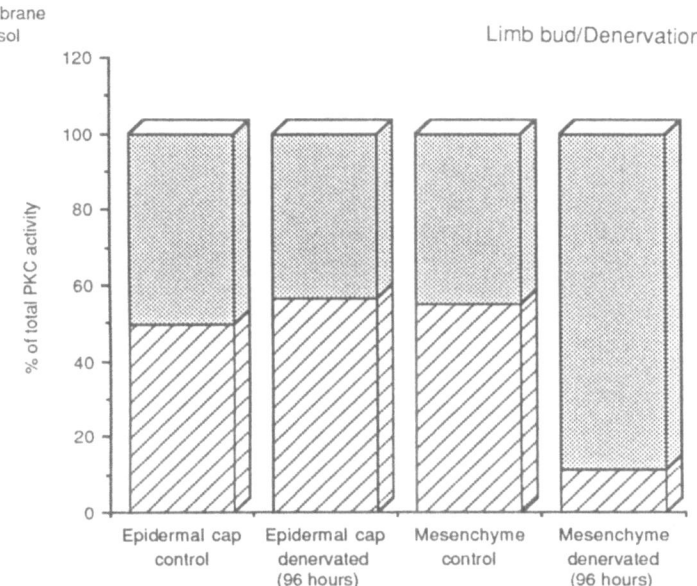

Fig. 3. Percentage of total PKC activity in membrane
and cytosol fractions of epidermal cap and
mesenchyme of mid bud stage blastemata of
<u>Pleurodeles</u> 96 hrs after denervation.

B) <u>PKC Activity in CNS During Limb Regeneration of
Pleurodeles</u>

    1) Compared to the change in PKC activity in blastemata,
specific PKC activity (125 pmol/mg/min, 337 pmol/mg/min,
respectively in spinal cord and brain of unamputated animals)
did not change greatly in either the spinal cord (1.7X
increase) or the brain (1.1X increase).

    But in both tissues total PKC activity increased in the
membrane and decreased in cytosol. This increase of PKC activity in the membrane compartment of nerve tissues appeared
earlier in spinal cord (7 days pa, Fig. 5) than in brain (14
days pa, Fig. 6) but the level of PKC activity in this cell
compartment was about the same (79% for spinal cord 7 days pa
versus 58% for the control; 75% for brain 14 days pa versus
53% for the control, Fig. 5 and 6).

    2) Denervation of limb without limb amputation
induced about the same response in nerve tissues as amputation. Specific PKC activity increased in a same manner
(1.6X, 14 days pa for spinal cord; 1.04X, 7 days pa for
brain) but the percentage of total PKC in membrane was
lower after denervation than after amputation (63% for spinal
cord versus 78% after amputation; 63% for brain versus 75%
after amputation, Fig. 7 and 8).

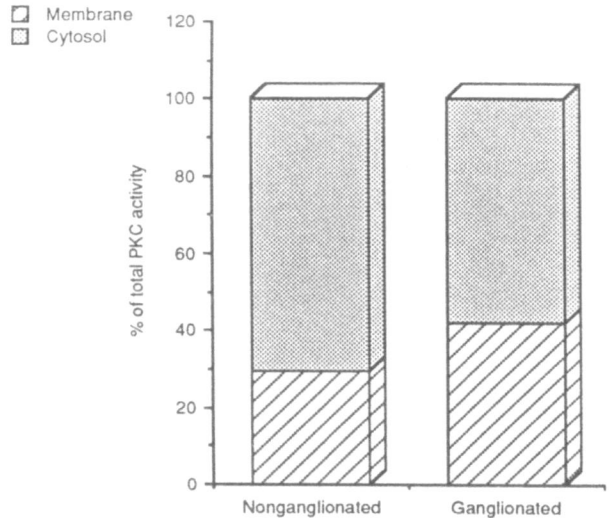

Fig. 4. Percentage of total PKC activity in membrane
and cytosol fractions of limb blastemata of
Pleurodeles cultured 2 days in the presence
(ganglionated) or in the absence (nonganglion-
ated) of a spinal ganglion. Control=blastema
just before culture.

C) <u>PKC Activity in CNS After Limb Amputation of Frogs</u>.

Limb amputation induced a light modification of specific PKC activity (1.05X,14 days pa for spinal cord;
1.1X, 14 days pa for brain). Similarly the percentage of
total PKC activity in the membrane was little modified at
least in spinal cord which had for its highest level 39%
of its total PKC activity in membrane fraction 14 days pa
versus 32% for the control (Fig. 9). On the other hand a
clear translocation of PKC to membrane was observed in
brain after limb amputation: 7 days pa 56% of the total
PCK activity was located in membrane fractions versus 38% for
the control (Fig. 10).

DISCUSSION

Our results show that limb regeneration in <u>Pleurodeles</u>
induces modifications of PKC activity both in blastemata and
in the CNS.

A) <u>Modifications of PKC activity in blastemata</u>. During
limb regeneration PKC activity increases and reaches its highest level 14 days pa (4X the level of control). In parallel
we observed a clear translocation of PKC from cytosol to membrane (which corresponds to an activation of the enzyme, Woodgett et al., 1987) the percentage of total PKC activity in
this cell compartment increasing from 28.7% in the control to
46% in 14 days old buds (i.e. an increase of 60%). At this
time, the blastema (mid-bud stage) shows a high proliferation

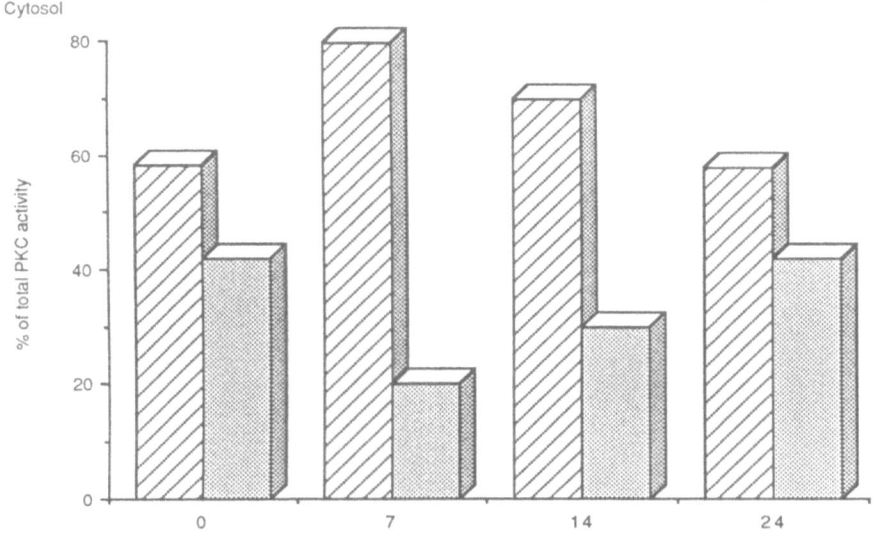

Fig. 5. Percentage of total PKC activity in membrane and cytosol fractions of spinal cord of <u>Pleurodeles</u> at 3 times after limb amputation: 7,14,24 days. 0 day=non amputated newts.

activity which is mostly under the control of nerves. Denervation of the limb 14 days pa which is known to depress the cellular proliferation of about 70%, 4 days after denervation (Boilly et al., 1985) also decreased the translocation of total PKC to membrane by about 20% for a 96 hours denervation; a fact which explains, why specific PKC activity increased in the cytosolic compartment after denervation. Conversely, translocation of PKC to membrane in cultured denervated blastemata increased in the presence of a spinal ganglion which is known to stimulate blastema cell proliferation. The decrease in the translocation of PKC after denervation and its increase in presence of nerve tissues might mean that the mitogenic nerve signal transduction is at least mediated by PKC which is known to be involved in cell growth (Woodgett et al., 1987). This idea is reinforced by the fact that the translocation of PKC to membrane occurred only in mesenchyme, the distribution of PKC in the cellular compartments of epidermal cells being unchanged 96 hours after denervation. This last result confirms the nerve independence of epidermal cap during regeneration as previously shown (Sidman and Singer, 1960; Géraudie and Singer, 1978; Boilly et al., 1985).

B) <u>Modifications of PKC activity in CNS</u>. During limb regeneration specific PKC activity was stimulated in spinal cord but did not increase significantly in brain. As in blastemata, in CNS cytosolic PKC was translocated to membrane which showed an increase of PKC activity of about 40%. These modifications of PKC activity in CNS during limb regeneration

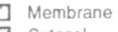

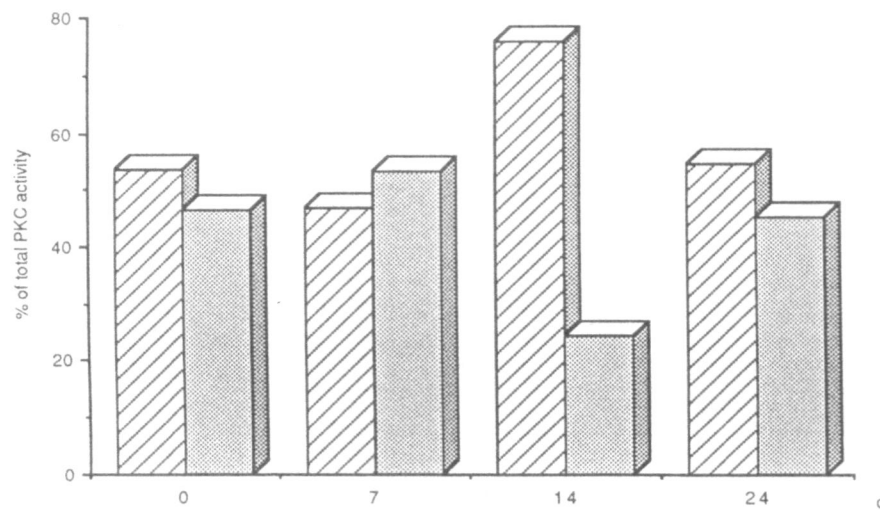

Fig. 6. Percentage of total PKC activity in membrane and cytosol fractions of brain of <u>Pleurodeles</u> at 3 times after limb amputation: 7, 14, 24 days. 0 day = non amputated newts.

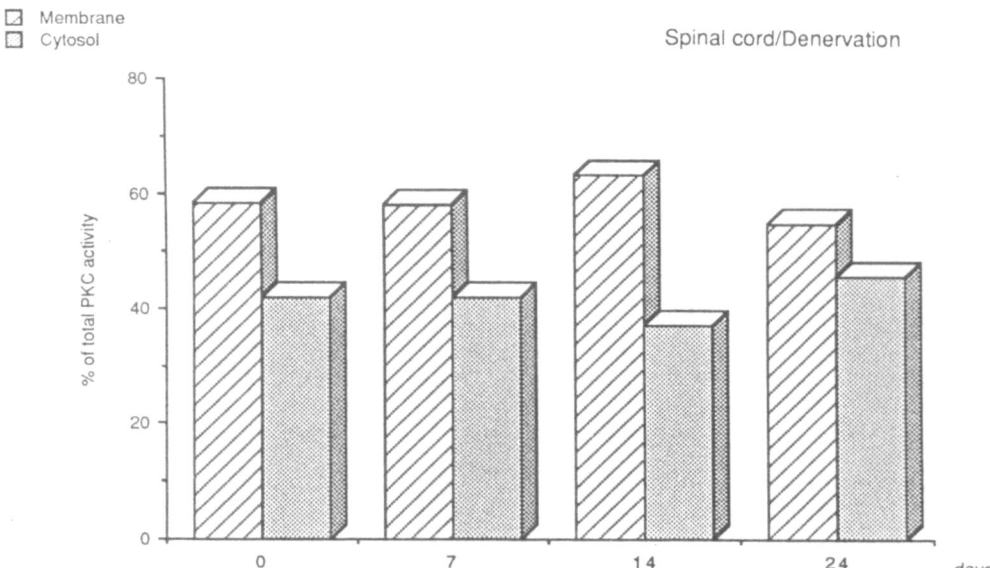

Fig. 7. Percentage of total PKC activity in membrane and cytosol fractions of spinal cord of <u>Pleurodeles</u> at 3 times after limb denervation: 7, 14, 24 days. 0 day = non denervated newts.

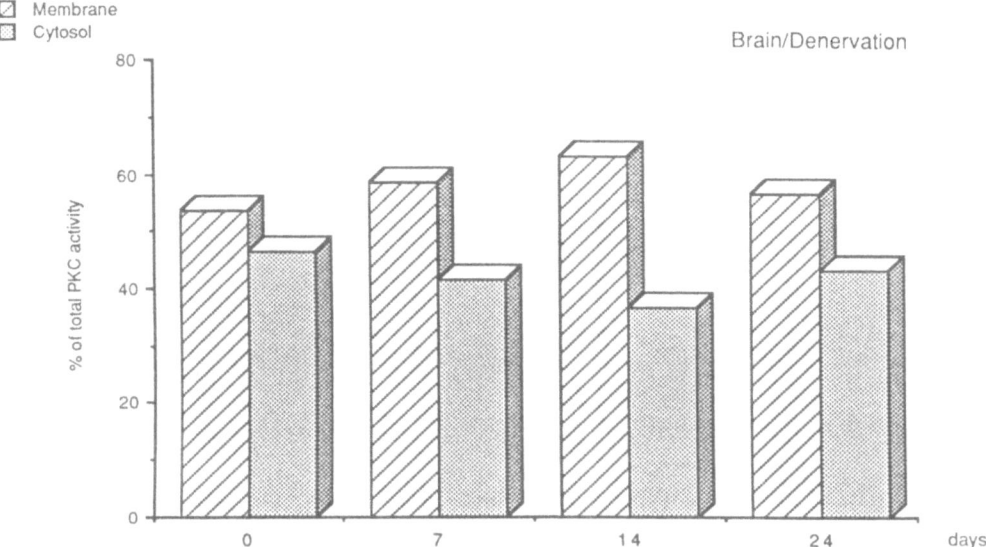

Fig. 8. Percentage of total PKC activity in membrane and cytosol fractions of brain of <u>Pleurodeles</u> at 3 times after limb denervation: 7, 14, 24 days. 0 day = non denervated newts.

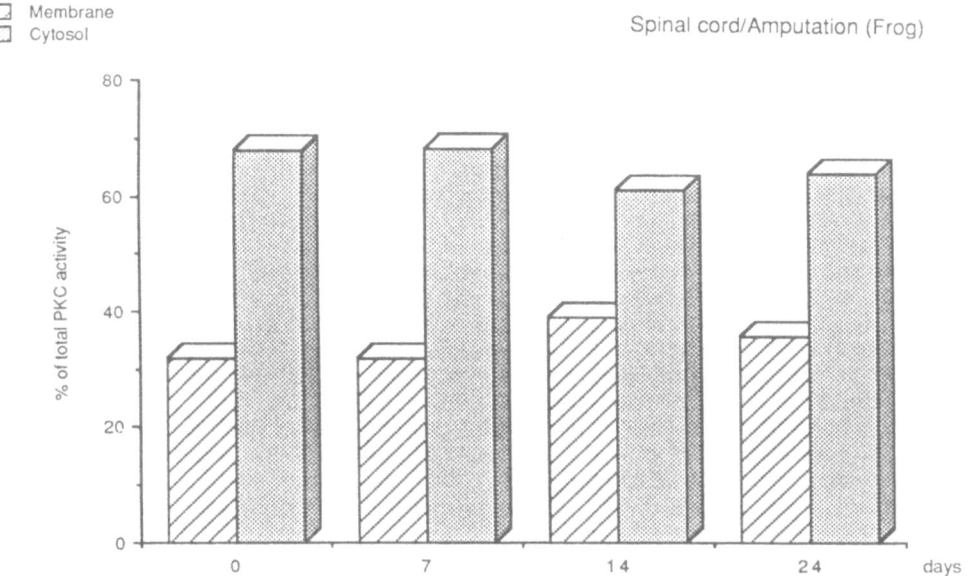

Fig. 9. Percentage of total PKC activity in membrane and cytosol fractions of spinal cord of frogs at 3 times after limb amputation: 7, 14, 24 days. 0 day =non amputated frogs.

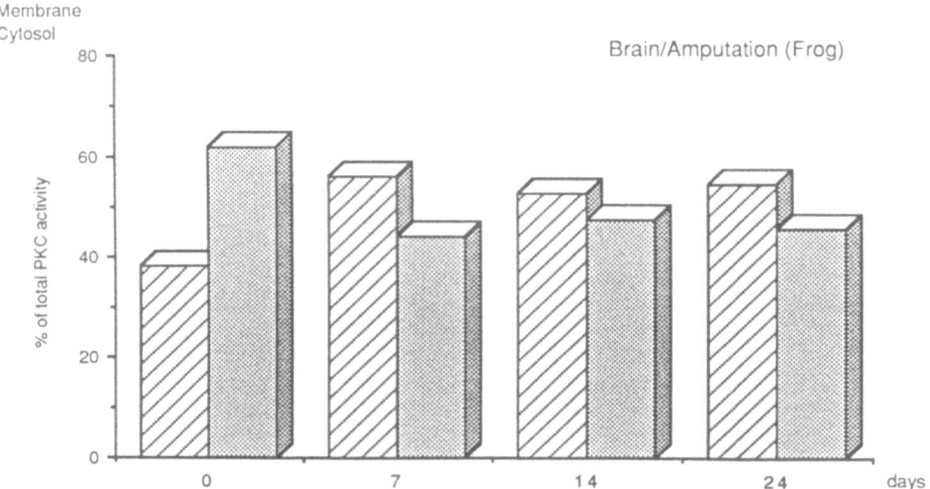

Fig. 10. Percentage of total PKC activity in membrane
and cytosol fractions of brain of frogs at 3
times after limb amputation: 7, 14, 24 days.
0 day = non amputated frogs.

seem to correlate with regeneration of nerves because we
obtained similar results after denervation of non amputated
limbs; that is, PKC activity increased in spinal cord but did
not change in brain after amputation; moreover, we observed
translocation of PKC to membrane after denervation but this
was less important after denervation than after amputation
(20% versus 40%). In frogs, a species which is unable to
regenerate its limbs and consequently does not regenerate limb
nerves, we observed different results. Contrary to newts, limb
amputation did not modify PKC activity in spinal cord and did
not induce translocation of PKC to membrane in this tissue.
On the other hand, we observed a translocation of PKC to membrane in the brain as in newts after limb amputation. We
hypothesize that the modification of PKC activity in spinal
cord is related to nerve regeneration; this would explain the
increase in protein synthesis in the whole spinal cord during
limb regeneration (Boilly and Scaps, 1988). Nevertheless, we
have to study how PKC activity is modified in spinal cord
after denervation of non amputated limbs of frogs in order to
verify this hypothesis. Moreover, we do not know why PKC activity is modified in the brain in both newts and frogs but it
may be correlated with reorganization of nerve connections in
brain as it was shown in adult monkeys after digit amputation
(Merzenich et al, 1983).

ACKNOWLEDGEMENTS

We thank Philip Brylski (UC Irvine) for improving the
english translation of the manuscript.

REFERENCES

Bao, C. Singer, M., and Ilan, J., 1986, Effects of forelimb amputation and denervation on protein synthesis in spinal cord ganglia of the newt, <u>Proc. Natl. Acad. Sci. USA</u>. 83:7971-7974.

Bauduin, B., Lassalle, B., and Boilly, B., The effect of regenerating blastema on the cultured spinal cord of the newt, <u>Pleurodeles waltlii</u>, <u>Differentiation</u>, (submitted).

Boilly, B., and Albert, P., 1988, Control of blastema cell proliferation during axolotl limb regeneration: <u>in vitro</u> cell culture study, in: "<u>Monogr. devl. Biol.</u>", H.J. Anton, ed., 21:1-8, Karger, Basel.

Boilly, B., and Bauduin, B., 1988, Production <u>in vitro</u> by spinal cord of growth factor(s) acting on newt limb regeneration: influence of regenerating nerve fibers, <u>Devl. Brain Res.</u>, 38:155-160.

Boilly, B., and Scaps, P., 1988, Biosynthetic activities in spinal cord during limb regeneration of axolotl, <u>in</u>: "<u>Monogr. devl. Biol.</u>", 21:89-95, Karger, Basel.

Boilly, B., Moustafa, Y., and Oudkhir, M., 1986, Image analysis of blastema cell proliferation in denervated limb regenerates of the newt, <u>Pleurodeles waltlii.</u>, <u>Differentiation</u>, 32:208-214.

Boilly, B., Oudkhir, M., and Lassalle, B., 1985, Role du systeme nerveux sur la proliferation cellulaire des blastemes de regeneration de membre de triton (<u>Pleurodeles waltlii</u> Michah), <u>Develop. Growth and Differ.</u>, 27:129-135.

Bradford, M.M., 1976, A rapid and sensitive method for the quantitation of microgram quantities of protein utilizing the principle of protein-dye binding, <u>Anal. Biochem.</u>, 72:248-254.

Couturier, A., Bazgar, S., and Castagna, M., 1984, Further characterization of tumor-promoter-mediated activation of protein kinase C, <u>Biochem. Biophys. Res. Com.</u>, 121:448-455.

Geraudie, J., and Singer, M., 1978, Nerve dependent macromolecular synthesis in the epidermis and blastema of the adult newt regenerate, <u>J. Exp. Zool.</u>, 203:455-460.

Merzenich, M.M., Nelson, R.J., Stryker, M.P., Cynader, M.S., Schoppmann, A., and Zook, J.M., 1984, Somatosensory cortical map changes following digit amputation in adult monkeys, <u>J. Comp. Neurol.</u>, 224:591-605.

Sidman, R.L., and Singer, M., 1960, Limb regeneration without innervation of the apical epidermis in the adult newt, <u>Triturus</u>, <u>J. Exp. Zool.</u>, 144:105-110.

Singer, M., 1978, On the nature of neurotrophic phenomenon in urodele limb regeneration, <u>Am. Zool.</u>, 18:829-841.

Tassava, R.A., Goldhamer, D.J., and Tomlinson, B.L., 1987, Cell cycle controls and the role of nerves and the regenerate epithelium in urodele forelimb regeneration: possible modifications of basic concepts, <u>Biochem. Cell Biol.</u>, 65:739-749.

Woodgett, J.R., Hunter, T., and Gould, K.L., 1987, Protein kinase C and its role in cell growth, <u>in</u>: "Cell Membranes: Methods and Reviews", Elson et al., ed., Plenum, New York.

PRODUCTION OF GROWTH FACTORS BY THE BLASTEMA DURING LIMB

REGENERATION OF URODELES (AMPHIBIA)

B. Boilly

Laboratoire de Biologie des Facteurs de Croissance, Université de Lille (France), Department
of Biological Chemistry and Developmental Biology
Center, University of California, Irvine, USA

SUMMARY

The blastema of regenerating limbs of Amphibia contains
neurotrophic factor(s) for spinal nerves and mitogenic fac-
tor(s) for blastema cells. The neurotrophic activity of
blastemata was shown by co-culture of spinal cord fragments
with blastemata by medium conditioned by blastemata or by
blastema extracts; only the mesenchymal component of blastemata exhibited neurotrophic activity which appeared to be
related to the proliferation rate of blastema cells. The
mitogenic activity of blastemata was shown by using blastema
extracts on cultured blastema cells; both the epidermal and
mesenchymal components of blastemata were mitogenic for these
cells. The nature of blastema growth factors has been
approached by testing mitogenic activity of heparin on
blastema cells, extraction of heparin-binding growth factors
from blastemata and immunolocalization of aFGF. Our results
show that both epidermal cap and blastema mesenchyme contain
aFGF. This growth factor is probably involved in blastema
cell proliferation during regeneration but not in the neurotrophic activity of blastemata.

INTRODUCTION

Amphibia (Urodeles and larvae of Anurans) are the only
vertebrates which are able to regenerate their limbs after
amputation. The spectacular nature of this phenomenon and its
uniqueness has led many scientists to try since the last century to elucidate the mechanism of limb regeneration in terms
of control of cellular proliferation and pattern formation.

Abbreviations: aFGF: acidic fibroblast growth factor; bFGF:
basic fibroblast growth factor; ECGF: endothelial cell growth
factor; EDGFII: eye derived growth factor II; EGF: epidermal
growth factor; GGF: glial growth factor; PDGF: platelet
derived growth factor; TGF: transforming growth factor, HPLH:
high pressure liquid chromatography; GAG: Glycosaminoglycan.

Limb amputation is followed by formation of a wound epidermis (cicatrisation) and accumulation of dedifferentiated
mesenchymal cells at the tip of the stump, just beneath the
wound epidermis. The first stages of limb regeneration are
characterized by a strong cell proliferation which allows the
formation of a blastema structure analogous to limb buds of
embryos; the blastema is formed by two major cellular components, an epidermis (epidermal cap) and a large population of
heterogenous mesenchymal cells (blastema cells). Early after
limb amputation severed brachial nerves regenerate towards
the tip of the stump and regenerating nerve fibers invade the
blastema (see Wallace, 1981 for review). Since blastema cell
proliferation is essential to limb regeneration, as was shown
by X-irradiation experiments (Lheureux, 1983), blastema cell
mitogens have been investigated for a long time. In fact,
these studies have focused essentially on mitogens produced by
nerves.

From the initial experiment of Todd (1823) it is known
that nerves are obligatory during the first stage of regeneration; denervation of a limb before amputation or before the
blastema reaches the late bud stage, prevents or stops regeneration respectively (see Carlone and Mescher, 1985 for
review). It was shown that nerves produce a diffusible factor, the so-called NTF (neurotrophic factor, Singer, 1978) or
NdBGF (nerve derived blastema growth factor, Boilly and
Albert, 1988a) the nature of which is still unknown (Singer,
1978). There has been less interest in mitogens produced by
the blastema itself. Several investigators have studied the
mitogenic effect of blastema extracts, with the idea that
blastema might contain on one hand, the factor delivered by
nerves (Deck and Futch, 1969, Burnett et al., 1971) during
regeneration and, on the other hand, substances which could
rescue irradiated blastema cells (Deck and Dent, 1970). Unfortunately these results are not definitive because similar
results were obtained with various tissues like living or dead
embryonic or young tadpole tissues (Malinin and Deck, 1958),
rat muscle or frog kidney (Polezhaev and Tuchkova, 1967) or
with homologous or heterologous liver extracts (Polezhaev,
1959; Polezhaev and Ermakova, 1960; Polezhaev et al., 1961,
1964; Smith and Crawford, 1969). Of the two cellular components of blastema the epidermal cap, which is known to be necessary for the growth of blastema (Polezhaev and Faworina,
1935; Goss, 1956; Stocum and Dearlove, 1972), received special
attention (Globus et al., 1980). These authors showed that
after removal of the epidermal cap, the blastema cells ceased
to cycle and differentiated precociously to cartilage nodules;
they concluded that the epidermal cap produced a "division
signal" which decreased in concentration in the proximal part
of the blastema and so allowed proliferation in the distal
mesenchyme and differentiation in the proximal one. Nevertheless, as restoration of proliferation in epidermis-free
blastemata by epidermal cap extracts has not yet been accomplished, we do not know the nature of this "division signal".
Although blastemata are only suspected of producing mitogenic
factors which influence blastema cell proliferation, it was
shown that they contain factors which promote neurite outgrowth from the spinal cord (Richmond and Pollack, 1983; Bauduin et al., 1988), a process which is known to enhance production of mitogens from severed nerves for blastema cells
(Boilly and Bauduin, 1988). Thus blastemata seem to contain

or/and produce both mitogenic and neurotrophic factors during limb regeneration. The present paper provides further information about these growth factors.

MATERIALS AND METHODS

Neurotrophic and mitogenic activities were studied with whole limb blastemata or from each of their two major components (epidermal cap, mesenchyme) in two species (_Pleurodeles waltlii_ and _Ambystoma mexicanum_).

1) _The neurotrophic activity of blastemata_ was studied on cultured fragments of spinal cord of adult (8-9 cm length) _Pleurodeles_ (which contrary to larvae or axolotls, do not spontaneously grow neurites from explanted spinal cord). Transverse slices (0.5 mm thick) of spinal cord were cultured in 35 mm dishes (Falcon Primaria) in N1 medium (Bottenstein et al.,1980) as described elsewhere (Bauduin et al., 1988) in the presence of forelimb blastemata (mid-bud stage, or palette) of the same species, in culture medium conditioned by blastemata (2 blastemata during 48 hours), or with crude blastema extracts (10000g, 1 hour, 4°C) (Bauduin, 1986). The neurotrophic activity of blastemata was measured by counting neurite outgrowth from the spinal fragment.

2) _The mitogenic activity of blastemata_ was studied using cultured blastema cells of axolotls. The cells were harvested from mid-bud stage forelimb blastemata of 9-12 months old axolotls and cultured in Petri dishes (35 mm diameter, Falcon Primaria) in diluted MEM (Seromed) serum-free during 16 days (Albert and Boilly, 1986). Crude extracts obtained after homogenization (Potter) and centrifugation (25,000 g, 1 hour, 4°C) of whole blastemata (mid-bud stage), epidermal cap or mesenchyme of blastemata of the same stage. The supernatants were added to the culture medium on the 10th and the 14th day of culture; on the 14th day, colchicine (20 µg/ml, Serva) was added together with blastema extracts. The mitogenic activity of these extracts was checked by measuring the mitotic index on the 16th day.

3) _The nature of blastema growth factors_ was approached by using three different methods: a) testing mitogenic activity of heparin on blastema cells, b) extraction of heparin-binding growth factors from blastemata, c) immunolocalization of FGF in blastemata. All were performed on axolotls (and Pleurodeles for immunolocalization).

a) _Mitogenic activity of heparin_ (Choay, 0.01-10 µg/ml was tested on cultured blastema cells in the presence or absence of bovine aFGF (50 pg/ml) or bFGF (5 pg/ml), by using the protocol described in 2). aFGF and bFGF were prepared by Courty et al (1985) from adult bovine retinas after ammonium sulfate precipitation, heparin Sepharose affinity chromatography and reverse phase HPLC.

b) _Extraction of heparin binding growth factors_ from blastemata. Different sets of blastemata (mid-bud stage) were extracted as follows. About 0.3 to 1.5 g of blastemata were homogenized (Potter) in 2 ml distilled water; an aliquot of 10 mM Tris HCl (pH 7.2) containing 3 M NaCl was added to obtain a

final concentration of 0.3 M NaCl. The homogenate was soni-
cated and centrifuged (43,000g, 1 hour, 4°C) and the supernatant (i.e. crude extract) applied to a heparin Sepharose (6B
Pharmacia) column (bed volume: 250 μl) preequilibrated at room
temperature with 10 mM Tris-HCl pH 7.2 containing 0.6 M NaCl.
The column was sequentially eluted with the same buffer containing 0.6 to 3 M NaCl (8 ml for each molarity of NaCl). The
eluates were concentrated on Centricon 10 microconcentrators
(10,000 cutoff, Amicon). In some cases, eluates with higher
salt concentrations (2 M, 3 M) were diluted in order to have a
final concentration of 1.1 M NaCl and in these cases, the
microconcentration was continued until the volume of the
retentate reached about the same volume as the 0.6 M and 1.1 M
samples (about 500 μl from 2 ml of eluates). This procedure
was used for the first 3 sets of blastemata:
First set: 60 blastemata (0.29 g), crude extract 0.3 M;
Second set: 113 blastemata (0.52 g), elution 2 M;
Third set: 358 blastemata, elutions 0.6 M, 1.1 M, 2 M, 3 M;
respectively from epidermal caps (1.6 g) and mesenchyme
(1.0 g).

The presence of heparin binding growth factors in the
different eluates from the heparin Sepharose column was tested
on PC12 cells. PC12 cells were cultured in Dulbecco's Modified Eagle's Medium (Irvine Scientific) supplemented with 10%
fetal calf serum and 5% horse serum. The test was conducted
with 24 well plates (Linbro). The eluates were added (5 to 20
μl per well with 500 μl of medium) every day or every two
days, 24 hours after plating in the presence or absence of
heparin (50 μl/ml). In some cases the medium was changed just
before adding eluate in order to avoid high salt concentration. At the same time, the response of cells in control
wells which received NGF or bFGF (5 or 50 ng/ml) and a volume
of buffer equal to that of eluates was checked. The response
of PC12 cells to these growth factors was determined by
observing neurite outgrowth. Eluates were electrophoresed on
17% sodium dodecyl sulfate polyacrylamide gels; the gels were
silver stained (Bio-Rad).

c) Immunolocalization of a FGF in blastemata. Immunolocalization of aFGF was performed on paraffin sections (7 μm
Bouin fixation) or frozen sections (10 μm 4% paraformaldehyde
fixation) of mid-bud blastemata. We used polyclonal anti-
bovine aFGF raised from rabbit (a gift of J.P. Carruelle,
Paris XII University, France) and FITC or peroxidase anti-Ig
rabbit from goat (Pasteur).

RESULTS

I. Presence of growth factors in the blastemata

1) Mitogenic factors. The presence of mitogenic factors
in blastemata was studied by testing different crude extracts
from blastemata of axolotls on cultured blastema cells of the
same species (Fig. 1). Crude extracts from whole mid-bud
blastemata and from each of the two components of blastemata
(mesenchyme and epidermal cap) are mitogenic for cultured
blastema cells. Epidermal cap extracts possess the highest
mitogenicity (11 times the mitotic index of the control for the
lowest concentration of protein (3.5 μg/ml). Extracts from

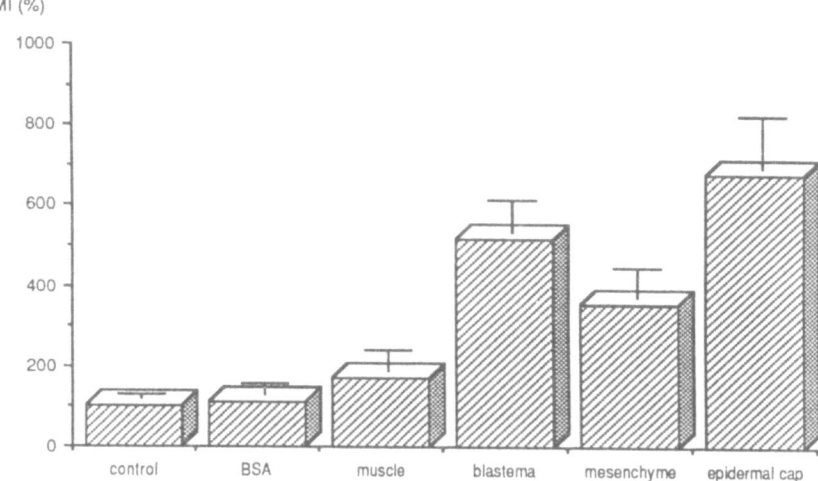

MI (%)

Fig. 1. Influence of crude extracts (7 µg protein/ml)
from blastemata on the mitotic index (MI) of
blastema cells. The results are expressed in
percentage of the control (i.e. no addition
of extracts nor exogenous protein: MI=100).
blastema: extract from whole blastemata;
mesenchyme: extract from the mesenchymal component of blastemata; epidermal cap: extract
from the epidermal component of blastemata;
muscle: extract from muscle; BSA: bovine
serum albumin. 500-1000 cells/dish examined
in 7 dishes for each point.

mesenchyme are about 2 times less potent (x 5.2) than epidermal cap extracts even with a 4fold higher concentration of
protein (14 µg/ml); nevertheless the mitogenic potency of
mesenchyme is still higher than that which we observed earlier
with nerve extracts (Albert and Boilly, 1986; Boilly and
Albert, 1988b). On the other hand, whole blastema extracts
(i.e. epidermal cap and mesenchyme) stimulate the mitotic
index at a value similar to that obtained for mesenchyme but
with a 2-fold lower concentration of protein (x 5.2 to 7 for 7
µg protein/ml).

2) Neurotrophic factors. The neurotrophic activity of
blastemata was shown by co-culture of blastemata from Rana
pipiens tadpoles (Richmond and Pollack, 1983) or from adult
Pleurodeles waltlii (Bauduin, 1986; Bauduin et al., 1988) and
homologous explants of spinal cord. The bioassay consisted of
counting neurites growing from the spinal cord explant in the
presence of different target tissue (Fig. 2). It was shown in
these two species that 1) a limb blastema had a stimulatory
effect on spinal fiber outgrowth and neuron survival, 2) the
neurotrophic influence of blastemata was provided by only its
mesenchymal component, 3) the neurotrophicity of blastemata is

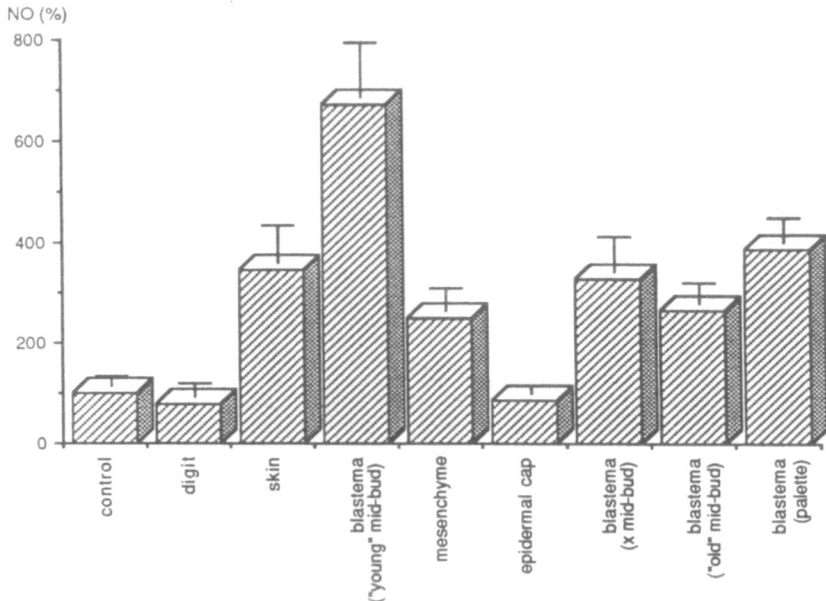

Fig. 2. Influence of blastemata on neurite outgrowth
(NO) from spinal cord explant in co-culture.
The results are expressed in percentage of
the control (i.e. spinal cord cultured alone:
NO=100). blastemata ("young" mid-bud), (X
mid-bud), ("old" mid-bud): whole mid-bud
blastemata respectively from young animals,
X-irradiated young animals, old animals;
blastema (palette): whole palette blastemata
from young animals; mesenchyme, epidermal
cap: respectively mesenchymal and epidermal
components of mid-bud blastemata from young
animals; digit: tip of an uninjured digit;
skin: piece of uninjured skin. 16 dishes
examined for each point.

correlated with the regenerative ability of the limb (Richmond
and Pollack, 1983) or the level of blastema cell proliferation
(Bauduin et al., 1988).

Richmond and Pollack (1983) noted that "the greatest
nerve growth enhancement effect appears to reside in blastemata taken from limbs capable of complete regeneration". In
our experiments we showed that blastemata with low proliferation rates were less neurotrophic for spinal cord explants
than highly proliferating blastemata; blastemata proliferating at a low level, which were obtained from old animals

(which regenerate more slowly than younger ones), X-irradiated blastemata or differentiating blastemata (palette stage), were about 2-times less potent than mid-bud stage blastemata of young animals (Fig. 2). Moreover the neurotrophic activity of blastemata seemed to be mediated by a diffusible factor(s) because 1) the outgrowth of neurites can be elicited without contact with the blastema, 2) because a culture medium conditioned by a blastema is able to stimulate nerve growth from spinal cord explants, and 3) blastema extracts are neurotrophic for the spinal cord (Bauduin, 1986),

II. Nature of growth factors contained in blastemata

We obtained some information concerning the nature of growth factors contained in limb blastemata by using the three following approaches.

1) Mitogenic effect of heparin on cultured blastema cells. Heparin is a GAG which presents a strong affinity for FGFs, a biochemical property which allows rapid purification of these growth factors (Shing et al., 1984) and lead Lobb et al. (1986) to propose a new name for these growth factors, namely HBGF (heparin-binding growth factors). Heparin is well known also because it potentiates the activity of aFGF (and analogous growth factors like ECGF, EDGFII) but not that of bFGF (Thornton et al., 1983; Maciag et al., 1984; Schreiber et al., 1985a and b; Orlidge and D'Amore, 1986; Lobb et al, 1986; Rosenbaum et al., 1986; Uhlrich et al., 1986a and b; Wagner and D'Amore, 1986; Inamura and Mitsui, 1987; Neufeld et al., 1987). On the other hand, heparin, when used alone inhibits proliferation of vessel wall-derived cells like endothelial cells (Rosenbaum et al., 1986; Gospodarowicz et al., 1986) and smooth muscle cells (Clowes and Karnovsky, 1977; Guyton et al., 1980; Hoover et al., 1980; Castellot et al., 1981; Majack et al., 1986; Reilly et al., 1986; Orlidge and d' Amore, 1986), skeletal muscle (Kardami et al., 1988) rat cervical epithelial cells (Wright et al., 1985) or transformed cardiac cells (Balk et al., 1985). The results we obtained showing the mitogenicity of aFGF and bFGF on cultured blastema cells (Albert et al., 1987) prompted us to use heparin in association with these growth factors. As expected, heparin (800 ng/ml) enhanced the stimulatory effect of aFGF on blastema cells, but did not modify the effect of bFGF on these cells (Table).

Surprisingly, we observed that heparin, when used alone, enhanced the mitotic index of blastema cells (Table) in a dose-dependent manner, up to 1 µg/ml (Fig. 3). As heparin potentiates aFGF and inhibits cell proliferation when used alone, it is possible that the stimulatory effect of heparin we have observed on blastema cells results from the potentiation of an aFGF present in our cell culture; as we use a serum-free medium, this growth factor, if any, might be provided by the cultured material (blastema cells and/or the associated extracellular matrix).

Table. Mitotic Index (%) of Blastema Cells Grown Without or With Heparin in Presence or Absence of aFGF (`) or bFGF(*).

FGF su/ml	0	0	0.1	
Heparin ng/ml	0	800	0	800
Mitotic index (%)	2.08±0.58 (7)	6.80±1.06	4.11±0.78` 5.75±1.13* (7)	9.55±1.93` 5.25±1.35* (7)

In parentheses, number of Petri dishes studied. 1 su of aFGF= 50 pg/ml, 1 su of bFGF=500 pg/ml for BEL (bovine epithelial lens) cells.

2) Extraction of heparin-binding growth factors from blastemata. PC12 is an established cell line from a transplantable rat adrenal pheochromocytoma which, when cultured in the presence of NGF, ceases to multiply and extends processes (neurites) similar to those produced by sympathetic neurons in primary cell culture (Greene and Tischler, 1976). Recently, it was shown that the PC12 responses elicited by NGF can also be obtained with commercially prepared FGFs (Togari et al., 1983, 1985), purified aFGF (Wagner and D'Amore, 1986; Rydel and Greene, 1987; Neufeld et al, 1987) or bFGF (Schubert et al., 1987; Rydel and Greene, 1987; Neufeld et al., 1987). Moreover, heparin was shown to potentiate neurite outgrowth induced by aFGF (Wagner and D'Amore,1986; Neufeld et al, 1987) and NGF (Neufeld et al., 1987); on the other hand, heparin both partially inhibited or stimulated neurotrophic activities of bFGF, depending upon the concentration of bFGF used (1 ng/ml, 35 ng/ml respectively) (Neufeld et al., 1987).

Because PC12 cells respond both to a neurotrophic factor (NGF) and to factors well known for their mitogenicity (FGFs), we used them in bioassays intended to test extracts from blastemata which have neurotrophic and mitogenic activities during regeneration as shown above. Crude extracts from the first set of blastemata was not effective on PC12 cells; only large amounts of extracts (40 µl) evoked any response and then only after two additions; as expected, it appeared highly toxic. 2 M eluates from the second set of blastemata gave (after three additions of the eluate, one addition each 2 days) a positive response for about 5 to 7% of the cells with 10 µl of the two first ml of the elution (20 µl of the same sample was toxic when the medium was not changed before addition of the eluate; but when the medium was changed, 20 µl of this sample gave the best response obtained with the eluate in the presence of heparin with 10% of the cells responding positively). 10 µl from the 3rd and 4th ml elution gave a negative response (contrary to the two first ml elution); but in presence of heparin the response was positive for about 5% of the cells. Thus it appears that the 2 M eluate, although relatively toxic (because of the high level of salts) was active on PC12 cells and that this activity was potentiated by heparin; this activity was eluted principally in the first two ml of elution.

Eluates from extracts from the two last sets of blastematata were added only two times and gave very clear responses as follows:

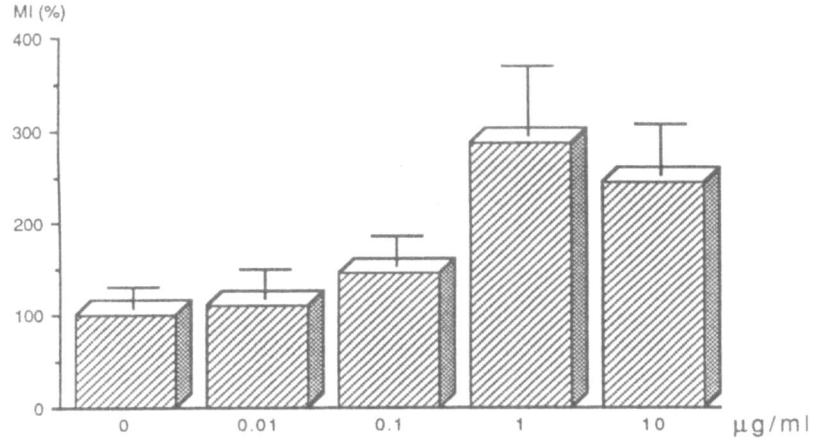

Fig. 3. Influence of heparin (µg/ml) on the mitotic index (MI) of blastema cells. The results are expressed in percentage of the control (i.e. no addition of heparin: MI=100). 500-1000 cells/dish examined in 7 dishes for each point.

Third set of blastemata: 1) fraction 0.6 M was inefficient on PC12 cells; 2) fraction 1.1 M gave a very good response in presence of heparin (25% of the cells treated two times with 10 µl of the eluate grow neurites versus 10% for cells treated two times with 50 ng/ml NGF) while in absence of heparin the response was low (around 5%); this activity was maximum in the first two ml of elution and disappeared from the 7th ml; 3) fraction 2 M, or 3 M even partially desalted and reconcentrated have no positive response.

Fourth set of blastemata: As for the third set of blastemata, the 2 M fractions did not stimulate PC12 cells to grow neurites, and heparin revealed the activity of 1.1 M fraction. Both epidermal cap and mesenchyme 1.1 M fractions gave a positive response on PC12 cells (32% and 15%, respectively); but because we extracted these 1.1 M fractions from different amounts of tissues it is not possible to know with precision if epidermal cap is more potent than mesenchyme on PC12 cells. Electropherogrammes of 1.1 M fractions from epidermal cap and mesenchyme extracts show a band with about the same molecular weight as FGF.

3) <u>Immunolocalization of aFGF in blastemata</u>. The polyclonal antibody against bovine aFGF produced a selective staining pattern in the section of blastemata. Basal membrane, when present, was heavily stained; this staining decreased rapidly at the base of the blastema and disappeared in the stump. Epidermal cap showed immunoreactivity in the median layers of cells (basal and apical cells were not stained) around the cells but also in cytoplasmic area as a diffuse staining. Mesenchyme exhibits high immunoreactivity in extracellular matrix and diffuse staining in the cytoplasmic area of the cells. The density of aFGF immunoreactivity was greatest just under the epidermal cap and specifically at the apex of blastema. It appears that the pattern of aFGF immunoreactivity in blastemata seems to change during regeneration from early to late bud stage blastemata; this pattern of immunoreactivity is now under investigation.

DISCUSSION

Our results show that limb blastemata contain growth factors with neurotrophic and mitogenic activities on spinal cord and blastema cells respectively. The neurotrophic growth factor(s) seems to be produced (or contained) only in the mesenchyme while both mesenchyme and epidermal cap are mitogenic for blastema cells, epidermal cap being more potent than mesenchyme. Results obtained from experiments using heparin potentiation of aFGF, extraction of heparin binding growth factors and anti-aFGF immunocytochemistry suggest that at least one of the growth factors contained in blastemata is an aFGF-related growth factor. Moreover, NGF is clearly not involved in limb regeneration, because anti-NGF which inhibits neurotrophic activity of NGF on PC12 cells, was ineffective on blastema extracts (data not shown); this result is consistent with the fact that NGF has no neurotrophic activity on spinal cord (Bauduin et al., 1988). Similarily, bFGF seems not to be involved in neurotrophic and mitogenic activities of blastemata because 2 M eluates (even partially desalted) from blastema extracts were inactive on PC12 cells.

The fact that blastemata contain aFGF (or a closely related molecule) is interesting to consider because this growth factor, contrary to bFGF, which is widely distributed, is localized only in a few tissues like brain and retina (see Baird et al.,1986, for review), bone (Hauschka et al., 1986), kidney (Gautschi-Sova et al., 1987), human synovial fluid (Hameerman et al., 1987), and vascular smooth muscle (Winkles et al., 1987). We think that aFGF is not concerned with neurotrophic activity of blastemata because conditioned medium by epidermal cap, a tissue which seems to contain more aFGF than mesenchyme, has no neurotrophic activity on spinal cord, while conditioned medium by mesenchyme has one (Bauduin et al., 1988). Nevertheless, we have still to check the activity of epidermal cap extracts and aFGF on spinal cord in order to be sure of this point.

Our results also show that the mitogenicity of epidermal cap and blastema mesenchyme can be explained by the presence of aFGF in these tissues and that this growth factor might represent the "division signal" hypothesized by Globus et al.,(1980). As we found aFGF in both epidermal cap and mesenchyme, we do not know if this growth factor is produced by these two tissues or only by one of them and released into the other. Several observations such as the chronology of immuno-reactivity and the location of aFGF in mesenchyme underlying the epidermal cap, suggests that aFGF is produced first in epidermal cap, and then transferred to mesenchyme. However, cytoplasmic localization of aFGF in mesenchyme suggests that at least some of this growth factor is provided by mesen-chymal cells. New investigations (i.e. appearance of immuno-reactivity during limb regeneration and especially hybridiza-tion in situ) are needed in order to answer this question.

Another question is to understand the function of this growth factor during limb regeneration. Since we know that about 60 to 70% of blastema cell proliferation is controlled by nerves (Boilly et al.,1985), we can expect that at least a part of the blastema cell population is under the control of aFGF; these cells might be concerned with angiogenesis, a well known function of FGF, because we frequently observed many blood capillary vessels at the tip of the blastema mesen-chyme, a place which exhibited intense positive staining with polyclonal anti-aFGF. The crucial role of blood supply and angiogenesis in limb regeneration is well known (Smith and Wolpert, 1975; Revardel and Chapron, 1975), and the presence of aFGF in blastemata is important to consider in this regard. On the other hand, we must remember the fact that aFGF is probably only one of the growth factors involved in limb regeneration. Brockes (1984) and Brockes and Kintner (1986) reported the presence of GGF in blastemata, a growth factor provided by nerves during regeneration. In embryonic chick wing bud, a developmental model which is close to regenerating limb, Bell (1986) and MacLachlan et al. (1988) showed that growth factors like PDGF, EGF and TGF can be detected during development. But, as in limb regeneration, FGF seems to play an important role during limb development (Hideaki and Hiroy-uki, 1988; Seed and Hauschka, 1988). We also have no idea about the nature of the neurotrophic factor contained in limb blastemata.

Another problem is the origin of blastema growth factors. It is possible to approach this by suppressing the source of growth factors as it was done for GGF by Brockes (1984) and Brockes and Kintner (1986) with denervation of regenerating limbs, but this prevents consideration of the autocrine pro-duction of growth factors, a cell control mechanism well known in tissues which are able to undergo fast histogenesis like capillary endothelial cells (Moscatelli et al., 1986; Schweig-erer et al.,1987) and vascular smooth muscle cells (Winkles et al.,1987) by FGF, for example. Molecular probes of growth factors for hybridization in situ can clearly aid in resolving this problem and explaining resolution of the nature of the control of blastema cell proliferation at the molecular level.

ACKNOWLEDGMENTS

We would like to thank R.A. Bradshaw (UC Irvine) for reading the manuscript. Research supported by CNRS, PHS grant HD06002, NIH NS 19964 and Amer. Can. Soc. BC 273.

REFERENCES

Albert, P., and Boilly, B., 1986, Evolution _in vitro_ de cellules de blastèmes de régénération de membre d'Axolotl: influence de l'insuline et d'extraits nerveux sur la prolifération cellulaire, _Biol. Cell._, 58:251-262.

Albert, P., Boilly, B., Courty, J., and Barritault, D., 1987, Stimulation in cell culture of mesenchymal cells of newt limb blastemas by EDGFI or II (basic or acidic FGF), _Cell. Diff._, 21:63-68.

Baird, A., Esch, F., Mormede, P., Ueno, N., Ling, N., Bolhen, P., Ying, S.Y., Wehrenberg, W.B., and Guillemin, R., 1986, Molecular characterization of fibroblast growth factor: distribution and biological activities in various tissues, _Recent Prog. Horm. Res._, 42:177-205.

Balk, S.D., Riley, T.M., Gunther, H.S., and Morisi, A., 1985, Heparin-treated, v-_myc_- transformed chicken heart mesenchymal cells assume a normal morphology but are hypersensitive to epidermal growth factor (EGF) and brain fibroblast growth factor (b FGF); cells transformed by the v-Ha-_ras_ oncogene are refractory to FGF and b FGF but are hypersensitive to insulin-like growth factors, _Proc. Natl. Acad. Sci. USA._, 82:5781-5785.

Bauduin, B., 1986, Influence neurotropedes blastèmes de régénération du membre antérieur sur la moëlle épinière du triton, _Pleurodeles waltlii_. Etude _in vitro_. Thèse, Université de Lille.

Bauduin, B., Lassalle, B., and Boilly, B., The effect of regenerating blastema on the cultured spinal cord of the newt, _Pleurodeles waltlii_. _Differentiation_ (submitted).

Bell, K.M., 1986, The preliminary characterization of mitogens secreted by embryonic chick wing bud tissues _in vitro_, _J. Embryol. exp. Morph._, 93:257-265.

Boilly, B., and Albert, P., 1988a, Control of blastema cell proliferation during axolotl limb regeneration: _in vitro_ cell culture study, in: _Monogr. devl. Biol._, 21:1-8, Karger, Basel.

Boilly, B., and Albert, P., 1988b, Blastema cell proliferation _in vitro_: effects of limb amputation on the mitotic activity of spinal cord extracts, _Biol. Cell._, 62:183-188.

Boilly, B., and Bauduin, B., 1988, Production _in vitro_ by spinal cord of growth factor(s) acting on newt limb regeneration: influence of regeneration of the nerve fibers, _Dev. Brain Res._, 38:155-160.

Boilly, B., Oudkhir, M., and Lassalle, B., 1985, Control of the blastemal cell cycle by the peripheral nervous system during newt limb regeneration: continuous labeling analysis, _Biol. Cell._, 55:107-112.

Bottenstein, J.E., Skaper, S.D., Varon, S.S., and Sato, G.H., 1980, Selective survival of neurons from chick embryo sensory ganglion dissociated utilizing serum-free supplemented medium, _Exp. Cell. Res._, 125:183-190.

Brockes, J.P., 1984, Mitogenic growth factors and nerve depen-
        dence of limb regeneration, _Science_, 255:1280-1287.
Brockes, J.P., and Kintner, C.B., 1986, Glial growth and ner-
        ve-dependent proliferation in the regeneration in the
        blastema of Urodele Amphibian, _Cell._, 45:301-306.
Burnett, A.L., Kary, C.E., and Lagorio, A.M., 1971, Induction
        of growth in newt and frog limbs after perfusion with
        extracts from newt blastemas, _Nature_, 234:98-99.
Carlone, R.L., and Mescher, A., 1985, Trophic factors from
        nerves; _in_: "Regulation of Vertebrate Limb Regener-
        ation" (R.E. Sicard, ed.), pp. 93-105. Oxford Univ.
        Press.
Castellot, Jr., J.J., Addonizio, M.L., Rosenberg, R., and Kar-
        novsky, M.J., 1981, Cultured endothelial cells produce
        a heparin like inhibitor of smooth muscle cell growth,
        J. Cell. Biol., 90:372-379.
Clowes, A.W., and Karnovsky, M.J., 1977, Suppression by
        heparin of smooth muscle cell proliferation in injured
        arteries, _Nature_, 265:625-626.
Courty, J., Loret, C., Moenner, M., Chevallier, B., Lagente,
        O., Courtois, Y., and Barritault, D., 1985, Bovine
        retina contains three growth factors activities with
        different affinity to heparin: eye derived growth
        factor I, II, III, _Biochimie_, 67:265-269.
Deck, J.D., and Futch, C.B., 1969, The effects of infused
        materials upon the regeneration of newt limbs: I. Blas-
        temal extracts in denervated limb stumps, _Develop.
        Biol._, 20:332-348.
Deck, J.D., and Dent, J.N., 1970, The effects of infused mate-
        rials upon the regeneration of newt limbs: III Blas-
        temal extracts and alkaline phosphatase in irradiated
        limb stumps, _Anat. Rec._, 168:525-536.
Gautschi-Sova, P., Jiang, Z.P., Schroder, M., and Bohlen, P.,
        1987, Acidic fibroblast growth factor is present in
        non-neural tissue: isolation and chemical characteriza-
        tion from bovine kidney, _Biochemistry_, 26:5844-5847.
Globus, M., Vethamany-Globus, S., and Lee, Y.C.I., 1980,
        Effect of apical epidermal cup on mitotic cycle and
        cartilage differentiation in regeneration blastema in
        the newt _Notophthalmus viridescens_, _Develop. Biol._,
        75:358-372.
Gospodarowicz, D., Manoglia, S., Cheng, J., and Fujii, D.K.,
        1986, Effect of fibroblast growth factor and lipopro-
        teins on the proliferation of endothelial cells derived
        from bovine adrenal cortex, brain cortex and corpus
        luteum capillaries, _J. Cell. Physiol._, 127:121-136.
Goss, R.J., 1956, The regenerative responses of amputated
        limbs to delayed insertion into the body cavity, _Anat.
        Rec._, 126:283-297.
Greene, L.A., and Tischler, A.S., 1976, Establishment of a
        noradrenergic clone line of rat adrenal pheochromocy-
        toma cells which respond to nerve growth factor, _Proc.
        Natl. Acad. Sci. USA_, 73:2424-2428.
Guyton, J.R., Rosenberg,R.D., Clowes, A.W., and Karnovsky,
        M.J., 1980, Inhibition of rat arterial smooth muscle
        cell proliferation by heparin. _In vivo_ studies with
        anticoagulant and nonanticoagulant heparin, _Circ. Res._,
        46:625-634.

Hamerman, D., Taylor, S., Kirschenbaum, I., Klagsbrun, M., Rainer, E.W., Ross, R., and Thomas, K.A., 1987, Growth factors with heparin binding affinity in human synovial fluid, <u>Proc. Soc. Exp. Biol. Med.</u>, 186:384-389.

Hauschka, P.V., Mavrakas, M.E., Iafrati, M.D., Doleman, S.E., and Klagsbrun, M., 1986, Growth factors in bone matrix: isolation of multiple types by affinity chromotography on heparin-sepharose, <u>J. Biol. Chem.</u>, 261:12665-12674.

Hideaki, A., and Hiroyuki, I., 1988, A gradient of responsiveness to the growth-promoting activity of ZPA (zone of polarizing activity) in the chick limb bud. <u>Develop. Biol.</u>, 128:136-141.

Hoover, R.L., Rosenberg, R.D., Haering, W., and Karnovsky, M.J., 1980, Inhibition of rat arterial smooth muscle cell proliferation by heparin, <u>Circ. Res.</u>, 47:578-583.

Inamura, T., and Mitsui, Y., 1987, Heparan sulfate and heparin as a potentiator or a suppressor of growth of normal and transformed vascular endothelial cells, <u>Exp. Cell. Res.</u>, 172:92-100.

Kardami, E., Spector, D., and Srohman, R.C., 1988, Heparin inhibits skeletal muscle growth <u>in vitro</u>, <u>Develop. Biol.</u>, 126:19-29.

Lheureux, E., 1983, Replacement of irradiated epidermis by migration of non irradiated epidermis in the newt limb: the necessity of healthy epidermis for regeneration, <u>J. Embryol. exp. Morph.</u>, 76:217-234.

Lobb, R.R., Harper, J.W., and Fett, J.W., 1986, Purification of heparin binding growth factors, <u>Anal. Biochem.</u>, 154:1-14.

Maciag, T., Mehlman, T., Friesel, R., and Schreiber, A.B., 1984, Heparin binds endothelial cell growth factor, the principal endothelial cell mitogen in bovine brain, <u>Science</u>, 225:932-934.

Majack, R.A., Cook, S.C., and Bornstein, P., 1986, Control of smooth muscle growth by composants of the extracellular matrix: autocrine role for thrombospondin, <u>Proc. Natl. Acad. Sci. USA</u>, 83:9050-9054.

Malinin, T., and Deck, J.D., 1958, The effects of implantation of embryonic and tadpole tissues into adult frog limbs. I. Regeneration after amputation , <u>J. Exp. Zool.</u>, 139:307-328.

Mc Lachlan, J.C., Macintyne, J., Hume, D.D., and Smith, J., 1988, Direct demonstration of production of transforming growth factor activity by embryonic chick tissue, <u>Experientia.</u>, 44:351-352.

Moscatelli, D., Presta, M., Joseph-Silverstein, J., and Rifkin, D.B., 1986, Both normal and tumor cells produce basic fibroblast growth factor, <u>J. Cell Physiol.</u>, 129:273-276.

Neufeld, G., Gospodarowicz, D., Dodge, L., and Fujii, D.K., 1987, Heparin modulation of the neurotropic effects of acidic and basic fibroblast growth factors and nerve growth factors on PC12, <u>J. Cell. Physiol.</u>, 131:131-140.

Orlidge, A., and D'Amore, P., 1986, Cell specific effects of glycosaminoglycans on the attachment and proliferation of vascular wall components, <u>Microvasc. Res.</u>, 31:41-53.

Polezhaev, L.V., 1959, Restoration of regenerative ability of the extremities of axolotls after irradiation with roentgen rays, <u>Dokl. Akad. Nauk. SSSR.</u>, 127:713-716.

Polezhaev, L.V., and Ermakova, N.I., 1960, Restoration of regenerative capacity of the extremities of axolotls depressed by roentgen radiation, <u>Dokl. Akad. Nauk. SSSR.</u>, 131:209-212.

Polezhaev, L.V., and Faworina, W.N., 1935, Uber die Rolle des Epithels in den Anfanglichen Entwicklungsstadien einer Regenerationsanlage der Extremität beim Axolotl, <u>Wilhelm Roux Arch.</u>, 133:701-727.

Polezhaev, L.V., Teplits, N.A., and Ermakova, N.I., 1961, Restoration of the regenerative property of extremities in axolotls inhibited by X-ray irradiation, by means of proteins, nucleic acids and lyophilized tissues, <u>Dokl.Akad. Nauk. SSSR</u>, 138:477-480.

Polezhaev, L.V., Teplits, N.A., and Tuchkova, S.Ya., 1964, Restoration of regenerational capacity of extremities of axolotls after it has been suppressed by X-rays with the aid of nuclear acids, <u>Dokl. Akad. Nauk. SSSR.</u>, 159:682-685.

Polezhaev, L.V., and Tuchkova, S.Ya, 1967, Restoration of regenerative power of axolotl limbs inhibited by X-ray irradiation with tissue heterografting, <u>Dokl. Akad. Nauk. SSSR.</u>, 180:754-757.

Reilly, C.F., Fritze, L., and Rosenberg, R.D., 1986, Heparin inhibition of smooth muscle cell proliferation: a cellular site of action, <u>J. Cell. Physiol.</u>, 129:11-19.

Revardel, J.L., and Chapron, C., 1975, Influence de la vascularization sur la régénération des membres chez les larves d' Urodèles. Nouvelle interprétation du rôle du système nerveux. <u>C.R. Acad. Sc.</u>, 280:1409-1411.

Richmond, M.J., and Pollack, E.D., 1983, Regulation of tadppole spinal nerve fiber growth by the regenerating limb blastema in tissue culture, <u>J. Exp. Zool.</u>, 225:233-242.

Rosenbaum, J., Tobelem, G., Molho, P., Barzy, T., and Caen, J.P., 1986, Modulation of endothelial cells growth induced by heparin, <u>Cell. Biol. Int. Rep.</u>, 10:437-446.

Rydel, R.E., and Greene, L.A., 1987, Acidic and basic fibroblast growth factors promote stable neurite outgrowth and neuronal differentiation in cultures of PC12 cells, <u>J. Neurosc.</u>, 7:3639-3653.

Schreiber, A.B., Kenney, J., Kowalski, J., Thomas, K.A., Gimenez-Gallego, G., Rios-Candelore, M., Di salvo, J., Barritault, D., Courty, J., Courtois, Y., Moenner, M., Loret, C., Burgess, W.H., Mehlman, T., Friesel, R., Johnson, W., and Maciag, T., 1985ÿ, A unique family of endothelial cell polypeptide mitogens: endothelial cell growth factor, brain-derived acidic fibroblast growth factor and eye-derived growth factor II, <u>J. Cell. Biol.</u>, 101:1623-1626.

Schreiber, A.B., Kenney, J., Kowalski, J., Friesel, R., Mehlman, T., and Maciag, T., 1985ö, The interaction of endothelial cell growth factor with heparin: characterization by receptor and antibody recognition, <u>Proc. Natl. Acad. Sci.USA.</u>, 82:6138-6142.

Schubert, D., Ling, N., and Baird, A., 1987, Multiple influence of a heparin-binding growth factor on neuronal development, <u>J. Cell Biol.</u>, 104:635-643.

Schweigerer, L., Neufeld, G., Friedman, J., Abraham, J.A., and Fidder, J.C., 1987, Capillary endothelial cells express basic fibroblast growth factor, a mitogen that promotes their own growth, <u>Nature</u>, 325:257-259.

Seed, J., and Hauschka, S.D., 1988, Clonal analysis of vertebrate myogenesis. Fibroblast growth factor (FGF)-dependent and FGF-independent muscle colony types during chick wing development, _Develop. Biol._, 128:40-49.

Shing, Y., Folkman, J., Sullivan, R., Butterfield, C., Murray, J., and Klagsbrun, P., 1984, Heparin affinity: purification of a tumor-derived capillary endothelial cell growth factor, _Science_, 223:1296-1299.

Singer, M., 1974, Neurotrophic control of limb regeneration in the newt, _Ann. NY Acad. Sci._, 228:308-322.

Singer, M., 1978, On the nature of neurotrophic phenomenon in urodele limb regeneration, _Am. Zool._, 18:829-841.

Smith, St D., and Crawford, G.L., 1969, Initiation of regeneration in adult _Rana pipiens_ limbs by injection of homologous liver nuclear R.N.P., _Oncogenesis_, 23:299-307.

Smith, A.R., and Wolpert, L., 1975, Nerves and angiogenesis in amphibian limb regeneration, _Nature_, 257:224-225.

Stocum, D.L., and Dearlove, G., 1972, Epidermal-mesodermal interaction during morphogenesis of the limb regeneration blastema in larval salamander, _J. Exp. Zool._, 181:49-62.

Thornton, C.S., Mueller, S.N., and Levine, E.M., 1983, Human endothelial cells: use of heparin in cloning and long-term serial propagation, _Science_, 222:623-625.

Todd, T.J., 1823, On the process of reproduction of the members of the aquatic salamander, _Quart. J. Sci. Lit. Arts_, 16:84-96.

Togari, A., Baker, D., Dickens, G., and Guroff, G., 1983. The neurite promoting effect of fibroblast growth factor on PC12 cells, _Biochem. Biophys. Res. Com._, 114:1189-1193.

Togari, A., Dickens, G., Kuzuya, H., and Guroff, G., 1985, The effect of fibroblast growth factor on PC12 cells, _J. Neurosc._, 5:307-316.

Uhlrich, S., Lagente, O., Choay, J., Courtois, Y., and Lenfant M., 1986a, Structure activity relationships in heparin: stimulation of non vascular cells by a synthetic heparin pentasaccharide in cooperation with human acidic fibroblast growth factor, _Biochem. Biophys. Res. Comm._, 139:728-732.

Uhlrich, S., Lagente, O., Lenfant, M., and Courtois, Y., 1986b, Effect of heparin on the stimulation of non vascular cells by human acid and basic FGF, _Biochem. Biophys. Res. Comm._, 137:1205-1213.

Wagner, J.A., and D'Amore, P.A., 1986, Neurite outgrowth induced by an endothelial cell mitogen isolated from retina, _J. Cell. Biol._, 103:1363-1367.

Wallace, H., 1981, "Vertebrate limb regeneration", John Wiley & Sons, Chichester.

Winkles, J.A., Friesel, R., Burgess, W.H., Howk, R., Mehlman, T., Weinstein, R., and Maciag, T., 1987, Human vascular smooth muscle cells both express and respond to heparin-binding growth factor I (endothelial cell growth factor), _Proc. Natl. Acad. Sci. USA._, 84:7124-7128.

Wright, T.C., Johnstone, T.V., Castellot, J.J., and Karnovsky, M.J., 1985, Inhibition of rat cervical epithelial cell growth by heparin and its reversal by EGF, _J. Cell. Physiol._, 125:499-506.

SECOND MESSENGERS IN NEWT LIMB REGENERATION: cAMP AND cGMP LEVELS AND DISTRIBUTION

C.H. Taban and M. Cathieni

Laboratory of Neurobiology, Psychiatric Clinic
of the University of Geneva, 8 ch. Petit-Bel-Air
CH - 1225 Chene-Bourg, Switzerland

SUMMARY

cAMP and cGMP measured by RIA and observed by immunohistochemistry in unamputated limbs and in regenerating newt
limbs at early- and medium-bud stages showed cAMP and cGMP
cytoplasmic accumulations in blastemata and in stumps of the
regenerating limbs, with increases up to 1900% over unamputated controls. In blastemata the cAMP increase was greater
than that of cGMP while in stumps of the regenerating limbs
the increase of cGMP was of a much greater magnitude than that
of cAMP. These increases were associated with cell cycling
(mainly in blastemata), cell dedifferentiation and cell movements (mainly in stumps).

INTRODUCTION

After limb amputation and wound healing the prominent
histological changes observed in the stump of the regenerating
newt limb (that is in the tissues proximal to the amputation
level) are cellular dedifferentiation and blastema cells formation. The dedifferentiated cells accumulate in the distal
stump where they divide, forming the bulk of the regenerating
blastema (see Hay and Fischman, 1961; Schmidt, 1968). The
stump cell dedifferentiation process lasts for several weeks,
while the blastema itself grows beyond the amputation plane,
ultimately rebuilding the amputated limb. This epimorphic
regeneration depends on the presence of normally regenerating
nerves and of normal hormonal and immunological functions.
Impairment of any one of these prevents or impairs normal
regeneration (see Sicard, 1985). It should be noted that
dedifferentiating cells lose their cell to cell close contacts
at first and that the quality of the extracellular matrix
changes after amputation. These changes also depend on the
presence or absence of the nerve (Carlone ; Mescher, present
volume) and on hormonal levels (see Schmidt, 1968). Moreover

after the amputation, the distal stump is inflamed, oedematous and invaded by granulocytes, macrophages and mastocytes. These inflammatory reactions which certainly exert an influence on the repair process (Sicard, 1985), decline after the wound retraction observed at the late dedifferentiation stage. Submitted to these events the stump cells most probably modify their fitting of cell membrane receptors. We hypothesized that these new cell membrane receptors will find in their environment the specific ligands able to activate them and that such activation will result in the formation of intracellular second messengers measurable by RIA and observable with adequate immunohistological means. The present results show that indeed the two second messengers cAMP and cGMP are formed in abundance in cells of the distal stump and in cells of the blastema, thus sustaining the formulated hypothesis. We have already reported results of such investigations performed during the wound healing stage (Midwestern regional conference, Columbus, Ohio, 1987) showing that epithelial and inflammatory cells accumulated the two cyclic nucleotides.

It should be noted that cAMP and cGMP are not the only possible second messengers and that cell receptor activation may be transduced within the cell by a small number of other second messengers (10 or 12, see Lichstein and Rodbard, 1987).

The probability that cyclic nucleotide formation was necessary for the newt limb regeneration appeared also from the results of several previous studies (Foret and Babich, 1973; Sicard, 1975 a,b; Jabaily et al., 1975; Liversage et al., 1977; MacLaughlin et al., 1983; Globus et al., 1987). Taban et al., (1978) reported that cAMP levels fluctuate in blastemata during the various stages of regeneration and they correlated these fluctuations with the analogous ones observed in the responses of beta-adrenoreceptors to norepinephrin. Rathbone et al., (1980) confirmed this effect of norepinephrin on cAMP formation in blastemata in vitro, a formation however not associated with high rates of protein synthesis thus indicating that other neurotransmitters or factors and not only catecholamines were necessary for blastema formation, growth and proliferation.

Due to the different experimental methodologies used, direct comparisons between these previously reported studies are difficult, as noted by MacLaughlin et al., (1983). Therefore we will only mention that generally their interpretation was that cAMP formation could be related to cell differentiation and cGMP to cell dedifferentiation.

The present results in addition to the measurements by RIA of the cAMP and cGMP levels in amputated newt limbs at budding stages also give a direct immunohistological observation as yet not used for the study of cAMP and cGMP in the regenerating newt limb. The levels of the two cyclic nucleotides measured in stumps and blastemata will be compared with the levels found in the analogous portions of unamputated limbs. Moreover in six animals a comparison will be made between measures of the cyclic nucleotides in the stump of the amputated limbs of each animal.

An unsuspected dramatic cGMP increase was found in the stump of the regenerating limbs, at budding stages.

MATERIALS AND METHODS

Adult newts, <u>Triturus cristatus</u> bought from Italy (Barilli and Biagi, Bologna) were kept in tap water and fed weekly with minced beef or Tubifex. The animals were anesthetized with 0.1% MS 222 Sandoz. The newts were amputated under a binocular magnifying glass, at forearm, half-way between the elbow and the carpal elements. They were left to regenerate until the early- and medium-bud stages (Iten and Bryant, 1973).

Radioimmunoassays

After collection (between 14 and 15 p.m.) all specimens were rinsed in cold amphibian Ringer solution and plunged into a glass homogenizer containing 150 μl of ice cold 5% trichloroacetic acid (TCA), homogenized, then centrifuged at 2500g for 15 minutes at 4°C. The supernatant fluid was poured off and stored at -20°C until measurement of cyclic nucleotide concentrations. Bidistilled water was added to bring each sample of supernatant fluid to a vol of 1 ml, TCA was removed by four extractions with three volumes of water-saturated diethyl ether, and the solution freeze-dried. The residue was dissolved in 0.05M sodium acetate buffer (pH 6.2) for duplicate measurements of cAMP and cGMP by radioimmunoassay with (^{125}I) RIA Kits (New England Nuclear, Boston). Counting was carried out in an automated gamma counter (Kontron). The pellet obtained during centrifugation was used to determine protein concentration (Lees and Paxman, 1972). Statistical analyses were carried out with the Wilcoxon Mann Whitney non parametric U test.

Immunohistochemistry

Regenerating and intact limbs were sectioned through the distal third of the humerus, fixed for 4 hours in an ice-cold 4% p-formaldehyde solution in 0.1M phosphate buffer, pH 7.4. The tissues were then rinsed at least 24 hr in three changes of 5% sucrose in 0.1M phosphate buffer, quickly frozen with powdered carbon dioxide and cut into sections at -16°C with an automated cryostat (Frigocut 2700, Reichert-Yung). Alternate sections (10 μm) were treated with the cGMP monoclonal or with the cAMP monoclonal antibodies diluted 1:100 or with the same antibodies preabsorbed with an excess of cGMP or cAMP respectively. Monoclonal antibodies were generously supplied by Susan L. Wescott, NIH, Bethesda (Wescott et al., 1985). The FITC-conjugated anti-mouse IgG (Miles Yeda) were used for the second layer (Coons et al., 1958). The sections were rinsed in PBS, mounted in a mixture of PBS and glycerol and examined with a Zeiss fluorescence microscope equipped with a HBO 50 lamp with BP 485-20 nm excitation and LP 520 stop filters. Pictures were taken with 3M Ektachrome films (1000 ASA). For more technical details see also Taban and Cathieni, 1983.

RESULTS

Radioimmunoassays

Figures 1 and 2 indicate the changes in the concentration of cAMP and cGMP, at various limb parts.

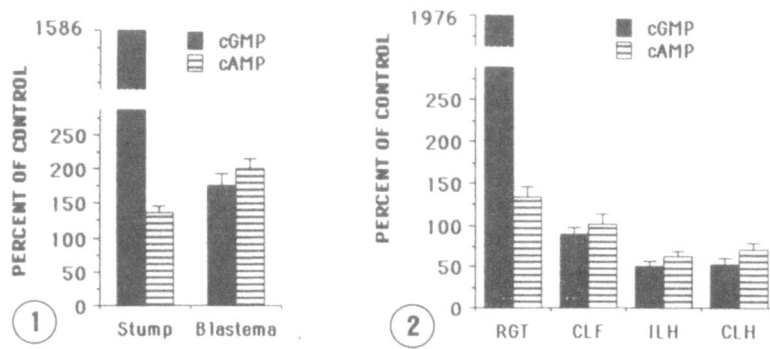

Fig. 1. Highly significant increase of 1586%±247 of cGMP, and cAMP in stumps (P<0.001, n=10) over control values: 100%=0.77±SEM 0.05 for cGMP; 4.47±0.16 for cAMP, pmol/mg prot; n=25. In blastemata the increase of cAMP is significant at P 0.001 and the slight increase of cGMP at P 0.05; n=12. Stump values were means of individually measured samples; blastema values were means of 5-7 pooled blastemata.

Fig. 2. Significant increase of cGMP and of cAMP (P<0.001) in the regenerating limb (RGT, n=6) over the 3 other limbs. Significant increase of cAMP over controls (P<0.001) in the contralateral forelimb (CLF). Low levels of cAMP and of cGMP in ipsilateral (ILH) and contralateral (CLH) hindlimbs (no control of unamputated hindlimbs).

Immunohistochemistry

In the unamputated non regenerating limbs, only a few structures were fluorescent, faintly in some zones of the epithelial basal layer, in some muscles and more marked in some individual cells of the skin glands (Fig. 3). In contrast, the majority and for some zones the totality, of the cells observed in stumps and blastemata of the amputated, regenerating limbs exhibited a very strong cytoplasmic fluorescence. In the blastemata practically all the cells reacted to one or the other antibody (or to both) with the exception of some vesicles filled with cell debris and macrophages and of lymphatic spaces (Fig. 4).

Practically all the blastemal cells reacted to both cAMP and cGMP antibodies, thus showing the simultaneous presence of both cyclic nucleotides in individual blastema cells. However in stump tissues not all the cells reacted to the cyclic nucleotide antibodies. Since we carried comparative staining of the two nucleotides in immediately following serial sections, but not on identical ones, it was not possible to ascertain the simultaneous presence of both cyclic nucleotides in individual stump cells. Moreover in stumps a proximodistal gradation was evident in the number of fluorescent cells. They were ubiquitous in the distal blastema, quite abundant in

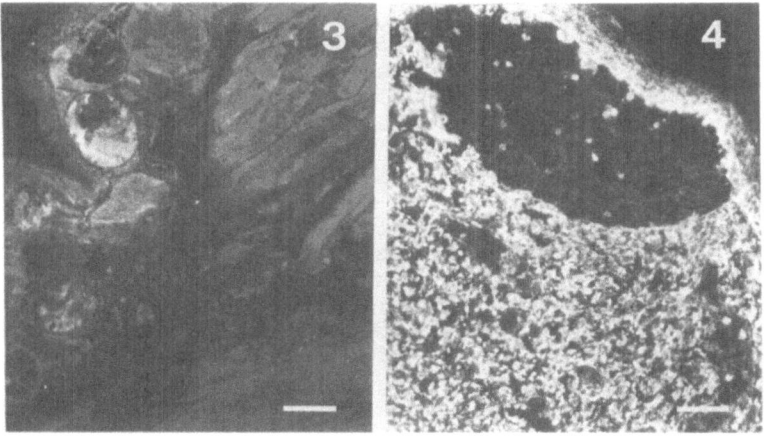

Fig. 3.   cAMP. Unamputated limb. Specific fluores-
          cence is seen only in some skin gland cells,
          epithelial cells of the basal layer, and in
          some muscles. Bar=100 μm.

Fig. 4.   cAMP in moderate early bud, heavy fluorescence
          in all blastema cells. Bar=80 μm.

the distal part of the forearm stump but were progressively
less abudant when we observed the more proximal stump (closer
to the elbow), and they become scarce in the arm. However,
through scarce, they were still more abundant than in the
analogous part of the unamputated limb. Muscles of the distal
stump showed fiber disruption and cell dedifferentiation.

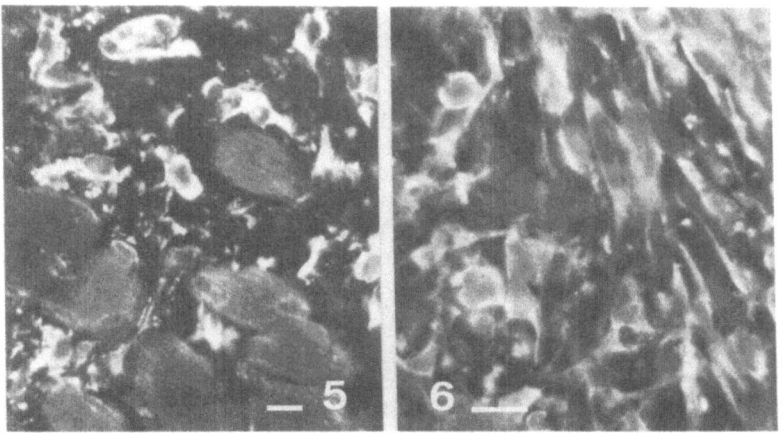

Fig. 5.   cGMP. Several highly fluorescent cells in
          dedifferentiating stump muscles, early bud
          stage. Bar=20 μm.

Fig. 6.   cGMP. Fluorescence in the cytoplasm of moving
          dermal cells. Bar=20 μm.

Dedifferentiating cells reacting to both antibodies moved out of the muscle compartment (Fig. 5). Several dermal elongated cells (fibroblasts or dedifferentiated cells) showed two cytoplasmic patches reacting to the two antibodies. These patches were located on the side of the nucleus and in cell extensions, and were placed along an axis giving the direction of the apparent cell movement, (Fig. 6).

Around the nerve almost all cells accumulated the cyclic nucleotides while within the nerve itself only occasional fibroblasts or Schwann cells exhibited some cytoplasmic fluorescence. Within blood vessels small and large monocytes often presented some fluorescence but the most intense reaction was noted in the endothelial cells of small vessels. In the stump epithelium the cells exhibiting a fluorescence were abundant in the zone immediately proximal to the blastema and diminished progressively in more proximal zones. In skin glands cyclic nucleotides accumulations were noted in individual cells, with great differences in the intensity of the reaction (from intense fluorescence to none). The pattern of this fluorescence was similar in unamputated and in regenerating limbs.

DISCUSSION

The present results concerned the observations of a fundamental intracellular transducing mechanism, and the study was conducted in normally regenerating animals, not submitted to additional experimental design other than plain limb amputation. They reflect therefore the usual biological game, adding a new dimension to the previous analyses carried out with classical histological staining, autoradiography and observations in vitro which have shown that the presence of the nerve and of hormones are necessary for regeneration but not precisely why (see Sicard, 1985 for reviews). The triggers and the triggered mechanisms responsible for cell dedifferentiation, cell cycle, cell movement and cell differentiation remain to be further defined and additional information on second messengers should help in this effort.

The low levels of cAMP and cGMP found in unamputated limbs and in the three other limbs of a one-limb amputated animal and the very high levels found in the regenerating limbs (Fig. 1,2) permit us to safely state that the increases observed in the levels of cAMP and of cGMP were essentially related to the cells directly and locally implicated in the limb regeneration process. We will therefore discuss only this kind of local relation.

At the early- and medium-bud stages three main cell biological events could be associated with cell accumulations of cAMP, of cGMP (or of both): cell cycle, cell dedifferentiation, cell movement. Observation of the presence of cAMP and cGMP in the basal epithelial cell layer, found in unamputated and in regenerating limbs could be associated with cycling cells since the cells of this layer are regularly dividing to insure cell renewal. According to their histological appearance and location these cells should be the same as the ones which were seen incorporating tritiated thymidine in comparable sections (Taban and Cathieni, unpublished). It is well known that almost all the cells of the blastema and some stump cells incorporate tritiated thymidine. We observed that their analogue were fluorescent in the sections treated with the two cAMP and cGMP antibodies. It was not surprising to find in rapidly proliferating cells, such as blastema cells, great reactions to the two cyclic nucleotides antibodies since in several other proliferating cell systems in vitro and in vivo studies have shown the existence of well defined fluctuations of the levels of cAMP and of cGMP related to cycling cells (see Goldberg and Haddox, 1977; Friedman, 1982; Boynton and Whitfield, 1983; Rixon and Whitfield, 1985). However, it was repeatedly observed that dividing cells are much less numerous in stumps than in blastemata (Chalkley, 1954; Hay and Fisch-man, 1961) therefore, the high level of cGMP found in stumps and which far exceeded the cGMP level measured in blastemata (Fig. 1,2) certainly could not be due solely to the presence of cycling cells. Instead of cell cycling, another cell pro-cess, cell dedifferentiation, could be related to these increased stump levels of cGMP. The characteristic of the stumps is the presence of abundant dedifferentiating cells. These cells reacted strongly with the two antibodies (Fig. 5,6). Moreover the RIA measures showed an extraordinary increase of cGMP over cAMP for stumps (Fig. 1,2). These observations strongly suggest a correlation between cell dedifferentiation and cGMP mobilisation. Under the micro-scope, dedifferentiating cells at the beginning of the process when they lose their intracellular differentiated structures, do not differ sharply from suffering, degenerating cells. It happens that increased levels of cGMP have been reported in degenerating retinal cells (Lolley and Lee, 1984). Our ensuing interpretation was that cell dedifferentiation in the amputated newt limb started with a phase of intracellular demolition of the previously differentiated intracellular structures, phase possibly analogous to a degenerating cell process, as observed in insect muscle and retinal cells. How-ever the increases of cGMP observed in the degenerating insect muscle and retinal cells is followed by cell death while in the amputated, non denervated newt limb, it is followed by the formation of living blastema cells. We further postulated the occurrence of a change in the number and quality of the cell membrane receptors presented by the dedifferentiating and the dedifferentiated cells allowing activation of these receptors by nerve-carried molecules and production of second messengers necessary for the modulation of the cell fate.

During dedifferentiation, dedifferentiating cells changed their original shape and this change was associated with the presence of the high levels of cGMP and of lower levels of cAMP (Fig. 1,2). These in vivo observations are in general agreement with the result gained in vitro by Yamada (1977) in studies on dedifferentiation and cell shape changes related to cyclic nucleotides in the iris epithelial cells of the newt. The patches of cyclic nucleotides seen in the apparently mov-ing cells (Fig. 6) could also be related to the cell shape changes necessary for the cell movements.

The small increase of cAMP and the slightly decreased level of cGMP observed in the contralateral forelimb of one-forelimb regenerating animals (Fig. 2) could be the result of a nerve transmitted modulation in the manner of the transneur-onal modulation described in the newt by Tweedle (1971) and known as the "Tweedle effect".

The magnitude of the increases in the levels of cGMP and the relations of the accumulation of this cyclic nucleotide with essential intracellular changes necessary for newt limb regeneration were unexpected and should be again emphasized.

REFERENCES

Babich, G.L., and Foret, J.E., 1973, Effects of dibutyryl cyclic AMP and related compounds on newt limb regeneration blastemas in vitro. II. ^{14}C-Leucine incorporation, Oncology, 28:89-95.

Boynton, A.L., and Whitfield, J.F., 1983, The role of cyclic AMP cell proliferation: a critical assessment of the evidence, Adv. Cyclic Nucleotide Res., 15:193-294.

Chalkley, D.T., 1954, A quantitative histological analysis of forelimb regeneration in Triturus viridescens, J. Morphol., 94:21-70.

Coons, A.H., 1958, Fluorescent Antibody methods, in: "General Cytochemical Methods", J.F. Danielli, ed., Academic Press, New York.

Foret, J.E., and Babich, G.L., 1973, Effects of dibutyryl cuclic AMP and related compounds on newt regeneration blastemas in vitro. 1. ^{3}H-Thymidine incorporation, Oncology, 28:83-88.

Friedman, D.L., 1982, Regulation of the cell cycle and cellular proliferation by cyclic nucleotides, in: "Cyclic Nucleotides II", Handb.Exp.Pharm. 58/2, J.W. Kebabian and J.A. Nathanson, eds., Springer-Verlag, Berlin.

Globus, M., Vethamany-Globus, S., and Kesick, A., 1987, Control of blastema cell proliferation by possible interplay of calcium and cyclic nucleotides during newt limb regeneration, Differentiation, 35:94-99.

Goldberg, N.D., and Haddox, M.K., 1977, Cyclic GMP metabolism and involvement in biological regulation, Ann. Rev. Biochem., 46:823-896.

Hay, E.D., and Fischman, D.A., 1961 Origin of the blastema in regenerating limbs of the newt Triturus viridescens. An autoradiographic study using tritiated Thymidine to follow cell proliferation and migration, Develop. Biol., 3:26-59.

Iten, L.E., and Bryant, S.V., 1973, Forelimb regeneration from different levels of amputation in the newt, Notophthalmus viridescens: length rate and stages, Wilhelm Roux's Arch., 173:263-282.

Jabaily, J., Rall, T.W., and Singer, M., 1975, Assay of cyclic 3'-5'-monophosphate in the regenerating forelimb of the newt, Triturus, J. Morph., 147:379-384.

Kintner, C.R., and Brockes, J., 1984, Monoclonal antibodies identify blastema cells derived from dedifferentiating muscle in newt limb regeneration, Nature, 308:67-69.

Lees, M.B., and Paxman, S., 1972, Modification of the Lowry procedure for the analysis of proteolipid protein, Anal., Biochem., 47:184-192.

Lichstein, D., and Rodbard, D., 1987, A second look at the second messenger hypothesis, <u>Life Sci</u>., 4:2041-2051.

Liversage, R.A., Rathbone, M.P., and McLaughlin, H.M.G., 1977, Changes in cyclic GMP levels during forelimb regeneration in adult <u>Notophthalmus viridescens</u>, <u>J.Exp.Zool</u>., 200:169-175.

Lolley, R.N., and Lee, R.H., 1984, Phosphodiesterase dysfunction, cyclic GMP accumulation, and visual cell degeneration in early-onset inherited blindness, <u>in</u>: "Advances in Cyclic Nucleotide and Protein Phosphorylation Research", vol. 17, P. Greengard, and J.A. Robinson, eds., Raven Press, New-York.

McLaughlin, H.M.G., Rathbone, M.P., Liversage R.A., and D.S. McLaughlin, 1983, Levels of cyclic GMP and of cyclic AMP in regenerating forelimbs of adult newts following denervation, <u>J.Exp.Zool</u>., 225:175-185.

Rathbone, M.P., Petri, J., Choo, A.F., Logan, M.D., Carlone, R.L., and Foret, J.E., 1980, Noradreline and cyclic AMP-independent growth stimulation in newt limb blastemata, <u>Nature</u> 283:387-388.

Rixon, R.H., and Whitfield, J.F., 1985, The Possible cyclic AMP-dependence of an early prereplicative event that determines mitosis in regenerating rat liver, <u>J. Cell. Physiol</u>., 124:397-402.

Schmidt, A.J., 1968, "Cellular Biology of Vertebrate Regeneration and Repair", University of Chicago Press, Chicago.

Schwartz, L.M., and Truman, J.W., 1984, Cyclic GMP may serve as a second messenger in peptide-induced muscle degeneration in an insect, <u>Proc. Natl. Acad. Sci. USA</u>, 81:6718-6722.

Sicard, R.E., 1985, "Regulation of Vertebrate Limb Regeneration", Oxford Univ. Press, New York.

Taban, C.H., Cathieni, M., and Schorderet, M., 1978, Cyclic AMP and noradrenaline sensitivity fluctuations in regenerating newt tissue, <u>Nature</u>, 271:470-472.

Taban, C.H., and Cathieni, M., 1983, Distribution of substance P-like immunoreactivity in the brain of the newt, <u>J. Comp. Neurol</u>., 216:453-470.

Thornton, C.S., 1968, Amphibian limb regeneration, <u>Adv. Morphog.</u>, 7:205-250.

Tweedle, C.D., 1971, Transneuronal effects on amphibian limb regeneration, <u>J. Exp. Zool</u>., 177:13-22.

Wescott, S.L., Nutman, T.B., Slater, J.E., and Kaliner, M.A., 1985, Production of monoclonal cGMP and cAMP antisera,, <u>J. Cyclic Nucleotide Prot. Phosph. Res.</u>, 10:189-196.

Whitfield, J.F., Boynton, A.L., Macmanus, J.P., Sikorska, M., and Tsang, B.K., 1979, The regulation of cell proliferation by calcium and cyclic AMP, <u>Mol. Cell. Biochem.</u>, 27:155-179.

Yamada, T., 1977, Control mechanisms in cell-type conversion in newt lens regeneration, <u>in</u>: "Monographs in Developmental Biology", 13, A. Wolsky, ed., S. Karger, Basel.

EPIMORPHIC REGENERATION AND THE IMMUNE SYSTEM

Raymond E. Sicard and Mary F. Lombard

Department of Pediatrics, Rhode Island Hospital
Providence, RI 02903, and Department of Biology
Regis College, Weston, MA 02913, USA

SUMMARY

Evidence for the suggested role of the immune system in epimorphic regeneration was sought by assessing effects on regeneration of immunological agents or active cellular immunity (allograft rejection) and by comparing immunological expression (e.g. mitogen responsiveness of splenocytes and interleukin 2 distribution) during regeneration with that following nonamputational trauma. There appeared to be no significant participation of humoral immunity in either epimorphic regeneration or repair of nonamputational wounds. However, expression of cellular immunity during regeneration differed from that observed during repair of nonamputational wounds. Moreover, expression of cellular immunity following amputation appeared to affect the progress of regeneration. This suggests that the nature of wounds can modulate expression of cellular immunity which, in turn, can affect mechanisms of wound repair, e.g. epimorphic regeneration vs fibroplasia and fibrosis.

INTRODUCTION

Characterizing the nature and mode of action of factors that regulate epimorphic regeneration in vertebrates has been the object of considerable interest. The notion that elements of the immune system participate in this capacity has appeared in several forms during past years. However, early postulates preceded comprehensive understanding of the immune system and were limited by ignorance of mechanisms of cellular interaction and expression that we take for granted today.

Early notions that blood cells gave rise to the blastema [Fritsch, 1911; Hellmich, 1930, 1931; Ide-Rozas, 1937] have been replaced by a realization that the blastema is of local origin [Butler, 1935; Brunst and Cheremetieva, 1936; Butler and O'Brien, 1942; Hay and Fischman, 1961]. However, blood cells, especially components of the immune syste, can contribute to regeneration in other ways. One current notion of the possible participation of the immune system in epimorphic regeneration derives from perceived analogies between the blastema and sarcoma and the proposed influence of the immune system as a modulator of tumor growth. Other notions relate to influence of the immune system on the expression of mechanisms of wound repair. For example, from studies of wound healing, Carrel [1921, 1922] discovered that lymphocytes exerted a growth-promoting effect on fibroblasts, at least in culture. Other studies have demonstrated that the immune system can affect the events of wound repair in mammals by regulating the functional expression of fibroblasts and macrophages [Wahl, 1981]. In addition, Liebman [1946, 1949] suggested that cells with lymphoid morphology provided a growth-promoting (trephocytic) influence in regeneration. The terminology used clearly implies that Liebman perceived that these cells contributed nutrients required for blastemal growth. More recently, drawing on parallels between the regeneration blastema and sarcomas, Prehn [Prehn, 1970-72; Prehn and Lappe, 1971] proposed that the immune system might exert growth-promoting influence on regeneration analogous to that experienced by certain sarcomas.

The immune system, thus, might affect the expression of epimorphic regeneration in several ways. For example, the immune system might interact with the wound site affecting the expression of repair mechanisms evoked, i.e. regeneration vs fibroplasia and fibrosis. Alternatively, the immune system might exert a growth-promoting influence on the dedifferentiated cells of the blastema. Undoubtedly other possibilities also exist.

The purpose of our investigations has been twofold. First, we have sought to determine the validity of the hypothesis that the immune system is an integral participant in epimorphic regeneration. Second, we have endeavored to ascertain more clearly the probable nature of interaction(s) between the immune system and the site of eventual regeneration. This has been accomplished by: (1) assessing influences of immunoactive agents on the progress of forelimb regeneration, (2) exploring reciprocal influences of active cellular immunity (rejection of skin allografts) (e.g. mitogen responsiveness of splenocytes and interleukin 2 [IL-2] distribution) following forelimb amputation and nonamputational wounding.

MATERIALS AND METHODS

General Procedures

Adult newts (Notophthalmus viridescens) used in our studies were obtained from C. Sullivan of Nashville, TN. Animals were maintained at 20 ± 2°C in large finger bowls containing either aged tap water or 10% Holtfreter's solution. Each bowl contained no more than fifteen animals and solutions were changed three times weekly.

Prior to surgical manipulations (e.g. skin grafting or removal of the humerus), newts were anesthetized in 0.1% aqueous methane tricaine sulfonate (MS-222; Sigma Chemical Co., St. Louis, MO). Postsurgical use of antibiotics was limited generally to cases in which infections occurred. Infected animals were isolated, treated topically with sulfadiazine

108

(Sigma Chemical Co.), and were excluded from further consideration.

Experimental Wounds

Wound consisted of either forelimb amputation or surgical removal of the humerus. Amputations were accomplished by unilateral transection through the distal third of the stylopodium. Any portion of the humerus found protruding from the wound site after retraction of soft tissues was carefully trimmed. In other cases, the humerus was removed through an incision in the shoulder. The progress of regeneration and wound healing were followed grossly. Identification of the stages of regeneration were based on the criteria of Iten and Bryant [1973].

Influence of Immunoactive Agents

Several substances were examined for their influence on the progress of regeneration. These substances were selected because of their ability to affect immunological expression in mammals, although their effects on amphibians were undetermined. The following substances were used and administered intraperitoneally (i.p.) every third day as specified: cyclophosphamide (Mead-Johnson Co., Evansville, IL; 50-100 µg); antilymphocyte serum (ALS, from Dr. A. Monaco, New England Deaconess Hosp., Boston, MA; 50-100 µg); thymic extract fractions (TE-H and TH-S, from Dr. A. Rule, Boston Coll.; 25-100 µg); cobra venom factor (CVF, Cordis Laboratories, Miami, FL; 2-4 U); interferon (IFN, Calbiochem Corp., La Jolla, CA; 1000 U); IL-1; (10 µg), IL-2 (50 µg), IL-3 (10 µg), and tuftsin (100 µg) (Sigma Chemical Co.). In addition three control series were employed in these experiments. They consisted of: (1) Holtfreter's solution (used to dilute all agents used), (2) bovine serum albumin (BSA [fraction V], Sigma Chemical Co.; 100 µg), and (3) inactivated thymic extract fractions (100 µg; heated to 95 °C for 30 min).

Effects of Challenge by Skin Allografts

Skin allografts consisted of small (2x2 mm) pieces of full thickness skin reciprocally transplanted between ventral and dorsal sites of paired newts. This provided simultaneously a color marker for the skin graft and a means of qualitatively assessing variations in histocompatibility among the animals used. In addition, each wound site created served as a transplantation site. Controls consisted of animals in which autografts were similarly performed. Following surgery, animals recovered overnight at 4°C in individual petri dishes. The progress of allograft rejection was subsequently followed with the aid of a dissecting microscope and related to the sequence of events identified by Cohen [1966] and Hightower and St. Pierre [1972].

Three experimental conditions were established: (1) amputation and grafting performed on the same day, (2) amputation followed by grafting, and (3) grafting followed by amputation. These conditions enabled effects of epimorphic regeneration on the expression of cellular immunity (condition 2) and the effects of activated cellular immunity on the expression of regeneration (condition 3) to be assessed. Progress of both allograft rejection and forelimb regeneration were evaluated relative to their respective controls (newts subjected to one or the other manipulation only).

Mitogen Responsiveness of Newt Splenocytes

During regeneration, or following wounding, spleens were removed at selected times. Splenocyte suspensions were prepared in Leibovitz L-15 medium (Difco Co., Detroit, MI) using pooled spleens from three newts in each group. Using the procedure of Horton et al. [1980], incorporation of [^{3}H]-thymidine (TdR*; Dupont Corp., Boston, MA) by splenocytes (10^5 cells/well) was assessed in the presence or absence of added mitogen.

Incorporation of TdR* in the absence of mitogen reflected endogenous proliferative activity of splenocoytes at a particular time after injury. Incorporation of TdR* in the presence of bacterial lipopolysaccharide (LPS, Sigma Chemical Co.; 2 mg/ml) reflected response of B-cells; while incorporation of TdR* in the presence of concanavalin A (con A, Sigma Chemical Co.; 2.5 µg/ml) reflected response of T-cells. Responses to mitogens, determined as incorporation in the presence of mitogens relative to incorporation in their absence, was designated the stimulation index.

Interleukin 2 Distribution

Distribution of IL-2 in newts during regeneration and following non-amputational trauma was accomplished using receptor-grade [^{125}I]-IL-2 (Dupont Corp.). Radiolabelled IL-2 (0.5 µCi/animal) was injected i.p. in 50 µL Holtfreter's solution. Thirty minutes later, animals were sacrificed and tissues removed for analysis. IL-2 content, assessed in major hemopoietic organs (liver, spleen, and thymus) and limb tissues (dehumerized limbs, amputated stumps or blastemata), and contralateral zeugopodia). IL-2 content was determined as counts per minute (cpm) per unit tissue mass and related to similar incorporation in tissues from newts subjected to no prior trauma. IL-2 distribution in tissues from animals undergoing regeneration was compared to incorporation in comparable tissues from newts repairing nonamputational wounds at equivalent times after trauma.

RESULTS AND DISCUSSION

If the immune system plays a role in regeneration, two predictions can be made: (1) Immune status of animals engaged in regeneration will be altered during regeneration. And (2) manipulation of immunological status of animals engaged in regeneration should alter its progress.

Effects of Immunoactive Agents

Initial investigations focused on determining effects on regeneration of amphibian forelimbs produced by administration of substances known to influence immunological expression in mammals. A variety of substances were tested [Table 1] which invariably adversely affected regeneration, although not necessarily in the same way [Sicard, 1981a, b; Schotté and

Sicard, 1982; Sicard and Laffond, 1983]. However, in all cases a reduced number of animals completed regeneration successfully. In addition a greater number of the blastemata formed were abnormal.

Although these experiments sought to determine if immunological agents could affect regeneration, immunological status of the animals treated was not assessed. However, neither BSA nor heated thymic extract fractions as foreign proteins, produced comparable results [Sicard, 1981b; Sicard and Laffond, 1983], suggesting that these agents exerted their effects on regeneration through biologically-specific actions rather than through nonspecific toxicity or antigenicity.

Since these results were consistent with immunological influence on regeneration, we examined immunological status during regeneration focusing on two parameters: (1) cellular immunity as observed during allograft rejection and (2) general immunological competence as measured by responses of splenocytes to mitogens. The former assumed that delayed graft rejection will occur under conditions that diminish immune responsiveness and that more rapid rejection will be correlated with immune stimulation; while the latter assumed that immunological status varies directly as a function of the number of proliferation-competent immunocytes responding to B-and T-cell mitogens.

Interactions between Allograft Rejection and Forelimb Regeneration

The first strategy used to identify possible immunological influence on regeneration was through challenge with skin allografts. This presented a challenge known to stimulate cellular immunity [Cohen, 1966], then assessed its impact on regeneration. When allografts were presented prior to amputation, regeneration was either delayed or accelerated, depending upon the time since grafting [Table 2]. A two week interval favored blastema formation and initial growth; however, a one week interval appeared to antagonize blastema formation and growth. We interpret this to mean that active cellular immune response, during its expansion phase, creates an environment disadvantageous to early events of regeneration. Perhaps destruction of blastemal cells occurs or fibroblasts are stimulated to a greater extent than might otherwise occur. When amputations preceded challenging by skin allografts, rejection was accelerated [Table 2]. This suggests stimulation of cellular immunity in association with epimorphic regeneration. However, presenting the challenges of allografts and regeneration simultaneously did not present an advantage to either process. In fact, there appeared to be an overall adverse effect that was especially evident as suppressed graft rejection among 40% of the newts subjected to simultaneous allograft challenge and amputation [Table 2]. Unfortunately the experimental design does not permit analysis of the complex interactions that likely arise under these conditions.

Splenocyte Responsiveness to Mitogens

Study of responsiveness of newt splenocytes to B-cell (LPS) and T-cell (con A) mitogens during regeneration and following nonamputational wounding was used to assess general

Table 1.   Effects of Immunoactive Agents on Amphibian Limb
           Regeneration[1]

Substance	Actions in mammals	Effects on Regeneration
Immune suppressants		
Cyclophosphamide	alkylating agent	complete suppression, no blastema formation
Cyclosporin	suppress activation of TH cells	suppression in 54% of cases
Antilymphocyte serum	"attack" native lymphocytes	retardation beyond blastema formation
Thymic extract fractions		
Helper	induce T_H cell differentiation	20-25% suppression; >50% abnormal limbs volumes decreased
Suppressor	induce T_S cell differentiation	12-15% suppression; >50% abnormal limbs volumes decreased
Lymphokines		
Interleukin 1	activate/promote	suppression in 25-58% of cases
Interleukin 2	differentiation of lymphocytes	
Interleukin 3		
Other Cytokines		
Interferon-beta	inhibit cellular proliferation	45% suppressed, 80% abnormal limbs
Tuftsin	macrophage chemotaxis/ migration	10% suppressed
Complement activator		
Cobra venom factor	activated properdin (or alternative complement pathway	retardation beyond blastema formation

[1]Details of certain of these data have been previously published,
see Schotté and Sicard, 1982; Sicard, 1981a,b; Sicard and Laffond, 1983.

Table 2. Reciprocal Effects of Graft Rejection and Regeneration[1]

Treatment	Time Required to Achieve[1] Rejection	Cone Stage	
Controls [18]	32 ± 3	28 ± 1	
Grafting + amputation [38]	41 ± 4[2]	33 ± 1*	
Grafting pre-amputation [13]	33 ± 3	41 ± 2*	(1 wk)
[6]		23 ± 2*	(2 wks)
Amputation pre-grafting [21]	22 ± 2*	22 ± 1*	

[1]Results = mean number of days ±1 s.e.m.; sample size is in brackets.
[2]Two responses occurred; 60% showed accelerated rejection (20±2 days, n=23) while 40% displayed apparent tolerance (73±4 days, n=15)
* p<0.05 vs corresponding control group

immunological activity during regeneration. The test system employed suspensions of isolated splenocytes as the source of immunocytes, since the spleen is the major lymphoid organ of adult newts [Hightower and St. Pierre, 1971]. The experiments assumed: (1) changes in TdR* incorporation in the presence of B- or T-cell mitogen reflect homeostatic activity of humoral or cellular immunity. (2) Differences in mitogenic responsiveness between wounded and control animals reflect changes in immunological homeostasis. (3) Differences in mitogenic responsiveness occurring among animals regenerating forelimbs, but not among newts reacting to nonamputational wounds, identified aspects of immunological responsiveness that might be significant for regeneration. In addition, times when changes might be expressed would provide insight into when immunological influence might be important to regeneration.

The results obtained [Table 3] revealed that proliferation in spleens of both regenerating and wounded animals decreased following trauma. However, this reduction was more dramatic among animals subjected to nonamputational trauma. During the acute phase of the inflammatory response (2 days post trauma) responsiveness to LPS was increased, suggesting possible enhancement of humoral immunity. However, no differences in responsiveness to LPS, compared to controls, were noted at other times. Thus, humoral immunity is not a significant factor in normal epimorphic regeneration or repair of uncomplicated nonamputational wounds. The pattern of responsiveness to T-cell mitogens was similar for animals subjected to nonamputational trauma. Among newts that were regenerating forelimbs, contrast with animals subjected to nonamputational trauma occurred only 12 days after amputation. At this time, responsiveness to con A was reduced to half that seen in controls.

Collectively, these observations suggest acute expansion in the numbers of activated B- and T-cells shortly after trauma. This is followed by emigration of activated lymphocytes from the spleen, as reflected by decreased TdR* incorporation. However, in animals engaged in epimorphic regeneration this was accomplished by apparent suppression of T-cell stimulation, reflected by decreasing responsiveness to con A. The consequence of these differences might be the presence of a substantial population of activated T-lymphocytes in association with nonamputated wounds but only a small population of activated T-cells in amputated stumps. Since the reparative events that occur under each of these conditions differ, one might speculate that the apparent differences in immunological activity (reflected in mitogenic responsiveness) are, in some way, instrumental in directing these events.

Distribution of Interleukin 2

This interpretation is reinforced by the distribution of [^{125}I] interleukin-2 (IL-2*) in adult newts during regeneration or following wounding [Table 4]. This lymphokine associates preferentially with activated T-cells and lymphokine-activated killer (LAK) cells and, to a lesser extent, with natural killer (NK) cells and macrophages. IL-2* was chronically reduced in hemopoietic tissues of regenerating newts,but reduced only transiently in newts subjected to nonamputational trauma. Moreover, IL-2* binding appeared to be decreased peripherally following trauma, as reflected by reduced IL-2* in nontraumatized limb tissues (data not shown). However, following removal of the humerus, IL-2* content of wound sites progressively increased from 2 to 21 days after trauma. This period coincides with the period of lowest proliferation index for this group [Table 3]. In contrast, IL-2* in stumps and blastemata decreased during this same period, suggesting reduced association of activated T-cells and LAK cells with regenerating limbs as compared to wounded limbs.

Interaction between the "Regeneration Zone" and the Immune System

Our studies into the possible role of the immune system as a source of regulatory influence on regeneration in amphibians have not yielded conclusive resolution of this question. Numerous substances with potential for affecting immunological activity clearly alter the progress of forelimb regeneration in adult newts. However, it cannot be concluded that these effects are the result of manipulation of immunological expression. On the other hand, experiments using allografts suggest that the immune system exerts influence on the expression and progress of epimorphic regeneration. Certain of these results indicate that alternation in cellular immunity can affect the normal progress of regeneration, while others suggest that cellular immunity is activated during regeneration.

The results of studies conducted with mitogens and with IL-2* provide additional support for possible participation of the immune system in regeneration. However, they also suggest that reciprocal interactions occur between the immune system and the wound site. Differences in proliferation index, mitogenic responsiveness of splenocytes, and IL-2* distribution between newts that were regenerating forelimbs and those healing nonamputational wounds suggest that the wound site itself communicates to central immune organs affecting their behavior. In turn, the nature of the immunological responses elic-

ited appear to affect the form of repair that is promoted in
the wound area. Thus, nonamputational wounds might produce
wound hormones that activate responses ensuring "healing". In
contrast, amputation might generate factors that prevent this
pattern of activation and, instead, evoke responses which
ensure that dedifferentiation and blastema formation occur.

We, therefore, propose the following: An amputated
appendage presents a distinctive wound environment that is
capable of evoking particular responses from the immune system
that favor epimorphic regeneration over fibroplasia and fibrosis with the eventual formation of cicatricial scar tissue.
This can be considered an "active regeneration zone". Denervation (and other manipulations) can render this "regeneration
zone" inactive (or incompetent), preventing it from generating
appropriate messages. The result is wound healing, but not
regeneration. Similarly, disturbing the immune system (e.g.,
following hypophysectomy) affects the ability to respond to
these messages and might yield similar results. Thus, we
envision dynamic interactions between the amputation site (or
"active regeneration zone") and the immune system that create
conditions permissive to epimorphic regeneration.

Future Directions

Our perception of the role of the immune system in regeneration leads us to anticipate three possibilities:
(1) Interaction will occur between the immune system and the
"regeneration zone" (or wound site) which is instrumental in
creating conditions favorable to epimorphic regeneration. This
notion derives, in part, from results of our studies and, in

Table 3.   Proliferation Indices and Mitogenic Responsiveness
           of Newt Splenocytes During Regeneration and Wound
           Healing

Time after trauma	Regenerating			Wounded		
	Prolif. index	Stim. index LPS	ConA	Prolif. index	Stim. index LPS	ConA
0 days	1.0	1.49	1.44	1.0	1.49	1.44
2 days	-	3.32*	4.32*	-	2.95*	3.48*
5 days	0.99	1.06	1.20	0.76	1.14	0.94
8 days	0.49*	1.08	1.01	0.48*	1.22	1.20
12 days	0.44*	1.47	0.59*[+]	0.21*[+]	1.09	1.27
16 days	0.40*	1.27	1.99	0.28*	1.38	0.94
20 days	0.62	1.00	1.15	0.44*	0.89	1.24
25 days	0.51*	1.19	1.04	0.88	0.92	0.86

Proliferation index = TdR* incorporated by experimental group
+ TdR* incorporated by controls;   Stimulation index = incorporation with mitogen + incorporation in absence of mitogen.
* $p < 0.05$ vs 0 days
[+] $p < 0.05$ vs other experimental group

Table 4. Interleukin 2 Distribution in Hemopoietic and "Wounded" Tissues of Newts During Regeneration and Wound Repair

Time after trauma (days)	Hemopoietic Tissue[1]		"Wounded" Tissue[1]	
	Dehumerized	Amputated	Dehumerized	Amputated
2	16.5^+	$23.1\pm 3.0^+$	11.8^+	$28.7\pm11.9^+$
8	$61.2\pm 6.3^+$	$40.8\pm 4.3*^+$	$45.4\pm 7.7^+$	$24.7\pm 2.8*^+$
15	95.3 ± 19.6	$55.2\pm 3.6*^+$	74.2 ± 13.9	$32.6\pm 3.6*^+$
21	96.6 ± 12.3	$36.7\pm 3.0*^+$	84.1 ± 8.0	$13.3\pm 0.6*^+$
29	N.D.	90.7 ± 16.2	N.D.	115.3 ± 30.8

[1] Values are presented as percent of control ·1 s.e.m.
$^+$ $p<0.05$ vs control; *, $p < 0.05$ vs dehumerized
N.D., not determined.

part, from investigations by others into mechanisms of wound repair [e.g., see reviews by Schmidt, 1968; Ross, 1970; Adair and Taban, 1971; Oppenheim et al., 1981; Sicard, 1985].
(2) Cytokines (growth factors?) will be elaborated and released by immunocytes which might promote blastemal growth and expression. We believe that this represents the "immune stimulation" hypothesis that we originally set out to explore.
(3) After differentiation and histogenesis are well-advanced, elements of the immune system will participate to ensure that tumor-like, dedifferetiated cells do not persist in the newly formed limb. This is an expression of the immunosurveillance function typically ascribed to and associated with the immune system.

Communication between a wound site and elements of the immune system has been known. A primary means through which this is accomplished is through changes in the concentrations of plasma proteins known as acute phase proteins. These substances modulate immunological expression and, in turn, affect progress of wound repair [Sipe and Rosenstreich, 1981]. However, we raise the question of whether wounds that lead to epimorphic regeneration present responses that can be readily contrasted from those associated with other wounds.

Our initial approach to this question has been through an effort to evaluate changes in plasma proteins following forelimb amputation or dehumerization. We have elected to examine plasma samples collected at varying times up to four weeks after trauma rather than focusing on the acute phase itself. Initial observations suggest that the protein composition of plasma varies as a function both of time after trauma and the nature of the trauma inflicted. We are currently endeavoring to confirm our preliminary observations and to identify the substances in question.

In addition, we are preparing to investigate the effects

of blastemal and wound (amputation) extracts and wound fluids (from nonamputational wounds) on the behaviors of macrophages, splenocytes, fibroblasts and blastemal cells in culture. These studies are intended to determine whether the "active regeneration zone" produces "wound hormones" capable of modulating the expression and behavior of cells that influence the nature of wound repair. Further studies will endeavor to understand (1) how these substances create conditions that favor epimorphic regeneration and (2) what factors are responsible for their production following amputation but not nonamputational injury.

ACKNOWLEDGEMENTS

The authors wish to express their gratitude to the numerous students whose technical assistance helped to generate the data presented in this report.

REFERENCES

Adair, L.C., and Taban,C.H., 1971, Comparison of wound healing and/or regeneration in mammals and urodeles, Arch. Sci. (Geneve), 24:499-516.

Brunst, V.V. and Cheremetieva, E.A., 1936, Sur la perte du pouvoir regenerateur chez le triton et l'axolotl par l'irradiation avec les rayons X., Arch. Zool. Exp. (Geneve), 79:57-67.

Butler, E.G., 1935, Studies on limb regeneration in x-rayed Amblystoma larvae, Anat. Rec., 62:295-307.

Butler, E.G. and O'Brien, J.P., 1942, Effects of localized x-irradiation on regeneration of the urodele limb, Anat. Rec., 84:407-413.

Carrel, A., 1921, Cicatrization of wounds. XII. Factors initiating regeneration, J. Exp. Med., 34:425-434.

Carrel, A. 1922. Growth-promoting function of leukocytes, J. Exp. Med., 36:385-391.

Cohen, N., 1966, Tissue transplantation immunity in the adult newt, Diemictylus viridescens, J. Exp. Zool., 163:157-190.

Fritsch, C., 1911, Experimentelle Studien über Regenerationsvorgänger des Gleidmassenskeletts, Zool. Jahrb. Abt. Physiol., 30:377-472.

Hay, E.D., and Fischman, D.A., 1961, Origin of the blastema in regenerating limbs of the newt Triturus viridescens. An autoradiographic study using tritiated thymidine to follow cell proliferation and migration, Dev. Biol., 3:26-59.

Hellmich, W., 1930, Untersuchungen über Herkunft und Determination des Regenerativen Materials bei Amphibien, Wilhelm Roux's Arch., 121:135-202.

Hellmich, W., 1931, Histology of regeneration in different species of adult and larval urodeles, Anat. Rec., 48:303-307.

Hightower, J.A., and St. Pierre, R.L., 1971, Hemopoietic tissues of the adult newt, Notophthalmus viridescens, J. Morph., 135:299-308.

Hightower, J.A. and St. Pierre, R.L., 1972, Skin allografts in the newt, Notophthalmus viridescens: cessation of blood flow as an indicator of rejection, J. Exp. Zool., 181:341-352.

Horton, J.D., Smith, A.R., Williams, N.H., Smith, A., and
    Sherif, N.E.H.S., 1980, Lymphocyte reactivity to "T"
    and "B" cell mitogens in Xenopus laevis: studies on
    thymus and spleen, Devel. Comp. Immunol., 4:75-86.
Ide-Rozas, A., 1937, Die Zytologische Verhältnisse bei der
    Regeneration von Kaulquappenextremitäten, Wilhelm
    Roux's Arch., 135:552-608.
Liebman, E., 1946, On trephocytes and trephocytosis: a study
    of the role of leukocytes in nutrition and growth,
    Growth, 10:291-330.
Liebman, E., 1949. The leukocytes in regenerating limbs of
    Triturus viridescens, Growth, 13:103-118.
Oppenheim, J. and Landy, M., 1983, "Interleukins, Lym-
    phokines, and Cytokines", Academic Press, New York.
Oppenheim, J., Rosenstreich, D.L., and Potter, M., 1981,
    "Cellular Functions in Immunity and Inflammation", Else-
    vier/North-Holland, New York.
Prehn, R.T., 1970, Immunosurveillance, regeneration, and
    oncogenesis, Progr. Exp. Tumor Res., 14:1-24.
Prehn, R.T., 1971, Percpectives on oncogenesis: Does immu-
    nity stimulate or inhibit neoplasia? J. Reticuloen-
    dothel. Soc., 10:1-16.
Prehn, R.T., 1972, The immune reaction as a stimulator of
    tumor growth, Science, 176:170-171.
Prehn, R.T. and Lappe, M.,1971, An immunostimulation theory
    of tumor development, Transplant. Rev., 7:26-54.
Ross, R., 1970, Wound healing, a review. in: "Chemistry and
    Molecular Biology of the Intercellular Matrix". Vol. 3
    E.A. Belasz, et al., eds. Academic Press, New York,
    pp. 1739-1751.
Schmidt, A.J., 1968, "Cellular Biology of Vertebrate Regener-
    ation and Repair", Univ. Chicago Press, Chicago.
Schotté, O.E. and Sicard, R.E., 1982, Cyclophosphamide-
    induced leukopenia and suppression of limb regeneration
    in the adult newt, Notophthalmus viridescens, J. Exp.
    Zool., 222:199-202.
Sicard, R.E., 1981a, The effects of putative immunological
    manipulations upon the rate of forelimb regeneration in
    adult newts, Notophthalmus viridescens, IRCS Med.
    Sci,. 9:962-963.
Sicard, R.E., 1981b, Morphologically aberrant limb regener-
    ation in adult newts induced by putative immunological
    manipulations, IRCS Med. Sci., 9:964-965.
Sicard R.E., 1985, Leukocytic and immunological influence on
    regeneration of amphibian forelimbs. in: "Regulation of
    Vertebrate Limb Regeneration", R.E. Sicard, ed., Oxford
    Univ. Press, New York, pp. 128-145.
Sicard, R.E. and Laffond, W.T., 1983, Putative immunological
    influence upon amphibian forelimb regeneration.
    I. Effects of several immunoactive agents on regener-
    ation rate and gross morphology, Exp. Cell Biol.,
    51:337-344.
Sipe, J.D., and Rosenstreich, D.L., 1981, Serum factors asso-
    ciated with inflammation. in: "Cellular Functions in
    Immunity and Inflammation", J.J. Oppenheim, D.L.
    Rosenstreich, and M. Potter, eds., Elsevier/North
    Holland, New York.
Sorg, C. and Schimpl, A., 1985, "Cellular and Molecular
    Biology of Lymphokines", Academic Press, New York.

Wahl, S.M., 1981. Inflammation and wound healing. in: "Cel-
    lular Functions in Immunity and Inflammation", J.J.
    Oppenheim, D.L. Rosenstreich, and M. Potter, eds,
    Elsevier/North Holland, New York.

IMMUNE SYSTEM AND REGENERATION

Anna Babaeva

Institute of Human Morphology
USSR AMS, Moscow USSR

SUMMARY

Splenocytes after partial hepatectomy, nephrectomy, resection of small intestine and massive haemorrhage acquire the ability to transfer regeneration information to non-operated syngenic recipients, that is, to increase proliferation in their corresponding organs. This phenomenon is biphasic, later the splenocytes become capable of suppressing proliferation. The unilateral (12-48 h) and bilateral nephrectomy (1-4 h) cause increase of immune response to the sheep's erythrocytes. These changes were also biphasic. Extirpation of the salivary gland (4-17 h) and spleen resection (17-168 h) reduce the number of antibody forming cells.

INTRODUCTION

According to conception of several investigators, the lymphoid system is responsible for the regulation of normal and reparative growth as well as of cell proliferation in certain organs (Burwell, 1963; Metcalf, 1964; Burch and Burwell, 1965; Babaeva, 1968, 1972, 1985). This point of view was subsequently confirmed by some experiments.

The purpose of the present report is to summarize briefly the results obtained in our laboratory which, as we hope, could help in understanding this problem.

In this paper we focus only on the morphogenetic function of the lymphoid cells, that is, their ability to transfer the regeneration information, and their immunological responsibility during various periods of different organ restoration.

MATERIALS AND METHODS

Young adult inbred male mice of CBA, C57BL/6J strains and hybrids (CBA x C57BL/6J) F-1 were used in the main part of the experiments. One experiment was done on male rats (August x Black Wistar) F-1.

Both the morphogenetic capacity and immunological responsibility of lymphoid cells were studied under conditions of adoptive transfer of splenocytes from operated, sham-operated or intact donors to syngenic recipients. All surgical procedures were performed under nembutal or ether anaesthesia.

Donor mice underwent one of the following operations: resection of the central lobe (25%) or two liver lobes (70%); resection of the kidney, unilateral or bilateral nephrectomy; resection of the spleen (70%), unilateral extirpation of submandibular salivary gland and corresponding sham operations. Donor rats were subjected to resection of half of the small intestine. Donors were killed by chloroform fumes or by cervical dislocation at different periods of time after surgery (donor interval). Lymphoid cells were isolated from the spleen in a glass homogenizer in the medium N 199, washed twice by centrifugation at 1500 rpm for 10 min. A dose of $60-70 \times 10^6$ viable nucleated cells in 1,0 ml medium $N_6 199$ was injected into a tail vein of each mouse (400×10^6 cells for each rat).

In a separate experiment splenocytes were injected after treatment by immune rabbit sera specific to mouse B or T lymphocytes.

Recipient mice were mainly non-operated, in some cases they were partially hepatectomized. All recipients were killed at different time after transfer of splenocytes (recipient interval). Their organs were fixed and processed routinely. The incidence of mitoses and the number of nuclei were determined by observation of 200-300 visual fields under oil immersion (900x). The mitotic index (MI) was expressed as the number of mitoses per 1000 nuclei. For the examination of the splenocytal capacity to immune response, lethally irradiated syngenic recipients were injected by spleen cells in a dose of 10^7 together with 2×10^8 sheep erythrocytes. The number of antibody forming cells (AFC) was determined by the method of Jerne and Nordin (1963) 7 days after injection.

RESULTS

In our early studies (Babaeva, 1968; Babaeva et al., 1969) we determined the optimal conditions for finding out the morphogenetic functions of lymphocytes. Initially time-response relationship was studied. Data on experiments of this kind are summarized in Fig. 1.

It can be seen that splenocytes removed from donors in 2-17 h after operation caused a singificant enlargement of the mitotic index both in hepatocytes and reticuloendotheliocytes of recipients' liver. Their ability to stimulation was completely lost 48 h after operation. It should be noted that different recipient interval is needed to find out this ability in different cells, i.e. 43 h for reticuloendotheliocytes, and at least 48 h for hepatocytes.

The results presented in Fig. 1 show clearly that morphogenetic activity of splenocytes appears long before the increase of mitotic activity in the regenerating liver. This indicates that splenocytes after operation can take part in the initiation of proliferation. Fig. 1 also shows that the enlargement of mitotic index reflects the true increase of

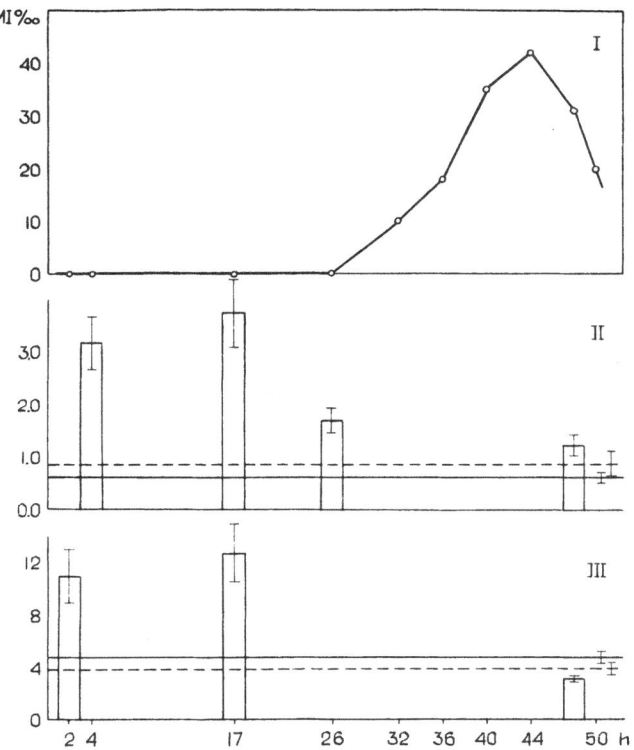

Fig. 1. Changes of the mitotic index in the cells of
regenerating (I) and recipients liver (II, III)
depending on the time interval after standard
hepatic resection (70%) and donor interval I -
in hepatocytes; II - in reticuloendotheliocytes (recipient interval - 43 h); III -in hepatocytes (recipient interval 48 h; colchemid
injected mice 4 h before sacrifice). Each column shows the mean ±SE.
Continuous line = MI in non operated mice.
Dotted line = MI in control recipients.

proliferative activity, as colchemid injection (4 h before
sacrifice) leads to accumulation of mitoses.

Fig. 2 demonstrates that splenocytes removed from intact
or sham operated donors do not stimulate proliferation in the
liver cells of intact recipients, as well as killed cells
removed from partially hepatectomized donors. At the same
time the activity of splenocytes from operated donors,
enriched by small lymphocytes, did not change.

Later it was found that mainly T-lymphocytes were responsible for transferring of regeneration information (Babaeva et
al., 1980). Pretreatment of splenocytes of partially hepatectomized mice by immune anti-T-serum caused a significant loss
of their ability to stimulate mitotic activity in liver cells
of recipients. Treatment of spleen cells by anti-B-serum was
not effective (Fig. 3).

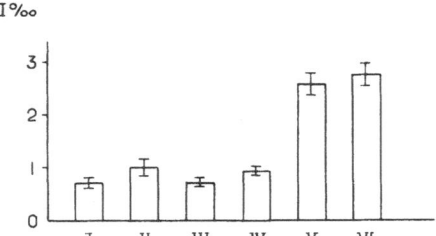

Fig. 2. Mitotic index in reticuloendotheliocytes of the intact mice liver (I), and recipient mice, that received splenocytes from donors; II - sham operated, III -intact, IV-VI - hepatectomized, (IV - killed splenocytes, V - enriched with small lymphocytes).

The surgery of some other organs as well as hepatectomy was accompanied by similar changes of splenocytic properties. Thus, splenocytes removed from unilaterally nephrectomized mice caused a significant increase of mitotic activity in renal tubular epithelial cells (Babaeva et al., 1973). Similar results were obtained after small intestine resection (Timashkevich et al., 1984) and massive haemorrhage (Babaeva, and Belan, 1988). Other regularities of changes in the splenocyte properties were also similar. The degree of morphogenetic activity was always positively correlated with the degree of tissue deficiency. Splenocytes isolated from unilaterally nephrectomized as well as from hepatectomized (70%) mice caused a greater increase of mitotic activity in cells of the corresponding organ than the 25% resection of renal or hepatic tissue (Fig. 4, Babaeva et al., 1973; Babaeva et al., 1982).

Numerous experiments have shown that the phenomenon of regeneration information transfer is only partially organ-specific. Splenocytes isolated from the spleen of unilaterally nephrectomized donors caused a significant increase of mitotic index in renal tubular cells. However, a slight effect was also observed in the reticuloendothelial cells of recipient's liver.

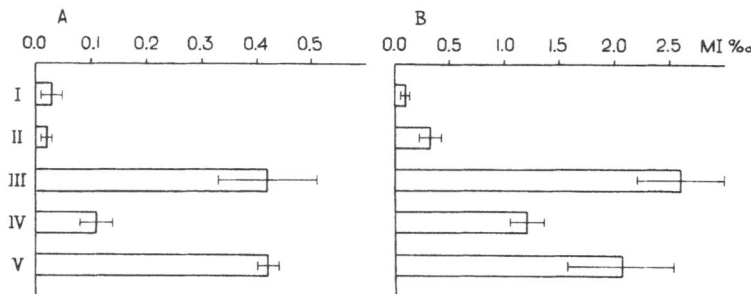

Fig. 3. Mitotic index in hepatocytes (A) and in reticuloendotheliocytes (B) of the liver from the intact mice (I), and recipients of splenocytes isolated from donors: intact (II), partially hepatectomized (III-V); III=whole suspension; IV=B-cells; V=T-cells; donor interval=17 h, recipient interval=48 h.

Similar data were obtained after small intestine resection: the splenocytes stimulated cell division in the oesophagus (Timashkevich et al., 1984).

As our last studies have shown, splenocytes not only initiate the proliferation, but they are also involved in its suppression (Babaeva and Zotikov, 1987). Splenocytes removed from partially hepatectomized donors 36 h after operation caused a significant suppression of mitotic activity in hepatocytes in regenerating liver (Fig. 5). A similar effect was obtained after transfer of splenocytes from regenerating spleen. In these cases the suppression was non-specific.

The splenocyte capacity to transfer regeneration information is biphasic: at the very early stages stimulates mitotic activity, whereas later its function is quite the reverse. The changes in the capacity of splenocytes for immune response after partial hepatectomy and unilateral nephrectomy in general were similar. They were also biphasic (Fig. 6). It was found that following injection of splenocytes removed from sham operated mice the immune response was the same as after injection of splenocytes from non-operated donors. This level of immune response was used as a control and expressed as 100%.

Fig. 6 demonstrates that surgery of liver and kidney is accompanied by an increase of immune response (Babaeva and Gimmelfarb, 1988), whereas operation on the salivary gland and spleen leads to immune response suppression (Yudina, 1979; Babaeva, 1985). There was a positive correlation between the degree of immune response and the amount of tissue removed (Fig. 7).

In contrast to unilateral nephrectomy, the injection of splenocytes from bilaterally nephrectomized donors leads to a more rapid response. The level of AFC was significantly less in the recipients of splenocytes from bilateral uretral ligation donors.

DISCUSSION

The data obtained in our laboratory confirm the general

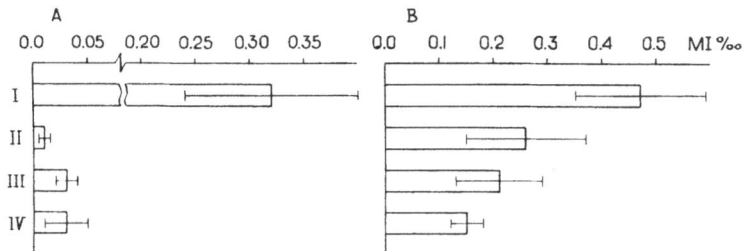

Fig. 4. Mitotic index of hepatocytes (A) and renal tubular epithelial cells (B) in recipients after the injection of splenocytes from 70% (I) and 25% (A-II) hepatectomized donors; BI - from 50%; BII - 25% nephrectomized donors; III - control recipients; IV - intact mice.

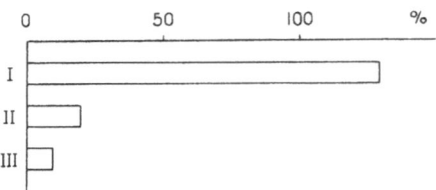

Fig. 5. Suppression of mitotic activity in hepatocytes
of regenerating liver (100%) after injection
of splenocytes from donors: I-intact;
II-partially hepatectomized (donor's interval
-36 h; recipient's interval -24 h);
III-partially splenectomized (donor's interval
-48 h; recipient's interval - 48 h).

idea that the lymphoid system plays an important role in regulation of reparative growth and, according to Burwell (1963)
it is responsible for establishing and maintaining morphosta-
sis for many tissue types.

All our findings and conclusions were confirmed by experiments of other investigators. The ability of splenocytes to
transfer regeneration information after hepatectomy and nephrectomy was clearly demonstrated by Pliskin and Prehn (1975).
They showed that spleen cells after these operations caused
the increase of DNA synthesis in the cells of corresponding
organs of non-operated and operated recipients. They also
observed both organ-specific and non-specific stimulating
effects of lymphoid cells.

Fox and Wahman (1968) and later Radosevic-Stasic and
Rukavina (1979) showed that lymphocytes transferred from unilaterally nephrectomized recipients caused an enhancement of

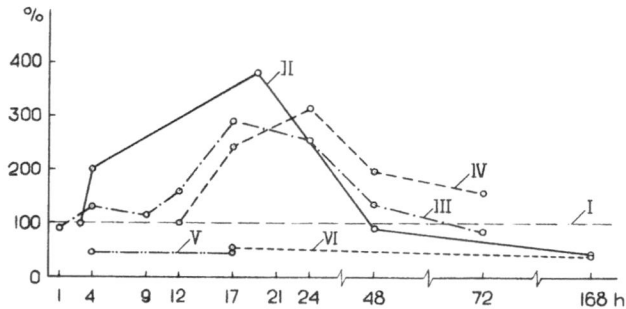

Fig. 6. Changes of the number of AFG in murine spleen:
I - sham operated (100%), II - partially hepatectomized, III - unilaterally nephrectomized,
IV - unilaterally ureter ligated, V - unilaterally sialadenectomized, VI - partially splenectomized.

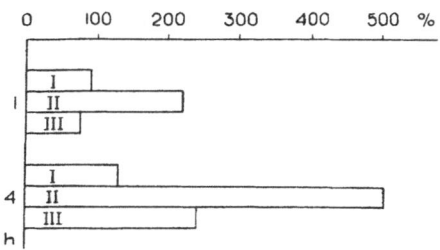

Fig. 7. Changes in the number of AFC in murine spleen
after 1 h and 4 h post operation:
I - unilateral nephrectomy, II - bilateral
nephrectomy, III - bilateral ureter ligation.

kidney growth. Both these findings indicate that there was a
true stimulation of mitotic activity in our experiments. It
has been well known that partial hepatectomy caused a stimulation of antibody-forming capacity (Sakai et al., 1970). But
the same stimulation of antibody-forming capacity occurs after
uni- and bilateral nephrectomy. A significant rapid increase
in the number of AFC in the spleen after bilateral nephrectomy
and the delay of this response after unilateral nephrectomy as
well as after bilateral ligation of the ureter allowed us to
suggest that the alteration of the immune response at an early
stage is the result of the immune system reaction to organ
antigen deficiency. The change of the immune system function
occurs not only in response to the introduction of foreign and
changed "self" antigens, but also to the amount of "self"
antigens. Reduction of organ mass probably leads to the disturbance of natural immunological tolerance, which is accompanied by a change in the ratio of immunoregulatory cells. We
suppose that morphogenetic activity is a cell-mediated phenomenon. Stimulating or suppressing effects are produced by
different lymphocyte populations (maybe T-helpers and T-suppressors). Lymphokins may play an important role in the realization of their morphogenetic function. Further investigations will help to elucidate these questions.

REFERENCES

Babaeva, A.G., 1968, Immunological reaction in the normal and
        reparative growth, in: "Proceedings of the 5th All
        Union Conference of Regeneration and Cell Division",
        Meditsina, Moscow.
Babaeva, A.G., 1972, "Immunological mechanisms in the regula-
        tion of the reparative processes", Meditsina, Moscow.
Babaeva, A.G., 1985, "Regeneration and the system of immu-
        nogenesis", Meditsina, Moscow.
Babaeva, A.G. and Zotikov, E.A., 1987, "Immunology of the pro-
        cesses of the adaptive growth, proliferation and their
        disorders", Science, Moscow.
Babaeva, A.G. and Belan, E.I., 1988, Changes in the pattern of
        differentiation of erythroid shoot of the bone marrow
        in mice during transplantation of peritoneal cells from
        the anemized donors, Cytologia, 19:125-131.

Babaeva, A.G. and Gimmelfarb, E.I., 1988, Antibody formation in
        the case of experimental renal deficiency and at the
        early stages of reparative renal growth in mice,
        Ontogenez, 19:42-47.
Babaeva, A.G., Kraskina, N.A., and Liosner, L.D., 1969,
        Enhancement of mitotic activity of liver cells of non-
        operated mice by the transfer of lymphoid cells of par-
        tially hepatectomized donors, Cytologia, 11:1512-1530.
Babaeva, A.G., Kraskina, N.A., and Liosner, L.D., 1973, Effect
        of lymphoid cells of unilaterally nephrectomized mice
        on proliferative activity of the kidney and liver of
        non-operated syngenic recipients, Bull. Exp. Biol.
        Med., 75:78-80.
Babaeva, A.G., Kraskina, N.A., and Yudina, N.V., 1980, Stimula-
        tion of the mitotic activity of hepatocytes and Kupf-
        fer cells in the liver of non-operated mice by T- and
        B-lymphocytes from partially hepatectomized syngenic
        donors, Bull. Exp. Biol. Med., 89:69-71.
Babaeva, A.G., Nesterenko, V.G., and Yudina, N.V., 1982,
        Increased capacity of lymphocytes of mice after
        repeated liver resection to transfer regeneration
        information, Bull. Exp. Biol. Med., 93:97-99.
Burch, R.A., Burwell, R.G., 1965, Self and not-self. A clonal
        induction approach to immunology, Quart. Rev. Biol.,
        40:252-279.
Burwell, R.G., 1963, The role of lymphoid tissue in morphosta-
        sis, Lancet, 2:69-70.
Fox, M., and Wahman, G.E., 1968, Etiology of the compensatory
        renal response: observation on the role of the lymphoid
        system, Invest. Urol., 5:521-538.
Jerne, N.K., and Nordin, A.A., 1963, Plaque formation in agar
        by a single antibody-producing cell, Science,
        40:405-410.
Metcalf, D., 1964, Functional interaction between the thymus
        and other organs, in: "The Thymus", Philadelphia.
Pliskin, M.E., and Prehn, R.T., 1975, Stimulation of liver
        regeneration and compensatory kidney hyperplasia by
        passive transfer of spleen cells, RES, J. Reticuloen-
        dothel. Soc., 17:290-299.
Radosevic-Stasic, B., and Rukavina, D., 1979, Renotropic acti-
        vity of lymphatic tissue, Periodiucum Biologorum,
        81:151-152.
Sakai, A., Muller-Berat, G.N., and Dabray-Sachs, M., 1970,
        Effect of partial hepatectomy on immune response to
        sheep red cells, Acta path. et microbiol. Scand.,
        78:652-656.
Timashkevich, T.B., Kharlova, G.V., and Yudina N.V., 1984, The
        lymphoid regulation of proliferation during regener-
        ation of intestine and oesophagus in rats, Ontogenez.,
        101-103.
Yudina, N.V., 1979, Peculiarities of the lymphoid tissue reac-
        tion in compensatory-reparative growth of salivary
        glands in mice, Bull. Exp. Biol. Med., 88:712-714.

DEPENDENCE OF ERYTHROBLAST PROLIFERATION AND DIFFERENTIATION
MANNER IN MICE HEMOPOIETIC TISSUE REGENERATION ON LYMPHOID-
MACROPHAGAL SYSTEM CELLS (*)

A.G. Babaeva and E.I. Belan

Institute of Human Morphology, USSR AMS
117418 Moscow, Tsurupa St. 3, U.S.S.R.

By adoptive transfer of T-lymphocytes from donors with
regeneration process to intact syngeneic recipients, ability
of such lymphocytes to transfer the "regeneration information"
has been discovered. Lymphocytes of operated donors caused
cell proliferation alterations in intact recipients similar to
those that have been observed in operated animals. In connection with restoring inner organs in mammals after damage we
continued the investigation of morphogenic properties of lymphocytes and macrophages at hemopoietic tissue regeneration.

The aim of the present work has been to investigate three
problems: 1) can a cell population with a great number of
macrophage transfer the "regeneration information"? 2) Do
lymphocytes and macrophages differ in the ability to transfer
to "regeneration information"? 3) Does this lymphoid-
macrophagal system control the processes of cell differentiation ?

The questions put forward have been investigated on the
model of hemopoietic tissue regeneration after massive haemorrhage since massive haemorrhage in pigeons, rats and mice,
initiates a particular type of erythroid cells differentiation, the "reserve erythropoiesis", in bone marrow. Polychromatophilic erythroblast proliferation increases and basophilic
erythroblast proliferation decreases since the latter misses
the stages of polychromatophilic and oxyphilic erythroblasts
transforming into immature reticulocytes. Oxyphilic erythroblasts which don't divide in normal conditions begin to proliferate intensively.

CBA/6J male mice have been used as donors and recipients
in the adoptive transfer of spleen cells or induced by inflammation peritoneal cell (PC) suspension. PC and spleen cells
have been obtained from intact and anemized donors (haemorrhage value was 2.5% of whole body weight). PC and spleen
cells have been obtained at different times after haemorrhage
and recipients have been injected different doses. Mitotic
indices in basophilic, polychromatophilic and oxyphilic erythroblasts and erythro-leucocytic indices in bone marrow and
blood reticulocytes number have been counted 4 days later.

129

PC containing 60-70% of macrophages and 25% of lymphocytes are demonstrated to acquire the ability to initiate the "reserve erythropoiesis" in intact recipients bone marrow already after 1/2 h and to initiate the increasing reticulocytosis in blood, if the dose is 10^7 cells.

This PC property is maximal 4 h later and remains during 48 h after haemorrhage. The "reserve erythropoiesis" initiation in intact recipients after PC adoptive transfer to them from anemized animals is not connected with the transfer of stem cells forming 11 and 8 days spleen colonies nor committed to erythroid lineage progenitor cells forming 5 days spleen colonies.

After adoptive transfer of 10^7 PC from intact donors to intact recipients the "reserve erythropoiesis" signs are not discovered. High mitotic indices of basophilic and polychromatophilic erythroblasts are revealed simultaneously with decreasing the erythro-leukocytic indices and blood reticulocyte number. Mitoses in oxyphilic erythroblasts are not discovered. Most likely in this case it is the erythropoiesis suppression by means of erythroblasts mitoses time-lenghtening.

Anemized donors PC in adoptive transfer initiate the "reserve erythropoiesis" already in doses of 10^6 cells. Maximal effect is achieved in doses of $5 \cdot 10^6$ cells and it is maintained in doses of 10^7 cells.

Anemized donors' spleen cells injected into intact recipients in doses of $5 \cdot 10^6$ and $6 \cdot 10^7$ cells initiate the "reserve erythropoiesis" less than PC in dose $5 \cdot 10^6$ cells. Only single mitoses in recipients' oxyphilic erythroblasts appear after transfer $6 \cdot 10^7$ spleen cells from intact donors.

Thus it is demonstrated that the cell population enriched by macrophages acquires the ability to initiate the hemopoietic tissue regeneration of the "reserve erythropoiesis" type in answer to massive haemorrhage. This morphogenic activity is expressed in PC more markedly than in spleen lymphoid cells. Furthermore in haemorrhage conditions spleen cells and especially PC are able not only to change erythroid cell proliferation intensity but also to initiate the particular type of erythroid cell differentiation to quick recovery of the cell number.

INSULIN LOCALIZATION IN THE PANCREAS AND IN FORELIMB

REGENERATES OF THE ADULT NEWT, <u>NOTOPHTHALMUS VIRIDESCENS</u> (*)

Ramsey A. Foty and Richard A. Liversage

Department of Zoology, University of Toronto
Toronto, Ontario, Canada M5S 1A1

INTRODUCTION

Metabolic hormones play an essential role in the regeneration of urodele appendages. Indeed, the involvement of insulin is essential as organ-cultured adult newt forelimb blastemata attain maximum growth and differentiation only when insulin is present in combination with thyroxine, hydrocortisone, and either growth hormone (1) or prolactin (2). In experimentally induced diabetic adult newts, forelimb blastemata fail to accumulate the critical mass of cells required for the continuation of regeneration. Also, cartilage formation as well as growth and differentiation of the tissues are markedly reduced (3,4). Given the importance of insulin to the process of regeneration, it was of interest to attempt localization of this hormone at its source (presumably the pancreas) and at its site of action in the regenerating forelimb.

MATERIALS AND METHODS

They include aldehyde fuchsin (AF) staining, peroxidase-antiperoxidase (PAP) immunolocalization for light microscopy (LM) and immuno-gold localization for scanning electron microscopy (SEM) in conjunction with backscattered electron imaging (BEI) of vibratome sectioned newt pancreatic tissue, and transmission electron microscopy (TEM) in conjunction with BEI for the localization of immunogold label in limb regenerates. Due to the scattering properties of protein A-gold, gold colloid is easily detected by TEM. And, because of the material-dependent signal, preferentially backscattered electrons are used in SEM (5,6). Backscattered electrons (BEs) are generated when incident beam electrons collide with atomic nuclei in the specimen. Such collisions frequently alter the electron's direction in a manner so as to reflect the electrons back to the surface of the specimen. Since the production of BEs increases with the mean atomic number (Z) of the material irradiated, as a specimen is scanned, more dense or higher Z areas (accumulation of gold label) will produce more backscattering than lower Z, less dense (e.g. carbon) regions (7). This analytical method utilizes the ten-fold difference in generated backscattered electron yield between gold and carbon to image areas in the pancreas of the newt that are immunoreactive to a specific insulin antibody conjugated to protein A-gold. This technique, in conjunction with secondary electron imaging (SEI), provide for an optimal correlation between labelled sites and ultrastructure (8).

RESULTS AND DISCUSSION

The hormone-secreting cells of the endocrine pancreas in the adult newt are arranged in a manner similar to those in many mammals. Insulin, glucagon and somatostatin-secreting cells are found within discrete islet-like cell clusters. The insulin secreting cells are centrally located within an islet. The glucagon- and somatostatin-immunoreactive cells are found at the islet periphery (9). In mammals, the islets are 100 to 200 µm in diameter (10,11). In the adult newt, islets range in size from 100 to 300 µm diameter. The islets of Langerhans in amphibia are usually small, contain mainly insulin secreting (B) cells and, in some urodeles, glucagon-secreting (A) cells are lacking (10). In the adult newt the islets are large, they contain mainly B cells, and they also contain A cells as well as somatostatin-secreting (D) cells (9). The current investigation demonstrates that the mammalian antibody used to detect insulin reacts in a cyto-immunologically specific manner. The insulin antibody utilized (courtesy of Dr. C.C. Yip, Banting and Best Department of Medical Research, University of Toronto), displays little non-specificity as demonstrated through homologous and heterologous antigen immunoabsorbance assays. Elliot and Youson utilized a similar antibody, also supplied by Dr. C.C. Yip, to localize insulin in the pancreas of the sea lamprey, Petromyzon marinus (12). Similarly, they found it to be highly specific. Having established that the adult newt pancreas contains insulin, and that a mammalian antibody against insulin provides a specific probe, it was of interest to determine whether insulin could be detected in newt pancreas by immunological techniques employing scanning electron microscopy. In this study, an electron-dense marker, protein A-gold, when conjugated with insulin antibody, was readily detected by BEI in the newt pancreas and its localization corroborated our results from the PAP-stained adjacent sections. Moreover, protein A-gold appears to bind specifically to the antibody as deletion of the antibody during incubation resulted in little or no gold signal BEI. Transmission electron microscopy in conjunction with immunogold labelling demonstrates the binding of insulin to cells of adult newt forelimb regenerates. In 14-21 day blastemata, immunoreactive insulin (IRI) was limited to cell membranes; in many cases being most prevalent at a cell membrane site nearest the nucleus. Evidence of insulin receptor aggregation, presumably prior to endocytosis, exists. An apparent increase in insulin binding occurs in regenerates undergoing differentiation, between 28 and 35 days postamputation. In our SEM and TEM preparations, BEI and energy dispersive X-ray microanalysis confirmed the presence of gold label. Controls for antibody specificity, including the incubation of sections in pre-immune serum or the deletion of the protein A-gold label, gave negative immunoreactivity.

CONCLUSIONS

a) The pancreas of the adult newt contains clusters of cells within the endocrine islets which are immunoreactive to insulin, glucagon and somatostatin and the arrangement of these cells is similar to that found in mammals; (b) insulin is detected by both PAP immunochemistry as well as immunogold localization by SEM in conjunction with BEI; c) the mammalian antibody probes employed in this study react specifically to the newt antigens; and d) IRI is detectable in regenerating limb tissues and there appears to be an increase in insulin binding during differentiation.

Cell membrane binding studies are currently being conducted, in order to determine whether cells of the regenerate undergo changes in receptor affinity for insulin during the regenerative process. Preliminary experiments indicate that increased IRI binding occurs in regenerates undergoing differentiation, comparable to the increased binding which occurs during differentiation in other systems (13,14).

ACKNOWLEDGEMENTS

Supported by NSERC of Canada Grant A-1208 to R.A.L.

REFERENCES

1. Vethamany-Globus, S. and R.A. Liversage, 1973a, J. Embryol. exp. Morphol., 30:397-413.
2. Liversage, R.A. et al., 1984, Roux's Arch. Dev. Biol., 193:379-387.
3. Vethamany-Globus, S. and R.A. Liversage, 1973b, J. Embryol. exp. Morphol., 30:415-426.
4. Vethamany-Globus, S. and R.A. Liversage, 1973c, J. Embryol. exp. Morphol., 30:427-447.
5. Walther, P. et al., 1984, Scanning Electron Microscopy, III:1257-1266.
6. De Harven, E. and D. Soligo, 1986, Am. J. Anat., 17:277-287.
7. Becker, R.P. and J.S. Geoffroy, 1981, Scanning Electron Microscopy, IV:195-206.
8. Hodges, G.M. et al., 1987, Scanning Microscopy, I:301-318.
9. Foty, R.A. et al., 1988, Tissue and Cell, (submitted).
10. Bentley, P.J., 1982, "Comparative Endocrinology", Cambridge University Press, London.
11. Pictet, R. and W.J. Rutter, 1972, Handbook of Physiology, Sect. 7 Endocrinology, volume 1 Endocrine Pancreas. American Physiological Society, Washington D.C.
12. Elliot, W.M. and J.H. Youson, 1986, Cell Tissue Res., 243:629-634.
13. Watanabe, N. et al., 1986, Diab. Res. Clin. Pract., 2:283-289.
14. Paireault, J. and F. Lasnier, 1987, J. Cell. Physiol., 132:279-286.

LENS AND LIMB REGENERATION IN AMPHIBIA UNDER THE SPACE

FLIGHT CONDITIONS (*)

V.I. Mitashov, E.N. Grigoryan, S.Ya. Tuchkova,
I.E. Malchevskaya

Institute of Developmental Biology
U.S.S.R. Academy of Sciences
26 Vavilov St., Moscow 117808, U.S.S.R.

It has been shown that lens and limb regeneration in lower vertebrates is dependent on various factors such as temperature, light, irradiation, age and species of animals. But nothing is known about the role of the space flight conditions in this process. On the other hand information of this kind can give us an opportunity to prognose the restoration processes in mammals and human. This idea has encouraged us to investigate lens and forelimb regeneration in newts during and after space flight.

Adult spanish newts (<u>Pleurodeles waltlii</u>) were used in ths spaceflight experiment. The lesn and forelimb were removed on the 5th and 7th days before space flight. The biosatellite flight times were seven and fourteen days after the two biosatellite launchings (on the landing days), and on the 9th, 14th, 135th days after the landing.

Necessary control experimjents were made under ground gravitation conditions. Influence of the transportation (animals were sent by train to the biosatellite starting place and returned back) and temperature were investigated in these controls, as well as the influence of vibration and mcirogravity by emans of vibrator and centrifuge.

Lens and forelimb regeneration were assessed on morphological stages of forming regenerates investigated by routine histology. In addition to this, radioautographic analysis of proliferation of the cells involved in regeneration process and measurement of the regenerate sizes were carried out.

It was found that the lens regeneration rate had no changes under weightlessness. On the landing days the majority of the newforming lenses reached similar stages in the experiment as in the control groups. However on the 9th and 14th days after the landing the synchronization of the lens regeneration and 1,5-2 times size enlargement of the advanced lenses were discovered in the spaceflight experiment but not in the control. In the experiment these events took place simultaneously with doubling of the number of ^{3}-thymidine labeled cells.

Similar results were obtained for limb regeneration. There were no changes in the forelimb regeneration rate in newts which had been into the space flight. In the experiment as well as in the control, on the landing days limb regenerates reached stage II (early bud, 7day flight) and stages II-III (early bud-bud, 14 day flight). But after the landing synchronization of the regeneration process was observed in the experiment, while asynchronization took place in the control. In addition to this, most of the limb regenerates were significantly promoted in their development as compared to the control group.

The 14-day stay of animals in the space flight did not disturb the morphogenetic process during forelimb regeneration: on the 135th day after the landing all regenerates had normal structure.

Therefore our first results suggest that the 7-day and 14-day stays of the newts under the spaceflight conditions affect the process of regeneration in the post flight time. This influence manifests itself in the increase of proliferative activity of the cells involved in the restoration, speeding up and synchronization of the regenerative processes and in enlargement of the regenerate size.

IN VITRO EFFECTS OF A MITOGENIC FACTOR

ON XENOPUS LAEVIS FORELIMB REGENERATION (*)

Catherine Tsilfidis and Richard A. Liversage

Ramsay Wright Zoological Laboratories
25 Harbord Street, University of Toronto
Ontario, Canada, M5S 1A1

INTRODUCTION

The essential role of nerves in the support of amphibian forelimb regeneration is well documented (Singer, 1952; 1954). In vitro experiments (Globus and Vethamany-Globus, 1977; Tomlinson et al., 1981) suggest that the neurotrophic agent is a diffusible proteinaceous nerve growth factor of low molecular weight, which is inactivated by heat and trypsin (Singer, 1976). The current investigation was designed to test the effects of a chick brain growth factor (CBGF) on Xenopus laevis forelimb regeneration; specifically its role as a mitogenic factor for mesenchyme-like cells in the regenerate. CBGF is a low molecular weight (500-2,000), relatively heat stable peptide (Carlone et al., 1988).

MATERIALS AND METHODS

The experiment consists of four series containing 15 regenerates each: a) controls (no mitogen), b) 100 ng, c) 1 ug and d) 10 µg CBGF/ml medium. Bilateral amputations were performed on 30 Xenopus laevis froglets which were anesthetized in 0.05% MS222 adjusted to pH 7.0. After a period of four weeks, regenerates were removed, decontaminated in 1% chloramine for 90 seconds followed by 3 rinses in double distilled H_2O and trimmed of extraneous stump tissue for explantation. Explants were placed at the air-medium interface on stainless steel grids in culture dishes containing Eagle's Minimum Essential Medium (Gibco). The medium was altered to frog serum osmolarity (240 ± 10mOsm/kg H_2O) and fortified with 10% fetal bovine serum, 0.05 I.U. bovine insulin and 100 I.U. penicillin G potassium per ml medium. Cultures were gassed with 5% CO_2 in 95% sterile air and maintained at $22\pm1°C$ for 24 hours. A fresh change of culture medium was concomitant with the addition of one of the three mitogen concentrations and 3 uCi/ml of [methyl-^{3}H] thymidine (spec. act. 56 Ci/mmol). Twenty-four hours later, explants were sampled for either liquid scintillation counting or autoradiography. The epithelium of the explants was removed prior to scintillation counting.

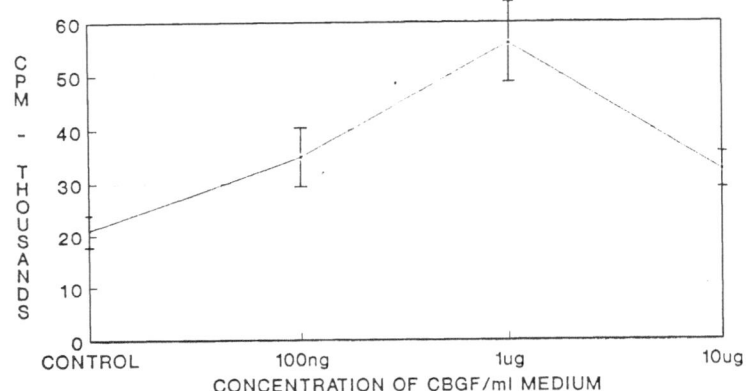

Fig. 1. Incorporation of [methyl ^{3}H] thymidine into
mesenchyme-like cells of the <u>Xenopus</u> regenerate. Using Duncan's multiple range test, significant differences were found between 1 μg
CBGF/ml medium and the other two series and
also compared with controls.

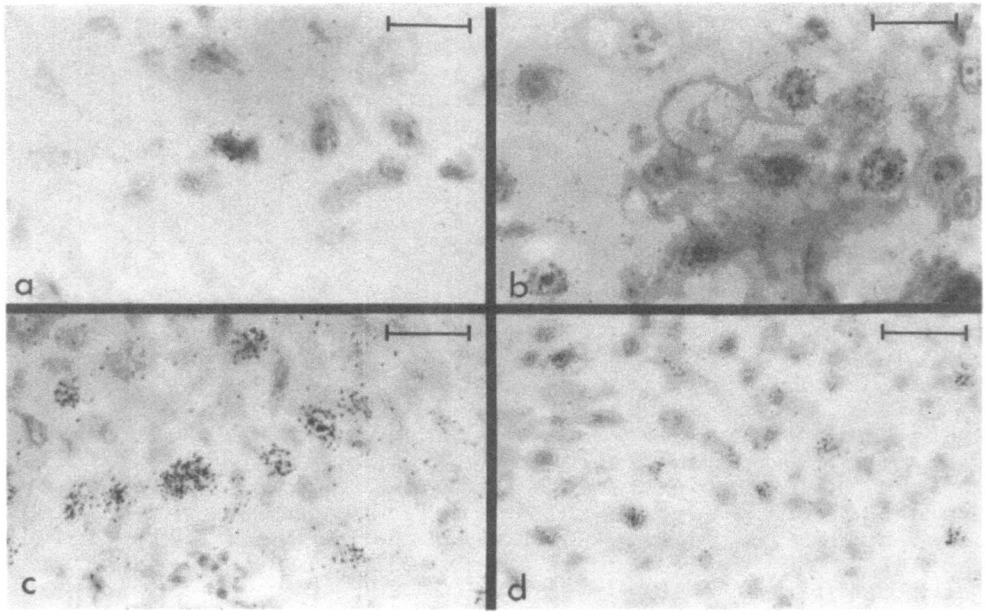

Fig. 2. Autoradiographs of typical sections through a
<u>Xenopus</u> regenerate: a) control, b) 100 ng
CBGF/ml, c) 1 μg CBGF/ml, d) 10 μg CBGF/ml
medium.   Autoradiographs confirm results
obtained by liquid scintillation counting
showing increased incorporation at 1 μg
CBGF/ml medium. Bar represents 10 μm.

RESULTS

Our results demonsrate that CBGF is mitogenic for mesen-chyme-like cells of a _Xenopus_ limb regenerate. In the pres-ence of the mitogen incorporation of [methyl-^{3}H] thymidine increased to a concentration of 1 μg CBGF/ml of medium but the cells showed less incorporation at 10 μg CBGF/ml (Fig. 1). Apparently, an increase in mitogen concentration above the optimum results in a decrease of responsiveness by the regen-erate cells. Statistical analysis using Duncan's Multiple Range Test shows significant differences between 1 μg CBGF/ml medium and the other two series and also compared with con-trols. Autoradiography confirms the results obtained by liq-uid scintillation counting (Figs 2a-d).

DISCUSSION

There have been several agents proposed as candidates for the "neurotrophic factor" in urodele forelimb regeneration including glial growth factor (Brockes and Kintner, 1986), fibroblast growth factor (Gospodarowicz and Mescher, 1980), substance P (Globus et al., 1983) and transferrin (Mescher and Munaim, 1984). However, the identity of the neurotrophic agent remains undetermined. Our results suggest a possible role for CBGF as a mitogenic agent in amphibian forelimb regeneration. It is a non-species specific molecule which has previously been shown to be mitogenic for chick astrocytes, Swiss mouse 3T3 fibroblasts and newt blastema cells (Carlone et al., 1988). Our experiments demonstrate that CBGF is also mitogenic for mesenchyme-like cells of the _Xenopus_ forelimb regenerate.

ACKNOWLEDGEMENTS

We wish to thank Dr. R.L. Carlone, Brock University, Ontario, Canada for the CBGF mitogen. This work was supported by NSERC of Canada grant A-1208 to R.A.L.

REFERENCES

Brockes, J.P. and Kintner, C.R., 1986, Glial growth factor and nerve-dependent proliferation in the regeneration blastema of urodele ammphibians, _Cell_, 45:301-326.

Carlone, R.L., Waters, B.W., Leonard, S.M. and Vijh, K.M., 1988, A low-molecular weight chick brain-derived growth factor is mitogenic for cultured astroglia from the chick embryo, _Dev. Brain Res._, 39:97-104.

Globus, M. and Vethamany-Globus, S., 1977, Transfilter mito-genic effect of dorsal root ganglia on cultured regen-eration blastemata in the newt, _Notophthalmus virides-cens_, _Dev. Biol._, 56:316-328.

Globus, M., Vethamany-Globus, S., Kesich, A., and Milton, G., 1983, Roles of neural peptide substance P and calcium in blastema cell proliferation in the newt, _Notophthal-mus viridescens_, _in_: "Limb Development and Regeneration (Part A)", J.F. Fallon and A.I. Caplan, eds., Alan R. Liss, New York.

Gospodarowicz, D. and Mescher, A.L., 1980, Fibroblast growth factor and the control of vertebrate regeneration and repair, Ann. N.Y. Acad. Sci., 339:151-174.

Mescher, A.L. and Munaim, S.I., 1984, "Trophic" effect of transferrin on amphibian limb regeneration blastemas, J. Exp. Zool., 230:485-490.

Singer, M., 1952, The influence of the nerve in regeneration of the amphibian extremity, Quart. Rev. Biol., 27:169-200.

Singer, M., 1954, Induction of regeneration of the forelimb of the post-metamorphic frog by augmentation of the nerve supply, J. Exp. Zool., 126:419-471.

Singer, M., Maier, C. and McNutt, W., 1976, Neurotrophic activity of brain extracts on forelimb regeneration in the urodele Triturus, J. Exp. Zool., 196:131-150.

Tomlinson, B.L., Globus, M. and Vethamany-Globus, S., 1981, Promotion of mitosis in cultured newt limb regenerates by a diffusible nerve factor, In vitro, 17:167-172.

Commentary

by Roy A. Tassava, on

CELLULAR AND IMMUNOLOGICAL ASPECTS OF REGENERATION

An important feature of epimorphic limb regeneration is the formation of a blastema, a mass of undifferentiated mesenchyme-like cells covered by a wound epithelium. The investigations in this section are concerned with basic problem areas relevant to the cellular composition of the blastema, the roles of wound epithelium, extracellular matrix, growth factors, and other effector molecules in blastema development, and the possible involvement of the immune system in limb and organ regeneration.

It is now quite clear from data presented by Wallace, and previously published by others, that cells of the blastema display considerable heterogeneity in their cycling characteristics. While some are actively progressing through the cycle (generally in 40-46 hrs), others are in a state of quiescence. That this quiescence is transient is shown by a continuous increase in labeling index with exposures to ^{3}H-thymidine longer that the combined lengths of G_2, M, and G_1. Various views were presented to explain the presence of transiently quiescent cells including, (1) recently dedifferentiated cells, (2) cells lacking proper growth stimulation, perhaps depleted of mitogen receptors, and (3) cells of a certain lineage which exhibit different growth characteristics.

With monoclonal antibodies (mAbs) it has been shown that blastema cells exhibit both heterogeneity and homogeneity in antigenicity. Brockes and collaborators have shown that a subpopulation of blastema cells, identified by reactivity to the 22/18 antibody, is particularly dependent on nerves for continued cycling. On the other hand, Tassava showed that essentially all blastema cells react to the 9G1 antibody; from Western blot data, the 9G1 antigen seems to be a microtubule associated protein. It will be of interest to know the lineage of 9G1 cells, particularly as to whether they are strictly from stump fibroblasts (fibrocytes), since fibroblasts are perceived by some to be the major source of positional information in the blastema.

It is noteworthy that the 22/18 and 9G1 antigens (both intracellular) and the 4G3 extracellular matrix antigen (tentatively identified as tenascin) are all persistent after denervation. These results are consistent with data in the literature and lead to the concept that amputation initiates many similar synthetic events in both regenerating limbs and in denervated, non-regenerating limbs. Forthcoming data will probably also show many similarities between flank wound healing and limb regeneration. It remains for investigators of regeneration to elucidate the events unique to regenerating limbs.

Based on studies by Tassava and collaborators with mAb WE3, it can now be stated with some degree of certainty that while the wound epithelium originates from skin epidermis, it subsequently expresses ("turns on") the WE3 antigen. The WE3 antigen is not present in the vast majority of epidermal cells and therefore a legitimate question is what causes wound epithelium to express this antigen. The WE3 antigen represents the only known antigenic difference between epidermis and wound epithelium. While circumstantial evidence argues that WE3 reactive cells are important to the function of the wound epithelium, exactly what that function is remains to be determined. To date, the WE3 antigen is resisting extraction and characterization efforts. Using the monoclonal antibody approach, efforts are continuing in the search for developmentally regulated antigens of regenerating limbs.

But why not search the blastema for molecules/factors already known and described in other systems? Interesting and potentially significant in this regard is the finding by Boilly and collaborators, with immunocytochemistry, that aFGF is present in the blastema mesenchyme and wound epithelium. The fact that aFGF is also in epidermis of skin need not detract from its possible role in regeneration. Continuing studies designed to determine the cellular origin of FGF, its target cells, and its possible relevance to the role of nerves and wound epithelium will no doubt be forthcoming. The possibility was raised that FGF and maybe other growth factors might be associated with components of the extracellular matrix. In fact, the pattern of immunoreactivity of FGF Ab in regenerates of <u>Pleurodeles</u> correlated with the pattern of reactivity of monoclonal antibody 4G3 in <u>Notophthalmus</u> presented by Tassava, but this correlation may merely be coincidental.

The mechanisms by which putative growth factors, neurotrophic factors, and wound epithelial mitogens act on cells of the regenerating limb are of great interest. In this regard, Boilly and collaborators investigated protein kinase C in cytosolic and membrane fractions of blastema cells and showed that increased protein kinase C levels in membrane fractions seem to be correlated with the neurotrophic stimulation of regeneration. These results are consistent with the view that protein kinase C plays a role in the membrane and responds to effector molecules acting on blastema cells.

Taban and collaborators measured cAMP and cGMP levels in unamputated and regenerating limbs by biochemical and immunocytochemical procedures. The presence of these cyclic nucleotides in cells of the regenerate cannot be disputed. Effort should now go into determining the relationship of growth factors and other effector molecules to protein kinase C, cyclic nucleotides, Ca^{++} levels (published papers of Globus and collaborators) and possibly other second messengers. It would be hoped that results here would relate to the ever growing body of knowledge concerning 2nd messengers and provide clues relevant to control/regulatory factors in regeneration.

Sicard presented data from his own laboratory and from the literature suggesting possible interactions between the immune system and the regenerating limb. In this scheme, such interactions could determine whether regeneration or repair occurs, could determine whether blastema antigens are tolerated, and could influence the behavior of blastema cells.

Implicating the immune system in the developmental program of regeneration is fine but supporting data are tenuous at best. Certain components of the immune system do appear to change during regeneration, both systemically and locally. Also, in general, suppression of the immune system occurs early in regeneration but no cellular or biochemical species have been correlated precisely with this suppression. Effects of immunoreactive substances, some stimulating, others inhibiting, need to be correlated with the immunological status of the animal in question. While data have emerged which are compatible with involvement of the immune system the relevant molecules and cells of both the immune system and limb have not yet been identified. The same is true for involvement of the immune system in the regeneration of the liver and other internal organs.

Babaeva showed that lymphocytes from partially hepatectomized rats could stimulate liver regeneration whereas lymphocytes from partial spleenectomized rats suppressed liver regeneration. Babaeva's suggestion of involvement of T-suppressors and self- vs non-self antigens in "lymphocytic morphogenic activity" is interesting and potentially relevant to both organ and limb regeneration.

These are exciting times in the field of regeneration. It seems that with traditional "tried and true" methods as well as with more modern monoclonal antibody and molecular approaches, we are very close to resolving some of the most basic problem areas but at the same time new problem areas are only now being uncovered. There is a need for many more clever experiments; stay at the laboratory bench !

REGENERATION

OF

SKELETAL AND CARDIAC MUSCLE

SOME ASPECTS OF REGENERATION IN SKELETAL AND CARDIAC MUSCLE

Bruce M. Carlson

Department of Anatomy and Cell Biology
University of Michigan
Ann Arbor, Michigan 48109 USA

SUMMARY

Although there are superficial similarities between skeletal and cardiac muscle, the differences between their regenerative response to injury appear to outweigh the similarities. The regeneration of skeletal and cardiac muscle is compared with respect to cellular origins, vascular and neural relations, morphogenesis and aging influences.

INTRODUCTION

The regeneration of skeletal and cardiac muscle is a study in contrasts that points out some profound differences in the biological strategies for dealing with tissue loss and replacement. At a superficial histological level, both tissues are classified as striated muscle designed for powerful contractions, but at that point the resemblance ceases. These two tissues have remarkably different embryological origins, adult structures and functions. They are also exposed to different types of injury during the normal course of an individual's life cycle. How these tissues respond to injury will be the subject of this review.

That skeletal muscle can regenerate has been established without a doubt, and many aspects of skeletal muscle regeneration are understood at least to a certain extent. Research on cardiac muscle, however, is still in its infancy. The fundamental question is whether or not the production of a local defect in the heart by ischemia, mechanical injury or other causes is followed by a regenerative response. What constitutes a regenerative response is a significant issue in the case of cardiac muscle.

ADULT STRUCTURE

From the standpoint of regeneration, the fundamental structural unit of skeletal muscle is a complex, consisting of 1) the multinucleated skeletal muscle fiber itself, 2) an enveloping basal lamina and 3) a population of mononuclear satellite cells (Mauro, 1961) situated between the muscle fiber and the basal lamina. Although larval urodele amphibians and neotenic axolotls possess satellite cells, they are not found in postmetamorphic adults (Popiela, 1976; Hay, 1979). The skeletal muscle of postmetamorphic newts, however, contains a population of cells, called postsatellite cells, which are similar to satellite cells except that a sheet of basal lamina is interposed between them and the muscle fiber (Cherkasova, 1982). Like the satellite cell, the postsatellite cell has been proposed to be a cellular source of regenerating muscle fibers. Skeletal muscle fibers possess one, or sometimes more than one neuromuscular junctions. The essential components of the neuromuscular junction are 1) the nerve terminal, with associated Schwann cells, 2) the postsynaptic surface, sometimes called the motor endplate, on the muscle fiber and 3) an intervening basal lamina, which is continous with the basal lamina that surrounds the rest of the muscle fiber. Acetylcholine receptors are located on the postsynaptic folds of the muscle fiber, and acetylcholinesterase molecules are embedded in the intervening basal lamina.

Cardiac muscle cells (cardiomyocytes) are superficially similar to skeletal muscle fibers in that they are cross-striated, but beyond this the differences seem to outweigh the similarities. Cardiac muscle is arranged in a three-dimensional meshwork formed from individual cellular units, which are connected to one another by structurally complex intercalated discs. The mammalian cardiomyocyte, in particular, is rarely a diploid cell. The centrally located nucleus is commonly polyploid, or the cardiomyocyte may by binucleated (each nucleus may be diploid or polyploid) as the result of mitosis without cytokinesis (Brodsky and Uryvaeva, 1985). Except for certain crabs, satellite cells are not associated with cardiac muscle, and most cardiomyocytes are not directly innervated.

EMBRYOLOGICAL ORIGINS AND POSTNATAL ONTOGENY

Virtually all skeletal muscle is derived from somatic mesoderm, specifically the somites or somitomeres. In the early embryo, premuscle cells migrate from the somites into the limb buds and collect into dorsal and ventral premuscle masses. From this point two developmental processes occur simultaneously. One is the fusion of myoblastic cells into multinucleated myotubes and their subsequent differentiation into muscle fibers. On the basis of evidence collected in vitro, mainly on avian muscle, there is reason to believe that there are several different subpopulations of myoblasts, which replace one another almost in the manner of developmental isoforms of proteins (Hauschka et al., 1982). There is also evidence from avian systems that populations of myoblasts may be committed to forming fast and slow muscle fiber types even in the absence of interactions with motor nerve fibres (Stockdale and Miller, 1987).

In the limb, cells of the myogenic lineage migrate from the somites to the early limb bud and become arranged into common dorsal and ventral premuscle masses (Kieny et al., 1986). Then, through a mechanism that is still very poorly understood, the premuscle masses split into the primordia of individual muscles. After they have been initially sculpted, the early muscles become connected with tendons that are derived from cells originating in the limb bud. Nerve fibers also grow into the limb bud at a very early stage (Landmesser, 1984) and make connections with the muscle fibers as they are forming.

Cardiac muscle is derived from splanchnic, rather than somatic, mesoderm in the embryo. Relatively little is yet known about the cell biology of early cardiac muscle differentiation. The possibility of cell lineages has received considerably less attention in cardiac muscle than in skeletal muscle. One important characteristic of cardiac muscle histogenesis is the ability of cross-striated cardiomyocytes to undergo mitotic division during embryogenesis (rev. Rumyantsev, 1982).

CELLULAR ORIGINS AND PROLIFERATION DURING REGENERATION

Skeletal Muscle

At the level of the muscle fiber, most experimental evidence points to the satellite cells as the chief, if not sole, precursor cells for regenerating skeletal muscle (Snow, 1977b). In ischemic mammalian muscle, at least, the myonuclei themselves die after a few hours and do not seem to be involved in the regenerative process (Snow, 1977a). Recent experimental work (McGeachie and Grounds, 1985) has not supported earlier suggestions that circulating cells can participate in myogenesis.

In the epimorphic regeneration of muscle in amphibians, the cellular origins of myoblasts are still not known, the two major options being 1) the formation of myoblasts from nucleated fragments that break off from the ends of the damaged muscle fibers and 2) the formation of new muscle fibers from reserve cells or from myogenic precursor cells (e.g., the postsatellite cells) (Mauro et al., 1970). Despite the increasing sophistication of more recent attempts to resolve this issue, the question remains open. For example, Kintner and Brockes (1984) used a double labelling technique involving a muscle-specific monoclonal antibody and an antibody (22/18) that was thought to be unique to blastema cells. However, with the later recognition that the labelling of cells with the 22/18 antibody also occurs in muscle regenerating in the absence of a blastema by the tissue mode (Griffin et al., 1987), further experiments are required to support the conclusion of the Kintner and Brockes (1984) study.

At a different level, the question of cell origins is a topographic one. There is increasing evidence that satellite cells, like myogenic cells in the embryo, are capable of migrating. Both Schultz et al. (1986) and Phillips (1987) have shown the migration of myogenic cells from normal muscle into a tightly adherent piece of frozen muscle. Schultz et al. (1985) have also shown that satellite cells can migrate along the length of muscle fibers toward the site of a focal lesion. In a free muscle graft, no viable myogenic cells are found in the ischemic center after a few hours (Phillips et al., 1987). Yet, in time the central region of the graft is repopulated with regenerating muscle fibers, showing that regenerating muscle has extensive powers of penetration into formerly necrotic areas. However, if the muscle graft is large or the process of revascularization is slow, the center of the graft becomes filled with a core of dense connective tissue, which forms before regenerating muscle can penetrate into the area (Markley et al., 1978). The dense connective tissue core is surrounded by a rim of regenerated muscle fibers.

Cardiac Muscle

There is little evidence for the outgrowth of muscle cells from the wound margin into a lesion in the heart. Rather, the damaged area is filled in by connective tissue. Therefore, one must look for a "regenerative" response in the remainder of the heart. The fundamental question during the past two decades of research on cardiac muscle has been whether or not cardiomyocytes are capable of dividing in response to damage to the heart. A second question, which is just beginning to be discussed, is whether or not there might be some hitherto unrecognized precursor or progenitor cell capable of producing new cardiomyocytes in the stable adult. Because of the lack of information of any sort about the latter question, this section of the review will concentrate on the issue of the ability of cardiomyocytes to proliferate in response to trauma.

The early studies of Rumyantsev (1961) and Oberpriller and Oberpriller (1974) demonstrated without a doubt that in amphibians, ventricular myocytes can respond to a traumatic stimulus by dividing mitotically. The dividing cells are not located directly at the wound margin, but rather they are found principally in a zone surrounding the site of injury. In contrast to the cellular response to skeletal muscle injury, which is a rapid one, the peak of the proliferative response in the heart occurs during the third week after injury (Oberpriller et al., 1979). Electron microscopic studies (rev. Rumyantsev, 1979, 1982) showed what was interpreted as a partial dedifferentiative response of the dividing cardiomyocytes. This response consisted of the partial dismantling of the array of contractile filaments in the dividing cardiomyocytes.

In mammals, the situation differs considerably from that in amphibians. Thymidine labelling studies on injured rat ventricles have shown only low labelling indices (typically 6% or less) even after multiple injections of H^3-thymidine. In contrast, after infarction of the left ventricle in the rat, as many as 85% of the atrial myocytes are labelled (Rumyantsev, 1979). What is the basis for the difference in proliferative response between the injured amphibian and mammalian ventricle and the mammalian atrium and ventricle?

One possible solution to this enigma may be in the degree of ploidy of the cardiomyocytes in there tissues. At birth, the ventricle of the rodent heart is made up almost entirely of diploid cells. However, during the first three weeks of postnatal life standard mitotic divisions are increasing replaced by incomplete divisions, leaving up to 95% of the ventricular myocytes binucleated or polyploid mononuclear cells. In contrast, the ventricle of the adult newt contains over 90% diploid myocytes (Oberpriller et al., 1988).

With descriptive data like those quoted above, one can hypothesize that there is an inverse relationship between the degree of polyploidy (or non-diploidy) in a ventricle and its ability to mount a proliferative response in response to injury. A first step in testing the hypothesis is to see if there is a correlation between the percentage of diploid myocytes in the heart and the ability of the cardiomyocytes to respond to injury by DNA synthesis and/or proliferation.

The remaining myocytes in the left ventricle of the rat mount only a weak proliferative response to damage, with 6% or less of the cells being labelled, even after multiple doses of thymidine (Rumyantsev, 1979). In rats of the age used in this study, well over 90% of the ventricular myocytes are non-diploid, being either binucleated or polyploid (Katzberg et al., 1977). In contrast, the ventricle of the newt, which undergoes an intense amount of ^{3}H-thymidine uptake after injury (Oberpriller and Oberpriller, 1974) is overwhelmingly diploid (over 95%) in the undisturbed state (Oberpriller et al., 1988).

Even within the rat heart, there is a correlation between the level of diploidy and DNA synthetic activity after cardiac njury. In a still poorly understood experimental model, an experimentally induced infarction in the left ventricle is followed by a tenfold higher rate of incorporation of ^{3}H-thymidine into left atrial myocytes (62% labelling index) than in the remaining left ventriculocytes (Rumyantsev, 1979). Correlated with this is a significantly higher percentage of diploidy (close to 50%) in the left atrium (Oberpriller and Oberpriller, 1985).

A second means of testing the cellular polyploidy hypothesis of mitotic inhibition of mitosis in cardiomyocytes is to devise means of increasing the proportion of diploid cells in the mammalian ventricle and then expose the cells to a stimulus that could lead to their division, or at least DNA synthesis. The first step has been accomplished by transplanting pieces of ventricle of newborn rats, which are almost all diploid, beneath the kidney capsule of syngeneic hosts (Brodsky et al., 1988). These grafts become well integrated into their new site and resume beating a few days after transplantation. However, the number of diploid cells in the grafts remains considerably higher than normal (from 40-70%). With this as a model, the next step is to determine the ability of the transplanted myocytes to incorporate ^{3}H-thymidine after an appropriate traumatic stimulus. This work has just begun.

REVASCULARIZATION AND MUSCLE REGENERATION

Except for certain specific myotoxic substances, such as local anesthetics (Foster and Carlson, 1980), most injuries to skeletal muscle involve injury to the local vascular bed, as well as the muscle fibers themselves, and regeneration of the local vascular bed is a prerequisite for successful muscle fiber regeneration. The phenomenon of revascularization is easiest to study in free muscle grafts or in minced muscles (Burton et al., 1987). In the strict sense of the word, a muscle is freely grafted when it is completely removed from its bed, with all connections with the host severed, and then replaced into its own bed or another site. Reintegration of a free muscle graft with the host must rely upon spontaneous revascularization and reinnervation. (Recent surgical literature has adopted the term free muscle graft in cases where the vasculature of graft and host are surgically anastomosed).

After a rat muscle has been freely grafted, the first vascular connections between graft and host are seen after about two days. Thereafter, new blood vessels, which reach the graft from many sites, grow into the graft toward its ischemic center (Hansen-Smith et al., 1980). Around the tips of the growing vessels is a characteristic association of numerous macrophages and the early appearance of activated satellite cells. Phillips (1987) has demonstrated that both normal and ischemic skeletal muscle posseses angiogenetic properties. One can hypothesize that a putative muscle-derived angiogenic factor causes the initial sprouting of blood vessels into the muscle graft and that once blood-borne macrophages enter the ischemic graft, an angiogenic factor produced by the oxygen-stressed macrophages (Polverini et al., 1977), acting alone or in concert with muscle-derived factors, maintains the progress of revascularization.

Although certain histochemical properties of capillaries in regenerating muscle return to normal (Grim et al., 1986), the structure of the regenerated vessels is often abnormal (Hansen-Smith et al., 1980). White et al.(1981) have also shown that blood flow in early muscle grafts is also abnormal.

Much more research must be done before we develop a real understanding of the restoration of the vascular bed in regenerating muscle. Major questions are 1) the exact nature of the angiogenetic stimulus, 2) the ability of vascular smooth muscle and endothelial cells to survive prolonged ischemia, 3) the role of macrophages in vascular regrowth and 4) the restoration of physiological control of the regenerating vasculature.

In studies of cardiac muscle regeneration, vascular relations have been viewed in a different light. Devascularization of a portion of the heart produces an infarct, which is presumed to be a permanent lesion, in which beating muscle dies and is replaced by connective tissue. Yet, as transplantation studies have shown, small pieces of neonatal cardiac muscle grafted beneath the kidney capsule become well vascularized. Virtually nothing, however, is known about mechanisms of revascularization of transplanted cardiac muscle.

REINNERVATION OF REGENERATING MUSCLE

The subject was recently reviewed (Carlson, 1988) and will not be repeated here except to note that reinnervation is essential for functionally successful skeletal muscle regeneration. In contrast to cardiac muscle, which is not dependent upon innervation for its function, regenerating skeletal muscle fiber must become reinnervated in order to go through the final stages of functional differentiation.

From the clinical standpoint, the reinnervation of muscle is perhaps the most difficult problem in restoring a damaged or grafted muscle to full function. At one level the problem of reinnervation is a topographical one - how can nerve fibers growing into a grafted or regenerating muscle be directed to all of the non-innervated muscle fibers? Even with the most careful neural anastomoses, reinnervation of muscles is normally incomplete. The other problem is a temporal one. Often months or even years elapse before regenerating nerve fibers reach a denervated or regenerating muscle in humans. There is currently considerable debate about whether long-term denervated muscle is capable of becoming reinnervated. Many clinicians feel that after 18 months the ability of non-innervated muscle to become successfully reinnervated is drastically reduced. Recently, Carlson and Faulkner (1988) have shown that grafts of rat extensor digitorum longus muscles that had been previously denervated for 22 months become reinnervated if grafted into an innervated leg.

MORPHOGENESIS

Studies conducted on minced muscle regeneration have shown that both gross form and internal architecture of skeletal muscle regenerating by the tissue mode can be accounted for almost entirely on the basis of mechanical conditions -pressure and tension - applied to the muscle tissue (Carlson, 1972). The experimental evidence points to the regenerating tendons and the intramuscular connective tissue as the mediators of tension. The morphogenesis of muscle regenerating by the epimorphic mode (Carlson, 1979) follows the same course as muscle forming in the embryo. Very little is yet understood about the mechanisms underlying the formation of muscles in either the embryonic or the regenerating limb.

An experiment by Bader and Oberpriller (1978) showed that a certain degree of morphogenesis is possible in the healing of severly damaged myocardial tissue in the newt. They amputated the tip of the ventricle in newt hearts, minced the amputated tip into small pieces and implanted these back into the pericardial cavity. The minced fragments coalesced and formed a mini-ventricle that was attached to the remainder of the ventricle. The myocytes in the mince showed a peak mitotic and thymidine-uptake index 16 days after mincing. Stable minced ventricles possessed a trabecular network and a structurally coherent outer wall which beat asynchronously with the main ventricle.

AGING AND MUSCLE REGENERATION

Grafting experiments have shown that muscles in old rats regenerate much more poorly than muscles in young rats (Gutmann and Carlson, 1976; Carlson and Faulkner, 1988). This is correlated with reduced numbers of satellite cells in old animals (Snow, 1977c; Gibson and Schultz, 1983). In addition, Schultz and Lipton (1982) have reported a reduced proliferative capacity in vitro of satellite cells derived from old rats.

A significant question is whether the age-related decline
in the success of regeneration reflects an intrinsic property
of the muscle itself or if there is an interaction between the
muscle and its environment. This was tested in a cross-age
transplantation experiment, in which muscles from old rats
were grafted into young rats and vice versa (Carlson and
Faulkner, 1987). This procedure resulted in a striking reversal in the success of regeneration in each case. Old muscles
grafted into young rats regenerated as well as young muscles
grafted into young rats, and young muscles grafted into old
hosts regenerated as poorly as old muscles grafted into old
hosts. These results show that the major determinant of the
success of skeletal muscle regeneration is the environment
provided by the host and not some factor(s) intrinsic to the
muscle itself. The nature of the age-related host influence
on the success of skeletal muscle regeneration has not yet
been identified. Virtually nothing is known about the influence of age on the regenerative capacities of cardiac muscle.

FUTURE DIRECTIONS

The fields of skeletal and cardiac muscle regeneration
are in quite different stages of development, and the respective natures of the systems lend themselves to different
experimental approaches. At the level of tissues and cells,
many of the basic tissue interactions involved in skeletal
muscle regeneration have been delineated, but still little
work has been done at the level of cell-matrix interactions,
for example. The burgeoning field of the molecular genetics
of muscle is just beginning to spill over into the area of
regeneration. A fundamental question is whether the sequence
of cellular and molecular isoforms in regenerating muscle
recapitulates that of embryonic development or if there are
significant adaptive changes from the normal ontogenetic
sequence. Very little is still known about the biology of the
satellite cell. Part of the problem lies in the difficulty in
identifying and isolating unquestionably pure populations of
satellite cells. After several years of rapid progress in
understanding the cell biology of nerve-muscle interactions
during regeneration, the pace of understanding them at the
molecular level has slowed. At anything beyond the gross
descriptive level, our understanding of any facet of the epimorphic regeneration of muscle is minimal.

Cardiac muscle regeneration is still at the stage of
seeking definition and acceptance. In mammals, understanding
the basis for the lack of proliferation in ventricular cells
is still the critical issue. In amphibians or in the mammalian atrium, a principal question remains how the proliferative
response is connected to the injury that presumably stimulated
it.

ACKNOWLEDGMENTS

Original research described in this review was supported
by FAO grants DE 07687 and EY 05813.

REFERENCES

Bader, D., and Oberpriller, J.O., 1978, Repair and reorganization of minced cardiac muscle in the adult newt (Notophthalmus viridescens), J. Morph., 155:349-358.

Brodsky, V. Ya, Carlson, B.M., Arefieva, A.M., and Vasilieva, I.A., 1988, Polyploidization of cardiac myocytes as an intracellular program, Cell Differen., (in press).

Brodsky, V. Ya., and Uryvaeva, I.V., 1985, "Genome Multiplication in Growth and Development: Biology of Polyploid and Polytene Cells", Cambridge Univ. Press, Cambridge.

Burton, H.W., Carlson, B.M., and Faulkner, J.A., 1987, Microcirculatory adaptation to skeletal muscle transplantation, Ann. Rev. Physiol., 49:439-452.

Carlson, B.M., 1972, The Regeneration of Minced Muscles. "Monographs in Developmental Biology", Vol. 4, Karger, Basel.

Carlson, B.M., 1979, The relationship between the tissue and epimorphic regeneration of muscle, in: "Muscle Regeneration", A. Mauro, ed., Raven Press, New York, pp. 57-71.

Carlson, B.M., 1988, Nerve-muscle interrelationships in mammalian skeletal muscle regeneration, in: "Control of Cell Proliferation and Differentiation during Regeneration", H.J. Anton, ed., Karger, Basel.

Carlson, B.M., and Faulkner, J.A., 1987, Age of the host is an important factor in determining the success of skeletal muscle grafts between young and old animals, Anat. Rec., 218 (1):20A.

Carlson, B.M., and Faulkner, J.A., 1988, Reinnervation of long-term denervated rat muscle freely grafted into an innervated limb. Exp. Neurol., (in press).

Cherkasova, L.V., 1982, Postsatellites in muscular tissue in adult tailed amphibia (Russian), Dokl. Akad. Nauk SSSR, 267:1235-1236.

Foster, A.H., and Carlson, B.M., 1980, Myotoxicity of local anesthetics and regeneration of the damaged muscle fibers, Anesth. Analog., 58:727-736.

Gibson, M.C., and Schultz, E., 1983, Age-related differences in absolute numbers of skeletal muscle satellite cells. Muscle & Nerve, 6:574-580.

Griffin, K.J.P., Fekete, D.M., and Carlson, B.M., 1987, A monoclonal antibody stains myogenic cells in regenerating newt muscle, Development, 101:267-278.

Grim, M., Mrazkova, O., and Carlson, B.M., 1986, Enzymatic differentiation of arterial and venous segments of the capillary level during the development of free muscle grafts in the rat, Am. J. Anat., 177:149-159.

Gutmann, E., and Carlson, B.M., 1976, Regeneration and transplantation of muscles in old rats and between young and old rats, Life Science, 18:109-114.

Hansen-Smith, F.M., Carlson, B.M., and Irwin, K.L., 1980, Revascularization of the freely grafted extensor digitorum longus muscle in the rat, Am. J. Anat., 158:65-82.

Hauschka, S.D., Rutz, R., Linkhart, T.A., Clegg, C.H., Merrill, G.F., Haney, C.M., and Lim, R.W., 1982, Skeletal muscle development. I. Developmental changes in muscle colony-forming cell type and location during vertebrate limb development. in: Disorders of the Motor Unit, D.F. Schotland, ed., Wiley, New York.

Hay, E.D., 1979, Reversibility of muscle differentiation in newt limb regeneration, in: "Muscle Regeneration", A. Mauro, ed., Raven Press, New York.

Katzberg, A., Farmer, B. and Harris, R., 1977, The predominance of binucleation in isolated rat heart myocytes, Am. J. Anat., 149:489-500.

Kieny, M., Pautou, M.P., Chevalier, A., and Mauger, A., 1986, Spatial organization of the developing limb musculature in birds and mammals, Bibl. Anat., 29:65-90.

Kintner, C.R., and Brockes, J.P., 1984, Monoclonal antibodies identify blastemal cells derived from dedifferentiating muscle in newt limb regeneration, Nature (London), 308:67-69.

Landmesser, L., 1984, The development of specific motor pathways in the chick embryo, Trends in Neurol. Sci., 7:336-339.

Markley, J.M., Faulkner, J.A., and Carlson, B.M., 1978, Regeneration of skeletal muscle after grafting in monkeys, Plastic Reconstr. Surg., 62:415-422.

Mauro, A., 1961, Satellite cells of skeletal muscle fibers, J. Biophys. Biochem. Cytol., 9, 493-495.

Mauro A., Shafig, S., and Milhorat, A.T., 1970, "Regeneration of Striated Muscle and Myogenesis", Excerpta Medica, Amsterdam.

McGeachie, J.K., and Grounds, M.D., 1985, Can cells extruded from denervated skeletal muscle become circulating potential myoblasts, Cell Tissue Res., 242:25-32.

Oberpriller, J.O., Bader, D.M., and Oberpriller, J.C., 1979, The regenerative potential of cardiac muscle in the newt, Notophthalmus viridescens, in: "Muscle Regeneration", A. Mauro, ed., Raven Press, New York.

Oberpriller, J.O., and Oberpriller, J.C., 1974, Response of the adult newt ventricle to injury, J. Exp. Zool., 187:249-259.

Oberpriller, J.O., and Oberpriller, J.C., 1985, Cell division in cardiac myocytes, in: "Cardiac Morphogenesis", V.J. Ferrans, G. Rosenquist and C. Weinstein, eds., Elsevier New York.

Oberpriller, J.O., Oberpriller, J.C., Arefysva, A.M., Mitashov, V.I., and Carlson, B.M., 1988, Nuclear characteristics of cardiac myocytes following the proliferative response to mincing of the myocardium of the adult newt, Cell and Tiss. Res., (in press).

Phillips, G.D., 1987, Limiting factors in the successful regeneration of skeletal muscle: Myogenic cell survival, myogenic cell migration, and revascularization, Ph.D. dissertation, University of Michigan.

Phillips, G.D., Lu, D., Mitashov, V.I., and Carlson, B.M., 1987, Survival of myogenic cells in freely grafted rat rectus femoris and extensor digitorum longus muscles, Am. J. Anat., 180, 365-372.

Polverini, P.J., Cotran R.S., Ginbrone, M.A., Unanue, E.R., 1977, Activated macrophages induce vascular proliferation, Nature (London), 269:804-806

Popiela, H., 1976, Muscle satellite cells in urodele amphibians: Facilitated indentification of satellite cells using Ruthenium red staining, J. Exp. Zool., 198:57-64.

Rumyantsev, P.P., 1961, Evidence of the regeneration of a significant part of frog myocardial fibers after traumatization (Russian), Arkh. Anat. Gistol. Embriol., 40(2):65-74.

Rumyantsev, P.P., 1979, Some comparative aspects of myocardial regeneration, in: "Muscle Regeneration", A. Mauro, ed., Raven Press, New York.

Rumyantsev, P.P., 1982, "Cardiomyocytes in Processes of Reproduction, Differentiation and Regeneration", Nauka, Leningrad.

Schultz, E., Jaryszak, D., Gibson, M., and Albright, D., 1986, Absence of exogenous satellite cell contribution to regeneration of frozen skeletal muscle, J. Muscle Research Cell Motil., 7:361-367.

Schultz, E., Jaryszak, D., and Valliere, C., 1985, Response to satellite cells to focal skeletal muscle injury, Muscle & Nerve, 8:217-222.

Schultz, E., and Lipton, B.H., 1982, Skeletal muscle satellite cells: Changes in proliferation potential as a function of age, Mech. Aging Devel., 20:377-383.

Snow, M.H., 1977a, Myogenic cell formation in regenerating rat skeletal muscle injured by mincing. I. A fine structural study, Anat. Rec., 188:181-200.

Snow, M.H., 1977b, Myogenic cell formation in regenerating rat skeletal muscle injured by mincing. II. An autoradiographic study, Anat. Rec., 188:200-218.

Snow, M.H., 1977c, The effects of aging on satellite cells in skeletal muscles of mice and rats, Cell and Tiss. Res., 185:399-408.

Stockdale, F.E., and J.B., Miller, 1987, The cellular basis of myosin heavy chain isoform expression during development of avian skeletal muscles, Devel. Biol., 123:1-9.

White, T.P., Maxwell, L.C., Sosin, D.M., and Faulkner, J.A., 1981, Capillarity and blood flow of transplanted skeletal muscles of cats, Am. J. Physiol., 241:H630-H636.

IMAGE ANALYSIS OF PROLIFERATION AND FUSION

IN CULTURED SATELLITE CELLS

Albert Le Moigne[1], Bernard Lassalle[2]
Isabelle Martelly[1] and Jean Gautron[1]

[1]Laboratoire de Recherche sur la Myogénèse et la Régénération Musculaire, UFR de Sciences et Technologie, Université Paris-Val de Marne, 94010 Creteil, France. [2]Laboratoire de Morphogénèse Animale, UA 685 CNRS, Université des Sciences et Techniques de Lille, 59655, Villeneuve d'Ascq France

SUMMARY

Myogenic cells were isolated from adult rat skeletal muscles and cultured in vitro. Cell proliferation was analyzed between days 1 and 14 after plating and the histological characteristics of the cultures were defined. The cell cycle phases determined by examining Feulgen-stained cultures with a cell image processor SAMBA 200. Nuclei were automatically analysed with two different methods which made it possible to analyse the nuclei of the myogenic cell populations which were either involved in each phase of the mitotic cycle, or left out of the cycle after fusion into myotubes. After 3 hr of culture 10% of the cell population was involved in the cell cycle. In the presence of fetal calf serum, this percentage increased until day 3 after plating. Histogram analysis showed that, at that time, the DNA content of 28,2% of the cell population was higher than 3C, and image analysis showed that 42% of the S and G_2 phase cells were involved in DNA synthesis. From day 4, the proliferation rate gradually slowed down until day 8, when less than 5% cells had a DNA content higher than 3C. After day 8, when numerous myotubes differentiated, the image analysis method revealed that the percentage of S and G_2 phase cells involved in DNA synthesis had diminished to between 3 and 8%. The percentage of nuclei in G_0 increased when the first myotubes differentiated around day 5. When horse serum was added to the culture medium on day 4 to enhance myotube differentiation, significant cell proliferation was observed for the first time in satellite cells before cell fusion. These methods of analysis gave the first daily pattern of myogenic cell proliferation and fusion in a cell population isolated from adult muscular tissue. This now makes it possible to study the effects of growth-promoting substances in vitro.

INTRODUCTION

The skeletal muscles of Vertebrates include a population of mononucleated cells located close to the plasmalemma inside the basal laminae of the myofibres, known as satellite cells. These cells, first described in frogs (Mauro 1961) have been reported in many species (Rev. in Campton, 1983). In undamaged muscle, they have the characteristic features of resting state cells. It has been shown that they can be activated during growth and regeneration. They proliferate and either fuse to give new myotubes which differentiate into myofibres, or they are incorporated into preexisting fibres (Bischoff 1986, Bischoff and Holtzer 1969, Lipton and Schultz 1979, Moss and Leblond, 1971).

The control of these myogenic processes which start with the activation of previously non dividing cells is not well understood. In vitro culture of these cells after isolation from muscular tissue makes it possible to study the factors involved in myogenic processes. However, little is known about satellite cell characteristics in vitro. Authors generally study defined myogenic strains, or myoblasts. A study of cell cycle under basic culture conditions was first necessary to permit investigation of the role of growth substances in activating proliferation or delaying differentiation. Indeed, differentiation, expressed by the muscle form of creatine phosphokinase (M-CPK) in BC_3H1 cells was observed in cells arrested in G_0. If proliferation was stimulated by a fibroblast growth factor, cells entered in G_1 and concomitantly M-CPK was repressed. Accordingly, mononucleated cells from striated muscles were cultured on glass slides and samples were analyzed with a SAMBA 200 (System for Analytical Microscopy in Biological Applications) microscopic image processor (Thomson TITN, France) which detects and computes densitometric, textural and morphometric cell features that allow cell cycle phase recognition.

METHODS

Isolation of Satellite Cells

Satellite cells from 8 weeks rat White Wistar muscles were isolated as described in Hantai et al. (1985), following a procedure which enabled us to eliminate most of the satellite cell contaminants such as fibroblasts, endothelial and Schwann cells, which usually remain after one step dissociation. In all experiments, the cell suspension was diluted with complete culture medium to obtain a final plating density of 2.000 cells/cm^2. Cells were plated on glass slide chambers (4 chambers per slide) each of which was filled with 0.5 ml of cell suspension. The culture were incubated at 37°C in a 5% CO_2 humidified atmosphere. The Dulbecco medium supplemented with fetal calf serum was renewed every three days. In some experiments 10% horse serum (Gibco) was substituted for calf serum on day 4. We checked that no satellite cells had been removed from the muscle fibres during the first trypsin and subsequent collagenase treatments. This was done by verifying that cloning the cells isolated after these two treatments did not reveal any alignment or fusion characteristic of myogenic cells.

Verification of the Purity of the Myogenic Cell Population

The dissociated cells were plated in 96 multiwell dishes at a concentration of 1 cell/well. Each well was filled with Dulbecco's modified Eagle medium (DMEM) supplemented with 10% fetal calf serum (proliferation medium) for the first 4 days; the medium was then replaced with DMEM containing 10% horse serum (differentiation medium). At day 11, each culture was washed with PBS, fixed and stained with May-Grunwald Giemsa. We examined each well to detect myotubes. After 11 days of culture, we considered that wells containing mononucleated cells but no myotubes had been initially plated with non myogenic cells.

Cell Image Analysis

Feulgen-stained cultures were analysed using a SAMBA 200 microscopic processor (Brugal and Chassery, 1971); the hardware and software of the system were described earlier (Brugal et al., 1979). Cell populations were analysed out on samples of 200-450 cell nuclei in slide regions where nuclear overlapping liable to cause misinterpretation was sparse. Their nuclei were automatically analyzed by calculating 18 parameters relating to the densitometry and texture of the chromatin as well as to the shape of each nucleus (Brugal 1984, Moustafa and Brugal, 1984). In studies of contiguous nuclei in myotubes, they were selected by a digitalizing tablet.

Cell Cycle Phase Identification

We applied the unsupervised recognition method of cell kinetic analysis (Emptoz et al., 1978) to determine the number of putative classes in proliferative cell populations 3 to 5 days after plating. Proliferation was estimated by microscope observation of mitotic phases and by DNA histograms of satellite cell populations. Our aim was to discriminate between cell cycle phases and we therefore used integrated optical density (IOD) as the major parameter since it measures the nuclear DNA content which shows the variations characterizing the mitotic process. The variance of the other 17 parameters was tested against the IOD variance from cell to cell. The classification obtained by this method was tested against the results obtained by stepwise linear discriminant analysis, as applied to the study of cell kinetics, using a SAMBA 200 system (Giroud, 1982). To visualize cell distribution during each cycle, the multi-parametric image featuring the 18-dimensional space for the 18 parameters measured was projected into a 2-dimensional factorial plane by canonical transformation. The classes determined by this method were used as learning sets for a stepwise linear discriminant analysis, in order to assign the nuclei of a cellular population aged from 0 to 14 days to the different phases of the cell cycle.

RESULTS

Cell Population Increase and Myotube Morphogenesis

Our isolation technique eliminated most non myogenic cells, and we counted in multiwell dishes seeded with single cells that they only constituted about 12 to 14% of the total plated population. A few hours after plating, cells adhered to glass slides (Fig. 1). Most non myogenic cells were fibroblasts. Giemsa-staining permitted easy detection of spindle shaped myogenic cells with a small nucleus and little cytoplasm. From day 0, when cells had just adhered to the substrate, to day 3 or 4, the cell population doubled every day. Past this time, the growth rate slowed down, and from days 4 to 10, the number of cells only doubled once. From days 2 to

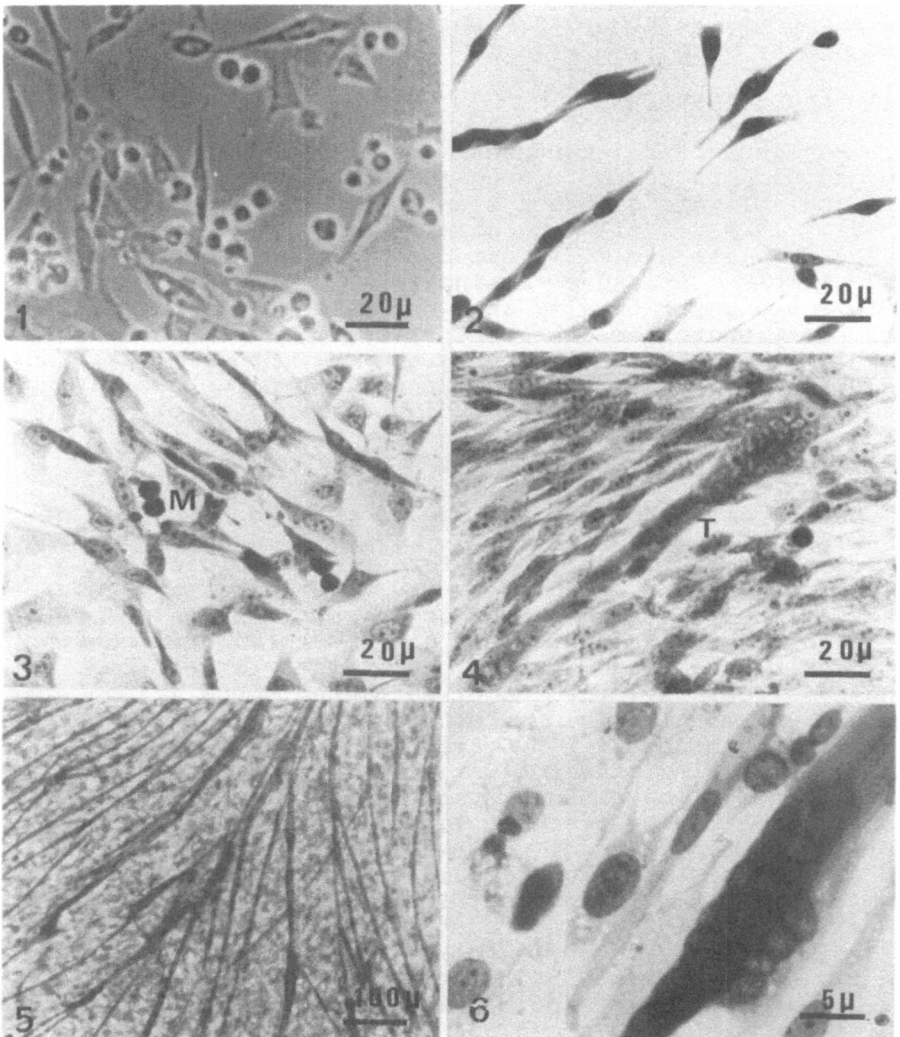

Fig. 1. Phase contrast micrography of a 4 hr rat myogenic cell culture.
Fig. 2. Two day myogenic cell culture.
Fig. 3. Four day myogenic cell culture. M: mitosis.
Fig. 4. Myotube (T) differentiated in a six day culture.
Fig. 5. A 14 day culture showing a network of myotubes.
Fig. 6. Detail from a multinucleated myotube in a 14 day culture

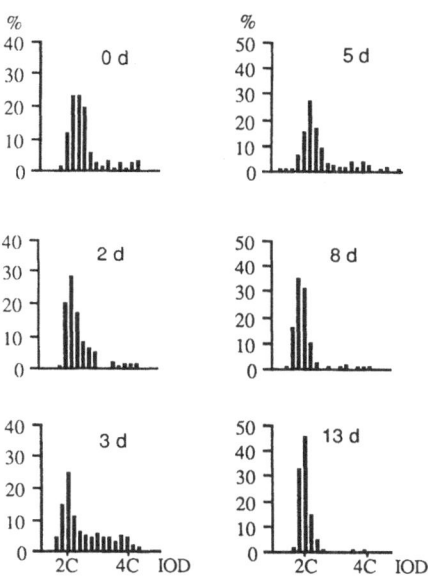

Fig. 7. Distribution of nuclear DNA content observed by integrated optical density (IOD) in satellite cell cultures between 0 and 13 days after plating. The percentage of nuclei containing 4C DNA increased from day 0 to day 3 and then declined with time.

to 4, cultured cells progressively colonized the whole slide area, during a rapid proliferation phase (Fig. 2-3). At about day 4, some of the spindle shaped cells began to align and to come into contact. In the absence of·contact inhibition, these alignments frequently intersected, and spindle shaped cells overlapped with flattened cells. In the presence of fetal calf serum, the proliferation rate began to decrease on day 4, and the number and size of cellular alignments was seen to increase on day 5. Contacts between cells were established through the lateral membrane of cell processes that displayed characteristic features of myogenic cells. These contacts were a necessary preliminary to cell fusion, which resulted in slender multinucleated myotube formation. Myotubes afterwards thickened and their nuclei aggregated in several rows along the myotube axis (Fig. 4). When the myotubes had enlarged, the nuclei migrated laterally (Fig. 5). Myotubes finally formed a network which covered the cultures (Fig. 6). Several mononucleated cells remained present among the growing myotubes and continued to divide. Under our experimental conditions, we counted that the fibroblast population gradually diminished.

Cell Distribution According to Nuclear DNA Content

The histograms of cultured cell distribution according to nuclear DNA content (IOD) show that the percentage of cells

Table 1. A Confusion Matrix Obtained by Unsupervised Cell-
Phase Recognition (A) and Discriminant Analysis (B)

B \ A	G_0	G_1	S	G_2	Met	Tel	Total cell numbers
G_0	**32**	4	0	0	0	0	36
G_1	7	**143**	0	0	0	0	150
S	0	5	**78**	7	0	0	90
G_2	0	0	0	**57**	0	0	57
Met	0	0	0	0	**8**	0	8
Tel	0	0	0	0	0	**14**	14

The percentage of cells identically classified by both methods
ranged between 86 and 100%. The numbers of cells classified
into the same class by both methods (in bold type) form a
diagonal line.

undoubtedly involved in the cycle (Fig. 7 values between 3C
and 4C) changed with the age of culture. Three hours after
plating, cells adhered to the culture support, and 10% of them
had already synthetized DNA. This proportion increased,
reached its peak at day 3 after plating (28.2% 3C), and thereafter regularly declined. Eight days after the beginning of
the culture, this proportion wasnever higher than 5%.

Cell Cycle Phase Recognition

The M and G_0 phases were recognized by the cytologist and
the G_1, S and G_2 phases were determined by the unsupervised
method on an actively proliferating cell population. This analysis, carried out after discarding the mitotic phases, led to

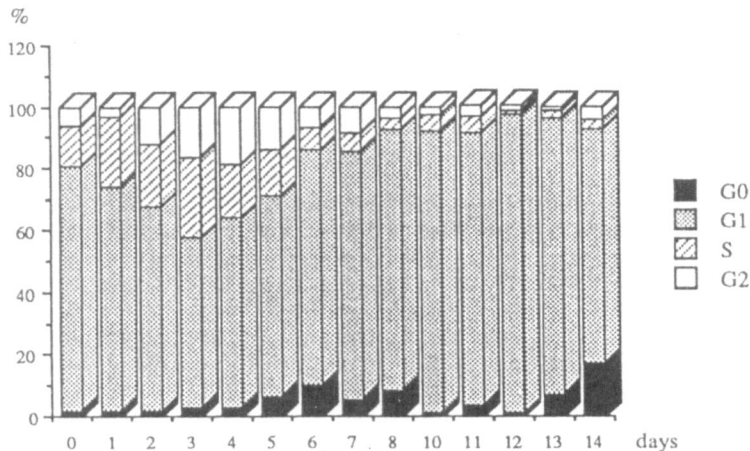

Fig. 8. Variation in the percentage of cells in G_0, G_1,
S and G_2 phases during 14 days of satellite
cell culture with fetal calf serum.

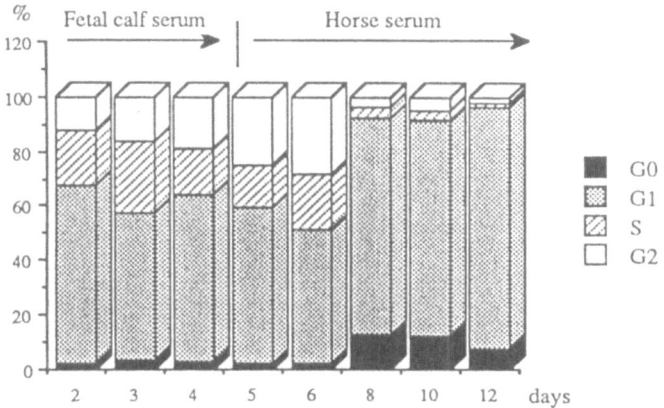

Fig. 9. Variation in the percentage of cells in G_0, G_1, S and G_2 phases during 12 days of satellite cell culture in medium in which fetal calf serum was replaced with horse serum 4 days after plating.

the detection by the program of 3 classes: G_1, S and G_2. Another 3 classes were detected by the cytologist: metaphase, telophase and G_0. The G_0 phase nuclei were collected in 10 day cultures among the myotubes nuclei, which at that time were numerous. As these nuclei were contiguous, the image processor was not able to discriminate between them, and they were separated by a digitalizing tablet.

Discriminant analysis applied to the six classes determined by the above methods, ranked the parameters according to their discriminatory power. Integrated optical density (IOD) was ranked first, thus confirming the importance of this parameter in cell phase recognition. Variance of optical density (VOD) which measures the densitometric heterogeneity of the nucleus, was ranked second. This indicated that the level of chromatin condensation accurately characterized each phase of the cell cycle.

We compared the results of classifying cultured cells into 6 classes by using the discriminant analysis program of Giroud (1982) with the results for the unsupervised-cell phase recognition of Moustafa and Brugal (1984). Confusion matrix (Table 1) showed that the percentage of cells identically classified by both methods ranged from 86 to 100%. We therefore concluded that the 6 classes detected were valid. In particular, the rate of confusion between the G_0 class and the others was very low, and only involved cells belonging to the adjacent class (G_1). We analysed 56 nuclei included into myotubes, in 13 and 14 days old cultures. The nuclei at the extremities of myotubes resulted from recently fused cells and were largely euchromatic; they were classified in G_0 or G_1. The nuclei included in the median part of myotubes resulted from the first fusions; they were heterochromatic and exclusively classified as G_0. In the same cultures, nuclei from mononucleated cells were in G_1.

Evolution of the Cell Cycle During Culture in Fetal Calf Serum

The discriminant analysis applied to the cellular population from 0 to 14 days of culture led us to evaluate daily cell distribution among the G_0, G_1, S and G_2 phases. The mitotic phases were excluded on account of their small number. Three to six slides were examined for each day. Fig. 8 shows the evolution of cultured cell populations. The proportion of S+G_2 cells increased until day 3, when up to 42% of the total cell population was involved in DNA synthesis. The importance of these two phases then decreased from day 4, dropped to a low steady state at day 8 (between 3 and 8%) and remained at this level until the end of the experiment. Conversely, the level of G_0 cells began to rise from day 5 when we observed the first myotubes, but this increase was not uniform and fairly large fluctuations were observed after day 6.

Effect of Replacing Fetal Calf Serum with Horse Serum on the Cell Cycle

Horse serum is known to enhance cell differentiation and myogenesis when substituted for fetal calf serum in the culture medium at day 4. Cell cycle analysis in such cultures showed that, between days 4 and 12, i.e. after adding horse serum to the medium, a recovery of proliferation preceded differentiation (Fig. 9). We checked that renewing fetal calf serum at day 4 had no effect on proliferation. With horse serum, the percentage of S+G_2 cells rose up to day 6, when it reached 48%. On and after day 8, cell distribution between the different cell cycle phases was similar to that observed in cells cultured with fetal calf serum, with a few cells in S+G_2. The percentage of G_0 cells in the samples analysed rose to 14% at day 8, and thereafter apparently declined with time.

DISCUSSION

We isolated mononucleated cells - chiefly satellite cells - from rat skeletal muscles and studied their proliferation and differentiation into myotubes in two different media. During the period from plating to day 4, cell population on an average doubled every 24 hours and the number of cells involved in DNA synthesis gradually increased. When, at days 4-5 the first satellite cells aligned and initiated fusion into myotubes, the growth rate of the cell population slowed down and then remained constant until the end of culture on day 14.

Analysis of the Feulgen-stained cell population using a SAMBA 200 microscopic processor provided results which would not have been distinguishable with conventional techniques of histological observations. Firstly, as soon as 2-3 hours after plating, 10% of the total cell population were in the S+G_2 phases. This proportion was higher than in intact adult muscle. In growing animals, the proportion of activated satellite cells is estimated at only 3% (Moss and Leblond 1971). In our cultures, DNA synthesis might have been initiated early by the effects of trypsin on the matrix and cell coat during dissociation. A second interesting observation was that, under our experimental conditions, a significant population of cells remained involved in the S and G_2 phases even when myogenesis was very advanced. Such continued division of satellite cells was also shown in 18-day Xenopus myogenic cell cultures (Franquinet et al., 1988). The present cells were not fibroblasts, since they were not arrested by inhibitory contact, and several cell layers were frequently observed. Thus, different subpopulations of satellite cells might possibly be engaged very late in the cell cycle or might be able to continue proliferation and delay fusion.

We could have expected a gradual increase in the number of cells arrested in G_0 after day 5, from the moment when the first myotubes were differentiated. In fact, however, we observed that the number in G_0 fluctuated, which revealed some of the limits of the image analysis technique used here. As this analysis was programmed to take account of isolated nuclei only, most of the nuclei in myotubes were discarded as they generally were close together. Thus, in the random samples we studied in 5-7 day cultures, an unknown number of myotube nuclei was eliminated. As a result, the population of nuclei in G_0 was underestimated in these cultures. Another original result obtained with this method of analysis was the demonstration that the nuclei of growing myotube were not homogeneous (Table 1). This heterogeneity supported the idea that after the arrest of nuclear divisions and fusion into myotubes, nuclei gradually pass from G_1 in recently fused nuclei to G_0 in the oldest part of the myotube, concomitantly with initiation of the myogenic program.

Cell fusion into myotubes was routinely enhanced by substitution of horse serum for fetal calf serum in the medium on day 4. Our methods of analysis showed for the first time in satellite cell cultures that before cell fusion, horse serum either initiated one or several further cell cycles, or triggered the proliferation of previously resting cellular subpopulations, as mentioned above. A parallel could be drawn between our observations on the mitotic effects of horse serum and the results obtained on L6 myoblasts (Pincet and Whalen, 1985) and avian myoblasts (Devlin and Koenigsberg, 1983). These cells also undergo one or more cycles before fusing if they are transferred to a fusion-promoting medium containing serum.

Our results defined the normal conditions of proliferation in a primary culture of mononucleated cells isolated from muscular tissue and enriched in myogenic cells. They now make it possible to compare the effects on in vitro myogenesis of substances acting on proliferation and/or differentiation, such as growth factors, hormones and oncogenes.

REFERENCES

Bischoff, R., 1986, Proliferation of muscle satellite cells in intact myofibers in culture, Dev. Biol., 115:129-139.
Bischoff, R. and Holtzer, J., 1969, Mitosis and the processes of differentiation of myogenic cells in vitro, J. Cell. Biol., 44:188-200.

Brugal, G., 1984, Image analysis of microscopic preparations, in: "Methods and Achievements in Experimental Pathology", Jasmin, G. and Porschek, L., eds, Karger, Basel.

Brugal, G. and Chassery, J.M., 1971, Un nouveau systeme d'analyse densitometrique et morphologique des preparations microscopiques, Histochemistry, 52:241-258.

Brugal, G., Garbay, C., Giroud, F. and Adelh, D., 1979, A double scanning microphotometer for image analysis: hardware, software and biomedical applications, J. Histochem. Cytochem., 27:144-152.

Campion, D.R., 1983, The muscle satellite cell: a review, Int. Rev. Cytol., 87:225-250.

Devlin, B.H., and Koenigsberg, I.R., 1983, Re-entry into the cell cycle of differentiated skeletal myocytes, Dev. Biol., 95:175-192.

Emptoz, H., Terrenoire, M. and Tounissoux, D., 1978, Indetermination measure for a sequential identification process, Proc. 4th Int. Conf. Pattern Recogn., :262-264

Franquinet, R., Aamiri, A., Gautron, J. and Le Moigne, A., 1988, In vitro proliferation and differentiation of myogenic cells from adult Xenopus, Biol. Struct. Morph., 1:84-88.

Giroud, f., 1982, Cell nucleus pattern analysis; geometric and densitometric featuring, automatic cell phase identification, Biol. Cell., 44:177-188.

Hantai, D., Tassin, A.M., Gautron, J. and Labat-Robert, J., 1985, Biosynthesis of laminin and fibronectin by rat satellite cells during myogenesis in vitro, Cell Biol. Int. Rep., 9:647-654.

Lipton, B.H., and Schultz, E., 1979, Development Fate of skeletal muscle cells, Science, 205:1292-1294.

Mauro, A., 1961, Satellite cells of skeletal muscle fibers, J. Biophys. Biochem. Cytol., 9:493-495.

Moss, F.P. and Leblond, C.P., 1971, Satellite cells as a source of nuclei in muscles of growing rates, Anat. Rec., 170:421-436.

Moustafa, Y. and Brugal, G., 1984, Image analysis of cell proliferation and differentiation in the thymus of the newt Pleurodeles waltlii Michah, by SAMBA 200 cell image processing, Roux's Arch. Dev. Biol., 193:139-148.

Pinset, C. and Whalen, R.G., 1985, Induction of myogenic differentiation in serum-free medium does not require DNA synthesis, Dev. Biol., 108:284-289.

REGENERATION OF SKELETAL MUSCLE FIBERS BY

IN VITRO MULTIPLIED AUTOLOGOUS SATELLITE CELLS (*)

Hala Alameddine

INSERM U. 153, 17 rue du Fer-à-Moulin
75005 Paris, France

Repopulating a partially destroyed muscle by myogenic stem cells able to reform differentiated and functional muscle fibers is the aim of the work we have developed in our laboratory.

Extensor digitorum longus (EDL) muscles of adult male rats were injured by autotransplantation according to the procedure of Carlson and Gutmann, 1975 (Anat. Rec., 183:47-62) modified by Gulati et al., 1982 (Anat. Rec., 204:175-183). This procedure determines muscle ischemia and is characterized by the survival of a peripheral rim of 3 to 5 muscle fibers thick which surrounds central necrotic fibers. Degeneration, phagocytosis, and regeneration proceed centripetally following revascularization. The most central fibers regenerate by the end of the second postoperative week.

The spontaneous regeneration that follows injury can be inhibited by X-ray irradiation. By administration of three different doses (i.e. 1500, 2500, 35000 rad) 24 hours before or after ischemia, we demonstrate that inhibition of regeneration is dose dependent and more effective when X-irradiation is performed before injury. A single dose of 2500 rad administered before ischemia completely inhibited spontaneous regeneration. When associated to autotransplantation, it resulted in the formation of a cystic structure formed by the peripheral rim of surviving muscle fibers surrounding a central space where no regenerating fibers could be identified even after long survival periods. This central space contains a loose connective network with few cellular elements. We assumed this space was suitable for the development of implanted autologous satellite cells.

Satellite cells enzymatically liberated from contralateral tibialis anterior and EDL muscles were cultured in a proliferation favoring medium composed of MEM 199 supplemented with 15% fetal calf serum and 2% embryo extract. Incubation of the cells in medium containing fluorescent labels such as FITC-latex beads and/or 1,1'-dioctadecyl-3,3,3',3' tetramethylindocarbocyanine perchlorate resulted in the internalization of either one or both labels if used together. Myoblasts carried these labels into myotubes when differentiation was allowed _in vitro_. Cells to be grafted were maintained in a mononucleated state by trypsinization and sequential passaging. Dense suspensions of labeled satellite cells were injected into the central space of X-irradiated ischemic EDL muscles at the 14th postoperative day. The fate of grafted cells was followed on morphological basis in cryostat transverse sections stained for routine morphology, histochemistry, and fluorescence microscopy.

Immediately after cell grafting, dense cellular masses could be found within the central space of X-irradiated ischemic muscles. The first myotubes were identified at the 3rd day postgrafting. Their number and diameter increased with time. Fluorescence observation showed the label to be confined to the central space. Most of it was observed within regenerating myotubes restricted to the central space. Few labeled mononuclear cells were present in the interstitial space.

Myofibers became histochemically differentiated between the 2nd and the 4th postgrafting week and fiber type grouping was then observed. Regenerated muscle fibers belonged to both fiber types. The central space was completely reorganized and the muscle had a fascicular pattern. Regenerated fibers could be distinguished from surviving muscle fibers by the presence, at the boundary, of a "no fiber land" devoid of muscle fibers. This appearance persisted for longer survival periods. Only the cross sectional area of the grafted muscles and hence, that of muscle fibers increased with time. Wet weight follow up of the irradiated-ischemic muscles which received cell grafts showed a progressive increase like the one observed in control animals whereas that of irradiated-ischemic muscles remained stationary. By the end of the 3rd postgrafting month, the wet weight of cell grafted muscles is 3 times that of ungrafted EDLs whereas it only reached 1:3 that of normal animals.

To study the role that muscle basement membrane could play in this reorganization, we studied the renewal of some of its well known components such as fibronectin, laminin, and type IV collagen by immunocytochemistry using polyclonal antibodies in X-irradiated ischemic muscles which received or did not receive cell grafts. Immunofluorescence showed reactivity to all three components in the central space of X-irradiated ischemic muscles examined at day 14 after ischemia. Type IV collagen, laminin, and fibronectin antibodies stained the remnants of old basement membranes. Fibronectin stained also all the interstitial space of the cystic structure. Mononuclear cells labeled with FITC-latex beads and grafted into that space, showed no reactivity to any of these components at early regenerative stages. For longer periods, it could be observed surrounding myotubes containing latex beads and their distribution is similar to that of normal muscles.

An electrophysiological study was undertaken to follow functional recovery of such grafted muscles by measuring latency periods, time to peak, twitch and tetanic tension in three groups of animals. The first formed by normal animals, the second by animals in which EDL muscles were X-irradiated and autotransplanted, and the third by animals in which EDL received the same treatment as the 2nd group plus satellite cells grafts. The electrophysiological parameters were measured at different time intervals until the 3rd month after cell grafting. The same increase in twitch and tetanic tensions was observed. They reached successively 4 and 5 times those measured for control X-irradiated ischemic EDL. However, latency period of the cell grafted muscles remained higher than in normal animals. Correspondingly, the forces developed by the grafted muscles remained below that of normal individuals.

To investigate in vivo proliferation of implanted satellite cells as well as the time interval at which they stop multiplying, we injected intraperitoneally ^{3}H-TdR at days 0, 1, 3, 7, and 14 after cell grafting. ^{3}H-TdR was also injected to animals which were submitted to injury after X-irradiation but did not receive cell grafts. Satellite cells of the peripheral surviving muscle fibers were never labeled in either ungrafted or cell-grafted muscles. In the latter, surprisingly, we could not identify labeled myotubes when ^{3}H-TdR was injected immediately after cell grafting. However some labeled fibroblasts were encountered not only in the central space, but also in the interstitial space between surviving muscle fibers. In rats which received ^{3}H-TdR injection between days 1 to 7 after cell grafting, labeling was found in both mononuclear cells interspersed into the central space of irradiated-ischemic muscles and multinucleated myotubes. Labeled myotubes are present only in muscles which received cell grafts. They could be detected in animals which received the ^{3}H-TdR injection between days 1 and 7 suggesting that cell proliferation occurs until the 7th postgrafting day.

The model we present allowed us to demonstrate that autologous satellite cells multiplied in vitro retain their myogenic potentiality. We give evidence that they are able to proliferate in vivo for at least 1 week after being implanted. They are able to form differentiated muscle fibers on morphological and histochemical criteria, and to improve functionality by means of all measured electrophysiological parameters. The regenerated muscle fibers renew their muscle basement membrane and reorganize among each other in a manner reminiscent of normal muscles.

This work was supported by grants from AFM (Association Francaise de lutte contre les myopathies) and INSERM.

"IN VITRO" POTENTIALITIES OF MYOGENIC CELLS

ISOLATED FROM ADULT URODELE MUSCLES (*)

Raphaël Franquinet

Laboratoire de Biologie Animale, UFR de Sciences
et Technologie, Université Paris XII
Avenue du Général de Gaulle, 94010 Creteil Cedex
France

When urodele limbs regenerate, it was formerly accepted that cells of regenerates resulted from the dedifferentiation of stump tissues, where multinucleated myofibres might break into mononucleated cells which proliferate and fuse anew. Another interpretation arose since Mauro (J. Biophys. Biochem. Cytol., 9:493-495, 1961) pointed out the existence of myogenic stem cells so-called "satellite-cells". Such cells are present in urodele muscles and it is tempting to image that they are involved in limb regeneration. Careful study of the myogenic potentialities of Urodele statellite cells requires in vitro cultures separately from myotubes. These urodele myogenic cells have never been isolated and cultured in vitro. We describe here a method of primary and secondary cultures of mononucleated cells after isolation from muscle fibres.

Myogenic cells were isolated from larval and adult Pleurodeles waltlii and from larval and adult Ambystoma mexicanum. Muscles were dissociated for 1 h in 0.1% pronase in culture medium. The culture medium contained 40% Eagle's MEM with Hank's salts, 40% HAM F12, 20% distilled water, 2.5 g/l NaHCO$_3$ (pH 7.35 at 25°C), 5% fetal calf serum or horse serum, 200 IU/ml penicillin, 200 µg/ml streptomycin, 0.56 IU/ml insulin. CaCl$_2$ was adjusted to 1.8 mM. Culture dishes were precoated with 0.5% gelatin. In some experiments 21 days primary cultures were treated with 0.1% trypsin, rinsed in PBS, and seeded in Petri dishes at the cell concentration of 1 cell/dish. Replating was repeated in the same conditions each 14 day during 3 months. In some experiments cultures were filtered, myofibres were recovered on nylon grids, after separation from mononucleated cells, rinsed and cultured in the same conditions as primary cell cultures.

Myofibres survived for as long as 30 days without fragmentation into myogenic cells and no mononucleated cells could be seen in these cultures.

In primary cell cultures, myogenic cells adhered to the support after 6 days and proliferated. In adults the labelling index augmented until day 15 and then reached a plateau. In larvae it rose until day 15 and then sharply diminished to a lower level from day 20. The first fusion into myotubes cells were seen around day 7 for _Ambystoma_ and day 12-15 for _Pleurodeles_. No fusion of axolotl mononucleated cells were observed in our _in vitro_ culture conditions.

In secondary cell cultures of adult _Pleurodeles_ myogenic cells, the cellular population doubled each 3 day. After several sub-cultures mitotic activity was still noticeable. In each culture the total number of cells obtained from a single replated cell decreased progressively. We calculated that after 4 successive replatings (one per 2 weeks) from a primary culture of 21 days we might obtain 32 cell generations (cells). We also calculated the mean volume of a _Pleurodeles_ multinucleated and replated during 3 months might give 1 cm^3 of muscle tissue, that is more than one _Pleurodeles_ leg volume.

In clonal replated cultures, we recognized four morphological myogenic cell types, these morphological differences were constant. These types appeared in either clonal sub-cultures and, at a lower rate, in long term primary cultures. We observed: 1) Little mononucleated cells, typical myogenic proliferating cells. 2) Elongated mononucleated cells, typical myotube cells. 3) Large mononucleated cells. 4) Large multinucleated cells.

In this study we analysed the potentialities of myogenic stem cells from urodele muscles to proliferate and fuse into myotubes. These cells showed abilities to proliferate during more than 3 months giving at least 32 cell generations ($4x10^9$ cells). The fusion potentialities reached levels which differed with species and ages of animals. These results allowed us to propose that proliferating and fusing abilities of the myogenic stem cells, so-called "satellite cells", are sufficient to explain the neoformation of muscles during urodele limb regeneration, without the need of an hypothetic dedifferentiation.

by Bruce M. Carlson, on

MUSCLE REGENERATION

Much of the research on the regeneration of striated muscle has taken place in a different conceptual framework from research on the regenerating limb, but with significant advances in both fields it is now more easily possible to compare the systems and to identify common problems. Research on mammalian skeletal muscle began with the question "Can it regenerate?" When that was answered affirmatively, two main lines of research arose. One concerned the cellular origin of regenerating myoblasts; the other was directed toward the more quantitative question of how much muscle can regenerate at one time and how well does it function. As research along these lines evolved, new data revealed important questions that are of general biological significance.

The origin of regenerating myoblasts continues to intrigue many investigators. Although the role of the satellite cell as the major (if not only) source of regenerating myoblasts in the mammal seems secure, many questions about their properties remain to be answered. In particular, there are tantalizing bits of evidence to suggest that these cells have a remarkable ability to adapt to particular conditions in the host musculature. This can be an adjustment of their numbers to correspond to the muscle fiber types with which they are associated or the apparent ability of satellite cells to migrate into areas of ischemic or degenerated muscle. A striking example of this was provided by Alameddine, who injected cultured satellite cells into grafted muscles of the rat and found that they penetrated the basal laminae of control muscle fibers that had been destroyed by ischemia but did not enter into association with intact original muscle fibers within the graft. Such experiments have profound implications in the possible treatment of diseased muscles by cellular engineering techniques.

A number of French laboratories (e.g., Lemoigne, Franquinet) have devised methods for isolating satellite cells from both amphibians and mammals and studying their behavior in vitro. As these techniques are refined, it will soon become possible to study directly the population characteristics and growth dynamics of satellite cells by direct methods in vitro instead of the laborious and often indirect in vivo methods that have often been necessary in the past. Particularly with the use of clonal cultures and some novel marking methods, it may finally be possible to resolve the decades-old question of the origin of muscle in the regenerating limb.

At a higher level of organization, there are now abundant examples of interactions and influences between regenerating muscles and the surrounding tissues of the host. For example, there is now persuasive evidence that ischemic muscle exerts an angiogenetic influence on the surrounding host musculature, causing the ingrowth of new blood vessels into areas lacking a blood supply. The extent to which degenerating or regenerating muscle stimulates neural ingrowth vs. regenerating nerves taking advantage of an empty, nerveless territory is still not well understood, but the influence of the regenerating nerve supply on the character of the regenerating musculature is great, as is the ability of regenerating muscle to adapt to changing neural or mechanical circumstances.

Grafted skeletal muscle has shown a surprising ability to adapt to the host in terms of age-related characteristics. The recovery of regeneration in old muscles grafted into young hosts has shown that despite a reduction in number and possibly the proliferative capacity of satellite cells in old animals, old muscles still possess enough intrinsic developmental potential to regenerate as well as young muscles when grafted into a young host. This implies that an old muscle has a regenerative potential greater than that which is normally expressed during its regeneration in situ. On the other side of the coin, despite an apparently great capacity to regenerate, young muscle is not allowed to reach its full regenerative potential when grafted into an old host. From the practical standpoint, this leads to the inescapable conclusion that it is better to have a young body than and old one! How the age-related environment provided by the body influence the success of muscle regeneration, and whether or not this applies to the regeneration of other organs, as well, remains a large problem for the future.

The study of cardiac muscle regeneration remains in its infancy. Although there is now convincing evidence that cardiac myocytes in amphibians can divide, there is little direct evidence of effective cell division in mammalian ventricular myocytes. As was pointed out in the meeting, the vast majority of mammalian ventricular myocytes are non-diploid, whereas the ventricular myocytes of the newt, which can divide, are mostly diploid. It would be interesting to determine if mammalian cardiomyocytes which are tricked into remaining diploid will have the competence to divide after injury. Until the problem of cell division is solved, the other major question at the cellular level is why in the mammal the atrial myocytes respond to deficiencies created at the ventricular level. Specific questions are: 1) how do the atrial cells know that something has happened to the ventricle and 2) by what mechanism do they respond? The relationship between mechanical conditions and cellular behavior in the heart remains an enigmatic one.

In looking at striated muscle in general, one of the most intriguing questions remains why a nucleus, once it is part of a multinucleated cell, seemingly loses its ability to participate in DNA synthesis or mitosis. This remains a challenge to the technique and conceptual framework of contemporary molecular biology.

From the practical level, there are many clinical conditions, affecting both skeletal and cardiac muscle that could benefit greatly from the controlled application of regenerative phenomena. One of the great challenges of the future is to understand enough about the basic biology of regeneration to put it to use for the betterment of mankind.

REGENERATION

AND THE

NERVOUS SYSTEM

MECHANISMS CONTROLLING DIRECTED AXON REGENERATION IN THE
PERIPHERAL AND CENTRAL NERVOUS SYSTEMS OF AMPHIBIANS

Nigel Holder, J.D.W. Clarke, Steve Wilson[+]
Kim Hunter and D.A. Tonge[*]

Anatomy and Human Biology and Physiology[*],Subject
Groups, Division of Biomedical Sciences
King's College, Strand, London WC2R 2LS
[+]Present address: Biology Department
University of Michigan
Ann Arbor, Michigan 48109, USA

SUMMARY

We describe a series of experiments designed to elucidate
mechanisms which guide axons to their appropriate targets in
the central and peripheral nervous systems of amphibia. In
both systems it is evident that pathway guidance cues exist in
the environment through which axons grow. In an initial
attempt to examine guidance cues in the CNS at the cellular
and molecular level we are focusing upon the role of radial
glia and are examining neurite outgrowth within organ cultures
and from explanted axolotl spinal cords.

GENERAL PRINCIPLES

As a general rule lower vertebrates (anamniotes - fish
and amphibia) are able to regenerate organs and tissues more
effectively than higher vertebrates (amniotes - reptiles
upwards). Nowhere is this distinction more evident than in
the peripheral and central nervous system. Following
transection of the spinal cord, for example, higher vertebrates are permanently paralysed in those body parts innervated
by nerves emanating caudal to the lesion (Puckula and Windle,
1977). In contrast, following such damage an axolotl will
functionally recover within a few weeks and lead an apparently
normal existence (Clarke et al., 1988). In recent years we
have been examining, using a multi-disciplinary approach, the
mechanisms controlling directed axon regrowth in amphibians in
an attempt to elucidate the general principles and cellular
interactions which underlie this process. We have suggested
that continuous growth in a particular area of the nervous
system may be correlated with the ability of that area to
functionally repair a lesion (Holder and Clarke, 1988). This
correlation holds as well for higher vertebrates as it does
for fish and amphibia, the crucial difference being that

neurogenesis in higher vertebrates ceases early in life; in mammals for example, virtually all CNS neurons are born before birth, whereas many lower vertebrates have continuous mode of growth. It may be that continuously growing animals which generate neurons throughout life (see Holder and Clarke, 1988) can be considered as continuously developing forms which maintain the developmental mechanisms necessary to build a functional nervous system. Following damage many of the mechanisms required to recreate the correct neuronal connections may be intact and regeneration occurs as a consequence. In amniotes such mechanisms are discarded once development is complete and directed axon growth cannot occur.

In this review of our work we will discuss experiments designed to examine these developmental mechanisms in the peripheral and central nervous systems (PNS and CNS) of the axolotl and in the CNS of the tadpole of the frog, Rana temporaria.

SPECIFIC AXON REGENERATION IN THE AXOLOTL PERIPHERAL NERVOUS SYSTEM

In urodeles functional recovery of limb movements follows either limb amputation and regeneration or peripheral nerve section and axon regrowth. The conditions for axon regrowth in these two situations are different; after limb amputation all target tissues distal to the amputation plane are missing and growth cones extend neurites into dedifferentiated blastema tissue. In contrast, when a nerve is cut or misrouted in an intact limb, denervated target muscles and an isolated nerve stump are present distally. We have studied both situations and will describe the results separately.

Specific reinnervation of regenerated limbs; the existence of local pathway cues

To demonstrate specificity of neuromuscular connections after limb regeneration, motor pool positions in the brachial and lumbar spinal cord were mapped following horseradish peroxidase (HRP) retrograde transport from identified limb muscles (Stephens and Holder, 1985). Although motor pools in the axolotl cord are not as clearly defined as those in higher vertebrates (see, for example, Landmesser, 1978; Hollyday, 1980; McHanwell and Biscoe, 1981) they are adequate to distinguish populations of motor neurons. Using the motor pool assay we have demonstrated that the great majority of neuromuscular connections are correct following fore- or hindlimb regeneration (Stephens and Holder, 1987a; Wilson et al., 1988). The question is, how do regenerating motor axons locate and innervate their appropriate muscles? Two main mechanisms have been discussed in the literature. First, axons are guided to their targets by pathway guidance cues located within the limb tissues. This notion places little emphasis on the role of the target. Secondly, axons innervate limb tissue randomly, innervating muscles as they make contact; the correct pattern of connections subsequently emerges as a result of competitive interactions between appropriate and inappropriate connections at the target site, resulting in withdrawal of inappropriate motor axons. Competitive interactions have been demonstrated in urodele

limb muscle following nerve implant and denervation experiments (reviewed by Mark, 1980; see also Holder et al., 1982)
and after collateral sprouting from an adjacent nerve into the
denervated territory of a muscle (Bennett and Raftos, 1977;
Wigston, 1980).

To distinguish between these two mechanisms we have
performed a number of experiments the results of which
indicate clearly that pathway guidance cues exist in the
axolotl limb and that the target muscle plays, at most, a
minor role in the initial establishment of neuromuscular
connections. Indirect evidence for pathway guidance cues came
from analysing motor pool position for the biceps muscle of
limbs serially duplicated following regeneration in the
presence of vitamin A (Stephens et al., 1985; Stephens and
Holder, 1987b). In such limbs, the "duplicated" biceps was
invariably innervated by inappropriate, distal limb nerves
suggesting that the target muscle was not capable of attract-
ing biceps specific motor neurons across inappropriate distal
limb territory. This result was achieved irrespective of
whether the limb nerves were cut to allow axon regeneration,
or whether or not the normal, proximal biceps was in place
(Stephens and Holder, 1987b).

A second piece of evidence which argues strongly against
a role for target site competition between appropriate and
inappropriate motor nerves comes from experiments in which the
motor pool maps of limb muscles are examined at different
stages during limb regeneration. By placing HRP into muscles
as soon as they differentiate, it is possible to establish
that motor pools are correct from the time that the first
neuromuscular contacts are made (Wilson et al., 1988). If
inappropriate motor axons were making contact with muscles and
subsequently being withdrawn, motor pools would become refined
as regeneration proceeded, but this was not observed.

Direct evidence for pathway cues has been obtained by
examining axon relationships in limb nerves of axolotl larvae
with fully formed limbs. By applying HRP to nerves in
positions from spinal roots through the plexus to distal limb
nerve branches we have built up a picture of axonal trajecto-
ries which reveals the presence of sorting regions where axons
change their neighbours (Fig. 1; Wilson and Holder, 1988).
These areas, termed "decision regions" by Tosney and Landmesser (1985) in the chick, are interspersed with nerve
lengths in which axons run in parallels and do not change
their neighbours. In the axolotl, a number of decision
regions have been identified; the major sorting area is the
limb plexus, but axons also sort at more distal positions with
axons from one motor pool only finally coming together at the
terminal muscle branch (Wilson and Holder, 1988). This
picture is consistent with that described first in the chick
by Landmesser and her coworkers (reviewed by Landmesser, 1984;
Tosney and Landmesser, 1985). The similarities between the
chick and the axolotl are striking and argue strongly that
axon guidance cues have evolved early as mechanisms which
underlie neuromuscular specificity in vertebrates.

Evidence for pathway cues has also emerged from experiments in which limb nerves are cut and re-routed in intact
limbs (Grimm, 1981; Cass and Mark, 1975). However, as

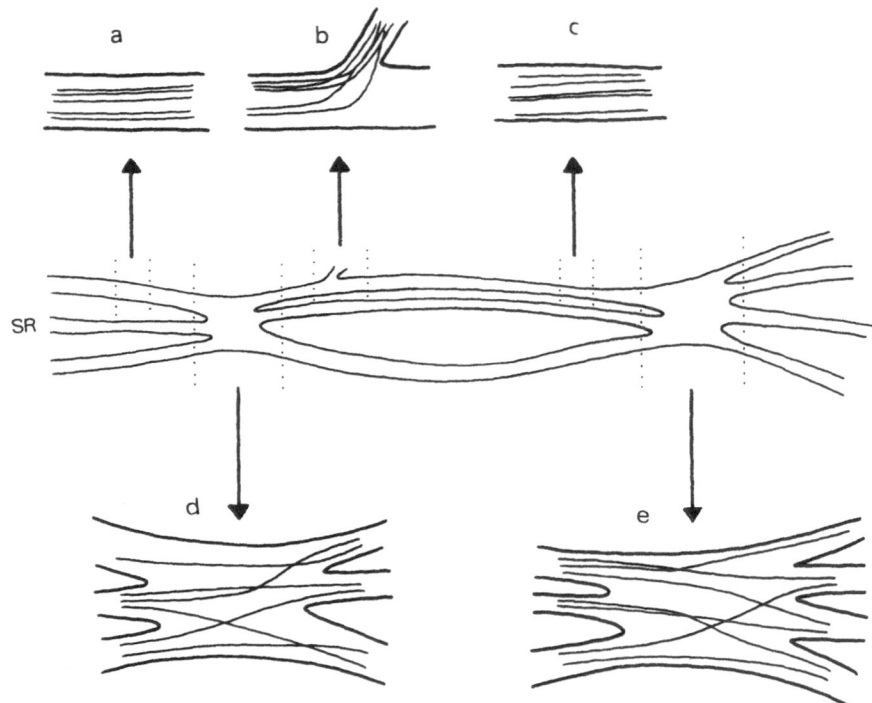

Fig. 1. Schematic view of the arrangements of axons
within the peripheral nervous system of the
axolotl (adapted from Wilson and Holder,
1988). Axons run in parallel in non-decision
regions such as the three spinal roots, SR
(a), and the main limb nerves (c), but sort
out from one another at final nerve muscle
branches (b), and proximal (d) and distal (e)
limb plexi.

mentioned above, distal denervated limb muscles and a distal
nerve stump remain in this experimental situation. Following
a nerve re-route, sprouted axons regrow from within the
proximal nerve and exit the rerouted piece somewhere along its
length, from the position at which it leaves its normal
pathway to the cut end (Holder et al., 1984; Holder et al.,
1982; Wilson et al., unpublished results). Once axons have
left the proximal nerve they invariably grow distally and will
eventually innervate their appropriate target (Holder et al.,
1982). The path taken by the regrowing axons is related to
the original pathway with fascicles running distally in
parallel. As yet it is unclear whether the distal directed
growth of these axons is caused by local environmental pathway
cues or by the presence of more long range attraction from a
signal emanating from the denervated muscle or isolated distal
nerve stump. We are currently using _in vivo_ and _in vitro_
methods to clarify these points.

DIRECTED AXON REGENERATION IN THE SPINAL CORD

It has been known for a long time that frog tadpoles and salamanders will recover function following spinal cord damage (Hooker, 1925; Platt, 1955). However, little attention has been paid to the extent and precision of axon regrowth in the amphibian CNS. In an attempt to discover how functional recovery is attained, our first aim was to establish the extent and precision of axon regrowth. Using tadpoles of common frog, Rana temporaria, we have examined the ability of a particular group of axons to grow across a completely transected spinal cord. The axons we chose to study were the ascending central axons of the lumbar dorsal root ganglion cells. These dorsal column (DC) axons offer a number of advantages over other spinal cord axons; since they alone enter the cord from the dorsal roots they can be selectively labelled with HRP, and are easily stimulated or recorded from using conventional electrophysiological techniques; the position of their cell bodies is well known, and they can be counted; they form a compact fasciculus close to the surface, adjacent to the dorsal midline of the spinal cord.

The growth of these axons across a spinal cord transection was found to be dependent upon the stage of development of the tadpole (Clarke et al., 1986). If the cord is cut before developmental stage VII, many dorsal column axons are able to cross the injury and grow along their usual pathway to the medulla. Between stages VII and X this ability is largely lost as fewer axons cross the injury and then do not extend more than 500 μm along their pathway and, in those cords transected after stage X, virtually no dorsal column axons cross the transection. This fall-off in growth across the transection appears to be specific to the DC axons since we have electrophysiological evidence that many axons within the ventral funiculus of the spinal cord are still able to cross the injury even after stage X.

The population of dorsal root ganglion (DRG) cells in frogs is born over a protracted developmental period (Prestige, 1965; Bidd, 1978). In R. temporaria our own data suggest new DRG cells are being born up to about developmental stage XII. This raises the possibility that the axon growth seen across a transection may consist of newborn as well as regenerating axons. It is also likely that the stage-related repair process reflects the timing of neurogenesis, a proposition described earlier in this review (Holder and Clarke, 1988).

In a second study of the consequences of complete spinal transection we concentrated on a more diverse group of axons, the descending projection from the brain to the lumbar spinal cord in the axolotl (Clarke et al., 1988). This projection arises from several nuclei in the medulla, midbrain and diencephalon. They can easily be studied using retrograde transort of HRP applied to the lumbar spinal cord. We analysed these projections in normal axolotls and axolotls of various sizes whose thoracic spinal cord had been transected 3 to 23 months earlier. The results showed that, given enough time, the regeneration of the descending axons was excellent. - after 23 months the number of neurones with axons reaching the lumbar cord was aproaching the control values, and all of the usual nuclei were represented. This regenerative potential did not appear to be dependent upon the axolotl's size (and hence its age). Perhaps the most surprising feature of the response was how slowly the number of axons reaching the lumbar cord returned to normal. At 3 to 4 months after the transection only 10% of the axons had reached the lumbar cord, whereas morphologically the repair at the transection site (as judged by cord diameter) is advanced at 3 months and hardly improves over the next 20 months.

If we now compare the results obtained from the axolotl with those from the frog tadpole dorsal columns and those from other amphibian systems, a rather complex picture emerges. Clearly, not all amphibian CNS axons have the same regenerative potential; in some systems this potential is greater early in development than later (e.g. dorsal column axons in R. temporaria, Clarke et al., 1986), in others there may be critical periods when regeneration is most efficient (e.g. during metamorphosis of R.catesbeiana, Forehand and Farel, 1982), and in other systems regeneration is not restricted to any one period (e.g. the retinotectal projection of frogs, Sperry, 1944; Gaze 1959; and descending spinal projections in axolotls, Clarke et al.,1988). It appears also that even homologous axon pathways in closely related species can behave quite differently; for example, Forehand and Farel (1982) found no restoration of dorsal column axons in R. catesbeiana tadpoles, whereas we find good restoration of this pathway in R. temporaria, albeit in a stage-dependent manner. Obviously, the generalization that axon regeneration in the amphibian CNS is better than that seen in higher vertebrates cannot be extrapolated to include all CNS axons in all amphibians. The factors which regulate axon regeneration appear to be restricted locally to particular axon pathways, rather than being ubiquitous within the CNS environment (see also Lyon and Stelzner, 1987). Again, this may reflect the periods of neurogenesis of cell bodies from which particular axon tracts arise (Holder and Clarke, 1988).

Evidence for axonal pathway cues in spinal cord

In both the frog tadpole and axolotl spinal cord axon regeneration studies we have only indirect evidence that axons grow back precisely to their normal sites of innervation: thus, in the axolotl, lumbar HRP fills resulted in labelling of all of the expected nuclei in the hindbrain and midbrain (Clarke et al., 1988). In order to examine directly the path-finding abilities of axons negotiating a spinal lesion we rotated thoracic spinal cord segments dorsal to ventral in stage VI or VII R. temporaria tadpoles and demonstrated dorsal column axons some weeks later with HRP applied to the hindlimb bud (Fig. 2; Holder et al., 1987). At the first transection where dorsal column axons encountered ventral cord tissue they showed one of several responses. Some stopped growing, other switched to the contralateral dorsal column and grew caudally, back down to the lumbar cord, but many grew downwards at the graft-host junction and located the dorsal column within the graft, growing indiscriminately down right or left tract but always in a rostral direction. In one case axons negotiated the rostral graft-host junction and they returned to the host dorsal side and grew on to their normal terminal site, the dorsal column nucleus.

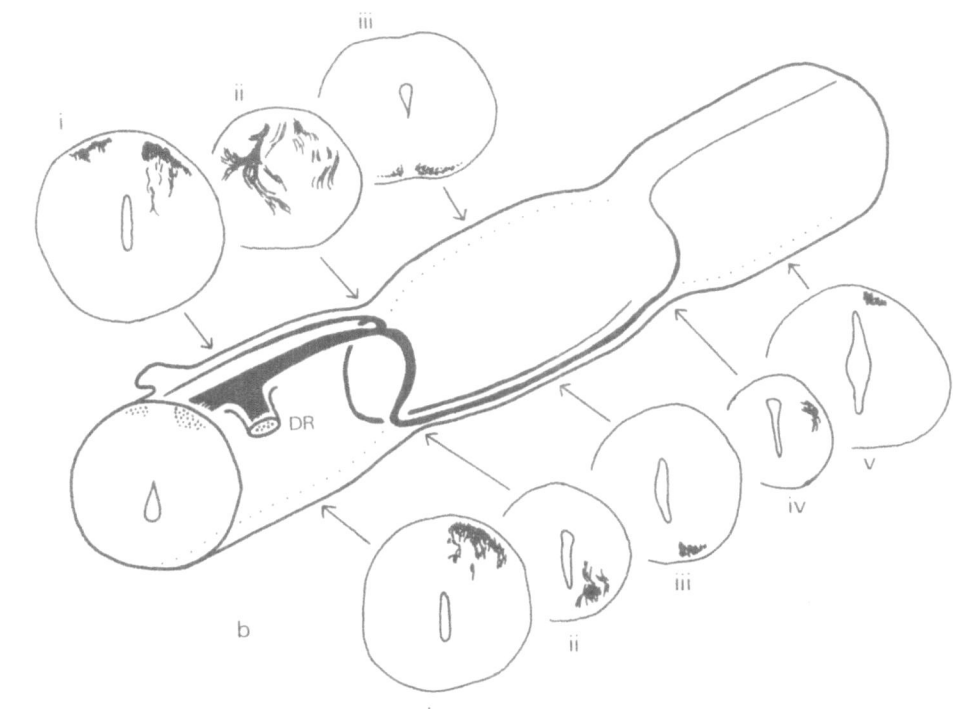

Fig. 2. The response of axons in the dorsal column to
dorsal-ventral rotation of the thoracic spinal
cord in R. temporaria tadpoles (adapted from
Holder et al., 1987). Schematic reconstruction and camera lucida drawings from two
specimens, (a) and (b). Axons, filled with
HRP via a lumbar dorsal root (DR), either
cross the midline at the first transection and
grow caudally in the contralateral dorsal
column (ai), "dive" into the dorsal side of the
graft (aii) and run within the dorsal columns
but peter out before reaching the rostral
graft margin (aiii); or, in a single case,
return to the dorsal side of the brachial cord
(biv and v) and project to the dorsal column
nucleus. In each situation axons run exclusively within the dorsal funiculus. The
ventral midline is shown dotted in the
reconstruction.

These results tell us that axons do pathfind within cord
tissue and that they have an affinity with a particular tract,
in this case the dorsal funiculus. They also revealed that
axons readily grow rostrally or caudally and on right or left
sides within this tract. The identity and location of pathway
guidance cues in the cord are unknown; however, some evidence
suggests an important role for radial glial cells. As the CNS
develops, these are one of the first cell types to appear;
They have a very characteristic morphology - their cell bodies

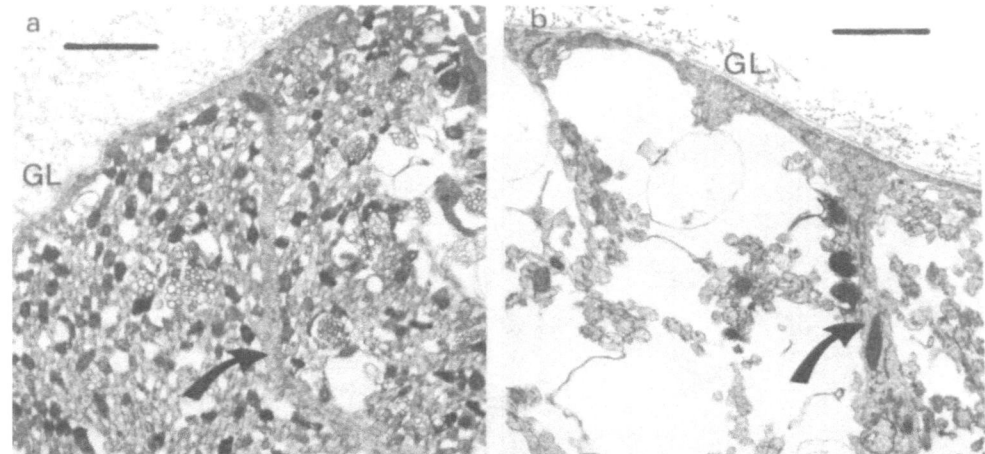

Fig. 3. Electron micrographs of transverse sections
within the dorsal funiculus of <u>Rana</u> <u>temporaria</u>
tadpole spinal cords. (a) is from a normal
cord, (b) was taken 13 days after spinal cord
transection. In both cases radial glia
processes (arrowed) can be seen traversing the
funiculus and forming the glia limitans (GL),
but the axons have largely disappeared from

are closely associated with the ependymal layer around the CSF
and they have one main process which stretches out towards the
surface of the spinal cord or brain. This process often
divides one or more times and the endings of these processes
form flattened swellings (end-feet) which interdigitate
together to form the glia limitans. Within the brain there is
considerable evidence that the surfaces of these cells are
important substances for the migration of neurons (Rakic,
1988) and some evidence that the outermost processes of these
cells are important substrates for the initial growth and
guidance of axons (Silver and Rutishauser, 1984). In mammals
these radial glial cells are transitory structures and most of
them have disappeared, soon after birth. This is not the case
in amphibians however as here radial glial cells are present
throughout life (Miller and Liuzzi, 1986). Given their
apparent importance in the initial development of the CNS, it
seems a reasonable hypothesis to suggest that their presence
in mature amphibian CNS may be one of the factors that make it
permissive for axon regeneration. In support of this notion,
preliminary experiments have shown that radial glia remain
intact with their spoke-like arrangement following spinal cord
transection or crush in <u>R. temporaria</u> and axolotls. Axon
regeneration is rapid, and within a few days the white matter
is a scaffold or radial glia with large spaces between them
(Fig. 3). We are currently studying radial glial cells in
more detail using an anatomical and immunohistochemical
approach to establish if these cells represent a homogeneous
or heterogeneous population.

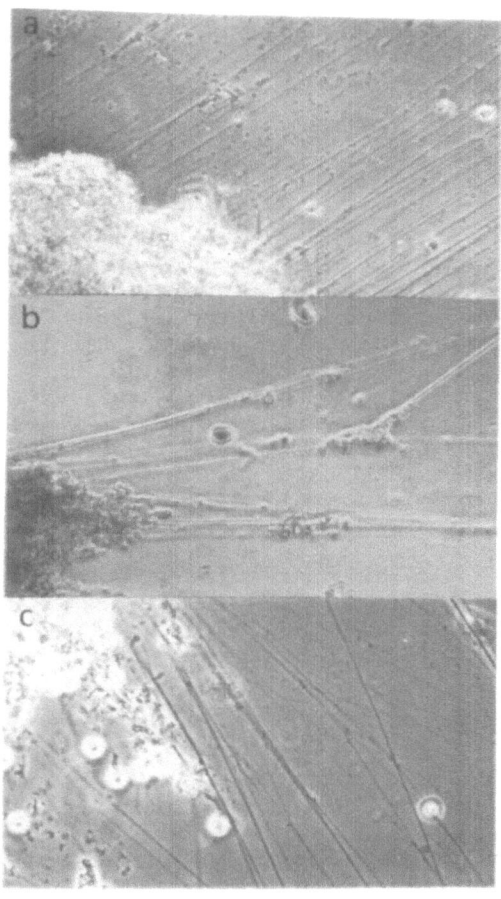

Fig. 4. Patterns and extent of neurite outgrowth from explanted axolotl spinal cord are different depending on the matrix component forming the substrate for growth. In these three examples the substrates are laminin (a), collagen (b) and fibronectin (c).

Studies of spinal cord regeneration in culture

One of our goals in studying axonal regeneration is to understand the cellular and molecular mechanisms controlling growth and navigation of neurites. Our studies to date in intact animals have provided a solid basis of descriptive knowledge but work in situ is restricted because of limitations on the degree to which the environment of growing axons can be manipulated. Therefore, we have developed organ and explant culture methods with axolotl spinal cord tissue. Both methods use a medium modified from that described by Olsen and Bunge (1986).

Organ cultures are prepared by exposing the spinal cord via a dorsal laminectomy and removing a length of cord cradled in the ventral half of the vertebral column with some surrounding axial muscle. Lengths of cord approximately 1.5 cm long were isolated and crushed midway along using watchmaker's forceps. Such preparations have been left for up to two weeks during which time axon regeneration across the lesion occurs. Neurite outgrowth has been assessed by crushing HRP onto the dorsal side distal to the crush and examining filled axons in sections. In such preparations growth cones can be observed near the lesion and these advance through the crush site and beyond in a time sequence and in numbers comparable to that seen following a crush in vivo (Tonge et al., 1988). This preparation holds considerable promise for further studies of axon regeneration because it allows administration of drugs and toxins that are known to affect neuronal behaviour and antibodies which block function of candidate molecules which may play a role in this process.

The second culture method successfully developed in our laboratory is explant culture of axolotl spinal cord. This system has been used primarily to examine neurite outgrowth on surfaces coated with various materials including laminin, fibronectin, collagen, polyornithine and polylysine. Such experiments have been performed previously using neurons from chick and mammals (see, for example, Pierce et al., 1988; Rogers et al., 1987). Using these materials as substrates axon outgrowth from axolotl cord explants occurs within 24 h in vitro. However, each substrate promoted a different type of outgrowth with respect to axon length and fasciculation pattern (Fig. 4; Hunter et al., 1988). It was also found that outgrowth is greatly enhanced if explants are placed within a collagen gel to create a three-dimensional substrate for neurite growth. These studies are continuing in parallel with immunocytochemical analysis of extracellular matrix components in the damaged and normal cord in situ.

CONCLUSIONS

There is considerable evidence for the existence of pathway guidance cues responsible for guiding growing or regrowing axons to their appropriate targets in the central and peripheral nervous systems of amphibia. We have further defined the timing and characteristics of axon regeneration in these two parts of the nervous system and have now embarked on a series of experiments using organ and explant culture methods in an effort to further understand mechanisms controlling axon growth at the cellular and molecular level.

ACKNOWLEDGEMENTS

The work reported here has been supported financially by Action Research for the Crippled Child, Medical Research Council and Wellcome Trust.

REFERENCES

Bennett, M. and Raftos, J., 1977, The formation and regression of synapses during reinnervation of axolotl striated muscles, J. Physiol., 265:261-275.
Bidd, H.D., 1978, Neuronal death in the development of normal and hyperplastic spinal ganglia, J. Exp. Zool., 206:65-72.
Cass, D.T. and Mark, R.F., 1975, Reinnervation of axolotl limbs. 1. Motor nerves, Proc. Roy. Soc. Lond. B, 190:45-58.

Clarke, J.D.W., Tonge, D.A. and Holder, N., 1986, Stage-dependent restoration of sensory dorsal columns following spinal cord transection in anuran tadpoles, Proc. Roy. Soc. Lond. B, 227:67-82.

Clarke, J.D.W., Alexander, R. and Holder, N., 1988, Regeneration of descending axons in the spinal cord of the axolotl, Neurosci. Lett., 89:1-6.

Forehand, C.J. and Farel, P.B., 1982, Anatomical and behavioural recovery from the effects of spinal cord transection: dependence on metamorphosis in anuran larvae, J. Neurosci., 2:654-662.

Gaze, R.M., 1959, Regeneration of the optic nerve in Xenopus laevis, Q.J. Exp. Physiol., 44:290-308.

Grimm, L.M., 1981, An evaluation of myotopic respecification in axolotls, J. Exp. Zool., 178:479-496.

Holder, N., Mills, J. and Tonge, D.A., 1982, Selective reinnervation of skeletal muscle in the newt, Triturus cristatus, J. Physiol. 326:371-384.

Holder, N., Tonge, D.A. and Jesani, M., 1984, Directed regrowth of axons from a misrouted nerve to their correct muscles in the limb of the adult newt, Proc. Roy. Soc. Lond.B, 222:477-489.

Holder, N., Clarke, J.D.W. and Tonge, D.A., 1987, Pathfinding by dorsal column axons in the spinal cord of the frog tadpole, Development, 99:577-587.

Holder, N. and Clarke, J.D.W., 1988, Is there a correlation between continuous neurogenesis and directed axon regeneration in the vertebrate nervous system? Tr. Neurosci., 11:94-99.

Hollyday, M., 1980, Organization of motor pools in the chick lumbar lateral motor column, J. Comp. Neurol., 194:143-170.

Hooker, D., 1925, Studies on regeneration in the spinal cord. III. Re-establishment of anatomical and physiological continuity after transection in frog tadpoles, J. Comp. Neurol., 38:315-347.

Hunter, K., Tonge, D.A. and Holder, N., 1988, In vitro neurite outgrowth from axolotl neurons, Neurosci. Lett. Sup., 32:S68.

Landmesser, L., 1978, The distribution of motor neurons supplying chick hindlimb muscles, J. Physiol., 284:371-389

Landmesser, L.T., 1984, The development of specific motor pathways in the chick embryo, Tr.Neurosci., 7:336-339.

Lyon, M.J. and Stelzner, D., 1987, Tests of the regenerative capacity of tectal efferent axons in the frog, Rana pipiens, J. Comp. Neurol., 255:511-525.

Mark, R.F., 1980, Synaptic repression at neuromuscular junctions, Physiol. Rev., 60:355-395.

McHanwell, S. and Biscoe, T.J., 1981, The localisation of motoneurons supplying hindlimb muscles of the mouse, Phil. Trans. Roy. Soc. Lond. B., 293:477-508.

Miller, R.H. and Liuzzi, F.J., 1986, Regional specialisation of the radial glial cells of the adult frog spinal cord, J.Neurocytol., 15:187-196.

Olsen C.L. and Bunge, R.P., 1986, Requisites for growth and myelination of urodele sensory neurons in culture, J. Exp. Zool., 238:373-384.

Piatt, J., 1955, Regeneration of the spinal cord in the salamander, J. Exp. Zool, 129:177-207.

Pierce, S.T., Bishop, A.K. and Thompson, J.M., 1988, Developmental patterns of neurite outgrowth from chick embryo spinal cord and retinal neurons on laminin substrates, Dev. Brain Res, 40:213-222.

Prestige, M.C., 1965, Cell turnover in the spinal ganglia of Xenopus laevis tadpoles, J. Embryol. exp. Morphol., 13:63-72.

Puckula, E. and Windle, W.F., 1977, The possibility of structural and functional restitution after spinal cord injury. A review, Exp. Neurol, 55:1-42.

Rakic, P., 1988, Specification of cerebral cortical areas, Science, 241:170-175.

Rogers, S.L., Letourneau, P.C., Peterson, B.A., Furcht, L.T. and McCarthy, J.B., 1987, Selective interaction of peripheral and central nervous system cells to two distinct cell binding domains of fibronectin, J. Cell Biol., 105:1435-1442.

Silver, J. and Rutishauser, U., 1984, Guidance of optic axons in vivo by a preformed adhesive pathway on neuroepithelial endfeet, Dev. Biol., 106:485-499.

Sperry, R.W., 1944, Optic nerve regeneration with return of vision in anurans, J. Neurophysiol., 7:57-69.

Stephens, N. and Holder, N., 1985, A horseradish peroxidase study of motor neuron pools of the forelimb and hindlimb musculature of the axolotl, Proc. Roy. Soc. Lond. B., 224:325-339.

Stephens, N., Holder, N. and Maden, M., 1985, Motor neuron pools innervating muscles in vitamin A induced proximal-distal duplicate limbs in the axolotl, Proc. Roy. Soc. Lond. B, 224:341-354.

Stephens, N. and Holder, N., 1987a, Reformation of the pattern of neuromuscular connections in the regenerated axolotl hindlimb, Development, 99:221-230.

Stephens, N. and Holder, N., 1987b, The pattern of innervation in serially duplicated axolotl limbs: Further evidence for the existence of local pathway cues? Development, 100:479-487.

Tonge, D.A., Clarke, J.D.W and Hunter, K., 1988, Axolotl spinal cord in vitro supports axon regeneration, Neurosci. Lett. Sup., 32:S68.

Tosney, K.W. and Landmesser, L.T., 1985, Specificity of early motoneuron growth cone outgrowth in the chick embryo, J. Neurosci., 5:355-366.

Wigston, D.J., 1980, Supression of sprouted synapses in axolotl muscle by transplanted foreign nerves, J. Physiol., 307:355-366.

Wilson, S. and Holder, N., 1988, Evidence for axonal decision regions in the axolotl peripheral nervous system, Development, 102:823-836.

Wilson, S., Jesani, M. and Holder, N., 1988, Reformation of specific neuromuscular connections during axolotl limb regeneration: evidence that the first contacts are correct, Development, 103:365-377.

RESPONSE OF AMPUTATED <u>XENOPUS</u> <u>LAEVIS</u> FORELIMBS

TO AUGMENTATION OF THE NERVE SUPPLY

Catherine Tsilfidis and Richard A. Liversage

Ramsay Wright Zoological Laboratories
University of Toronto, Toronto
Ontario, Canada, M5S 1A1

SUMMARY

This study concerns the effects of nerve augmentation on
growth and development of forelimb regenerates in metamorphosed <u>Xenopus</u> <u>laevis</u> froglets. Augmentation was accomplished
by implanting spinal ganglia from sibling donors near the base
of the host regenerate. The results show that nerve augmentation increases the potential growth of regenerates and also
leads to perturbations of the normal cartilaginous outgrowth.
However, nerve augmentation does not appear to make a substantial contribution of the definitive pattern formation in the
regenerate.

INTRODUCTION

In the late tadpole and post-metamorphic stages <u>Xenopus</u>
<u>laevis</u> exhibits a heteromorphic forelimb regenerative out-
growth characterized by a cartilaginous spike and lack of
digit formation (Dent, 1962; Goode, 1967; Korneluk et al.,
1982; Korneluk and Liversage, 1984). Many investigations have
shown that nerves play an integral role in determining whether
regenerative ability is retained in amphibians. Singer et
al., (1967) demonstrated that <u>Triturus</u> <u>(Notophthalmus)</u> <u>viri-
descens</u> has more nerve fibres per amputation surface area than
<u>X. laevis</u> and <u>Rana</u> <u>pipiens</u>. However, when cros-sectioned area
was considered, Singer found that the axoplasmic area (trophic
factor availability) was significantly higher in <u>Xenopus</u> and
<u>Notophthalmus</u> than in non-regenerating <u>Rana</u>. Scadding (1982)
found that in the eight anuran species studied, regenerative
ability of the forelimb was directly correlated with nerve
fibre number and cross-sectional area. Singer (1954) obtained
a regenerative response in several non-regenerating species of
<u>Rana</u> when he augmented the nerve supply by deviating the
sciatic nerve into the ipsilateral forelimb.   Konieczna-
Marczynska and Skowron-Cendrzak (1958) also studied   the
effects of nerve augmentation in hindlimb regeneration of
metamorphosed <u>X. laevis</u>. They deviated the sciatic nerve
contralaterally into the right hindlimb followed by limb ampu-

tation;   only two experimental cases showed improved regenerates.

In the current investigation, we examined the effects of
augmenting the normal nerve supply of Xenopus forelimb regenerates by implanting spinal ganglia from siblings. This was
an attempt to stimulate further the growth potential of the
regenerative outgrowth.

MATERIALS AND METHODS

Xenopus laevis froglets were obtained by induced mating
of proven breeders from NASCO (Wisconsin). Froglets selected
were 4-5 months post-metamorphosis (snout-vent length 2.4-3.9
cm). Animals were kept in running dechlorinated tap water at
$22\pm1°C$ and fed chopped beef heart once weekly. Of the 112
sibling froglets used, 80 served as hosts and 32 as donors of
spinal ganglia and muscle. Our four groups contained 20 animals each; controls (no ganglion implant), muscle (sham control), single ganglion, and double ganglion implants. Frogs
were anesthetized in 0.05% MS-222 w/v aqueous solution
adjusted to pH 7.0 with 1.6g/l $NaHCO_3$. Right forelimb amputations were made through the mid-zeugopodium. One week postamputation, ganglia or muscle was implanted subcutaneously near
the base of the regenerate (see also Kamrin and Singer, 1959;
Globus and Liversage, 1975). A 2mm incision was made at the
base of the upper arm; the implant was inserted subcutaneously and pushed to within 2-3 mm of the amputation plane.
The second and third spinal ganglia (see Liversage et al.,
1987) from donour siblings were implanted. Latissimus dorsi
muscle implants were obtained from sibling donors. Muscle
tissue the size of a double ganglion implant was used.
Implants were allowed one week to adhere and become vascularized. Beginning at this time (time 0), limb regeneration was
monitored for up to 8 weeks using camera lucida drawings.
Limbs were sampled at time 0 and 1, 2, 3, 4 and 8 weeks. Then
host frogs were given 2 mm x 2 mm donour sibling skin grafts
which were placed in the lower back to prevent removal by the
host. After grafting, froglets were kept immobilized in 0.02%
MS222 for two hours to allow grafts to adhere. Graft
responses were monitored visually and bi-weekly for 4 weeks
for any reduction in size. Limbs were fixed in G-Bouins,
decalcified in Jenkins, stained in 1% Victoria Blue-B (BV-B)
and cleared in methyl salicylate. After recording the VB-B
stained cartilage patterns, limbs were embedded and sectioned
at 8 μm. Alternate sections were stained with hematoxylin and
orange-G-eosin (H and OG-E) or silver stained (Samuel, 1953).
Camera lucida drawings of limbs were analyzed using a planimeter. Areas obtained were divided by the square of the snout-
vent length of the froglet to correct for size differences
between animals.

RESULTS

Within two weeks donour sibling muscle implants were
resorbed. However, implanted donour ganglia adhere, become
vascularized and send fibres into the regenerate. Regardless
of ganglion orientation, nerve fibres re-route themselves and
grow towards/into the regenerate. Ganglia do not always

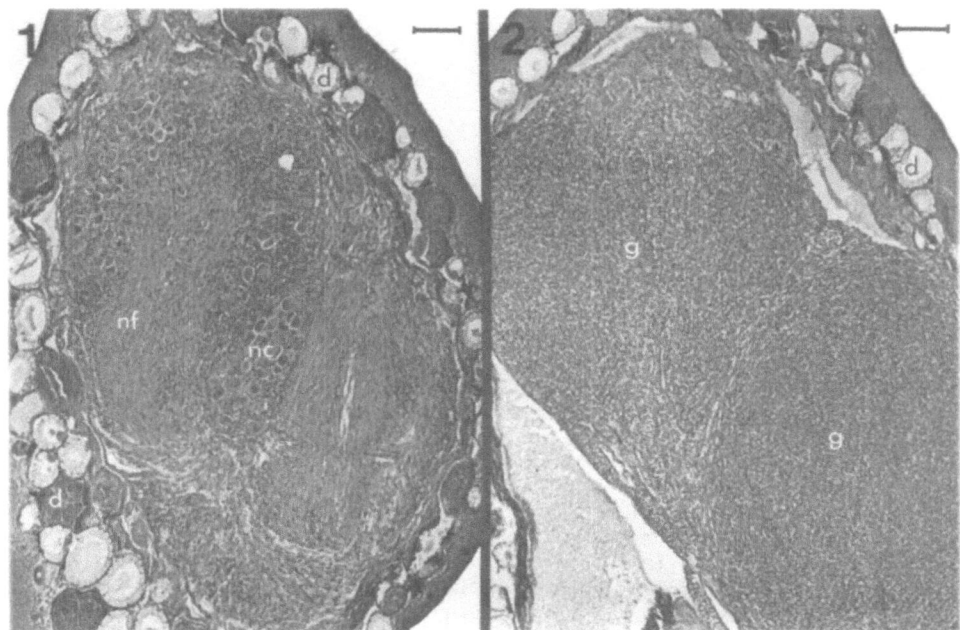

Fig. 1. Double ganglion implant at 8 weeks from time 0
         (9 weeks postimplantation). The cytoplasm of
         neuronal cell bodies (nc) is strongly stained
         with H and OG-E and the cellular and nuclear
         boundaries are clearly evident. Note: nerve
         fibre (nf) bundles originating from the neur-
         onal cell bodies, and large concentration of
         dermal glands (d) surroundings the implant.
         Bar represents 0.1 mm.

Fig. 2. Double ganglion implant at 4 weeks from time
         0. Implants (g) show later stages of rejec-
         tion. Cells are compacted and no neuronal cell
         bodies or fibres are evident. Dermal glands
         (d) are apparent. H and OG-E stained. Bar rep-
         resents 0.1 mm.

[Note: Long arrows indicate amputation level, short arrows
point out extent of dermal migration into the regenerate in
all photomicrographs].

remain at the site of implantation, they frequently become
rearranged and are found in the stylopodial region. Host
acceptance or rejection of ganglion and muscle implants was
determined by three criteria; visually, using skin grafts, and
histologically. A primary immune response was indicative of
first contact with a foreign antigen resulting in a delayed
rejection response. Skin graft depigmentation was not evident
until 2 to 3 weeks after grafting. Priming by an antigen from
the first graft led to a rapid rejection (secondary response)
of skin grafts. This was characterized by depigmentation
within 7-10 days. Muscle implanted limbs always reacted with
a primary response to their skin grafts. In ganglionated
cases, a size reduction of the graft resulted in a secondary

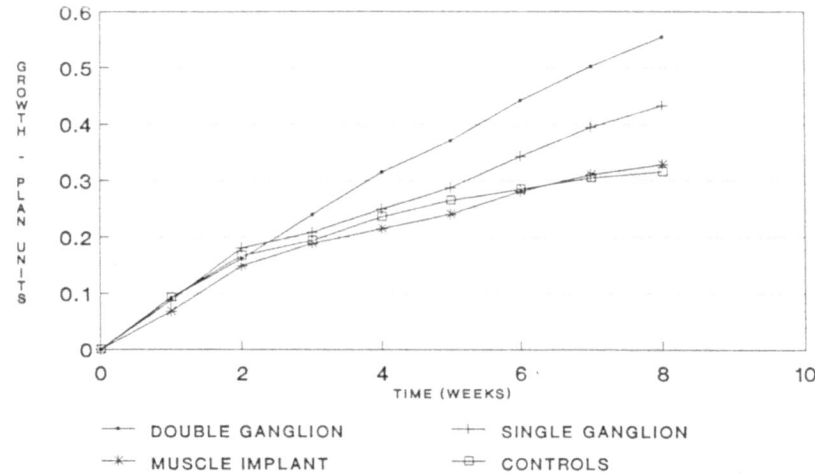

Fig. 3. Growth response of <u>Xenopus</u> forelimb regenerates
to nerve supply augmentation. Camera lucida
drawings were analyzed using a planimeter. Relative growth rates (in planimeter units) were
compared using Duncan's multiple range test.
Significant differences were found with single,
and particularly double ganglionated regenerates compared with controls.

response to the skin graft. Accepted ganglion grafts had distinctly different staining properties. The cytoplasm of neuronal cell bodies stained strongly with the silver technique
and also with OG-E (Fig. 1). Grafts in the early stages of
rejection were vascular in regions previously occupied by
neuronal cell bodies and were weakly stained. At later stages
of ganglion rejection, the cell bodies were not evident and
grafts had undergone cellular compaction (Fig. 2). Although
siblings were used, about 20-30% of ganglion implanted animals
experienced graft rejection. As a result, growth rates of
regenerates diminished and were comparable to controls. Frogs
experiencing graft rejection were not considered among our
histological or morphological results. Consequently, any
effects observed are attributed to the ganglion implants and
not to activation of the immune system. However, in determining effects on regenerate growth, all cases were considered.
Control and sham cases exhibited very similar results and are
considered jointly.

Analyses using Duncan's multiple range test showed significantly greater growth rates of double ganglionated regenerates from 3-8 weeks compared with control and sham cases;
and also greater growth rates of single ganglionated regenerates at six weeks compared with control (Fig. 3). Presumably,
by these times, implanted ganglia have sent out fibres into
the regenerate which contributed significantly to the level of
forelimb innervation. Furthermore, when our experiments were

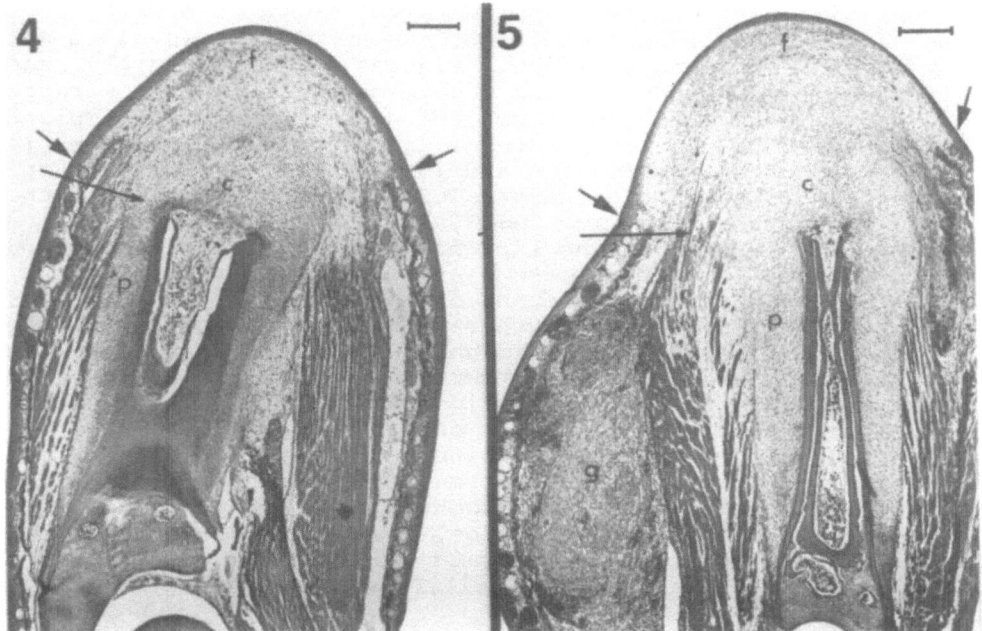

Fig. 4.   Time zero control regenerate. Thickening of
          the periosteum (p) in the stump region extends
          distally forming a small cartilage mass (c) in
          the regenerate. Note the small fibroblast (f)
          accumulation. The dermis has just begun to
          develop distally into the regenerate. H and
          OG-E stained. Bar represents 0.25 mm.

Fig. 5.   Time 0 double ganglionated regenerate. Gan-
          glion implant (g) is visible subcutaneously
          near the regenerate base. The limb character-
          istics are similar to those of the control
          (see Fig. 4 legend). H and OG-E stained. Bar
          represents 0.25 mm.

terminated, ganglionated regenerates displayed elevated growth
rates whereas in controls the rates had diminished considerably (Fig. 3).

        Control and experimental time zero regenerates exhibit
similar morphological and histological patterns. There is an
extensive thickening of periosteal cartilage along the length
of the radius-ulna diaphysis. This terminates distal to the
amputation surface of the bone in a cartilage mass. Further
distally and immediately subjacent to the wound epithelium is
a small accumulation of fibroblasts (see Figs. 4,5). In contrast, control regenerates contain a typical single continuous
cartilaginous rod from 4 to 8 weeks (Figs. 6,7). However,
considerable alteration of the cartilage pattern is apparent
in double ganglionated regenerates at 4 and 8 weeks (Figs.
8,9). These limbs have wider, more bulbous regenerates and
appear to have a diffuse cartilaginous matrix. In general,
the experimental limbs show perturbation of the typical rod-
like cartilage formations, they also contain loci of cartilage

condensation markedly interspersed with nerve fibres. Some muscle is also found distal to the amputation level in our control and experimental limbs. Figures 10-15 illustrate the difference between control and double ganglionated regenerates at 4 and 8 weeks. Fig. 16 shows the double ganglion implant corresponding to the regenerate of Fig. 15. Twenty percent (4 of 20) of double ganglionated and 10 percent (2 of 20) of single ganglionated regenerates exhibit a bifurcation of the distal regenerate tip (Fig. 12-13, 15, 17, 18). Bifurcation is only associated with the implantation of nervous tissue and is not evident in control cases. In cases with a bifurcation, implanted ganglia were found distal to the elbow joint and sufficiently close to the regenerate base that branching of the emerging nerve fibres did not occur until the developing nerve trunk(s) had insinuated themselves further distally into the regenerating tissues. Histological results further corroborate the growth rate data. In controls at 8 weeks, the dermis is nearly completely insinuated between the overlying epidermal and underlying mesodermal tissues, an indication that no further growth in the regenerate will ensue (Fig. 14). In double ganglionated regenerates, the dermis did not migrate to the distal tip of the regenerate (Fig. 15).

DISCUSSION

Siblings share a variety of histocompatibility antigens, therefore they are preferred donors in tissue transplantation experiments. However, unless siblings are mono-zygotic, differences will occur resulting in rejection responses. As our muscle implanted hosts responded with a primary response to skin grafts, either the implants had degenerated before an immune response could be initiated, or muscle and skin do not share any histocompatibility antigens. Thus, the host reacted with a primary response to each graft. In most ganglion implant cases, reduction of the original implant corresponded with a secondary response to the skin graft.

Previous studies show that the level of innervation in an amphibian forelimb determines its regenerative ability (Van Stone, 1964; Singer et al., 1967; Geraudie and Singer, 1977; Scadding, 1982). The regenerative capabilities of amphibians as well as other vertebrates have been improved by augmenting the normal nerve supply (Singer, 1954; Simpson, 1961; Mizell, 1968; Fowler and Sisken, 1982). In most cases, the results have been due to enhanced tissue regeneration. As well, in X. laevis forelimbs nerve (spinal ganglion) augmentation significantly increases the growth of a regenerate. Presumably, these regenerates retain a further potential for growth after the eight week period. This is supported by the slopes of our growth curves and the lack of dermal migration into the distal tip of the regenerate. An increased nerve supply also results in morphological changes. In some cases, there was a bifurcation of the distal tip of the regenerate yielding a "more normal" forelimb. We interpret this as disruption of regeneration of a continuous cartilaginous rod by the augmented nerve supply. This results in an irregular, diffuse mass of cartilage characterized by several loci of cartilage condensation. We also observed regenerated muscle fibres, presumably as a result of "sprouting of muscle" (Goss, 1983), in the proximal region of the ganglionated regenerates and also in controls.

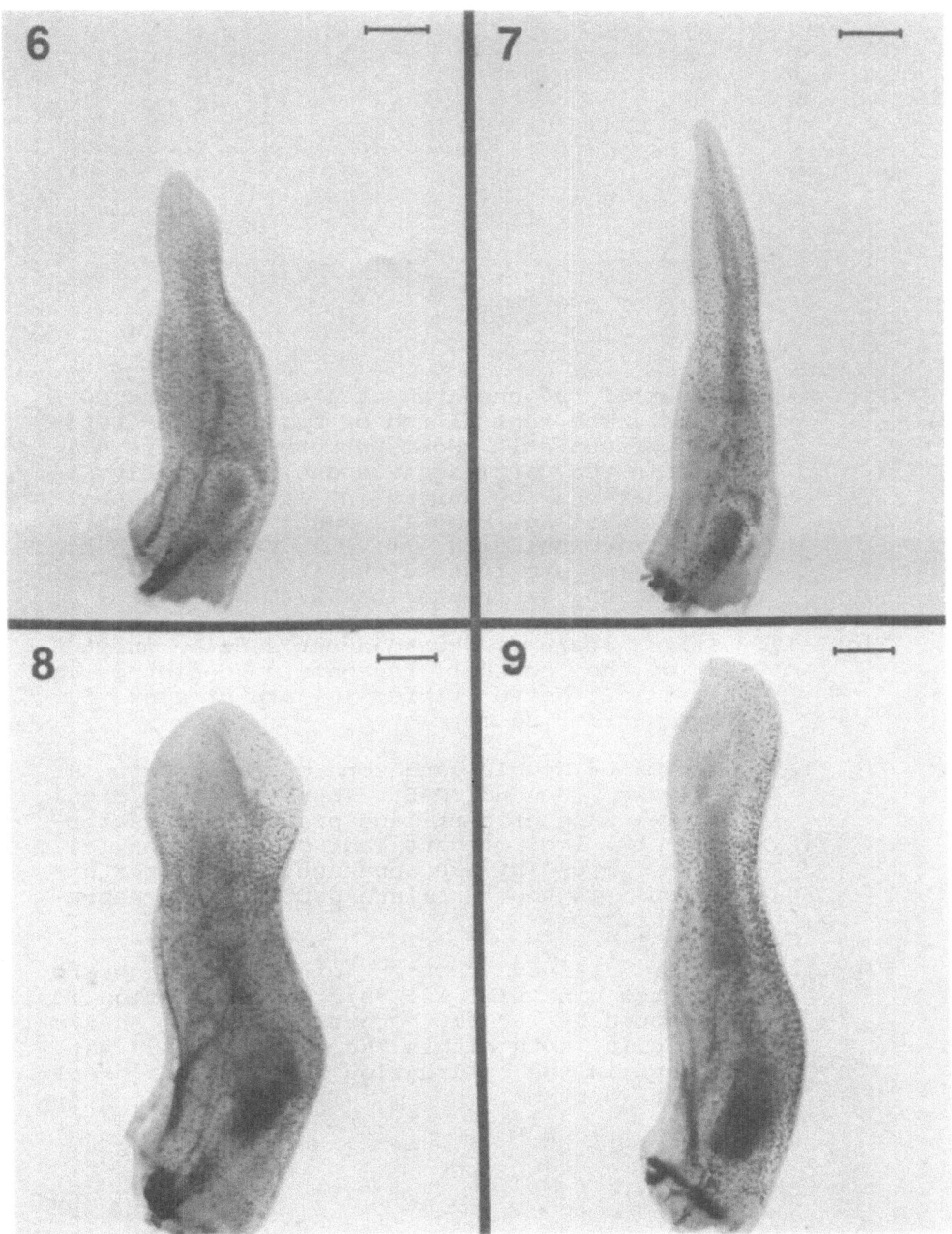

Fig. 6-9. Control regenerates at 4 (Fig. 6) and 8 (Fig. 7) weeks. VB-B stains the periosteum of the radius-ulna in the stump and the congiguous rod-like cartilage regenerate. Double ganglionated regenerates at 4 (Fig. 8) and 8 (Fig. 9) weeks. Regenerates are more bulbous compared with controls and exhibit atypical patterns of regenerating cartilage. Bar = 0.7 mm.

Fig. 10.   Control regenerate at 4 weeks (H and OG-E
            stain). The central rod of cartilage is typi-
            cal of controls. Some perturbations are evi-
            dent in the cartilage rod due, presumably, to
            insinuations by nerves (n) into the regener-
            ating cartilage mass. A small distal fibro-
            blast accumulation (f) is present. Muscle
            fibres (m) are found distal to the level of
            amputation. Bar represents 0.25 mm.

Fig. 11.   Silver stain of the adjacent section (next 8
            µm) of the control regenerate depicted in
            Fig. 10. Nerve fibres (n) are present. Bar
            represents 0.25 mm.

Fig. 12.   Bifurcated double ganglionated regenerate at
            4 weeks (H and OG-E). There is an atypical
            central mass of cartilage present character-
            ized by loci of cartilage condensation (c).
            Nerve fibres (n) are abundant as are pertur-
            bations in the cartilage pattern. Bar repre-
            sents 0.3 mm.

Fig. 13.   Silver stained section of the regenerate
            depicted in Fig. 12. This serial section is
            distanced by 160 µm. Note the numerous nerve
            fibres (n) both within the cartilaginous mat-
            rix and in the bifurcation regions. Bar rep-
            resents 0.25 mm.

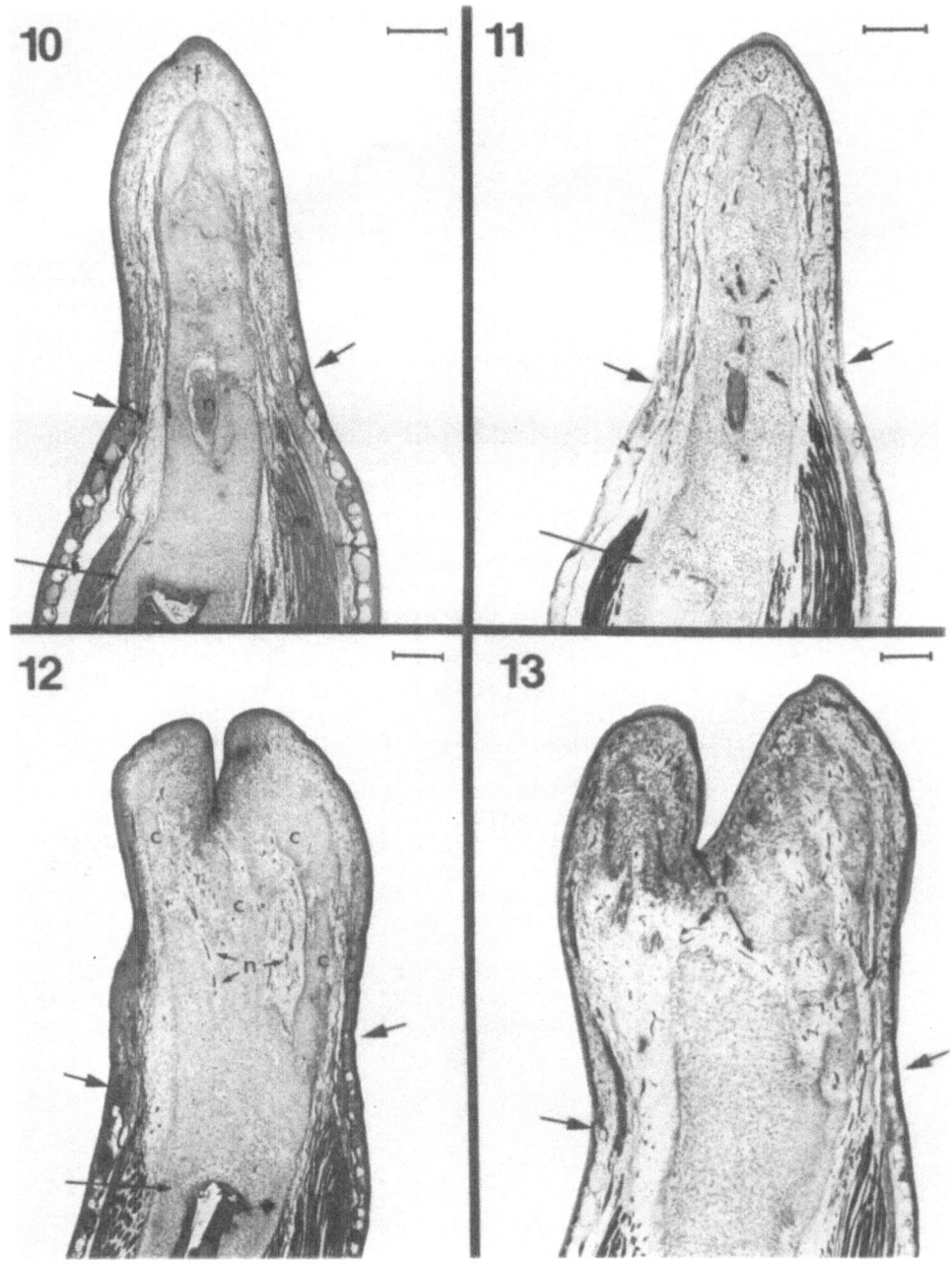

Fig. 14. Control regenerate at 8 weeks (H and OG-E stain). The regenerate has a single continuous cartilage rod (c) typical of controls. Dermis (d) has migrated over the distal regenerate tip indicative of the termination of growth. Bar represents 0.75 mm.

Fig. 15. Silver stained section of a double ganglionated regenerate at 8 weeks. The cartilaginous rod (c) is atypical, possessing a bifurcation distally. Note presence of nervous tissue (n) near the base of the bifurcation. Dermal glands are not evident in the distal regenerate tips which indicates a further potential for growth. Bar represents 0.75 mm.

Fig. 16. Silver stain of the double ganglion implant present in the regenerate in Fig. 15. Neuronal cell bodies (nc) are evident. Nerve fibres (nf) are also visible. Bar represents 0.2 mm.

Fig. 17. Single ganglionated regenerate at 2 weeks (H and OG-E). Implant (g) has moved proximally from the site of implantation and is located subcutaneously near the elbow joint. Note bifurcation of the distal tip of the regenerate. Bar represents 0.5 mm.

Fig. 18. Higher magnification of the regenerate in Fig. 17. Note prominent nerve trunk (n) near the bifurcation base. Bar represents 0.2 mm.

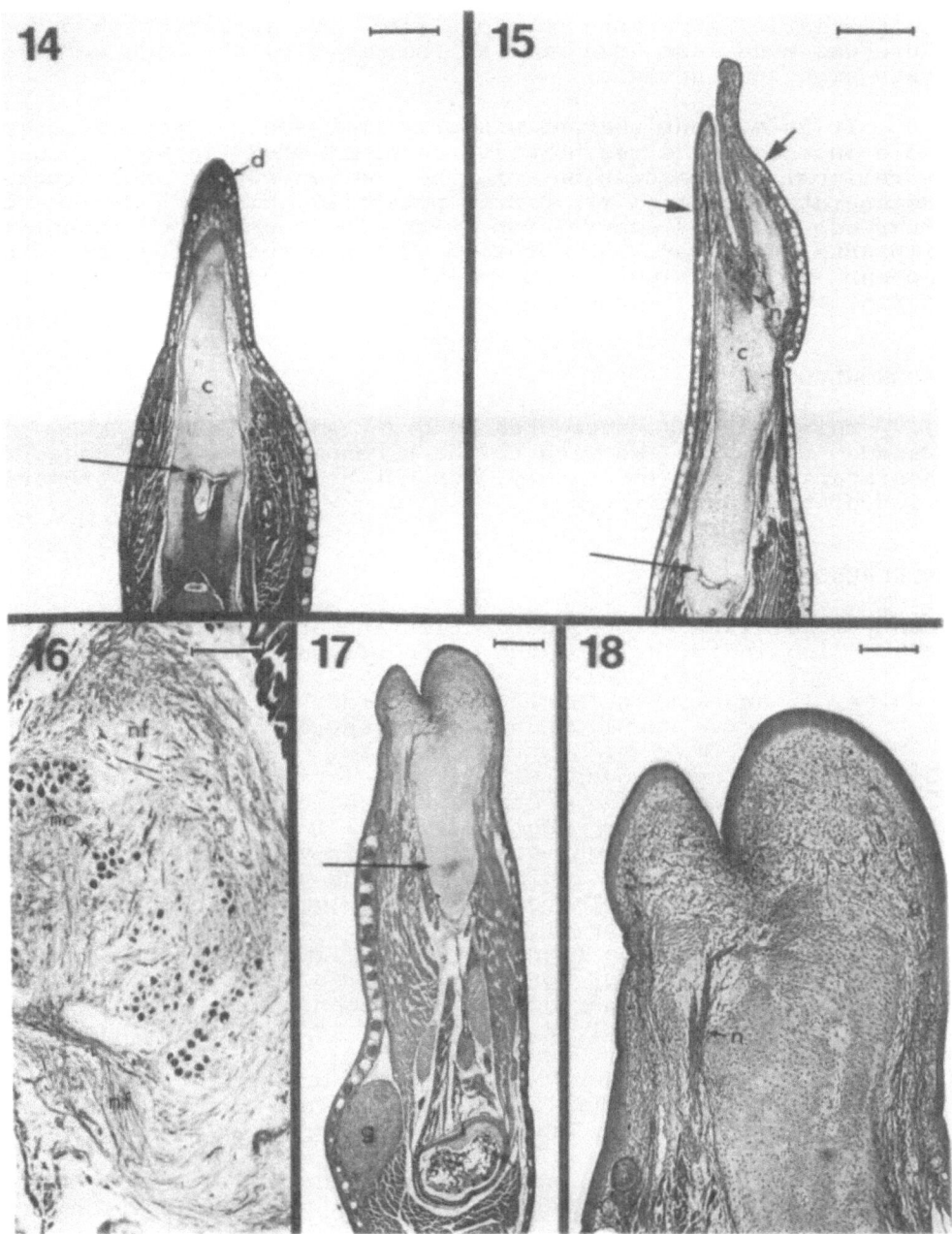

Konieczna-Marczynska and Skowron-Cendrzak (1958) observed a delay in the regeneration rate of their nerve augmented _Xenopus_ hindlimbs compared with controls, and attributed this to blood loss and extensive tissue destruction. Singer (1954) also had a high fatality rate when he augmented the nerve supply in several anuran species by ipsilateral deviation of the left sciatic nerve into the forelimb. Our implantation procedure was much less traumatic as indicated by the high survival rate among our animals.

It is evident that nerves play an essential proliferative role in the tissue regenerative response of _X. laevis_ froglet forelimbs. Augmentation of the nerve supply in a froglet regenerate increases its growth potential, and the degree of increase can be correlated with the number of ganglion implants. However, augmentation of the nerve supply does not appear to contribute substantially to the definitive pattern formation of the regenerate.

ACKNOWLEDGMENTS

This study was supported by grant A-1208 from NSERC of Canada to R.A.L. We wish to thank Donna R. Wheeler, Research Assistant in this laboratory, for her help with the preparation of the manuscript.

REFERENCES

Dent, J.N., 1962, Limb regeneration in larvae and metamorphosing individuals of the South African clawed toad, _J. Morph._, 110:61-77.

Fowler, I. and Sisken, B.F., 1982, Effect of augmentation of the nerve supply upon limb regeneration in the chick embryo, _J. Exp. Zool._, 221:49-59.

Géraudie, J. and Singer, M., 1977, Relation between nerve fibre number and pectoral fin regeneration in the teleost, _J. Exp. Zool._, 199:1-8.

Globus, M. and Liversage, R.A., 1975, _In vitro_ studies of limb regeneration in adult _Diemictylus viridescens_: Neural dependence of blastema cells for growth and differentiation, _J. Embryol. exp. Morph._, 33:813-829.

Goode, R.P., 1967, The regeneration of limbs in adult anurans, _J. Embryol. exp. Morph._, 18:259-267.

Goss, R.J., 1983, Chondrogenesis in Regenerating Systems. _in_: "Cartilage", vol. 3, Biomedical Aspects. B.K. Hall, ed. Academic Press, New York.

Kamrin, A.A., and Singer, M., 1959, The growth influence of spinal ganglia implanted into the denervated forelimb regenerate of the newt, _Triturus_, _J. Morph._, 104:415-440.

Konieczna-Marczynska, B. and Skowron-Cendrzak, A., 1958, The effect of the augmented supply on the regeneration in postmetamorphic _Xenopus laevis_, _Folia biol._, 6:37-46.

Korneluk, R.G., Anderson, M.J., and Liversage, R.A., 1982, Stage dependency of forelimb regeneration on nerves in postmetamorphic froglets of _Xenopus laevis_, _J. Exp. Zool._, 220:331-342.

Korneluk, R.G., and Liversage, R.A., 1984, Tissue regeneration in the amputed forelimb of <u>Xenopus</u> <u>laevis</u> froglets, <u>Can. J. Zool.</u>, 62:2383-2391.

Liversage, R.A., Anderson, M.J., and Korneluk, R.G., 1987, Regenerative response of amputated forelimbs of <u>Xenopus</u> <u>laevis</u> froglets to partial denervation, <u>J. Morph.</u>, 191:131-144.

Mizell, M., 1968, Limb regeneration: Induction in the newborn opossum, <u>Science</u>, 161:283-285.

Samuel, E.P., 1953, Towards controllabe silver staining, <u>Anat. Rec.</u>, 116:511-520.

Scadding, S.R., 1982, Can differences in limb regeneration ability between amphibian species be explained by differences in quantity of innervation? <u>J. Exp. Zool</u>, 219:81-85.

Simpson, S.B., 1961, Induction of limb regeneration in the lizard, <u>Lygosoma</u> <u>laterale</u>, by augmentation of the nerve supply, <u>Proc. Soc. Exp. Biol. and Med.</u>, 107:108-111.

Singer, M., 1954, Induction of regeneration of the forelimb of the post-metamorphic forg by augmentation of the nerve supply, <u>J. Exp. Zool.</u>, 126:419-472.

Singer, M., Rzehak, K. and Maier, C., 1967, The relation between the caliber of the axon and the trophic activity of nerves in limb regeneation, <u>J. Exp. Zool.</u>, 166:89-97.

Van Stone, J.M., 1964, The relationship of nerve number to regenerative capacity in the developing hind limb of <u>Rana</u> <u>sylvatica</u>, <u>J. Exp. Zool.</u>, 155:293-302.

NEURAL INFLUENCE ON THE EXTRACELLULAR

MATRIX DURING BLASTEMA FORMATION

Anthony L. Mescher[1] and Cheryl A. Cox[2]

[1]Medical Sciences Program, Indiana University
School of Medicine, Bloomington, IN 47405 USA
[2]Department of Biochemistry, University of
Birmingham, Birmingham, B15 2TT, U.K.

SUMMARY

By investigating reinnervation of denervated/amputated limbs of urodele (Ambystoma mexicanum) larvae, it is shown that nerves affect hyaluronate accumulation in the extracellular matrix as well as mitotic activity in cells of distal, dedifferentiated regions of the limb stump. Accumulation of hyaluronate is increased two-fold and mitotic activity is stimulated four-fold over that seen immediately prior to the reappearance of nerve fibers in the distal areas. Formation of a hyaluronate-rich matrix is accompanied by enlargement of the intercellular space and a decrease in cell density, which occur simultaneously with the appearance of a morphologically recognizable blastema on the tip of the limb stump. The increase in proliferative activity includes cells which entered the cell cycle in response to the amputation injury and incorporated ^{3}H-thymidine five days later. These results indicate that nerves promote blastema formation and growth both by stimulation cell division and by enhancing production of a large extracellular component important for expansion of intercellular space and tissue volume. Work by others with cultured mesenchymal cells and purified growth factors suggests that such factors stimulate both cell cycling and hyaluronate production as part of a single mitogenic effect.

INTRODUCTION

Occurring in the aftermath of amputation, the dedifferentiation phase of amphibian limb regeneration drastically modified physiological conditions and intercellular relationships within tissues of the limb. These changes set the stage for developmental events which not only repair the injured tissues, but continue through complete epimorphic replacement of lost structures. Although the molecular basis for the dedifferentiative changes has not been thoroughly investigated, histological studies have described many aspects of the process (see Wallace, 1981, for review). Such analyses have led to the general concept that removal of cell-associated material unique to specialized tissues results in the release of embryonic-like cells proliferating as populations that reform the missing tissues of the limbs (Wallace, 1981).

The extracellular matrix near the site of amputation is completely remodeled during dedifferentiation and several important aspects of this process have been demonstrated with studies of newt limb regeneration. Immunocytological studies reveal that laminin is completely removed as tissues with basement membranes are affected (Gulati et al., 1983). Fibronectin also disappears from these locations, but is apparently resynthesized by the undifferentiated cells and accumulates evenly in the interstitium of the early blastema (Repesh et al., 1982; Gulati et al.,1983). Collagenolytic enzyme(s), which degrade the major extracellular component of intact limbs, appear during dedifferentiation and become maximally active in the blastema (Grillo et al., 1968).

While these extracellular components of tissues are being broken down, the cells begin production of hyaluronate, a principal constituent of embryonic mesenchyme (Toole and Gross, 1971; Mescher and Munaim, 1986). This very large, polyanionic glycosaminoglycan (GAG) accumulates around dedifferentiated cells of the newt limb stump during the preblastemic period and can be demonstrated histochemically (Mescher and Munaim, 1986). Denervation at the time of amputation, which prevents regeneration, reduces hyaluronate production by approximately 50% (Smith et al., 1975; Mescher and Munaim, 1986). Conversely, dedifferentiating limb stump tissues placed in organ culture with dorsal root ganglia show twice as much hyaluronate synthesis as controls cultured in the absence of nerves (Mescher and Munaim, 1986).

Since matrices rich in hyaluronate are frequently associated with rapid cell proliferation,cell migration, and tissue expansion (Toole, 1981), all of which are important for the early phase of limb regeneration, the neural effect on formation of such a matrix may have considerable influence on the course of events during establishment of the blastema. We have examined the neural effect on hyaluronate production in denervated-amputated limbs of <u>Ambystoma</u> larvae, where reinnervation of the dedifferentiated tissue leads to rapidly increased mitotic activity and blastema formation (Olsen and Tassava, 1984). Microspectrophotometric examination of hyaluronate accumulation and determination of cell cycle activity indicate that these processes are stimulated concomitantly by reinnervation (Mescher and Cox, 1988).

MATERIALS AND METHODS

Larval <u>Ambystoma mexicanum</u>, 25 to 35 mm in length, were maintained at 18±1°C and fed newly hatched brine shrimp (<u>Artemia</u>) daily. The animals were anaesthetized in an aqueous solution of benzocaine (0.025%; Sigma) and forelimbs were amputated bilaterally, midway between the wrist and elbow. Left limbs were denervated at the time of amputation by transecting spinal nerves 3, 4 and 5 near the brachial plexus; right limbs remained innervated.

For determination of the hyaluronate content in limb stumps, three larvae were anaesthetized and fixed in Bouin's fluid containing 0.5% cetylpyridinium chloride (Sigma) each day from days 5 to 15 postamputation. Forelimbs were embedded in paraffin and sectioned longitudinally at 10 μm. The extent of reinnervation in each denervated limb was determined by staining sections from representative levels for nerve fibers by the Bodian method (Humason, 1979). Hyaluronate was measured microspectrophotometrically by the method of Derby (1978), which is based on the specific removal of this GAG from sections digested with <u>Streptomyces</u> hyaluronidase (SH). Deparaffinized sections were covered individually with SH (100 units/ml; Sigma) in 0.1 M acetate buffer, pH 4.8, and incubated at 37°C in a humid chamber for 3 h. Adjacent sections were treated similarly with buffer alone. Rinsed slides were then stained with 1% alcian blue 8GX (Sigma) in 3% acetic acid containing 0.025 M $MgCl_2$ and mounted. The concentrations of alcian blue in distal dedifferentiation tissue of the limb stumps were determined by measuring the optical densities in selected areas of sections with a Nikon P1 microspectrophotometer. Details of the method are given elsewhere (Mescher and Cox, 1988). The relative concentration of hyaluronate was obtained by comparing the mean absorption values of buffer- and SH-treated adjacent serial sections. The SH-induced decrease in alcian blue staining was analyzed statistically with Student's t-test and expressed as a percentage of stain present in the same region of the buffer-treated sections.

Another group of amputated-denervated larvae was used to examine mitotic activity during reinnervation of the denervated limb stumps. Five days after the operations, each animal was injected intraperitoneally with 0.25 uCi ^{3}H-methylthymidine. Three hours later and at 24 h intervals through day 15, two larvae were fixed in Bouin's fluid. Forelimbs were processed and reinnervation was examined as described above. Other sections from each limb were processed for radioautography and stained with hematoxylin and eosin. Labeled, unlabeled and overall mitotic indices were calculated from a total of 1000 to 4500 dedifferentiated or blastemal cells counted in representative sections of each limb.

RESULTS

The relative concentration of hyaluronate, represented by hyaluronidase-sensitive alcianophilic GAG, was approximately constant at 25-30% in the extracellular matrix (ECM) of innervated blastemata from days 5 through 15 postamputation, which included early bud stage blastemata through the onset of redifferentation. Denervated limb stumps contained similar concentrations of hyaluronate on days 5 and 6 postamputation, but on day 8 this percentage had decreased to 15% (Fig. 1A). By this time dedifferentiated cells in distal regions of denervated stumps were present at increased density around the ends of skeletal elements, a feature Olsen and Tassava (1984) termed a "pseudoblastema". Nerve staining confirmed the report of these authors that regenerating axons reached the distal tips of such limb stumps by day 9 or 10. Beginning with the return of nerve fibers, there was a two-fold increase in the percentage of SH-sensitive GAG distally in the stumps. As the concentration of hyaluronate in distal regions increased, pseudoblastemata were no longer observed and the limbs rapidly formed medium bud stage blastemata.

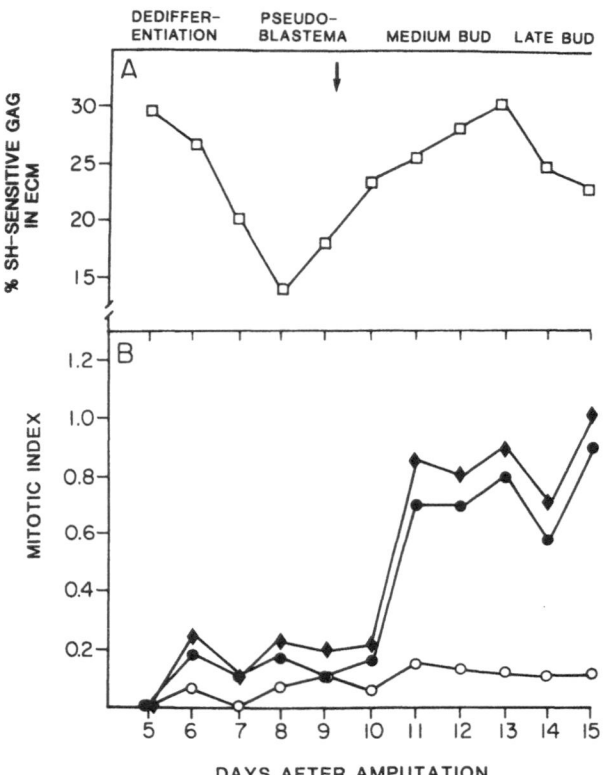

Fig. 1.   (A) Percentage of SH-sensitive staining material in denervated/reinnervated larval limb stumps at different times after amputation. Alcianophilic material was measured by microspectrophotometry in equivalent regions of adjacent sections treated with buffer or Streptomyces hyaluronidase (SH). Each point represents the percentage by which staining was decreased by enzyme digestion as described in Materials and Methods. For each case the difference between the mean absorptions of buffer-and SH-treated sections was significant (P<0.05). Arrow indicates approximate time of reinnervation. (B). Total mitotic index ($\blacklozenge$), mitotic index of ^{3}H-thymidine labeled cells ($\bullet$) and mitotic index of unlabeled cells (o) for denervated/ reinnervated limb stumps. Each point is the average of determinations from two limbs.

Examination of sections processed by radioautography showed that little proliferative activity occurred in the denervated limb stumps prior to reinnervation (Fig. 1B). Low numbers of mitotic figures were observed in both unlabeled cells and cells labeled by ^{3}H-thymidine on day 5. The total mitotic index seen each day from days 5 through 10 was similar to that of innervated, regenerating limbs during the early phase of dedifferentiation (Kelly and Tassava, 1973; Tassava et al., 1974; Maden, 1978).

Reinnervation of distal limb stump regions stimulated a rapid rise in proliferative activity (Fig. 1B). The total mitotic index on days 11 through 15 was approximately four times that of day 10 (Fig. 1B) and was similar to that of contralateral control limb stumps at the early and medium bud stages of regeneration (not shown). The increase in mitotic activity was seen two days after the apparent onset of hyaluronate accumulation in the reinnervated tissues (Fig. 1A). Unlike the observation of Olsen et al.(1984), the increased proliferation seen here primarily involved the labeled cell population (Fig. 1B). It is possible that some cells labeled here on day 5 degenerated during the extended period of dedifferentiation, releasing ^{3}H-thymidine which was then available to label cells during reinnervation and thereby augmenting the labeled mitotic index.

DISCUSSION

Removal of extracellular components from limb stump tissues is of major importance for the onset of regenerative growth. By this process cells are released from differentiated tissues and surrounded by new microenvironments in which they can proliferate and give rise to the cellular patterns that lead to replacement of amputated structures (Schmidt, 1968). Specific molecular or subcellular changes in the extracellular matrix that occur during dedifferentiation have not been investigated extensively, but certain general aspects of the process are apparent. Basement membranes of blood vessels, neural tissues, muscle fibers and epidermis are digested, as indicated by the loss of laminin from these structures (Gulati et al., 1983). Fibronectin is also removed from basement membranes and connective tissues early in dedifferentiation (Repesh et al., 1982; Gulati et al., 1983). The collagen content drops from approximately 25% of the total limb protein at amputation to less than 5% in cone-stage blastemata, after which there is a net accumulation of newly synthesized collagen during redifferentiation (Mailman and Dresden, 1976).

Neither the sources nor the biochemical characteristics of the degrading enzymes that carry out the early, catabolic phase of limb regeneration have been well-studied. It is known however that the collagenolytic activity in newt limbs is maximal during the third week after amputation in both regenerating limbs and denervated, nonregenerating limbs (Mailman and Dresden, 1979) and is optimal at neutral pH (Grillo et al., 1968). In mammals, both macrophages (see Werb, 1983, for review) and neutrophils (Senior and Campbell, 1983) secrete neutral proteinases which have specificities for all types of collagen, elastin, proteoglycans, and specific extracellular glycoproteins such as laminin and fibronectin. It is likely that similar enzymes from neutrophils and macrophages are responsible for the catabolic changes in the extracellular matrix during limb regeneration. Histological studies have shown that both neutrophils and macrophages are numerous within injured tissues during the dedifferentiative phase of newt limb regeneration, with increased numbers of macrophages persisting into the period of blastemal growth (Liebman, 1949; Schmidt, 1968).

Before the histolytic phase of limb regeneration is complete, dedifferentiated cells enter the cell cycle (Hay and Fischman, 1961) and simultaneously begin to produce extracellular matrix components such as fibronectin (Repesh et al., 1982; Gulati et al., 1983) and hyaluronate (Toole and Gross, 1971; Mescher and Munaim, 1986). Similar increase in hyaluronate synthesis upon mitogenic stimulation occur with cultured cells. Serum and several purified growth factors have been shown to stimulate hyaluronate production as well as mitotic activity in quiescent cells (e.g., Moscatelli and Rubin, 1975; Lembach, 1976; Toole, 1981; Hamerman and Wood, 1984; Hamerman et al., 1986). By manipulating the growth rate of cultured fibroblasts in various ways, Matuoka et al. (1987) found a tight association between the rate of proliferation and synthesis of hyaluronate. Evidence reviewed by these authors, from studies with a wide variety of cells and mitogens, suggests that hyaluronate production is an integral part of the proliferative response in mesenchymal cells.

The mitogens that stimulate cell-cycle re-entry and hyaluronate synthesis during the dedifferentiative response to amputation in amphibian limbs have not been identified. However, results from the much more thoroughly investigated process of mammalian tissue repair indicate the likely nature of the factors involved (Hunt and Van Winkle, 1979). These studies have revealed a principal role for the macrophage in regulating and promoting proliferative aspects of a tissue's response to injury (see Hunt and Van Winkle, 1979; Ross, 1980 for reviews). Macrophages arriving at the site of injury synthesize and release several polypeptide mitogens for endothelial and connective tissue cells (Rappollee et al., 1988). Macrophages have also been shown to secrete growth factors specific for granulocytes and monocytes as well as angiogenic factors and various chemotactic factors (reviewed by Werb, 1983; Rappollee et al., 1988). Although agents secreted by macrophages appear to be of paramount importance, various proteins and low molecular weight factors affecting the growth response in injured tissue may also be released from platelets, from other leucocytes, and from damaged cells (Ross, 1980; Boucek, 1984).

The close association between the proliferative response and hyaluronate production during dedifferentiation is indicated by the observation that neither activity is sustained if the amputated limb is denervated. Using larval <u>Ambystoma</u> limb stumps, Maden (1978) showed that cell cycling activity elicited by the injury begins to decline approximately 5 days postamputation in denervated limbs. The present study suggests that the concentration of hyaluronate in denervated larval limb stumps, though elevated initially in response to amputation, also decreases after 5 or 6 days. Comparable results indicating the importance of the neural effect for the maintenance of both hyaluronate synthesis (Smith et al., 1975; Mescher and Munaim, 1986) and mitotic activity (Singer and Craven, 1948; Mescher and Tassava, 1975) have been obtained from regenerating adult newt forelimbs.

Unlike newt limbs, amputated and denervated larval limbs begin to regenerate when axons grow back to the distal dedifferentiated region of the stumps (Butler and Schotte, 1941). Olsen et al. (1984) quantified proliferative activity of dedifferentiated cells in such limb stumps and found a sharp increase in the mitotic index following distal reinnervation, which was shown to occur approximately 10 days postamputation. Evidence was also presented that reinnervation brought about the "rescue" and renewal of cycling in cells which had entered the cycle several days earlier as a result of amputation but failed to maintain mitotic activity in the absence of nerve's trophic effect (Olsen et al., 1984; Barger and Tassava, 1985).

Data presented here confirm that the mitotic activity stimulated by regenerating axons includes cells which entered the cell cycle as a result of mitogenic influences released by the amputation injury and which incorporated ^{3}H-thymidine given on day 5 postamputation. Moreover, the renewal of cell proliferation is shown to be accompanied by production and accumulation of hyaluronate in the reinnervated regions of the stump. The effect of reinnervation is similar to that found in vitro, when newt blastemata denervated by explantation, are cultured with dorsal root ganglia. In this system, nerves stimulate a twofold increase in both mitotic activity (Globus and Vethamany-Globus, 1977) and hyaluronate synthesis (Mescher and Munaim, 1986) in the blastema cells. The polypeptide mitogen, Fibroblast Growth Factor, has been shown to stimulate threefold increases in both DNA (Mescher and Loh, 1981) and hyaluronate synthesis (Mescher and Munaim, 1986) in cultured newt forelimb blastemata. All these results are consistent with work discussed above indicating that hyaluronate synthesis and mesenchymal cell proliferation are coordinately regulated (e.g., Matuoka et al., 1987).

The observations that the growth-promoting influence of nerves on dedifferentiated cells in the limb stump involves enhanced hyaluronate production as well as mitotic activity add a new dimension to our understanding of the neural effect on blastemal growth. Individual chains of this GAG can have very high molecular weights and, being hydrophilic and polyanionic, attract large amounts of water. Upon interaction with water, molecules of hyaluronate occupy large domains and even at very low concentrations form hydrated gels which exert considerable swelling pressure in the extracellular matrix (Roden, 1980). In several developing systems localized production of hyaluronate has been correlated with tissue expansion and shape changes in organ rudiments (reviewed by Toole et al., 1984; Mescher and Cox, 1988). This indicates that neural stimulation of hyaluronate synthesis in the amphibian limb stump is likely to lead to expansion of the extracellular space in the dedifferentiated tissue and to the swollen, bud-like morphology of the early blastema. In the system studied here, the increase in hyaluronate concentration which follows distal reinnervation is accompanied by expansion of the limb tips (Olsen and Tassava, 1984) and by significantly decreased cell density in the dedifferentiated tissue (Mescher and Cox, 1988). The observations reviewed here suggest that the neural effect on cell cycling during the early phase of limb regeneration not only produces an increased number of cells, but also causes a concurrent expansion of the space between mesenchymal cells by stimulating accumulation of hyaluronate.

In addition to its effect on tissue volume, the presence of hyaluronate in the extracellular matrix of the blastema may influence various other aspects of the regeneration process as shown by its biological properties in other systems. For example, accumulations of hyaluronate have been found to accompany onset of cellular migration in many developmental processes (reviewed by Toole et al., 1984). Hyaluronate produced during dedifferentiation in amphibian limb stumps may facilitate the directed migration of fibroblasts from the dermis (Gardiner et al., 1986) and chondrogenic cells from the periosteum (Neufeld, 1985; Mescher and Munaim, 1986) which occur during blastema formation. It is also known that diffusion of solutes, especially large molecules, through the extracellular matrix is sterically hindered by the presence of hyaluronate and other GAGs (reviewed by Hruza, 1977; Meyer, 1986), suggesting that local concentrations of this component will affect delivery of nutrients and other factors to blastema cells.

Experiments with primary embryonic cell cultures show that hyaluronate in the cellular microenvironment can influence the onset of differentiation in myogenic and chondrogenic cells (reviewed by Caplan, 1986). Moreover, local accumulations of hyaluronate in the chick limb bud are apparently responsible for maintaining specific areas of avascularity during angiogenesis (reviewed by Solursh, 1984; Caplan, 1985). Growth of capillaries into the regeneration blastema may be affected by accumulation and removal of this extracellular material. Peadon and Singer (1966) analyzed the process of neovascularization during regeneration of adult newt forelimbs and showed that growing vessels do not begin to penetrate proximal regions of the blastema until the advanced early bud stage approximately two weeks after amputation. This corresponds to the time when hyaluronidase activity first becomes detectable in regenerating limbs (Toole and Gross, 1971), raising the possibility that capillary ingrowth to the blastema involves localized removal of hyaluronate. This suggestion is supported by the observation of West et al.(1985) that angiogenesis is augmented by oligosaccharides generated during hyaluronate degradation.

In summary, it is clear from the work reviewed here that renovation of the extracellular matrix is a very important part of the dedifferentiation phase of amphibian limb regeneration. The trophic effect of nerves on dedifferentiated cells stimulates not only mitotic activity, but also production of hyaluronate. This in turn apparently causes expansion of the intercellular space and tissue volume and may influence other physiological and developmental processes regulated by the extracellular matrix in the blastema.

REFERENCES

Barger, P.M., and Tassava, R.A., 1985, Establishment of a regeneration-specific bioassay for neurotrophic activity in denervated Ambystoma forelimbs, Experientia, 41:1405-1407.

Boucek, R.J., 1984, Factors affecting wound healing, Otolaryngologic Clin. North. Amer., 17:243-264.

Butler, E.G., and Schotte, O.E., 1941, Histological alterations in denervated non-regenerating limbs of urodele larvae, J. Exp. Zool., 88:307-341.

Caplan, A.I., 1985, The vasculature and limb development, Cell Diff. 16:1-11.

Caplan, A.I., 1986, The extracellular matrix is instructive, Bioessays, 5:129-132.

Derby, M.A., 1978, Analysis of glycosaminoglycans within extracellular environments encountered by migrating neural crest cells, Devel. Biol., 66:321-336.

Gardiner, D.M., Muneoka, K., and Bryant, S.V., 1986, The migration of dermal cells during blastema formation in axolotls, Devel. Biol., 118:488-493.

Globus, M., and Vethamany-Globus, S., 1977, Transfilter mitogenic effect of dorsal root ganglia on cultured regeneration blastemata in the newt, Notophthalmus viridescens, Devel. Biol., 56:316-328.

Grillo, H.C., LaPiere, C.M., Dresden, M.H., and Gross, J., 1968, Collagenolytic activity in regenerating forelimbs of the adult newt (Triturus viridescens), Devel. Biol., 17:571-583.

Gulati, A.K., Zalewski, A.A., and Reddi, A.H., 1983, An immunofluorescent study of the distribution of fibronectin and laminin during limb regeneration in the adult newt, Devel. Biol., 96:355-365.

Hamerman, D., and Wood, D.D., 1984, Interleukin 1 enhances synovial cell hyaluronate synthesis, Proc. Soc. Exper.Biol. Med. 177:205-210.

Hamerman, D., Sasse, J., and Klagsbrun, M., 1986, A cartilage-derived growth factor enhances hyaluronate synthesis and diminishes sulfated glycosaminoglycan synthesis in chondrocytes, J. Cell. Physiol., 127:317-322.

Hay, E.D., and Fischman, D.A., 1961, Origin of the blastema in regenerating limbs of the newt Triturus viridescens, Devel. Biol., 3:26-59.

Hruza, Z., 1977, Connective tissue, in: "Microcirculation", vol. I, G. Kaley and B.M. Altura, eds., University Park Press, Baltimore.

Humason, G.L., 1979, "Animal Tissue Techniques", W.H. Freeman and Co., San Francisco.

Hunt, T.K., and Van Winkle, W., 1979, Normal repair, in: "Fundamentals of Wound Management", T.K. Hunt and J.E. Dunphy, eds., Appleton-Century-Crofts, New York.

Kelly, D.J., and Tassava, R.A., 1973, Cell division and ribonucleic acid synthesis during the initiation of limb regeneration in larval axolotls (Ambystoma mexicanum). J. Exp. Zool., 185:45-53.

Lembach, K.J., 1976, Enhanced synthesis and extracellular accumulation of hyaluronic acid during stimulation of quiescent human fibroblasts by mouse epidermal growth factor, J. Cell. Physiol., 89:277-288.

Liebman, E., 1949, The leucocytes in regenerating limbs of
    Triturus viridescens, _Growth_, 13:103-118.
Maden, M., 1978, Neurotrophic control of the cell cycle during
    amphibian limb regeneration, _J. Embryol. exp. Morph._,
    48:169-175.
Mailman, M.L., and Dresden, M.H., 1976, Collagen metabolism in
    the regenerating forelimb of _Notophthalmus_ _viridescens_:
    synthesis, accumulation, and maturation, _Devel. Biol._,
    50:378-394.
Mailman, M.L., and Dresden, M.H., 1979, Denervation effects on
    newt limb regeneration: collagen and collagenase,
    Devel. Biol., 71:60-70.
Matuoka, K., Namba, M., and Mitsui, Y., 1987, Hyaluronate syn-
    thetase inhibition by normal and transformed human
    fibroblasts during growth reduction, _J. Cell Biol._,
    104:1105-1115.
Mescher, A.L., and Tassava, R.A., 1975, Denervation effects on
    DNA replication and mitosis during the initiation of
    limb regeneration in adult newts, _Devel. Biol._,
    44:187-197.
Mescher, A.L., and Loh, J.-J., 1981, Newt forelimb regeneration
    blastemas _in vitro_: cellular response to explantation
    and effects of various growth-promoting substances, _J.
    Exp. Zool._, 216:235-245.
Mescher, A.L., and Munaim, S.I., 1986, Changes in the extra-
    cellular matrix and glycosaminoglycan synthesis during
    the initiation of regeneration in adult newt forelimbs,
    Anat. Rec., 214:424-431.
Mescher, A.L., and Cox, C.A., 1988, Hyaluronate accumulation
    and nerve-dependent growth during regeneration of lar-
    val _Ambystoma_ limbs, _Differentiation_ (in press).
Meyer, F.A., 1986, Distribution and transport of fluid as
    related to tissue structure, _in_: "Tissue Nutrition and
    Viability", A.R. Hargens, ed., Springer-Verlag,
    New York.
Moscatelli, D., and Rubin, H., 1975, Increased hyaluronic acid
    production on stimulation of DNA synthesis in chick
    embryo fibroblasts, _Nature_, 254:65-66.
Neufeld, D.A., 1985, Bone healing after amputation of mouse
    digits and newt limbs: implications for induced regen-
    eration in mammals. _Anat. Rec._, 211:156-165.
Olsen, C.L., and Tassava, R.A., 1984, Cell cycle and histolog-
    ical effects of reinnervation in denervated forelimb
    stumps of larval _Ambystoma_, _J. Exp. Zool._, 229:247-258.
Olsen, C.L., Barger, P.M., and Tassava, R.A., 1984, Rescue of
    blocked cells by reinnervation in denervated forelimb
    stumps of larval _Ambystoma_, _Devel. Biol._, 106:399-405.
Peadon, A.M., and Singer, M., 1966, The blood vessels of the
    regenerating limbs of the adult newt, _Triturus_, _J.
    Morph._, 118:79-90.
Rappolee, D.A., Mark, D., Banda, M.J., and Werb, Z., 1988,
    Wound macrophages express TGF- and other growth factors
    in vivo: analysis by mRNA phenotyping, _Science_,
    241:708-712.
Repesh, L.A., Fitzerald, T.J., and Furcht, L.T., 1982, Changes
    in the distribution of fibronectin during limb regener-
    ation in newts using immunocytochemistry, _Differentia-
    tion_, 22:125-131.

Roden, L., 1980, Structure and metabolism of connective tissue proteoglycans, in: "The Biochemistry of Glycoproteins and Proteoglycans", W.J. Lennarz, ed., Plenum Press, New York.

Ross, R., 1980, Inflammation, cell proliferation, and connective tissue formation in wound repair, in: "Wound Healing and Wound Infection", T.K. Hunt, ed., Appleton-Century-Crofts, New York.

Schmidt, A.J., 1968, "Cellular Biology of Vertebrate Regeneration and Repair", Univ. of Chicago Press.

Senior, R.M., and Campbell, E.J., 1983, Neutral proteinases from human inflammatory cells, a critical review of their role in extracellular matrix degradation, Clin. Lab. Med, 3:645-666,

Singer, M., and Craven, L., 1948, The growth and morphogenesis of the regenerating forelimb of the newt Triturus following denervation at various stages of development, J. Exp. Zool., 108:279-308.

Smith, G.N., Toole, B.P., and Gross, J., 1975, Hyaluronidase activity and glycosaminoglycan synthesis in the amputated newt limb: comparison of denervated, nonregenerating limbs with regenerates, Devel. Biol., 43:221-232.

Solursh, M., 1984, Ectoderm as a determinant of early tissue pattern in the limb bud, Cell. Diff., 15:17-24.

Tassava, R.A., Bennett, L.L., and Zitzik, G.D., 1974, DNA synthesis without mitosis in amputated denervated forelimbs of larval axolotls, J. Exp. Zool., 190:111-116.

Toole, B.P., 1981, Glycosaminoglycans in morphogenesis, in: "Cell Biology of the Extracellular Matrix", E.D. Hay, ed., Plenum Press, New York.

Toole, B.P., and Gross, J., 1971, The extracellular matrix of the regenerating newt limb: synthesis and removal of hyaluronate prior to differentiation, Devel. Biol., 25:57-77.

Toole, B.P., Goldberg R.L., Chi-Rosso, G., Underhill, C.B., and Orkin, R.W., 1984, Hyaluronate-cell interactions, in: "The role of the Extracellular Matrix in Development", R.L. Trelstad, ed., Alan R. Liss, New York.

Wallace, H., 1981, "Vertebrate Limb Regeneration", John Wiley & Sons, Chichester.

Werb, Z., 1983, How the macrophage regulates its extracellular environment, Amer. J. Anat., 166:237-256.

West, D.C., Hampson, I.N., Arnold, F. and Kumar, S., 1985, Angiogenesis induced by degradation products of hyaluronic acid, Science, 228:1324-1326.

EFFECTS OF DENERVATION ON THE EXTRACELLULAR COLLAGEN MATRIX OF LIMB REGENERATE OF THE NEWT, PLEURODELES WALTLII

Stephane Vanrapenbusch and Bernard Lassalle

Laboratoire de Morphogenèse Animale, UA 685
Université des Sciences et Techniques de Lille
59655 Villeneuve d'Ascq Cedex, France

SUMMARY

We studied the effect of denervation on the extracellular matrix of a newt limb regenerate with a specific dye of collagen, Sirius red. Blastemata harvested on adult animals 4 days after denervation accumulated more collagen (+54%) than blastemata harvested just after denervation. On the other hand, no difference appeared when denervation was performed on young animals. The distribution of collagens of types I, III and IV was analyzed before and after denervation using immunocytochemical techniques. The results showed that the deposition of collagen in denervated blastemata concerned types I and IV. The basement membrane zone was considerably thicker in denervated regenerates. We hypothesize that the basement membrane that takes place after denervation could constitute a barrier to epidermal proteolytic enzymes and would result in an enlarged collagen deposition between mesodermal cells.

INTRODUCTION

Regeneration of the amputated limb in adult Urodeles is controlled by the nervous system (Wallace, 1981). Regeneration only occurs in the presence of a threshold number of nerve fibers at the amputation surface (Rzehak and Singer, 1986). On the other hand, if denervation occurs later, cellular proliferation is arrested but differentiation already initiated in the blastema continues. The regenerate forms a small limb (Singer and Craven, 1948). Owing to these results, it is usual to distinguish two phases in regeneration: a nerve-dependent stage, where blastemal cells actively proliferate, and a nerve-independent stage, where the differentiation occurs. The regenerate differentiation is marked by the synthesis of an extracellular matrix composed of collagen and glycosaminoglycans. Biochemically, it was shown that denervation led to a rapid decline of macromolecular synthesis, especially protein (Dresden, 1969; Lebowitz and Singer, 1970; Singer and Caston, 1972).

Singer (1978) argued that the denervated regenerate synthesizes as much protein as the innervated one. However he
suggested that the quantitative effect of denervation on the
rate of protein synthesis is not uniform among the protein
species; the rate of synthesis in some is increased. At this
time the only one protein studied is collagen. Nerve withdrawal results in a low rate of synthesis (Mailman and Dresden, 1979). However, microscopic observations of semithin
sections (Bryant et al., 1971) revealed an accumulation of
extracellular matrix between mesodermal cells of denervated
newt blastema. This matrix deposition was neither qualitatively nor quantitatively measured.

The apparent discrepancy between the biochemical (Mailman
and Dresden, 1979) and electron microscopic results (Bryant et
al., 1971) led us to reexamine the qualitative and quantitative evolution of collagen after denervation of a newt regenerate. The collagen was visualized and measured with a specific dye, Sirius red.

MATERIALS AND METHODS

Amphibian Urodele, <u>Pleurodeles waltlii</u> Michah were anaesthetized in MS 222 (Sandoz) and amputated through the zeugopod
near the elbow. The animals were maintained in tap water
at 20°C and fed twice a week with chironomid larva. About
fifteen days later, the regenerates reached the mid cone stage
when the denervation was performed. The denervation is carried out by sectioning the spinal nerves 3, 4 and 5 of the
right limb; the left limb was used as control.

Regenerates were fixed in Bouin-Holland's fluid, embedded
in paraffin, then serially sectioned at 7 µm.

<u>Histological Technique</u>

<u>Sirius red, a specific dye of collagen</u>. To study the
specificity of Sirius red, we used a palette-stage regenerate,
which was obtained 20 days after amputation. Some serial sections from this blastema were divided into 3 samples on a
slide. Each sample was treated in a different way. A control
group was treated with Hepes Buffer 50 mM, $CaCl_2$ 0.036 mM, pH
7.2. The second group was treated with 1 mM papain dissolved
in the same buffer. The third group was treated, first with a
mixture of 1 mM papain (Sigma) and 1 mM type VII collagenase
(Sigma) dissolved in the same buffer. After 6 hours of treatment at 37°C, the three samples were coloured with Sirius
red.

<u>Immunohistochemistry</u>. De-paraffinized sections were
immersed twice for five minutes in an agitated 10 mM phosphate
buffer (pH 7.2). Air-dried slides were incubated for 30 minutes at 37°C in a damp atmosphere with diluted (1:50) chicken
anti-collagens of types I, III and IV raised in rabbit (Institut Pasteur, Lyon). After several washes in the same buffer,
the slides were incubated for 30 minutes at room temperature
with diluted (1:100) fluorescent anti-rabbit IgG raised in
goat. Slides repeatedly washed in phosphate buffer were
stained with diluted (1:10000) Evans blue for 10 minutes, then
mounted in Fluoprep (Biomerieux).

Biochemistry

Sections obtained from a regenerate were analyzed to quantify collagen and total protein (Schwartz et al., 1985). The section ribbon was divided into four samples composed of an equal number of sections. The measurement of collagen was carried out on the two even samples and the measurement of total proteins on the two odd samples.

Collagen measurement. The sections used to measure the quantity of collagen were de-paraffinized then stained for 15 minutes in 0.1% Picrosirius solution. The stained sections were washed in 0.01 N HCl, dehydrated and dried at room temperature. Small aliquots of collagen solution (0 to 70 ug) were spread out on a glass slide, dried at room temperature then fixed for one hour with Bouin Holland's fluid. The slides were washed with tap water, stained with Picrosirius red then dried at room temperature. These slides were used to obtain a standard curve. Dried tissue sections and collagen aliquots were then scraped from the slide with a razor blade and transferred into tubes containing 1 ml of 0.1 NaOH. The tubes were immersed for 30 minutes in a boiler at 37°C and occasionally hand shaken. The measurement was made with a spectrophotometer (Pye-Unicam SP6-550) at 540 nm.

Total proteins measurement Sections to be analysed were placed in the bottom of polypropylene tubes (Nalgene). 0.15 ml 4 M NaOH was added to the tubes, which were capped and heated to 120°C for 120 min. After cooling, the hydrolysate was neutralised with glacial acetic acid, and then 0.25 ml ninhydrin reagent, freshly removed from the stock bottle, was added. The tubes were recapped, mixed and heated in boiling water for 20 minutes. After cooling on ice, 2,5 ml 50% ethanol was added to each tube and mixed vigorously. Absorption at 570 nm was determined with a spectrophotometer (Schwartz et al., 1985).

RESULTS

Standard Curve

The results showed a linear relationship between the amount of collagen and the optical density of Sirius red, (Fig. 1).

Specificity of Sirius Red

Control sections showed that only the extracellular matrix and the basal lamina were stained with Sirius red, (Fig. 2a). The extracellular matrix was composed of a dense network of fibers between the distal and peripheral mesodermal cells, (Fig. 2d). Proximal cartilagenous differentiation was clearly visible in the palette-stage regenerate. Sections treated with papain and stained with Sirius red showed similar features with control section, (Fig. 2b). A dense network of extracellular fibers and a basal lamina were yet visible but intensively red stained compared to the control section (Fig. 2e).

Sections treated with papain and collagenase showed a different aspect (Fig. 2c). The basal lamina usually heavily

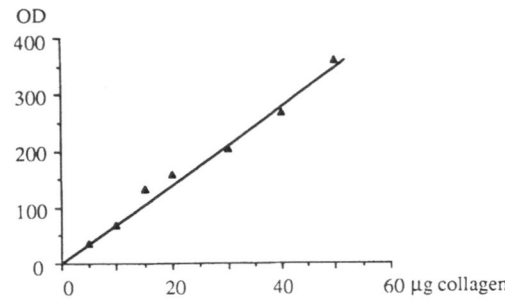

Fig. 1. Standard curve. There is a linear relationship between amount of collagen and optical density of Sirius red.

stained and the network of fibers were not clearly distinguishable (Fig. 2f). Similar results were obtained with papain on innervated or denervated mid-cone regenerate; in both cases, Sirius red heavily stained the basal lamina and the fibrillar matrix more intensively than the control sections.

Biochemical Results

The results showed that Sirius red was a specific dye for collagen. We used it to evaluate the amount of collagen in innervated or denervated limb regenerates. The denervation was performed on mature <u>Pleurodele</u> (8 cm). The results reported in Fig. 3 show that the blastema harvested on adult animals 4 days after denervation (Denerv. D4) accumulated more collage (+54%) than the blastema harvested just after denervation (Control D0). On the other hand, no difference appeared between denervated blastemata (Denerv. D4) and innervated blastemata harvested 4 days after the beginning of the experiment. At this time, these blastemata had reached the palette stage (Control D4).

Immunocytochemical Observations

Use of specific anticollagen of types I, III and IV, enabled us to investigate the distribution of the different types of collagen in the extracellular matrix of the denervated and non-denervated blastemata.

<u>Type I collagen</u>. With anti-type I collagen antibody, the fluorescent labelling in control sections was uniformly distributed between the mesodermal cells and underlined the basal lamina, (Fig. 4c). In denervated sections, the distribution of the label was similar, but highlighted, (Fig. 4d).

<u>Type III collagen</u>. Anti-type III collagen antibody labelling was uniform throughout the mesodermal tissue and underlined the epithelial-mesodermal junction in the control blastemata, (Fig. 4e). The distribution of labelling was very similar in denervated blastemata (Fig. 4f).

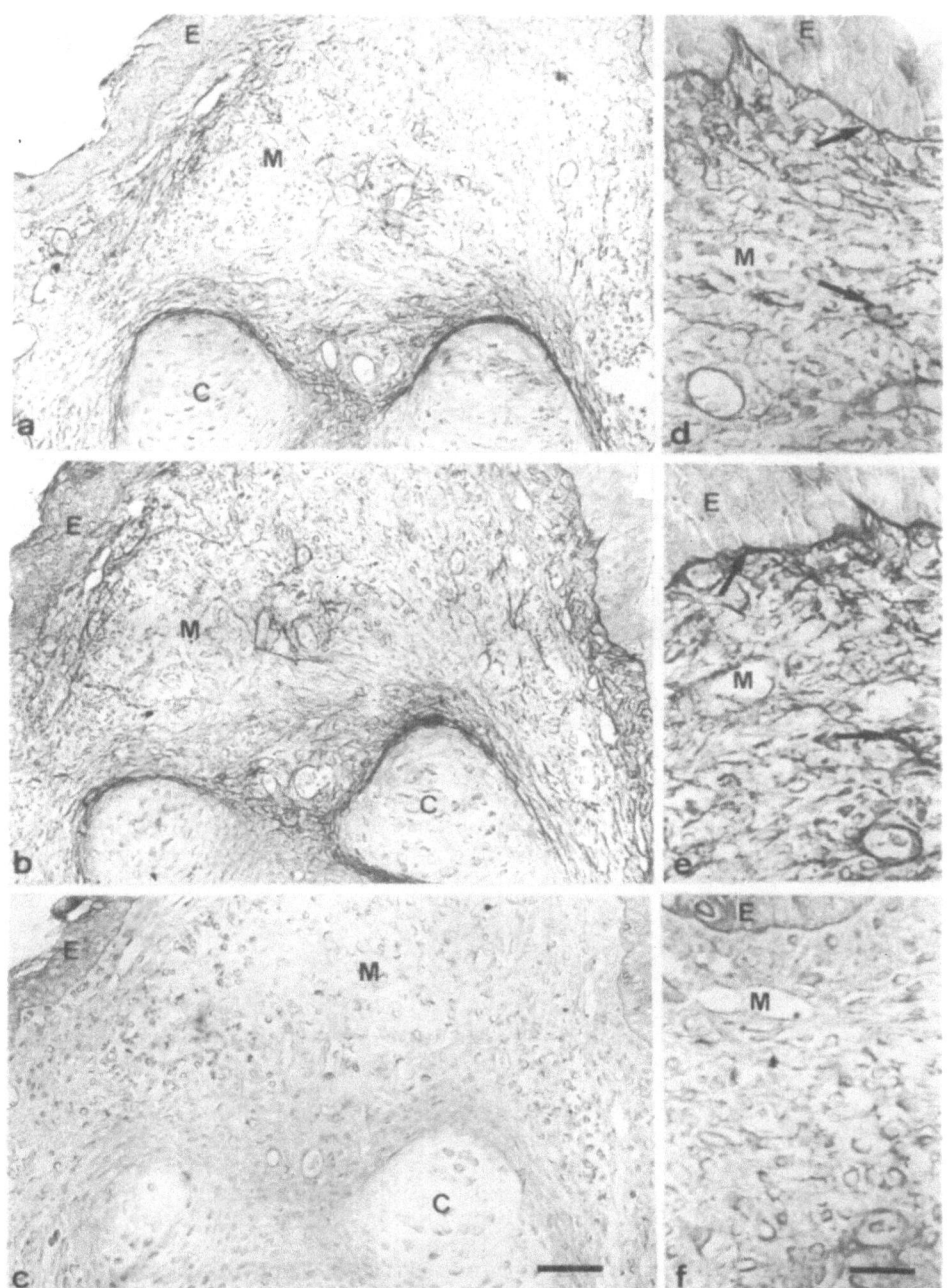

Fig. 2. Light micrographies of a palette-stage regenerate showing specificity of picro-sirius
staining. a,d=control sections. Note the presence of collagen fibers (arrows) at the basal
lamina and between the mesodermal cells.
b,e=sections treated with papain. Collagen
fibers are more intensively stained.

Continued on page 222

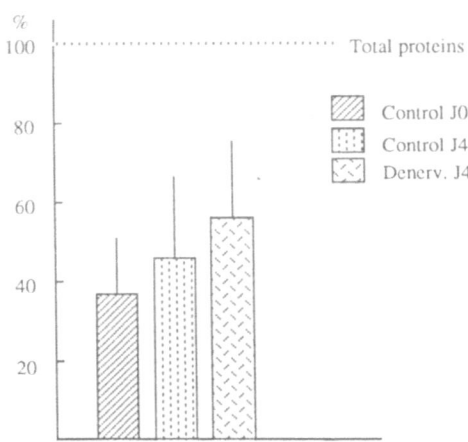

Fig. 3. Amount of collagen compared to total proteins
in innervated and denervated limb regenerate
of adult Pleurodeles. Blastemata harvested on
animals 4 days after denervation (Denerv. D4)
accumulated more collagen (+54%) than blastemata harvested just after denervation (Control
D0).

Type IV collagen. In control sections, type IV collagen
antibody revealed a weak surface fluorescence of the mesodermal cells and of the basement membrane zone, (Fig. 4g). In
denervated sections the latter was strongly labelled (Fig. 4h)
as also was the collagen network between mesodermal cells.
Note the intra-epidermal label in the basal layer of the epidermis. In all cases, when stained with phosphate buffer or
non-immune rabbit serum, the sections remained negative, (Fig.
4a and Fig. 4b).

DISCUSSION

Our observations showed that Sirius red was a specific
dye of collagen (Junqueira et al., 1979b). Indeed, the basal
lamina mainly composed of collagen (Burgeson, 1987), was
heavily stained with Sirius red, as was also the extracellular
matrix of the mesodermal part of the regenerate. The section
treated only with papain was more strongly stained than the
control section. This can be explained as the result of
enzyme cleavage of the bound between the proteoglycan and the
collagen fibers network. The digestion of these closely associated molecules of proteoglycan permit a larger access for the
Sirius red to the collagen sites. Junqueira's results
(Junqueira et al., 1979a) strongly suggest that the interaction between collagen and Sirius red is due mainly to the
reaction of its sulfonic acid groups with the basic groups of

Continued from page 221
c,f=sections treated with papain and collagenase. The basal lamina and the network of fibers are not clearly distinguishable.
E=epidermis, M=mesodermis, C=cartilage. Bars
represent 100 μm (a,b,c), 75 μm (d,e,f).

collagen. When both enzymes, papain and collagenase were used on the same section, collagen digestion is completed. In this latter case the section was not stained by Sirius red.

Owing to the histochemical results, we used Sirius red as a specific dye for the quantitative estimation of collagen, in the amphibian regenerates. The only molecule that makes exception to this specificity is the component Clq of the complement (Junqueira et al., 1979a).

The significant increase in the accumulation of collagen, obtained in denervated blastemata harvested from adult animals (Denerv. D4) compared to innervated mid-cone blastemata (Control D0), suggests that the nervous system controls the metabolism of that protein. This enhanced metabolism, which occurs in the absence of nerves, can be the result either of a higher collagenolytic activity in innervated blastemata or a better accessibility to collagen sites.

The former hypothesis cannot explain our results. Mailman and Dresden (1979), argued that collagenolytic activity is similar in innervated and denervated blastemata up to 21 days post-amputation. Neither does the latter hypothesis explain our results, because using Sirius red on papain treated sections of innervated and denervated mid-cone stage blastemata, leads to a heavy staining. Thus denervation does not result in a better accessibility to collagen sites.

The increase in the deposition of collagen in denervated blastemata must be the result of an enhancement of protein synthesis by mesodermal cells. Nerves indirectly control the structural protein synthesis rate. So, although, total protein synthesis rate is much lower in denervated blastemata than in innervated blastemata (Mailman and Dresden, 1979; Singer and Caston, 1972), collagen synthesis seems to be specifically and differently affected by denervation. Mailman and Dresden (1979) showed, at palette stage, a decrease in the proline uptake much higher in collagen (65%) than in non collagenous proteins (50%). However, the decrease of the collagen synthesis does not affect the differentiation of the denervated regenerate. According to these authors, a lower turnover in the denervated blastemata could explain the results.

Our immunocytochemical results led us to identify the different types of collagen synthetized by the denervated regenerate. Type IV collagen is strongly accumulated between the mesodermal cells and at the epithelial-mesodermal junction. The thicker basement membrane zone of the denervated regenerate shows a heavy fluorescent label. The setting of a thick basement membrane zone in denervated regenerates is concomitant with the blastema resorption (Bryant et al., 1971). The increase in the thickness of the basement membrane zone could constitute a barrier between the epidermis and the mesoderm.

The determining influence of epidermis on underlying cells in the beginning stages of regeneration of a limb has often been demonstrated (Wallace, 1981). It involves dedifferentiation and proliferation of the underlying cells, under the influence of wound epithelium (Revardel and Chapron,

Fig. 4.   Immunocytochemical detection of collagens.
a: innervated blastema stained with non-immune
rabbit serum. b: denervated blastema stained
with non-immune rabbit serum. c: innervated
blastema stained with antitype I collagen
antibody. Note the fluorescent labelling uniformly distributed between the mesodermal
cells and at the basal lamina (arrows).
d: denervated blastema stained with antitype I
collagen antibody. Note that the distribution
of the label at the epithelio-mesodermal junction is highlighted. e: innervated blastema
stained with antitype III collagen antibody.
f: denervated blastema stained with antitype
III collagen antibody. Note presence of label
at the epithelio-mesodermal junction and
between the mesodermal cells. g: innervated
blastema stained with antitype IV collagen
antibody. Note continuous labelling of under-
side of epidermis (arrows). h: denervated
blastema stained with antitype IV collagen
antibody. The epithelio-mesodermal junction is
labelled by a very thick continuous line.
E=epidermis, M=mesenchyme. Bar represents 0.75
μm.

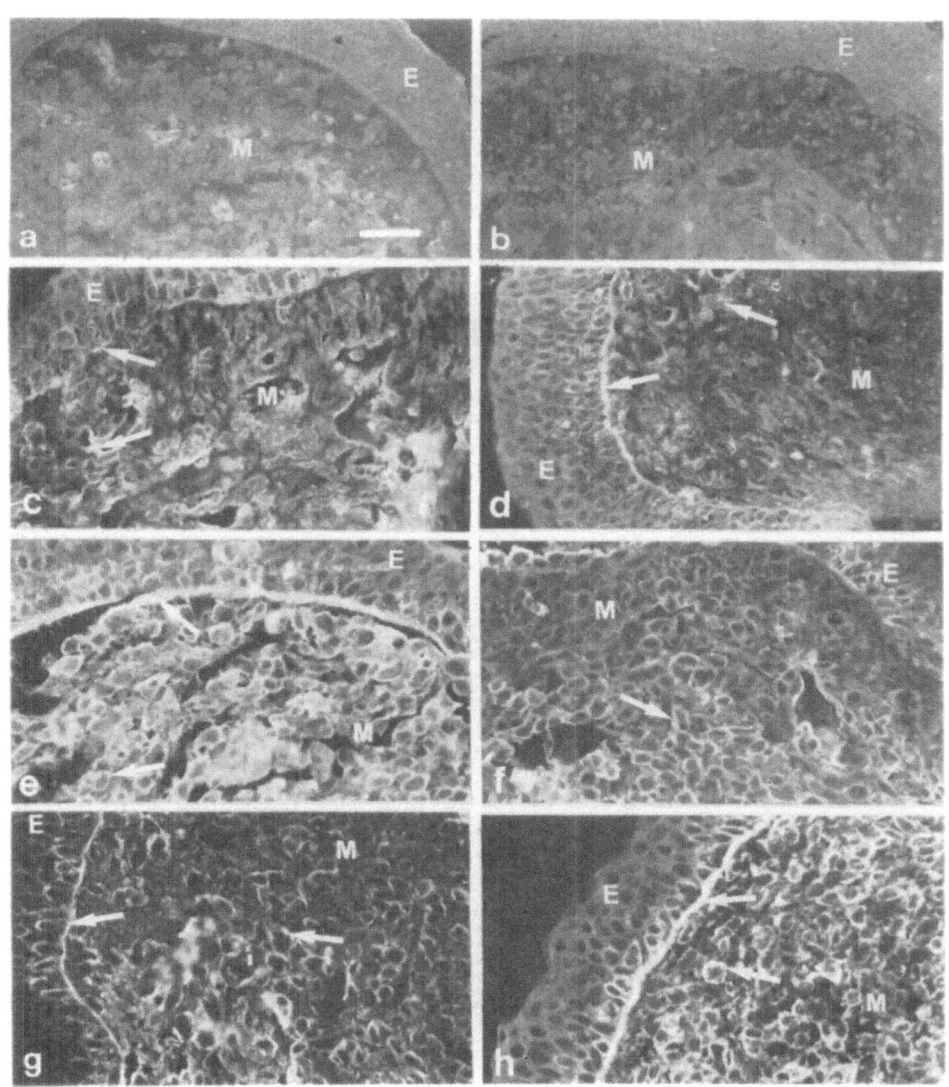

225

1984). This effect appears to be governed by enzymes acting on the extracellular matrix and cell surface. Apparently, the epidermis activity seems to depend on proteolytic enzymes of the same types as collagenases (Revardel et al., 1986).

Thus, we can hypothesize that the basement membrane zone that takes place in the denervated regenerate, could constitute a barrier to proteolytic enzymes and would result in an enlarged collagen deposition. However, it appears that this matrix deposition in denervated regenerates is concomitant with their resorption, when denervation is performed in adult newts.

REFERENCES

Bryant, S.V., Fyfe, D. and Singer, M., 1971, The effects of denervation on the ultrastructure of young limb regenerates in the newt, Triturus. Devel. Biol., 24:577-595.

Burgeson, R.E., 1987, The collagens of skin, Curr. Probl. Derm., 17:61-75.

Dresden, M.H., 1969, Denervation effects on newt limb regeneration. DNA, RNA, and protein synthesis, Devel. Biol.,19:311-320.

Junqueira, L.C.U., Bingolas, G. and Brentani, R.R., 1979a, A simple and sensitive method for the quantitative estimation of collagen, Anal. Biochem., 94:96-99.

Junqueira, L.C.U., Bingolas, G. and Brentani, R.R., 1979b, Picrosirius staining plus polarization microscopy, a specific method for collagen detection in tissue sections, Histochem. J., 11:447-455.

Lebowitz, P. and Singer, M., 1970, Neurotrophic control of protein synthesis in the regenerating limb of the newt, Nature (London), 225:824-827.

Mailman, M.L. and Dresden, M.H., 1979, Denervation effects on newt limb regeneration: collagen and collagenase, Devel. Biol., 71:60-70.

Revardel, J.L. and Chapron, C., 1984, Mise en évidence, au cours de la régénération des Urodèles, d'une influence dédifférenciatrice de l'épithélium cicatriciel (étude chez la larve de Salamandra Salamandra), C.R. Acad. Sc. (Paris), 299:671-676.

Revardel, J.L., Aoussi, S. and Chapron, C., 1986, Action d'extraits d'épithélium cicatriciel sur la régénération des doigts chez Rana ridibunda adulte: étude histologique et biométrique, Can. J. Zool., 64:2683-2689.

Rzehak, K. and Singer, M., 1986, Limb regeneration and nerve fiber number in Rana Sylvatica and Xenopus laevis, J. Exp. Zool., 162:15-22.

Schwartz, D.E., Choi, Y.Y., Sandel, L.L. and Manson, W.R., 1985, Quantitative analysis of collagen, protein and DNA in fixed, paraffin-embedded and sectioned tissue, Histochem. J., 17:655-663.

Singer, M. and Craven, L., 1948, The growth and morphogenesis of the regenerating forelimb of adult Triturus following denervation at various stages of development, J. Exp. Zool., 108:272-300.

Singer, M. and Caston, J.D., 1972, Neurotrophic dependence of macromolecular synthesis in the early limb regenerate of the newt, _Triturus_. _J. Embryol. exp. Morphol._, 28:1-11.

Singer, M., 1978, On the nature of the neurotrophic phenomenon in Urodele limb regeneration, _Amer. Zool._, 18:829-841.

Todd, T.J., 1923, On the process of reproduction of the members of the aquatic salamander, _Quart. J. Sci._, 16, 84-96.

Wallace, H., 1981, "Vertebrate Limb Regeneration", John Wiley and sons, London.

A QUANTITATIVE STUDY OF INNERVATION IN NEWT MULTIPLE LIMBS: NERVE FIBRE COUNTS IN MULTIPLE LIMBS PRODUCED FROM EITHER LIMB BUDS OR MATURE REGENERATING LIMBS

Chafika Zenjari and Emile Lheureux

Laboratoire de Morphogénèse animale
Université des Sciences et Techniques de Lille
F - 59655, Villeneuve d'Ascq, Cedex, France

SUMMARY

Multiple limbs were obtained from either limb buds or mature limbs. By counting the fibers of the main nerves at homologous levels of both multiple limbs and contralateral single limbs, it was shown that the innervation of each duplicated distal part of the limb was quantitatively similar to the innervation of an equal single limb. Only in the late induced multiple limbs from regenerating mature limbs, the excess of nerve fibers distal to the branching level of the multiple limbs, modified a little if any the fiber amount of the proximal common base. In the early induced multiple limbs from limb buds, a large percentage of the distal fiber excess was found in the nerves of the common base of the limb. These results suggest a possible additional recruitment of neurons only at an early stage of limb formation. They are discussed in connection with results obtained in other vertebrates.

INTRODUCTION

The limb of vertebrates has been extensively used to study how the neuro-muscular relationships are established. During the developmental events in mammals, birds and anurans, clusters of motoneuron cell bodies differentiate in the spinal cord. From each neuron, an axon grows, branches and connects with a few mature muscle fibers in the limb. An initial over-abundance of neurons in the spinal cord is subsequently regulated by cell death. In the chick embryo, about half of the motoneurons die between the 6th and the 9th day in ovo (Hamburger, 1975). Later, half of the functional synapses are removed. How the amount of neuron cell bodies is controlled has been investigated by altering the size of the target. It was shown that a supernumerary limb bud graft in the chick embryo allowed survival of some of the motoneurons destined to die (Holliday and Hamburger, 1976). Similar observations were done by Lamb (1979) in Xenopus. Conversely, removal of target reduced strongly the number of motoneurons (Hamburger, 1958; Prestige, 1967). In urodeles, little is known on quantitative

229

relations between motoneurons and limb muscle fibers. Clusters of neurons which innervate specific muscles were described in axolotl by means of horseradish peroxidase (Stephens and Holder, 1985). The number of motoneurons in urodeles is very low (Stephens and Holder, 1985) in comparison with that of motoneuron pools observed in chick (Hamburger, 1975) or in anurans (Prestige, 1967). If the neuron pool innervating the limb is similar to the number of nerve fibres entering the limb as it is in anurans (Prestige and Wilson, 1974), the assay of nerve fibre amount should give information on the possible extra recruitment of neurons for the multiple limb innervation. Multiple limbs produced before the nerves invade the limb bud might influence the motoneuron pool as did similar experiments carried out in higher vertebrates. Multiple limbs from developed limbs by means of regeneration is not possible in other vertebrates, and should bring new data. Our study consisted in counting nerve fibers proximally to and distally to the level of branching of the multiple limbs. Nerve fiber amounts entering the base of the limbs differed in both groups of multiple limbs, suggesting that a recruitment of additional neurons was possible during the initial processes of innervation of the limb.

MATERIALS AND METHODS

The experiments were performed on the newt, <u>Pleurodeles waltlii</u>. Larvae and adults were bred in the laboratory and reared separately in tap water. All the operations were done under anaesthetization in MS222 (Sandoz).

<u>Supernumerary limbs</u>. Two larval stages were used to induce supernumerary limb formation and two different methods were used.

<u>Supernumerary limbs induced in embryos and young larvae</u>. Multiple limbs were expected from limb rudiment exchanges in embryos (Harrison, 1921). Most of these limbs had separate bases and were not used for this study. Subsequently, a retinol palmitate treatment was used to induce transverse supernumerary limb at the moment of limb bud growth and differentiation. New-hatched stage 36 (Gallien et Durocher, 1957) larvae have a short forelimb bud. Amputated limb buds can regenerate as transverse duplicated limbs after a treatment of the animals by retinol palmitate for 11 days (Lheureux et al., 1986). These double limbs were used as <u>early induced multiple limbs.</u>

<u>Supernumerary limbs induced in older larvae</u>. Since the vitamin A treatment do not induce transverse limb duplications of amputated differentiated limbs (Lheureux et al., 1986), contralateral blastema transplantations were performed in stage 55 larvae to induce regenerating multiple limbs to develop. The latter were called <u>late induced multiple limbs.</u> After the limbs of the experimental animal were regenerated, they were studied for their innervation.

<u>Limb Histology and Nerve Fiber Counts</u>

Both multiple limbs and contralateral single limbs of each animal were fixed in Bouin's fluid, then decalcified and embedded in paraffin. Seven micrometer thick transverse sections were treated by Bodian's protargol method to stain nerve fibers. Fibers were counted in the main nerves of either stylopodium, namely, the two branches of the <u>nervus brachialis</u> and the two branches of the <u>n. extensorius</u>, or zeugopodium <u>n. dorsalis manus radialis</u>, <u>n. dorsalis manus intermedius</u>, <u>n. dorsalis manus ulnaris</u>, and <u>n. interosseus</u>). Since the nerve pattern of the limb of <u>Pleurodeles</u> has not yet been described, the terminology used in this report was that used by Francis (1934) on the salamander. Nerve sections were observed at a microscope magnification of 1000 times. The fibers were then counted through a squared reticle. The total number of the nerve fibers in one section is a quantitative evaluation of the local innervation. Measurements of the whole sections were carried out to be used as limb size values.

RESULTS

Production of Supernumerary Limbs

Contralateral transplantation of a limb blastema on a mature limb stump allowed supernumerary limbs to develop. Double or triple regenerating limbs were composed of two or three normal distal limbs with a common base. The axial limb regenerated from the graft kept its opposite handedness. Anterior and posterior supernumeraries limbs had normal handedness. Duplicated regenerating limbs obtained by retinoid treatment on stage 36 larvae were composed of a normal dorsal limb and a ventral, d-v reversed limb whose handedness was that of the contralateral limb. The only double limb induced in the embryo had similar pattern.

Nerve Fiber Amount in the Limb Parts Distal to the Branching Level of the Limb

All the distal parts of the duplicates and triplicates used for this study had normal structures. Transverse sections of these limbs showed normal pattern of skeleton, muscles and nerves. Duplicates induced by retinol palmitate treatment rose at the central or proximal level of the stylopodium. Duplicates and triplicates obtained from contralateral blastema transplantations branched at the distal stylopodium or at the elbow level. The assessment of the nerve fiber amount resulted then from the amount of the nerve fiber counts in the six main nerves of the zeugopodium. For a comparison, counts were done in sections corresponding to the same level of the contralateral single limbs. The nerve fiber amounts differed from one limb to the other, but the values per surface unit of the section were less different. The fiber amount differences may be due to different degrees of development of the limbs. Some of them may be errors bound to the selection of section which were not strictly homologous since the number of fibers decreases from the base of the limb to the tip, as shown by the zeugopodium levels. Obliquity of some sections may also false the area values. Nevertheless, these data (Table 1) suggest that both supernumerary and single limbs proportionally received an equal innervation. As a result, the total numbers of the nerve fibers in the two or three elements of the multiple limbs were always higher than

Table 1. Nerve Fibre Counts in Duplicate and Triplicate Parts of Multiple Limbs.

	Multiple limb type	Studied limb	N=Number of fibres	Operated/ Control (%)	A=amputa-tion surface (mm^2)	N/A
Late induced mult. limps	1 triple (Z)	control	565	242	6.99	81
		ant	412		4.75	87
		axial	375		3.84	93
		post	584		6.16	95
	2 triple (Z)	control	568	208	6.75	84
		ant	365		5.21	70
		axial	376		3.49	108
		post	445		2.24	105
	3 triple (Z)	control	588	191	6.76	87
		ant	530		5.45	97
		axial	594		7.69	78
		post*				
	4 triple (Z)	control	525	258	6.76	78
		ant	370		3.20	116
		axial	476		5.04	93
		post	512		5.28	97
	5 double (Z)	control	544	186	6.01	74
		ant	543		6.23	87
		axial	472		5.05	93
	6 double (Z)	control	456	160	6.14	74
		ant-vent	373		5.87	67
		axial	358		5.34	67
early induced mult.limbs	7 double (S)	control	635	252	3.60	193
		dorsal	956		4.04	237
		ventral	649		3.24	200
	8 double (S)	control	605	224	2.82	214
		dorsal	788		3.04	259
		ventral	570		2.36	241
	9 double (S)	control	1116	160	6.93	161
		dorsal	1066		6.81	157
		ventral	718		3.20	224
	10 double (S)	control	165	211	1.04	159
		dorsal	174		1.13	154
		ventral	175		0.98	179
	11 double (Z)	control	488	169	3.49	140
		dorsal	564		3.04	186
		ventral	264		2.08	127

Z (Zeugopodium) and S (Stylopodium) indicate the level of counts. The control limb corresponds to the contralateral single limb. Limb 11 results from embryo rudiment transplantation. (ant=anterior, post=posterior, vent=ventral. *=only the hand was developed).

Table 2. Nerve Fiber Counts in Proximal Common Base of Multiple Limbs

	Multiple limb type	Studied limb	N=Number of fibers	Operated/ Control (%)	A=Amputation surface (mm²)	N/A
Late induced mult. limbs	1 triple	control operated	1183 1287	108	4.92 5.23	240 246
	2 triple	control operated	1290 1572	121	6.82 7.22	189 217
	3 triple	control operated	1303 1893	145	6.75 7.69	193 246
	4 triple	control operated	1093 1285	117	5.21 7.04	210 183
	5 double	control operated	1010 1078	106	5.78 5.89	175 183
	6 double	control operated	1265 1401	110	6.51 6.83	194 205
Early induced mult. limbs	7 double	control operated	865 1565	180	4.48 6.30	193 248
	8 double	control operated	1560 2768	170	5.40 6.76	285 409
	11 double	control operated	783 1683	214	3.14 3.74	249 450

the nerve fiber amounts in the contralateral single limbs (Table 1). The percentages of extra fiber amounts ranged from 60% to 158% with a mean value of 107% in late induced multiple limbs and from 60% to 124% with a mean value of 89% in early induced multiple limbs. In order to know whether these additional fibers originated from branching of the nerve fiber tips at the amputation plan or were already present at a proximal level suggesting they belonged to additional neurons in the spinal cord, nerve fiber counts were carried out in the common base of the multiple limbs.

Nerve Fibre Counts in the Proximal Common Base of the Multiple Limbs

In the stylopodium, proximally to the level of the limb branching, nerve fibers were counted in the four main nerves cited above. The results differed according to the moment the multiple limbs were induced (Table 2).

In the case of the multiple limbs obtained from the differentiated limb stumps on which a contralateral blastema had been transplanted, the nerve fiber numbers were similar to or slightly above those at the same level of the single contralateral limb. The percentages of extra fibers were ranging from 6% to 45% with a mean value of 17% (Table 2). They were less than those expected from the total counts at the distal levels (Table 1).

In the case of the double limbs induced at an early stage of development, only three limbs were assayed. In the two other limbs of this series, the level of duplications was too proximal and a nerve pattern of single limb was not observed. Despite the limitations of the assay, it was shown that the amount of nerve fibers was markedly higher than those of the single limbs (Table 2). The percentage of extra nerve fibers ranged from 70% to 114% with a mean value of 88%. These values were equal to the expected value if the nerve fibers from the proximal levels did not branch to innervate the duplication distal parts. These results suggest that additional neurons might innervate the additional structures when the latter were induced early in the development, while only nerve fiber branching explains that the innervation of the late induced multiple limbs was quantitatively normal.

DISCUSSION

The quantitative analysis of the duplicates showed that the supernumerary limbs were innervated with number of fibers similar to those of normal limbs. The innervation of the duplicates seemed to be size dependent. This point will be discussed later. The excess of distal nerve fiber amounts was not found in the proximal common base of the multiple limbs developed from blastema transplantation in old larvae, but it was at least partly, in the proximal base of early induced multiple limbs. It was supposed that most of the nerve fibers found in the supernumerary limbs originated from branching of the fiber tips at the amputation area of the old larva limb stump while in the double limbs induced at the moment of the limb formation, the additional amount of fibers might be cor-related to an additional recruitment of neurons, since no branching of axon is supposed to occur in the ventral ramus of the spinal nerve, as shown in Xenopus (Prestige and Wilson, 1974). The supernumerary limbs would be differently recruited according to the moment of the multiple limb formation.

The light microscope counts of silver impregnated sec-tions gave an undervaluation of the nerve fiber amount (Egar and Singer, 1971), but the Bodian's method remains reliable for comparative studies.

An additional pool of neurons early in limb development would be likely since it was proved in higher vertebrates. To try to rescue some of the nervous cells destined to die, Holli-day and Hamburger (1976) grafted an extra limb bud onto a chick embryo before the beginning of the motoneuron cell death. However, a double target only saved a low percentage of the cells that normally would have died, ranging from 11% to 27%. These low rates may be due to inappropriate neuromus-cular junctions, since the graft could not be at the correct limb location. Incorrect neuromuscular connections can end in being withdrawn and the muscles atrophy (Butler et al., 1986). In Xenopus, Lamb (1970) studied a triple limb and observed an increase of the lumbar motoneuron pool equal to 18%, even though the extra mass of the muscle tissues was assessed to 100%. Holliday and Hamburger (1976) reported that Detwiler (1920) was unable to detect any effect of supernumerary limbs in the urodele Ambystoma punctatum. However, by limb bud excision in Ambystoma embryos, Detwiler and Lewis (1925) showed that branchial motor-horn cells and ventral roots suffered atrophy.

Precise data obtained in other vertebrates such as differentiation of motoneurons in excess and natural neuron death remain to be demonstrated in urodeles. However, during the early stages of the axolotl hindlimb development, counts of spinal axons were carried out in the dorsal and ventral rami at the limb level (Freeman and Davey, 1986). A large number of axons destined to innervate the limb bud were counted in the ventral ramus. A maximal number of fibers was reached in 26mm larvae at the stage of paddle limb bud. In the 2-digit limb bud stage which corresponds to 30mm larva, the axon number in the ventral ramus stopped increasing and even was slightly reduced. Because axons were growing, the corresponding limb neurons were differentiated and the limb neuron pools were supposed to be static before the limb bud was growing. Motor and sensory neuron pools cannot be assessed separately in urodeles. On the base of these results and hypothesis, our data on Pleurodeles are surprising. Either the limb neuron populations in the urodeles behave as do motoneuron pools in higher vertebrates, and in this case,half the neuron pools present in the larvae at the early stage of limb development would be destined to die and the early induction of double limbs in Pleurodeles would rescue a motoneuron pool equal to the normal one, or, according to a second hypothesis, no neuron death would occur after the limb morphogenesis ended, and the fiber excess would result from more proximal branching located between the spinal cord and the base of the limb. Such a possibility cannot be excluded since during the production of double limbs following a retinol palmitate treatment, the limb bud stump regressed and subsequently, the double limbs either had two separate bases or a common base branched distally (Lheureux et al., 1986). A quantitative study of the influence of multiple limbs upon the neuron populations by labelling neuron cell bodies would allow a choice between these hypotheses.

Variations in nerve fiber numbers were also observed in newts having different sizes as if nerve fiber amount still increased during growth. In this case, the urodele muscle innervation would be different from that of other vertebrates during the post-morphogenetic growth of the limb. In mammals, the number of muscle fibers remains stable after the birth (Rowe and Goldspink, 1969). Since there is only one neuromuscular junction per muscle fiber and no major changes occur in the motoneuron pools after the cell death period, the nerve fiber number does not vary during the postnatal growth of mammals. In Xenopus,no influence of growth on both motoneuron and muscle fiber pools was shown in giant frogs from experimental giant tadpoles when they were compared with normal frogs (Sperry and Grobstein, 1985). Additional information on muscle growth in urodeles together with nerve fiber variation during the growth must be obtained for a better understanding of interactions between neurons and muscles during multiple limb morphogenesis and growth.

REFERENCES

Butler, J., Cauwenbergs, P., and Cosmos, E., 1986, Fate of brachial muscles of the chick embryo innervated by inappropriate nerves: structural, functional and histochemical analyses, J. Embryol. exp. Morph., 95:147-168.

Detwiler, S.R., 1920, On the hyperplasia of nerve centers resulting from excessive peripheral loading, Nal. Acad. Sci., 6:96-101.

Detwiler, S.R., and Lewis, R.W., 1925, Size changes in primary brachial motor neurones following limb excision in Ambystoma embryos, J. Comp. Neurol., 39:291-300.

Egar, M., and Singer, M., 1971, A quantitative electron microscope analysis of peripheral nerve in the urodele amphibian in relation to limb regenerative capacity, J. Morph., 133:387-398.

Francis, E.T.B., 1934, "The anatomy of the salamander". Oxford University Press, London.

Freeman, J.M., and Davey, D.F., 1986, The precision of pathway selection by developing peripheral axons in the axolotl, J. Embryol. exp. Morph., 91:117-134.

Gallien, L. et Durocher, M., 1957, Table du developpement chez Pleurodeles waltlii, Michah., Bull. Biol. Fr. Belg., 91:97-114.

Hamburger, V., 1958, Regression versus peripheral control of differentiation in motor hypoplasia, J. Anat., 102:365-410.

Holliday, M., and Hamburger, V., 1976, Reduction of the naturally occurring motor neuron loss by enlargement of the periphery, J. Comp. Neurol., 170:311-320.

Harrison, R.G., 1921, On relations of symmetry in transplanted limbs. J. Exp. Zool., 32:1-136.

Lamb, A.H., 1979, Ventral horn cell counts in a Xenopus with naturally occurring supernumerary hind limbs, J. Embryol. exp. Morph., 49:13-16.

Lheureux, E., Thoms, S.D., and Carey, F., 1986, The effects of two retinoids on limb regeneration in Pleurodeles watlt and Triturus vulgaris, J. Embryol. exp. Morph., 92:165-182.

Prestige, M.C., 1967, The control of cell number in the lumbar ventral horn during the development of Xenopus laevis tadpoles, J. Embryol. exp. Morph., 18:359-387.

Prestige, M.C., and Wilson, M.A., 1974, A quantitative study of the growth and development of the ventral root in normal and experimental conditions, J. Embryol. exp. Morph., 23:819-833.

Rowe, R.W.D., and Goldspink, G., 1969, Muscle fiber growth in five different muscles in both sexes of mice. I. Normal mice, J. Anat., 104:265-283.

Sperry, D.G., and Grobstein, P., 1985, Regulation of neuron numbers in Xenopus laevis: Effects of hormonal manipulation altering size at metamorphosis, J. Comp. Neurol., 232:287-298.

Stephens, N., and Holder, N., 1985, A horseradish peroxidase study of motorneuron pools of the forelimb and hindlimb musculature of the axolotl, Proc. R. Soc. Lond. B, 224:325-339.

THE ORIGINS OF SPINAL GANGLIA IN THE AMPHIBIAN TAIL

Jacqueline Géraudie[1], Ruth Nordlander[2]
and Marcus Singer[3]

[1]Laboratoire d'Anatomie Comparée
 Université Paris 7, 2, Place Jussieu
 75005 Paris, France
[2]Department of Oral Biology, and
[3]Department of Anatomy, Case Western
 Reserve University, Cleveland, Ohio 44106, USA

SUMMARY

Sensory ganglia of the amphibian tail differ from those of the trunk with regard to developmental pattern and ability to regenerate. We set out to examine tail ganglion formation during normal development and regeneration. Ganglion ontogenesis and regeneration in the tail follow a similar sequence beginning with ventral root outgrowth, formation of anlagen whose cells divide and go on to form ganglion neurons and glia and the later appearance of dorsal roots. Because there is no classical neural crest the source of the anlagen cells is not entirely clear. Our findings suggest that cells forming the regenerating ganglion migrate from the ventrolateral part of the regenerating spinal cord, whereas those of the embryonic and larval tail probably come from the tail bud.

INTRODUCTION

Sensory ganglion formation at trunk regions follows a well documented pattern in vertebrates. A portion of the neural crest migrates as metameric streams along a stereotyped pathway until the cells reach an appropriate position between myotome and neural tube (Detwiler, 1937; Weston, 1970; Le Douarin, 1982, 1984; Bronner-Fraser, 1986; Teiller et al., 1987). There, they coalesce and develop into neural and glial cells of the ganglia. In the past decade, neural crest migration and differentiation have been extensively studied, especially in the chick (reviewed in Bronner-Fraser, 1986), but also in the amphibian (Krotosky et al., 1988).

Ontogenesis of segmental sensory ganglia at caudal regions is less well understood. The spinal cord at this level forms differently. At trunk levels the cord forms by primary neurulation, i.e. the rolling and dorsal closure of the neural plate with the exclusion of the neural crest. The tail spinal cord is formed by secondary neurulation (Schoen-

237

wolf, 1977; Nakao and Ishizawa, 1984). Here, the spinal cord arises from the tail bud, probably as a solid rod of cells that is continuous with the neural tube of the trunk and which later develops a lumen via cavitation. Neural crest analogues in this region may also arise directly from the tail bud (Schoenwolf, 1977; Schoenwolf and Nicholls, 1984; Schoenwolf et al., 1985). If so, cells forming the tail spinal ganglia should originate in the tail bud; however, there is still no hard evidence for this assertion.

Another intriguing problem is the development of spinal ganglia during tail regeneration in the urodele amphibia. Each regenerate is complete with pigmented skin, new metameric muscles, segmented vertebral components, a regenerated spinal cord, and new spinal ganglia (Goss, 1969). The origin of ganglia of the regenerate is puzzling since the regenerating tail shows neither classical neural crest nor histologically identifiable tail bud (Géraudie et al., 1988).

Several basic questions persist regarding formation of sensory ganglia in caudal regions of vertebrate embryos. Among the most fundamental problems are identification of the source of cells giving rise to their neurons and glia, and elucidation of the actual pattern of ganglion development. Special questions exist for anamniotes, amphibia and fish, in which ganglia originate under a variety of conditions. A comparison of ganglion formation in the amphibia (for example, in trunk versus tail regions, embryos versus regenerates, anurans versus urodeles) may help identify common features of ganglion formation under varied circumstances and help define the essential mechanisms that lead to the development of functional somatic sensory apparatus in vertebrates. We have begun to look at tail ganglion formation in this class. We report here progress of our examination of ganglion formation during normal development of urodelan and anuran tails and in the regenerating urodele tail.

MATERIALS AND METHODS

Species used for our studies include: Xenopus laevis embryos and larvae of stages 35-48 (Nieuwkoop and Faber, 1967), Pleurodeles waltlii embryos of stages 30, 32 and 38 (Gallien and Durocher, 1957), Ambystoma mexicanum of stages 31,34,35,38,39,40 (Schreckenberg and Jacobson, 1975), and 2-8 week tail regenerates of adult Notophthalmus viridescens.

Cytological aspects of tail ganglion development in Pleurodeles embryos and tail regenerates of Notophthalmus were studied in transverse and longitudinal sections of specimens embedded in plastic (see Géraudie et al., 1988). For this work we took advantage of the rostrocaudal developmental gradient to follow ganglion formation from the most primitive (caudal) levels to progressively more advanced (rostral) levels.

Development of tail ganglia in Xenopus embryos and larvae was studied in wholemount preparations with the aid of horseradish peroxidase (HRP) applied to peripheral sensory processes (Nordlander et al., 1988).

238

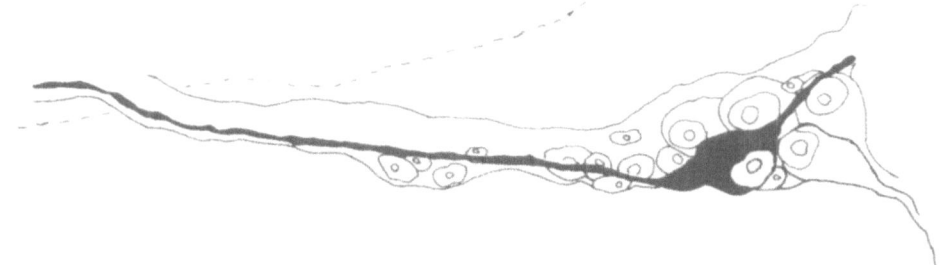

Fig. 1.   Camera lucida drawing of a tail spinal gan-
           glion of a stage 48 <u>Xenopus</u> larva.   Blackened
           ganglion cell was filled with HRP applied to a
           peripheral nerve. Note the acute angle between
           root and spinal cord.   Rostral is to the left,
           (600x).

        The distribution of HNK-1 (gift of Story Landis), an
antibody that recognizes neural crest cells as well as some
other cells (Tucker et al., 1984; Rickmann et al., 1986; Bronner-Fraser, 1986), was examined in the developing tails of
<u>Xenopus</u> , <u>Pleurodeles</u>, and <u>Ambystoma</u>.   Standard immunocytoche-
mical procedures (Beltz and Burd, 1986) were peformed on fixed

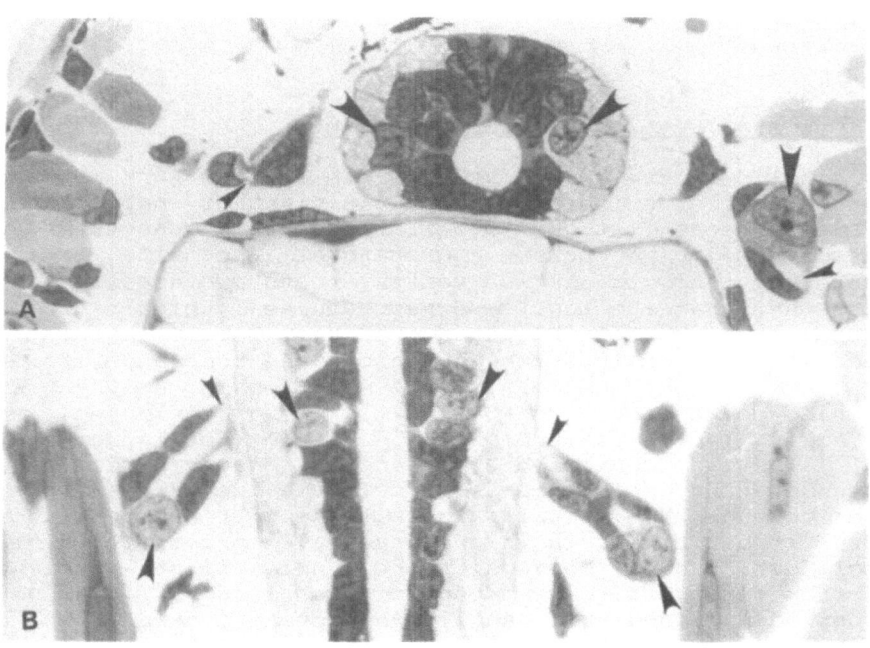

Fig. 2.   Transverse   (A)   and   longitudinal   (B)   views   of
           developing tail ganglia in <u>Pleurodeles</u> showing
           their symmetrical arrangement and relationship
           to   ventral   roots   (small   arrowheads).   Large
           arrowheads point to differentiating neurons in
           the   spinal   cord   (left)   and   ganglion   (right),
           (400x).

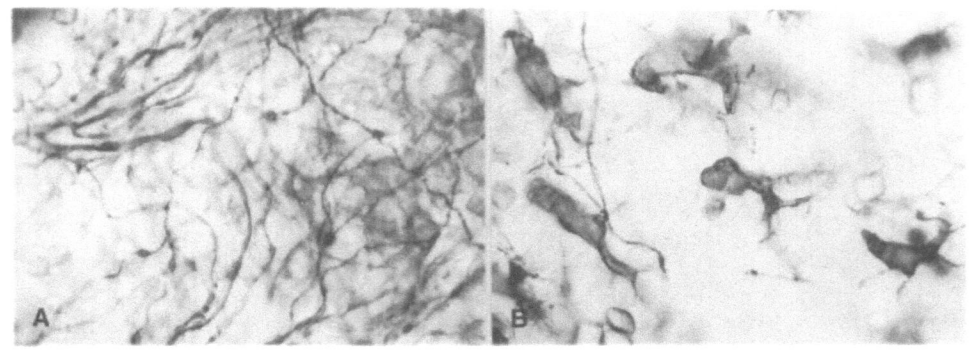

Fig. 3. HNK-1 immunolabeling of peripheral sensory
fibers at the epidermis (A) and of mesenchymal
cells of the dorsal fin (B) of a stage 32
<u>Pleurodeles</u> larva, (400 X).

mounts of whole embryos dissected to expose the tail bud,
nerve cord, and surrounding packing tissues. Antibody distribution was visualized using indirect immunofluorescence or the
ABC method (Vector Laboratories). Specimens were mounted in
glycerol and viewed at the light microscope.

RESULTS

<u>Tail Ganglion Development in an Anuran</u>

Ganglion development at the level of the 20th myotome in
<u>Xenopus</u> is first evident at about the time of hatching from
the jelly coat (stages 35-37). Clusters or sheets of cells
from surrounding mesenchyme aggregate at the site of motor
axon exit. At this time ventral roots are usually ribbon-
like rather than the tight nerve bundles typical of later
stages. Ganglion configuration gradually becomes more stream-
lined with some cells strung out along the root and its primary rami. Most neurons, however, remain at the primary
branching point of the root (Fig. 1). Separate dorsal roots
are lacking in the tails of <u>Xenopus</u> larvae (Nordlander et al.,
1988). Instead, central projections of sensory ganglion neurons enter the spinal cord directly via the ventral root and
travel through the lateral marginal zone to reach the dorsolateral fiber bundle in which they travel primarily rostrally
(Nordlander, 1987). Neurons of the ganglia first label with
HRP from peripheral nerves beginning at stage 39. Throughout
the period examined tail ganglion neurons are bipolar in configuration. Ganglion cell number eventually reaches
20-25/hemisegment.

<u>Tail Ganglion Development in the Urodele Embryo</u>

Spinal ganglia of the tailed amphibia (<u>Pleurodeles</u>) (Fig.
2) are first recognizable as small clusters of 2-3 cells surrounding the ventral roots at a rather uniform distance from
the neural tube. Soon two cell types can be identified in
these cell clusters on the basis of size and staining proper-

ties. The smaller and more intensely stained of these cells
enwrap axons of the root, whereas the larger and more pale
cells are probably neurons by virtue of their resemblance to
motor and interneurons differentiating within the neural tube.
Ganglia are arranged with nearly perfect bilateral symmetry
(see below).

The neural tube of the embryonic tail is entirely closed
and is surrounded laterally by the developing myotomes and
dorsally by loose mesenchyme which is continuous with that of
the dorsal fin. Mesenchymal cells are sparsely distributed in
the tail tissue. A careful study of serial sections of the
tail failed to show any evidence of cell migration from the
neural tube, either from its dorsal or ventrolateral aspect
(see below).   Ventral roots seem to be required for ganglion
formation becaue no cell aggregations appear anywhere among
the tail mesenchyme at levels where ventral roots have not yet
formed, and all aggregates there are associated with ventral
roots. Dorsal roots form at more rostral levels where ganglia
are already well established.

Distribution of HNK-1

We have recently applied the antibody HNK-1, a commonly
used neural crest marker, to wholemounts of developing embryos
of Ambystoma and Pleurodeles, and embryos and young larvae of
Xenopus.   HNK-1 marking appears on the surfaces of developing
neurons and axons of the central and peripheral nervous systems (Fig. 3A) (Nordlander, manuscript in preparation). At
cranial and trunk levels HNK-1 immunoreactivity follows a
rough rostrocaudal gradient.   In Xenopus labeled ganglion
cells are observed along the nerve roots and their primary
branches,again with labeling intensity showing a rostrocaudal
gradient. In urodeles marking of ganglion cells was less
obvious.   In addition, HNK-1 labels mesenchymal cells of the
dorsal fin (Fig. 3B), to a lesser extent those of the ventral
fin, and some mesenchymal cells near the tail bud. In young
Xenopus embryos most stellate cells in the tail region display
HNK-1 immunoreactivity. Differentiating pigment cells of the
tail region of all three species display HNK-1 reactivity.
With time, HNK-1 reactivity decreases rostrally in ventral and
lateral mesenchyme but persists in that of the dorsal fin.
Thus HNK-1 immunolabeling of most tissues seems to be transitory. In the specimens studied here, we have seen no evidence
of reactive cells moving from the tail cord at its dorsal edge
into the extracellular space surrounding it. Our HNK-1 preparations suggest that cells with neural crest-like properties
(i.e., HNK-1 antigenicity) are present in tail and that perhaps these cells may originate in the tail bud as suggested by
Schoenwolf (1977).

Ganglion Development in the Regenerating Tail

Steps in spinal ganglion reconstruction in the tail
regenerate have been followed in eight week regenerates of
Notophthalmus by study of a succession of rostrocaudal levels
(Géraudie et al., 1988). Regeneration begins with wound closure by overgrowing epidermis. This is soon followed by the
formation of a heavily pigmented tail blastema which will give
rise to the regenerated tail (Egar and Singer, 1972; Nordlander and Singer, 1978). The ependyma of the prelesion spi-

nal cord forms a closed tube, the tip of which is tightly adherent to the most distal extent of the blastema epidermis. Cells of this tube rapidly divide and the tube grows in length along with the blastema. This tube becomes the rudiment of the new tail spinal cord, and its cells give rise to new spinal neurons and glia.

Here, as in the embryo, the earliest signs of ganglion formation appear only after ventral roots have already formed (Fig. 4) (Nordlander et al., 1981; Géraudie et al., 1988). The rudimentary ganglion consists of a small collection of cells located around the ventral root at consistent distances from the neural tube. Beginning at about the same level that the cell clusters are first observed on the roots, small cells that we have designated as "M"-cells (because they appear to be migrating) are seen in an arrangement suggesting movement from the mantle layer of the cord and into the ventral root. Cells in this configuration were observed in a metameric pattern at all levels where ganglia are adding new cells and while these cells are still homogeneous in appearance. Ganglion size increases by means of cell divisions, but perhaps also by the addition of cells migrating from the ventral region of the neural tube. With cell differentiation, two cell types are distinguished. One cell type shows dense nuclear staining and is the smaller of the two types. These cells are considered to be developing glia of the ganglia. The other cell type stains more lightly and contains larger and more rounded nuclei. These cells closely resemble developing neurons within the nearby neural tube.

Dorsal roots of the regenerate, like those of the urodelan embryo (see above), develop later. Patterns of dorsal root entry are variable, ranging from entry at normal (dorsolateral) positions to entry anywhere along the lateral cord surface. Some segments lack dorsal roots altogether and occasionally only a single sensory root (rather than bilateral roots) enters the cord at the dorsal midline. Ganglia and roots of the regenerate are seldom bilaterally symmetrical.

In summary, none of the approaches we have used here have provided any evidence that cells are shed from the dorsal edge of the developing tail spinal cord in a pattern that mimics that of neural crest formation in the trunk. This applies to both normal and regenerative nervous system growth in the tail. Although we are unable to state with certainty the origins of cells giving rise to ganglia of the tail in either of these circumstances, our data suggest that cells are generated within the tail itself, either from the neural tube of the regenerate or from mesenchyme of the tail that may have been produced by the tail bud of the embryo.

DISCUSSION

The origin and development of spinal ganglia in caudal regions of the embryo are not well understood, nor is the process of tail ganglion replacement in those species where it occurs. Comparison of these processes in the amphibia is of special interest because it appears that ganglion formation and the achievement of the proper central connections can occur under a number of differing circumstances and using several different modes (Nordlander, 1987; Géraudie et al., 1988). We have described here our observations to date of the patterns of ganglion formation during normal development of several urodele species and one anuran species, and during tail regeneration in the newt.

Tail Ganglion Formation in the Embryo

In both urodele and anuran embryos tail ganglion formation seems to involve the condensation of mesenchymal-like cells upon developing ventral roots. Cells of the ganglion divide and subsequently go on to form the neurons and glia of the ganglion. In the urodele embryo, central projections of the ganglion neurons enter the spinal cord in classical dorsal roots that form relatively late in the developmental period. In the anuran species examined here (Xenopus) separate roots never form, but central sensory projections from the ganglia enter the cord via the venral roots and travel through the lateral marginal zone to reach their appropriate central distribution (Nordlander, 1987).

What is the nature of these mesenchymal cell? We have seen that mesenchymal cells of the tail environment where ganglia arise share some features with neural crest cells of the trunk. In particular, mesenchyme near the ventral roots of the tail also showed immunoreactivity with HNK-1, suggesting that these cells display surface properties in common with the neural crest (Bronner-Fraser, 1986). Although HNK-1 has been widely used as a marker for tracing the behavior of neural crest cells and some of their derivatives, this antibody also marks other cells including neurons within the CNS (Bronner-Fraser, 1986; Holley and Yu, 1987; Nordlander, in preparation), mature Schwann cells, and some cells of the immune system (Tucker et al., 1984). It is known to bind to the cell surface and to share an epitope with several cell adhesion molecules that play important roles in neural development (Kruse et al., 1984).

The source of the condensing mesenchymal cells giving rise to the ganglia of the tail is not clear. Our observations along with those of earlier workers suggest that the most likely source of ganglion cells in caudal regions of the embryo is the tail bud itself (Schoenwolf, 1977). Other possible sources include migrating neural crest cells from trunk levels into the developing tail (Bronner-Fraser, 1987). The earlier observations of Harrison (1898) and Fraisse (1885) that cells migrating from the ventrolateral region of the tail neural tube participate in the construction of the ganglia was not confirmed by our observations of embryonic tail ganglion development. Perhaps migration would have been observed if we had looked at later stages as Fraisse (1885) probably did. However, we have found that this pattern seems to hold during tail regeneration (see below; Géraude et al., 1988). To test these possibilities we continue our studies using HNK-1 and have begun to use transplant experiments similar to those used earlier for the study of neural crest movements (Weston, 1970; Le Douarin, 1984; Bronner-Fraser, 1986) and tail development (Harrison, 1898; Detwiler, 1936; Schoenwolf et al., 1985).

Ganglion formation in the amphibia needs to be examined separately for the many settings in which it occurs. For

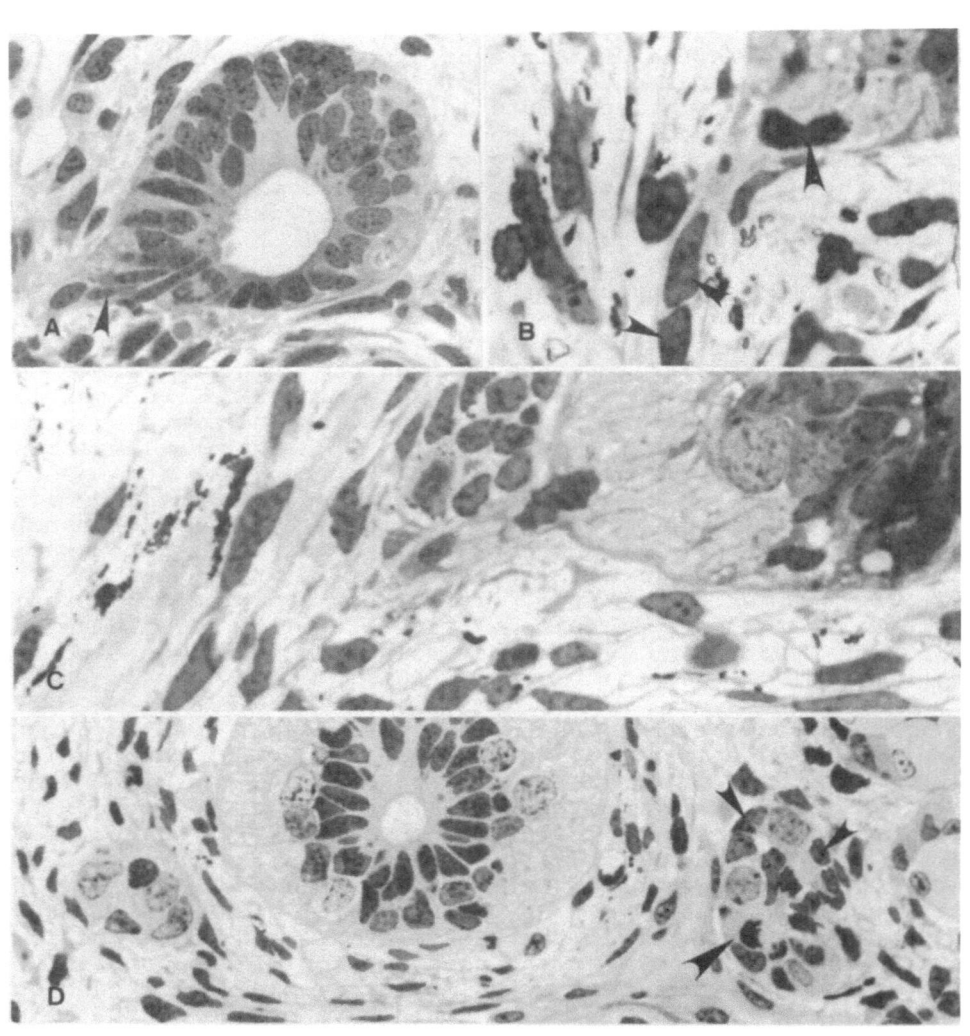

Fig. 4. Regenerating spinal ganglia of <u>Notophthalmus</u> as viewed in transverse sections. A. Exit point of a ventral root at a caudal (primitive) level of a 33 day regenerate. Several cells of the ependymal tube seem to be extending toward and into the root (arrowhead). Note the evidence of asymmetry here, the mitotic cell at the floor of the tube, and the absence of differentiating neurons. B. "M"-cell (large arrowhead) at more rostral ventral root of a 29 day regenerate. Here the cell is close to the root but not in it. Other cells with similar features (small arrowheads) are seen in the root. Dividing cells and differentiating neurons are apparent in C. "M"-cell bridging the spinal cord and ventral root compartments. This cell appeared at an even more rostral level of a 34 day regenerate where the marginal and mantle zones are more developed. D. Ganglia at a level comparable to that of C showing two asymmetric ganglia. Because of the angle of root exit (see Fig. 1), ventral root exit and ganglion cannot be seen in the same transverse section. This photograph shows clearly the similarity between the staining patterns of neurons differentiating within the cord and ganglia. In the ganglion to the right can also be seen a dividing cell (large arrowhead) and nuclei that we believe belong to glia of the ganglion (small arrowheads), (A,D: 420 X),(B,C: 860 X).

example, in anurans where the tail is transient, ganglion formation may follow different developmental patterns from those of the urodeles where tails are permanent. Indeed, one of us (Nordlander, 1987; Nordlander et al., 1988) has described some unique features of ganglion formation in the anuran tail (Xenopus) such as the lack of dorsal roots and the existence of diffuse sensory ganglia, different in structure from those of the trunk (Hughes and Tschumi, 1958).

Ganglion Formation in the Regenerating Tail

There are many parallels between ganglion development in the tail regenerate and in the embryo. The most obvious of these is the sequence which begins with ventral root formation, followed by the appearance of cell clusters upon the root, subsequent cell division within the cell cluster, and later cytodifferentiation within the ganglion. The most striking difference is the appearance of cells migrating from the ventrolateral part of the neural tube into the ventral root and perhaps contributing to the formation of the ganglia. On the basis of our work thus far, we hypothesize but cannot assume that these cells are involved in gangliogenesis, though experiments are underway to test this hypothesis using immunocytochemistry to identify the migrating cells and follow their fate. Alternatively, these cells may be destined to become sheath cells of the ventral root, as demonstrated during ontogenesis in trunk regions in the chick (review in Lunn et al., 1987). Still, as Harrison (1898) and Fraisse (1885) suggested long ago for normal ganglion development and we have suggested more recently for the tail regenerate, they may contribute to ganglion reconstruction. It may be more difficult to distinguish these two possibilities in the amphibia than in the chick because amphibian ganglia show a very intimate association with the ventral roots rather than the more segregated arrangement of the chick ganglia. If the migrating cells do incorporate into the ganglia, our observations would suggest that they divide before differentiating. It would be interesting to know whether they give rise there to both cell types or to one only.

Perhaps regenerated ganglia are a mosaic of cells from several sources. Other possible sources of cells forming the ganglion rudiment include mesenchyme of the prelesion tail or mesenchyme of the blastema. The source of blastema mesenchyme has been debated for some time but is thought to come from dedifferentiated cells of prelesion tissues (Holtzer et al., 1955; Bijtel, 1958). One other suggested source for ganglion precursor cells, the meninges of the regenerated cord (Anton et al., 1986), is not consistent with our observations.

In summary we present the following chart listing some of what is known and speculated about the development, regenerative capacities, and configuration of spinal ganglia and roots in the amphibia. It is based upon our observations of the embryos and larvae of Ambystoma, Pleurodeles, and Xenopus; our observations of tail regenerates of adult Notopthalmus; and other reports in the literature (see text and literature cited).

	Origin of ganglion cells	Ganglia regenerate ?	Ganglion configur.	D. root configur.
Urodeles				
Embryo				
trunk	neural crest	?	symmetrical	present
tail	? probably tail bud	yes	symmetrical	present
Adult				
trunk	------	no (Singer,1946)	symmetrical (Francis,1934)	present
tail regen.	? neural tube, ? blastema	yes	variable	variable
Anurans				
Embryo				
trunk	neural crest	?	symmetrical	present
tail	? probably tail	?	non classical	lacking
Adult				
trunk	-----	no	symmetrical	present

ACKNOWLEDGMENTS

The authors wish to recognize the skillful technical assistance of F. Allizard in Paris and Elizabeth Ullman in Cleveland. We also thank Story Landis for the monoclonal antibody HNK-1. This work was supported in part by Monsanto Corp., St. Louis, Missouri and NIH (NS-07403-15) to M. Singer, NIH (NS-18773) to R.N., and by CNRS (UA041137) to J.G.

REFERENCES

Anton, H.J., D. Weber and H. Doring, 1986, The restitution of spinal ganglia during tail regeneration in Triturus alpestris, Abst. Regeneration Club, Koln, 23-24.

Beltz, B.S., G.D. Burd, 1986, "Basic Immunocytochemical Techniques", Marine Biological Laboratory, Woods Hole, Mass.

Bijtel, J.H., 1958, The mode of growth of the tail in Urodele larvae, J. Embryol. exp. Morph., 6:466-478.

Bronner-Fraser, M., 1986, Analysis of the early stages of trunk neural crest migration in avian embryos using monoclonal antibody HNK-1, Devel. Biol., 115:44-55.

Bronner-Fraser, M., 1987, Perturbation of cranial neural crest migration by the HNK-1 antibody, Devel. Biol., 123:321-331.

Detwiler, S.R., 1936, "Neuroembryology: an Experimental Study", The MacMillan Co., New York.

Detwiler, S.R., 1937, An experimental study of spinal nerve
    segmentation in Ambystoma with reference to the pleuri-
    segmental contribution to the brachial plexus, J. Exp.
    Zool., 67:395-441.
Egar, M. and Singer, M., 1972, The role of ependyma in spinal
    cord regeneration in the urodele, Triturus, Exp.
    Neurol., 37:422-430.
Fraisse, P., 1885, "Die Regeneration von Geweben und Organen
    bei den Wierbelthieren, besonders Amphibien und Reptil-
    ien", Verlag von Theodor Fischer, Cassel und Berlin.
Francis, E.T.B. 1934, "The anatomy of the salamander", Oxford
    University Press, Oxford.
Gallien, L. and Durocher, M., 1957, Table chronologique du
    developpement chez Pleurodeles waltlii, Michah, Bull.
    Biol. Fr. Belg., 91:97-114.
Geraudie, J., Nordlander, R.H., Singer, M. and Singer, J.,
    1988, Early stages of spinal ganglion formation during
    tail regeneration in the newt, Notophthalmus virides-
    cens, Am. J. Anat., (in press).
Goss, R.J., 1969, "Principles of Regeneration", Academic
    Press, N.Y.
Harrison, R.G., 1898, The growth and regeneration of the tail
    of the frog larva, studied with the aid of Born's
    method of grafting, Arch. Entwickl., 7:430-485.
Holley, J.A. and Yu, R.K., 1987, Localization of glycoconju-
    gates recognized by the HNK-1 antibody in mouse and
    chick embryos during early neural development, Dev.
    Neurosci., 9:105-119
Holtzer, H., Holtzer, S., and G. Avery, 1955, An experimental
    analysis of the development of the spinal column. IV:
    Morphogenesis of tail vertebrae during regeneration,
    J. Morph., 96:145-168.
Hughes, A. and Tschumi, P., 1958, The factors controlling the
    development of the dorsal root ganglion and ventral
    horn in Xenopus laevis, J. Anat., 92:498-527.
Krotosky, D.M., Fraser, S.E. and Bronner-Fraser, M., 1988,
    Mapping of neural crest pathways in Xenopus laevis
    using inter- and intraspecific cell markers, Devel.
    Biol., 127:119-132.
Kruse, J., Mailhammer, R., Wenecke, H., Faissner, A., Sommer,
    I., Goridis, C., and Schachner, M., 1984, Neural cell
    adhesion molecules and myelin-associated glycoprotein
    share a common carbohydrate moiety recognized by mono-
    clonal antibodies L2 and HNK-1, Nature (Lond.),
    311:153-155.
Le Douarin, N.M., 1982, "The Neural Crest", Cambridge Univer-
    sity Press, Cambridge.
Le Douarin, N.M., 1984, Cell migrations in embryos, Cell,
    38:353-360.
Lunn, E.R., Scourfield, J., Keynes, R.J. and Stern, C.D. 1987,
    The neural tube origin of central root sheath cells in
    the chick embryo, Development, 101:247-254.
Nakao, t. and Ischizawa, A., 1984, Light- and electron-
    microscope observations of the tail bud of the larval
    lamprey (Lampetra japonica) with special references to
    neural tube formation, Am. J. Anat., 170:55-71.
Nieuwkoop, R.D. and Faber, J., 1967, "Normal Table of Xenopus
    laevis (Daudin)", 2nd ed., Elsevier, Amsterdam.
Nordlander, R.H., 1987, Establishment of a sensory pathway in
    the amphibian tail spinal cord, Abst. Soc. Neurosci.,
    13:468.

Nordlander, R.H., Awwiller, D.M. and Cook, H., 1988, Dorsal roots are absent from the tail of larval _Xenopus_, _Brain Res._, 440:391-395.

Nordlander, R.H. and Singer, M., 1978, The role of ependyma in regeneration of the spinal cord of the urodele amphibian tail., _J. Comp. Neurol._, 180:349-374.

Nordlander, R.H., Singer, J.F., Beck, R. and Singer, M., 1981, An ultrastructural examination of early ventral root formation in Amphibia. _J. Comp. Neurol._, 199:535-551.

Rickmann, M., Fawcett, J.W. and Keynes, R.J., 1985, The migration of neural crest cells and the growth cones of motor axons through the rostral part of the chick somite, _J. Embryol. exp. Morph._, 90:437-455.

Schoenwolf, G.C., 1977, Tail (end) bud contributions to the posterior region of the chick embryo, _J. Exp. Zool._, 201:227-246.

Schoenwolf, G.C., Chandler, N.B. and Smith, J.I., 1985, Analysis of the origins and early fates of neural crest cells in caudal regions of avian embryos, _Devel. Biol._, 110:467-479.

Schoenwolf, G.C. and Nicholls, D.H., 1984, Histological and ultrastructural studies on the origin of caudal neural crest cells in mouse embryos, _J. Comp. Neurol._, 222:496-505.

Schreckenberg, G.M. and Jacobson, A.G., Normal stages of development of the axolotl, _Ambystoma mexicanum_, _Devel. Biol._, 42:391-400.

Singer, M., 1946, The nervous system and regeneration of the forelimb of adult _Triturus_. IV. The stimulating action of a regenerated motor supply, _J. Exp. Zool._, 101:221-240.

Teillet, M.A., Kalcheim, C., and Le Douarin, N.M., 1987, Formation of the dorsal root ganglia in the avian embryo: segmental origin and migratory behavior of neural crest progenitor cells, _Devel. Biol._, 120:329-347.

Tucker, G.C., Aoyama, H., Lipinski, M., Curz, T. and Thiery, J.P., 1984, Identical reactivity of monoclonal antibodies HNK-1 and NC-1: conservation in vertebrates on cells derived from neural primordia and on some leukocytes, _Cell Differ._, 14:223-230.

Weston, J.A., 1970, The migration and differentiation of neural crest cells, _Adv. Morphogen._, 8:41-114.

THE ROLE OF NERVES IN ACCESSORY LIMB DEVELOPMENT

Margaret Wells Egar

Department of Anatomy, Indiana University
School of Medicine
Indianapolis, IN 46223 USA

SUMMARY

Accessory limbs are a type of ectopic supernumerary growth that may develop after trauma in urodeles. In the experiments described here, the removal of a patch of skin plus the reflection of a large nerve constituted the trauma that induced high proportion of growths in young axolotls. The regenerating axons of the reflected nerve supplied the necessary neurotrophic factor, and the Schwann cells of the same nerve may supply the mitotic cells for blastema formation.

INTRODUCTION

It has been known since 1925 from the pioneer work of Locatelli (Wallace, 1981) that the proximal stump of a cut nerve, still connected to the cord and ganglion but reflected to a more superficial position, could stimulate the growth of an ectopic limb at the site of the reflected nerve. Locatelli believed that the nerve itself specified the type of growth that resulted; i.e., an ectopic leg resulted from sciatic nerve reflection. She also established the need to reflect a large nerve. Subsequent work has verified the large nerve requirement in the production of these accessory outgrowths (Bodemer, 1960; Reynolds et al., 1983; Maden and Holder, 1984; Egar, 1988a,b).

The requirement for a large number of regenerating nerve fibers may be the factor behind implant or graft-induced accessory limbs that have been shown to be due to the severity and duration of the immune reaction or trauma. Many different types of trauma, with or without a reflected nerve, have been tested for the ability to stimulate accessory limbs (Breedis, 1952; Bodemer, 1958,1959,1960; Ruben, 1955,1960; Ruben and Stevens, 1963; Stevens et al., 1965; Thornton and Thornton, 1964; Carlson, 1967; Carlson and Morgan, 1967; Lheureux, 1977; Wallace, 1981; Reynolds et al., 1983). However, it has been shown that implant induced accessory limbs will not form on denervated limbs (Ruben and Frothingham, 1958).

Bodemer (1960) pointed out that it might not be the number of fibers per se that was important but rather the number of potential cells available for producing a blastema. He indicated that these cells might come from the nerve sheath cells, as well as from the traumatized muscle. In 1926, Guyenot and Schotte had also implicated the nerve itself as a source for accessory blastema cells (cf. Bodemer, 1960). Such possible metaplasia of Schwann cells has also been described in accessory limbs by Trampusch and Harrebommee (1969), and in regeneration at an amputation stump by Chalkley (1954), Wallace (1972), Wallace and Wallace (1973) and Maden (1977). Cartilage and other tissues of the stump have also been shown to be metaplastic (Chalkley, 1954; Wallace et al., 1974; Namenwirth, 1974; cf. Wallace, 1981).

The role of the nerve in regeneration has been well supported by previous work (cf. Singer, 1952). Much of the more recent experiments on supernumerary limb formation (excess tissue formed at an amputation stump) has focused on pattern regulation and the role of the regenerative territory with mismatched skin grafts (cf. Maden, 1981; Tank and Holder, 1981). Recent work on accessory limb induction has emphasized the need for deviating a large nerve but not the possible roles for this nerve in induction, focusing rather on the mismatched skin graft response and positional effects (Reynolds et al., 1983; Maden and Holder, 1984).

The role of the wound epidermis in implant-induced accessory limb development was explored by Ruben and Frothingham (1958). They found an important relationship between the position of the implant and the insertion wound. It was necessary for the insertion wound to be close to or distal to the implant. Wounds that were proximal to the implant were not successful in giving rise to accessory growths. They found that this proximo-distal feature was not related to a gradient that might exist on the host limb, for the accessory limbs developed with equal ease either at the elbow or at the wrist. It was thus emphasized that the antigenic reaction to the graft and the wound epithelium must be partners in this type of induction of accessory limbs.

The present experiments were designed to test the effects of skin grafts, the wound epithelium, the size of a reflected nerve and the site of the nerve cut on the resulting accessory growths. The results show that grafts are not essential in juvenile axolotls but that a large nerve and its association with a large wound epithelium are required. The site of the nerve cut seemed to determine the extent of proximo-distal tissue that developed in these ectopic limbs.

METHODS AND RESULTS

Axolotls (Ambystoma mexicanum) were selected in size and age matched groups of ten or more for experimental treatment. They were maintained in individual plastic containers in dechlorinated tap water at room temperature, fed strips of liver or salmon chow bi-daily and cleaned as required. Animals for surgery or examination were anesthetized with a 0.05% dilution of MS-222 (Finquel, Ayerst).

Histological examination was carried out on selected early stages by removal and fixation of the tissue at the proximal wound site, including the nerve and the underlying muscle. Glutaraldehyde fixation (2.5% in cacodylate buffer) was followed by fixation in 2% osmic acid in the same buffer. Dehydration in graded ethanols preceded critical point during for scanning electron microscopy and araldite embedment for light and for transmission electron microscopy. The plastic embedded tissue was cut and stained with toluidine blue for one-micron thick sections and with lead citrate for thin sections. Later stages of accessory limb growth were examined after whole body perfusion with saline through the heart followed by the buffered glutaraldehyde. The entire limb was then removed for Victoria blue staining of the skeletal elements. Serial one-micron sections were also studied with the aid of computer reconstruction.

In one series of experiments, skin patches were removed from the flank or stylopodium (or cuffs from the latter site) of upper (or lower) limbs of dark axolotls and grafted in the same position and orientation on white host animals. The reciprocal graft was also made. Two weeks later a large nerve (sciatic or superficial and deep brachial) was deviated to the site of the graft through the skin incision.

The homografts of skin (from siblings of the same spawning) were unsuccessful in inducing impressive accessory growths or even a high frequency of the unimpressive growths on lower limbs of older axolotls (late juveniles or adults). Such grafts made on upper limbs between young juveniles and adults of different spawnings were eventually rejected and no accessory growths developed. As expected, the young juvenile axolotls were found to be more responsive to accessory limb induction than older juveniles or adults or even larvae (hatchlings up to 3-4 months of age, which is prior to cryptic-metamorphosis, cf. Wallace, 1981).

Homografts made between the upperlimb stylopodia of differently pigmented younger juveniles from the same spawning also. failed to induce impressive growths, unless the end of the reflected nerve had retracted to the larger area of wound epidermis where the distal edge of the graft intersected with the skin incision made to free the nerve.

This observation led to the subsequent series of experiments involving large nerve reflection to the proximal site of a 1-2 millimeter wound area where full thickness skin had been removed. The results of this series indicated that a high proportion of accessory limbs could be induced in young animals without grafts. In larval animals, this same paradigm gave well-formed blastemata that were later resorbed. So far, only small, unimpressive accessory growths have been achieved in older animals (Egar, 1988a, 1988b).

Unlike an amputation stump blastema, an accessory blastema contained a central core of regenerating axons with accompanying Schwann cells and blood vessels from the reflected nerve. The axons and the mesenchymal components were continuous with the reflected nerve but there was no apparent contribution of cells from the underlying muscle that had received little or no damage. Although mitotic stages were

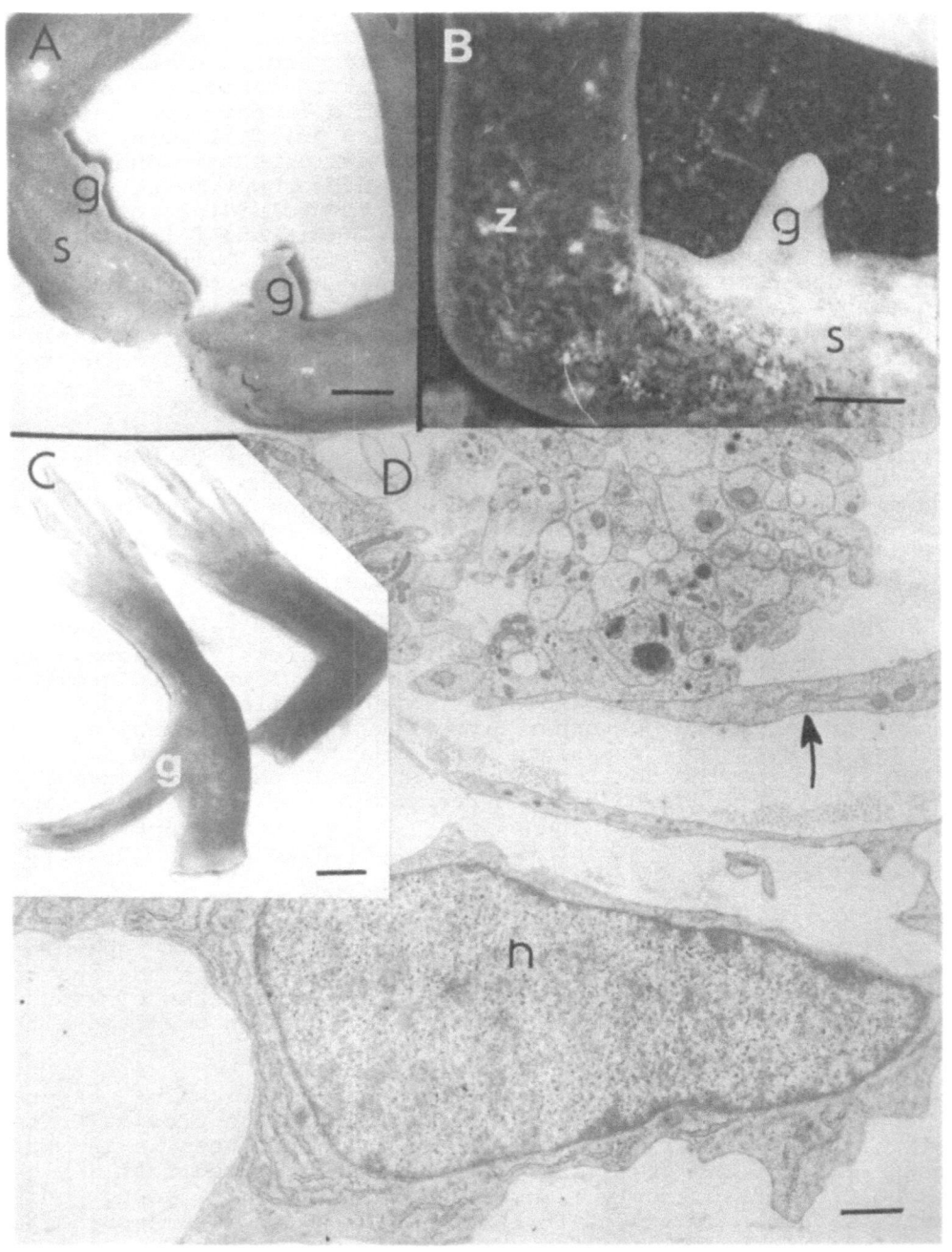

Fig. 1. A: accessory outgrowths (g) on stylopodia (s) from a small adult. Nerves cut at mid-brachium, bar=5 mm; B: a perpendicular accessory with a right-angle joint. Zeugopodium (z), bar=7 mm; C: both arms of a juvenile that had a well-formed left accessory from a joint level nerve cut, Bar=4 mm; D: electron micrograph from the core of a midbud stage accessory blastema. Arrow indicates Schwann cell process wrapping a bundle of axons; blastemal cell nucleus (n), bar=10^3 nm.

not found in association with the host muscle, they were characteristic of the mesenchymal-type cells which form a stream-like continuation with the reflected nerve. Similar to an amputation-stump blastema, the accessory blastema lacked a fully-formed dermis when it appeared to be in a still-growing stage. Those mounds showing a completed dermis in both types of blastemata were judged to have reached their maximum growth (Ruben and Frothingham, 1958; cf. Wallace, 1981).

The cells that compose both types of blastemata (accessory and amputation stump) have the same ultrastructural features but the central concentration of axons, characteristic of accessory blastemata, was not found in the more homogenous mesenchyme of stump blastemata (Geraudie and Singer, 1981). Another difference appeared to be the phagocytic role assumed by "vacuolar" cells of the accessory blastema (cf. Bryant et al., 1971; Egar, 1988a, Fig. 1).

In the previous experiments, the nerves (or nerve) were cut at the joint (either knee or elbow) and reflected through a continuous skin incision to the more proximal wound site. In no instance did tissue of the stylopodium or proximal zeugopodium develop in the accessory limbs (cf.Reynolds et al., 1983; Maden and Holder, 1984). Therefore, in the most recent experiments, the reflected nerves were cut at the mid-humeral level, just proximal to the exit of elbow branches. The initial results seem to indicate that, under these conditions, more proximal tissue may be represented in the accessory growth.

In a further attempt to elucidate the role of the Schwann cells in accessory blastemal formation, a graft of a segment of nerve from the opposite limb was positioned at the wound site adjacent to the reflected small nerve (deep brachial nerve alone, cf.Bodemer, 1960). The results were compared to growths achieved with no graft and those with a reflected large nerve (both superficial and deep brachial nerves).

None of the control series showed any growth at all (grafted nerve alone, small reflected nerve alone or no nerve involvement other than preparation of the wound site and freeing of the nerve trunks). The nerve graft increased the proportion of outgrowths induced by the small reflected nerve (35% growth in one series and 54% in another - total of 42 adult animals) but did not reach the 75% outgrowth (also adult animals) from the large nerve reflection alone (Egar, 1988b).

DISCUSSION

In the overall similarity between the nature and sequences of accessory limb development, as compared to normal regeneration at an amputation stump, the results of the present experiments agree with those of Carlson (1967). Carlson used frog kidney implants as the inducing agent that had been previously described by Ruben (1955; 1960) as being more highly inductive than other types of implanted tissue. In addition to many similarities, the present experiments showed a distinct difference in the overall timing, with accessory blastemata growing much more slowly and unpredictably. There was also a much greater variability in the position, orientation and completeness of accessory outgrowths. These variabilities also have been noted by others (cf.Reynolds et al., 1983; Maden and Holder, 1984).

Histological differences were seen in the early accessory limb blastema induced by a reflected nerve (Egar, 1988a). The initial dispersal of axons and uniformity of the overall pattern of mesenchymal cells that is so characteristic of amputation stump blastemata (Singer, 1952; Géraudie and Singer, 1981; Wallace, 1981), was not found in accessory blastemata. Rather, a central accumulation of axons, associated capillaries and debris was surrounded by blastemal cells. This core and the mesenchymal blastema, itself, showed continuity with the end of the reflected nerve. Strikingly similar features of the accessory blastema with that which forms at an amputation stump included the absence of dermis, no basal lamina (or a loosely formed one), an apical epidermal cap and the morphology of individual cell types (cf. Bryant et al., 1977).

The wound epithelium and nerve requirements are additional similarities between the two types of limb outgrowths (Ruben and Frothingham, 1958; Tassava and Mescher, 1975). The accessory limbs induced by frog kidney implants (Ruben, 1955; Carlson, 1967; Carlson and Morgan, 1967) have a high rate of induction and completenesss, which is still another similarity with normal limb regeneration.

Perhaps the most striking difference between the two types is the absence of proximal completeness of the accessory limbs produced by a reflected nerve, as related to the site of origin. If, instead of the site of attachment to the host, one considers the site at which the nerve was cut (in order to reflect it to a more proximal position), then this distinction between the two types of blastemata disappears. The proximal incompleteness is not seen in implant-induced (Bodemer, 1958; 1960; Carlson,1967) or injection-induced accessories (Breedis, 1952). In the latter work, the similarity in completeness with a normal regenerate was stressed as being related to the initial site of injection that produced the stimulus for outgrowth.

In the present simplified paradigm only three factors appear to be required for successful induction of an accessory outgrowth: (1) a large source for the neurotrophic factor, (2) a large area for wound epithelium at the same site and (3) a large source for proliferating cells. Requirements (1) and (3) appear to be supplied by the large reflected nerve, an observation also suggested by Bodemer (1958, 1960).

The third requirement differs slightly from that proposed by Reynolds et al. (1983) who emphasized the need for positionally mismatched limb tissue. It is possible that in the present experiments, the reflected nerve serves as mismatched tissue similar to a skin homograft. Nerve homografts were not as potent an inducing agent as skin (Bodemer, 1959; Ruben, 1959). However, the cut, reflected and misaligned large nerve appeared to satisfy all requirements when it was in contact with a large wound epithelium.

The wound epidermis allows for a dermis-free epithelial area to interact with underlying regenerating axons and with the mesenchymal components that will give rise to the accumulation of blastemal cells deep to the apical cap (cf. Wallace, 1981). This wound epidermis is a constant in all types of regeneration and in case of the accessory limb growth, its thickened derivative (apical cap) can actually give rise to an accessory growth when transplanted to the base of a stump blastema (Thornton and Thornton, 1964).

The first requirement (in the present paradigm), for accessory limb induction, a large source of the neurotrophic factor (and its continuing presence), is also a constant feature of all successful accessory limb inductions (Ruben and Frothingham, 1958; Reynolds et al., 1983; Maden and Holder, 1984). The quantitative aspect of this requirement has been the subject of much study related to the blastema formation at an amputation stump (cf. Singer, 1952; Wallace, 1981). This quantitative factor relates the induction rate with the fiber number (threshold hypothesis, Singer, 1952; cf. Brockes, 1984) and also the quantity of fibers with quality of growth (Egar, et al., 1982, 1982). This last feature was pointed out 30 years ago for accessory limbs by the work of Ruben and Frothingham (1958) and subsequently by Bodemer (1958, 1960).

Variability has been one of the principal characteristics of all accessory limbs reported (cf. Maden and Holder, 1984). It is seen in the amount of response (no growth, a blastema-like mound, a spike, a complete or incomplete distal limb); in the frequency of response (up to 100% in young axolotls in some of the present experiments, to too low to be used as an assay, Reynolds et al.,1983); in the specific site of attachment of the accessory growth (proximo-distal, dorso-ventral, anterior-posterior on the host limb segment), and in the orientation of the ectopic limb (parallel, perpendicular, oblique or opposite-handed to the host limb).

This variability may be due to the final position of the nerve, and therefore, its relationship to the wound epithelium, to the quantity of the neurotrophic factor available and to the number of potentially mitotic cells that could contribute to the blastema. Although the large demand for the neurotrophic factor and for potential blastemal cells, could be supplied by a large reflected nerve (cf. Wallace, 1973), this retraction factor cannot be so easily controlled. Should nerve retraction result in the return of the tip of the nerve to a more normal position, then the proximity of the large wound epithelium and the first two requirements are lost, as is also the possible stimulation due to mismatched tissue.

Lheureux (1977) emphasized that the position of the accessory was due to the position of the tip of the reflected nerve. He found that the surrounding tissues were influencial in the outgrowth and achieved more complete accessory limbs when tissue from opposite axes (either anterior and posterior or dorsal and ventral) were present at the site.

Distalization has been another characteristic feature of accessory limbs induced by a reflected nerve (Reynolds, et al., 1983; Maden and Holder 1984). In the present experiments, accessory limbs did not grow tissue characteristic of the limb segment more proximal than that of the site of the original nerve-cut, regardless of where on the host limb (or flank) the accessory actually developed. This evidence implies that the Schwann cells of the nerve sheath play a prominent role in the nature of the growth resulting from the reflected nerve.

Schwann cells react to cut axons by proliferation (Wallace, 1972; Singer and Steinberg, 1972; Maden, 1977; Bunge, 1983), they compose the larger of the two cell populations in a nerve (the other being connective tissue cells of the epi-, peri- and endoneural sheaths that comprise only 3% of the total, Egar and Singer, 1971), and their absence in large quantities may explain hypomorphic regeneration after partial denervation (Egar et al., 1982; Egar et al., 1982).

If the Schwann cells do indeed, contribute most of the mesenchymal cells for the accessory blastema, then it would be expected that those at the tip of the cut nerve are responsible for this feature. Normal tissues de-differentiating at an amputation plane to give rise to a stump blastema, don't loose sight of the location in the proximo-distal axis of the host limb (Singer, 1952; Both, 1970; Brockes, 1984). This prevents proximo-distal duplication of the re-established limb and insures that only that which was removed is regenerated.

If it is the Schwann cells at the site of the tip of the reflected nerve (or at the location of an implant, graft or injection) that are stimulated to de-differentiate and contribute to the resulting blastema, then the feature of distalization and the difference in completeness of accessories induced by different methods may be explained.

Nerve grafts in association with the reflection of a small nerve resulted in an increased percentage of accessory growths (47.6%) as compared to nerve grafts alone (0%) or small nerve reflection alone (0%), but did not equal the percentage of responses achieved in adult animals with the reflection of a large nerve (75%, cf. Egar, 1988b). However, a grafted nerve might also supply additional Schwann cells for blastema formation and thus, help supply one of the requirements when in association with the regenerating axons of the reflected small nerve tip. This could explain the beneficial effects seen in the present experiments.

Juvenile axolotls gave a higher percentage of response to accessory limb induction than did adult axolotls or larval animals (up to 100% in some experiments, Egar, 1988a). There was also less variability and a higher percentage of well-formed accessory limbs in these rapidly-growing axolotls that had passed the cryptic-metamorphic phase but had not yet reached sexual maturity. In the present experiments, young juveniles gave 100% response whereas the older juveniles (and adults) showed a high rate of resorption of the initial blastemata.

The results of the present experiments suggest that the reflected nerve may supply several of the requirements for accessory limb development: the neurotrophic factor, the mismatched tissue of a homograft stimulation, and the proliferative cell-population essential for the development of an accessory limb. Regenerating axons appeared to stimulate Schwann cell mitoses and to interact with the wound epithelium in the induction of the ectopic outgrowths. The actual final position of the cut-end of the reflected nerve is suggested as being the determining factor in the site and orientation of the ectopic limb and perhaps, its variable morphology as well. The position of the cut along the proximo-distal length of the nerve is suggested as determining the distalization of the accessory. The age of the axolotl and the size of the reflected nerve and wound also are suggested as being among the factors that determine the completeness of the ectopic growth.

ACKNOWLEDGMENTS

I gratefully acknowledge support for this work from the American Cancer Society, Ohio Division, Inc., Cuyahoga County Unit.

REFERENCES

Bodemer, C.W., 1958, The development of nerve induced supernumerary limbs in the adult newt, Triturus viridescens, J. Morph., 102:555-581.

Bodemer, C.W., 1959, Observations on the mechanism of induction of supernumerary limbs in adult Triturus viridescens, J. Exp. Zool., 140:79-99.

Bodemer, C.W., 1960, The importance of quantity of nerve fibers in development of nerve-induced supernumerary limbs in Triturus and enhancement of the nervous influence by tissue implants, J. Morph., 107:47-59.

Both, N.J. de, 1970, The developmental potencies of the regeneration blastema of the axolotl limb, Roux. Arch., 165:242-276.

Both, N.J. de, 1971, The regeneration territories - a critical note, Folia Morph., 19:177-181.

Breedis, C., 1952, Induction of accessory limbs and of sarcoma in the newt (Triturus viridescens) with carcinogenic substances, Cancer Res., 12:861-873.

Brockes, J.P., 1984, Mitogenic growth factors and nerve dependence of limb regeneration, Science, 225:1280-1287.

Bryant, S.V., D. Fyfe, and Singer, M., 1971, The effects of denervation on the ultrastructure of young limb regenerates in the newt, Triturus, Dev. Biol., 24:577-595.

Bunge, R.P., 1983, Recent observations on the control of Schwann cell functions, Anat. Rec., (Suppl. 1):3-26.

Carlson, B.M., 1967, Studies on the mechanism of implant-induced supernumerary limb formation. I. Histology, J. Exp. Zool., 164:227-241.

Carlson, B.M., and Morgan, C.F., 1967, Studies on the mechanism of implant-induced supernumerary limb formation. II. The effect of heat treatment, lyophilization and homogenization on the inductive capacity of frog kidney, J. Exp. Zool., 164:243-249.

Chalkley, D.T., 1954, A quantitative histological analysis of forelimb regeneration in Triturus viridescens, J. Morph., 92:21-90.

Egar, M.W., 1988a, Accessory limb production by nerve induced cell proliferation, Anat. Rec., 221:550-564.

Egar, M.W., 1988b, Does the nerve have dual roles in the pro-
    duction of accessory limbs? in: "Regeneration and
    Development", The Proceedings of the 6th International
    Marcus Singer Symposium, S. Inoue et al., eds., Okada
    Printing Company, Maebashi.
Egar, M.W., and Singer, M., 1971, A quantitative electron
    microscope analysis of peripheral nerve in the urodele
    amphibian in relation to limb regenerative capacity, J.
    Morph., 133:387-397.
Egar, M.W., McCredie, J., and Singer, M., 1982, New forelimb
    cartilage regeneration after partial denervation, Anat.
    Rec., 204:131-136.
Egar, M.W., Wallace, H., and Singer, M., 1982, Partial dener-
    vation effects on limb cartilage regeneration, Anat.
    Embryol., 164:221-228.
Géraudie, G., and Singer, M., 1981, Scanning electron microscopy
    of the normal and denervated limb regenerate in the
    newt, Notophthalmus, including observations on embry-
    onic amphibian limb-bud mesenchyme and blastemas of
    fish-fin regenerates, Am. J. Anat., 162:73-87.
Lheureux, E., 1977, Importance des associations de tissus du
    membre dans le developpement des membres surnumeraires
    induits par deviation de nerf chez le triton Pleuro-
    deles waltlii Michah, J. Embryol. exp. Morph.,
    38:151-173.
Maden, M., 1977, The role of Schwann cells in paradoxical
    regeneration in the axolotl, J. Embryol. exp. Morph.,
    41:1-13.
Maden, M., 1981, Morphallaxis in an epimorphic system: size,
    growth control, and pattern formation during amphibian
    limb regeneration, J. Embryol. exp. Morph., 65:151-167.
Maden, M., and Holder, N., 1984, Axial characteristics of
    nerve induced supernumerary limbs in the axolotl,
    Roux's Arch. Dev. Biol., 193:394-401.
Namenwirth, M., 1974, The inheritance of cell differentiation
    during limb regeneration in the axolotl, Dev. Biol.,
    41:42-56.
Reynolds, S., Holder, N., and Fernandes, M., 1983, The form
    and structure of supernumerary hindlimbs formed follow-
    ing skin grafting and nerve deviation in the newt Tri-
    turus cristatus, J. Embryol. exp. Morph., 77:221-241.
Ruben, L.N., 1959, The effects of implanting anuran cancer
    into non-regenerating and regenerating larval urodele
    limbs, J. Exp. Zool., 128:29-51.
Ruben, L.N., 1960, An immunological model of implant-induced
    supernumerary limb formation, Am. Nat., 94:427-434.
Ruben, L.N., and Frothingham, M.L., 1958, The importance of
    innervation and superficial wounding in urodele acces-
    sory limb formation, J. Morph., 102:91-117.
Ruben, L.N., and Stevens, J.M., 1963, Post-embryonic induction
    in urodele limbs, J. Morph., 112:279-301.
Singer, M., 1952, The influence of the nerve in regeneration
    of the amphibian extremity, Quart. Rev. Biol.,
    27:169-200.
Singer, M., and Steinberg, M., 1972, Wallerian degeneration: A
    re-evaluation based on transected and colchicine-
    poisoned nerves in the amphibian, Triturus, Am. J.
    Anat., 133:51-84.
Stevens, J., Ruben, L.N., Lockwood, D. and Rose, H., 1965,
    Implant-induced accessory limbs in urodeles: fresh,
    frozen and boiled tissues, J. Morph., 117:213-228.

Tank, P.W., and Holder, N., 1981, Pattern regulation in the regenerating limbs of urodele amphibians, Quart. Rev. Biol., 56:113-142.

Tassava, R.A., and Mescher, A.L., 1975, The role of injury, nerves and the wound epidermis during the initiation of amphibian limb regeneration, Differentiation, 4:23-24.

Trampusch, H.A.L., and Harrebomee, A.E., 1969, Dedifferentiation and the interconvertibility of the different cell-types in the amphibian extremity, Acta Emb. Exp., 1969:35-69.

Thornton, C.S., and Thornton, M.T., 1964, The regeneration of accessory limb parts following epidermal cap transplantation in urodeles, Experientia, 21:146-148.

Wallace, H., 1972, The components of regrowing nerves which support the regeneration of irradiated salamander limbs, J. Embryol. exp. Morph., 28:419-435.

Wallace, H., 1981, "Vertebrate Limb Regeneration", John Wiley, New York.

Wallace, B., and Wallace, H., 1973, Participation of grafted nerves in amphibian limb regeneration, J. Embryol. exp. Morph. 29:559-570.

Wallace, H., Maden, M., and Wallace, B., 1974, Participation of cartilage grafts in amphibian limb regeneration, J. Embryol. exp. Morph., 32:391-404.

RAPID NEURONAL AND GLIAL CHANGES IN THE PHRENIC NUCLEUS
FOLLOWING SPINAL CORD INJURY: A POSSIBLE MORPHOLOGICAL BASIS
FOR THE UNMASKING OF FUNCTIONALLY INEFFECTIVE SYNAPSES

Harry G. Goshgarian and Jose A. Rafols

Department of Anatomy and Cell Biology
Wayne State University School of Medicine
Detroit, MI 48201, U.S.A.

SUMMARY

The present study describes specific morphological alterations of the normal ultrastructure of the rat phrenic nucleus
which occur within four hours after an ipsilateral spinal cord
hemisection. Qualitative and quantitative morphometric analyses demonstrated a rearrangement of the glial compartment
and synaptic architecture in the phrenic neuropil. It is possible that the injury-induced neuronal and glial changes may be
related to the unmasking of a latent motor pathway which
restores function to muscle paralyzed by spinal cord injury.

INTRODUCTION

The primary objective of regeneration research is to
understand the mechanisms underlying the replacement of tissue
and organ systems that have been damaged by injury or disease.
In the case of the mammalian spinal cord, regeneration
research is directed toward establishing functional restitution by the growth of new axon pathways through the damaged
regions of the spinal cord. In spite of several recent
advances in this area, very little functional recovery has
been documented as a result of mammalian spinal cord regeneration research.

An alternative approach to functional recovery after spinal cord injury would be to activate already existing, functionally latent axon pathways which survive spinal cord injury
and are found in the nondamaged regions of the spinal cord.
Several examples of functionally latent pathways have been
demonstrated not only in the spinal cord (Basbaum and Wall,
1976; Guth, 1976; Goshgarian and Guth, 1977; Devor and Wall,
1981a, b; Devor, 1983; Seltzer and Devor, 1984), but also in
the visual system (Berman and Sterling, 1976), somatosensory
cortex (Wall and Egger, 1971a; Frank, 1980), thalamus (Wall
and Egger, 1971b; Rhoades et al., 1987), trigeminal complex
(Dostrovsky et al., 1982), and dorsal column nuclei (Dostrovsky et al., 1976; Millar et al., 1976; Merrill and Wall,

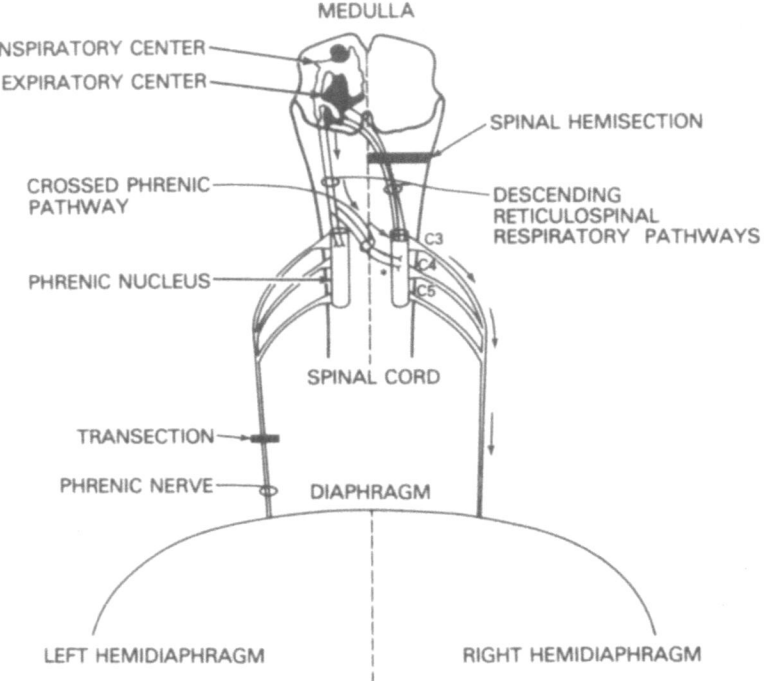

Fig. 1.   Diagram of the descending respiratory pathways
and the surgical procedures required to induce
the crossed phrenic phenomenon in rats (see
text).

1972). The pathways are latent because their synaptic connections are initially functionally ineffective in firing the
postsynaptic cell. Shortly after injury to the nervous system
or other various manipulations, however, the functionally
ineffective synapses are rapidly converted to ones which
become effective in initiating activity in the postsynaptic
cell. The mechanism for the unmasking of these synapses is as
yet unknown (Cragg and McLachlan, 1978; Devor et al., 1986).

Although the unmasking of ineffective synapses has been
demonstrated physiologically by other investigators on sensory
neurons in many regions of the CNS, our physiological studies
(Goshgarian and Guth, 1977; Goshgarian, 1979, 1981) are the
only ones that have ever related functionally ineffective synapses to a latent motor pathway in the injured spinal cord. We
use a simple respiratory reflex known as the "crossed phrenic
phenomenon" (CPP) as model to demonstrate neural plasticity in
the injured spinal cord. The CPP can be briefly described as
follows (Fig. 1). Spinal cord hemisection rostral to the
phrenic nucleus will paralyze the ipsilateral hemidiaphragm.
The paralysis is permanent and no spontaneous functional
recovery generally occurs (Aserinsky, 1961; Guth, 1976).   If
the phrenic nerve to the opposite functioning side of the
diaphragm is then cut (i.e., if the animal is subjected to
respiratory stress), functional recovery is achieved in the
hemidiaphragm paralyzed by spinal cord hemisection (Porter,
1895; Chatfield and Mead, 1948).

Our previous studies have shown that there is a delayed expression of the CPP in guinea pigs (Goshgarian and Guth, 1977) and young adult rats (Goshgarian, 1979). That is, if these animals are subjected to spinal hemisection followed immediately by contralateral phrenicotomy, the CPP is either not expressed at all or it is relatively weak and the animals usually die of asphyxia shortly after surgery. However, if several hours are allowed to pass after spinal hemisection before the contralateral phrenicotomy is carried out, the animals display a prominent CPP (Goshgarian and Guth, 1977; Goshgarian, 1979).

The above results imply that there are rapid, lesion-induced changes that occur in the spinal cord within hours after injury which unmask functionally ineffective connections and allow the animal to more fully express the CPP. Although the unmasking of ineffective synapses has been demonstrated physiologically in many regions of the CNS, and several physiological mechanisms for their unmasking have been hypothesized (Cragg and McLachlan, 1978; Devor et al., 1986), a morphological mechanism for the unmasking of the connections has not been suggested.

The present study describes specific morphological changes that occur in the phrenic nucleus as early as 4 hours after spinal cord injury. These changes should be of interest to investigators in the field of regeneration because such changes have never before been shown to occur so rapidly after injury to the spinal cord. Moreover, evidence will be presented which suggests that these changes could represent the morphological basis for the unmasking of functionally ineffective synapses in our spinal cord injury model.

MATERIALS AND METHODS

Twenty young adult (6-14 weeks old) male and female Sprague Dawley rats were evenly divided into 5 groups: a normal control group and 4 spinal cord hemisected groups. All animals were anesthetized with 4% chloral hydrate (400mg/kg, i.p.) prior to surgery and perfusion.

All of the morphological and surgical techniques employed by this study have been described previously (Goshgarian and Rafols, 1981, 1984). Briefly, horseradish peroxidase (HRP) was applied to the transected left phrenic nerve two days before sacrifice in all animals to retrogradely label phrenic motoneurons. At 4 hours, 1 day, 2 days, and 4 days prior to sacrifice, 4 rats at each time interval were subjected to a left cervical spinal cord hemisection at C_2.

Following surgery, all rats were sacrificed under anesthesia by bilateral pneumothorax and transcardial perfusion with isotonic saline followed by a mixture of 2.5% glutaraldehyde and 0.5% paraformaldehyde in 0.1M phosphate buffer. The cervical spinal cord was exposed by laminectomy and then post-fixed in situ for 24 hrs in the aldehydeperfusate mixture. After removing the spinal cord from the vertebral canal, the C_3 to C_5 segments were embedded in 7% agar and sectioned transversely at 100 μm with a tissue chopper. The sections were collected serially and then stained for peroxidase acti-

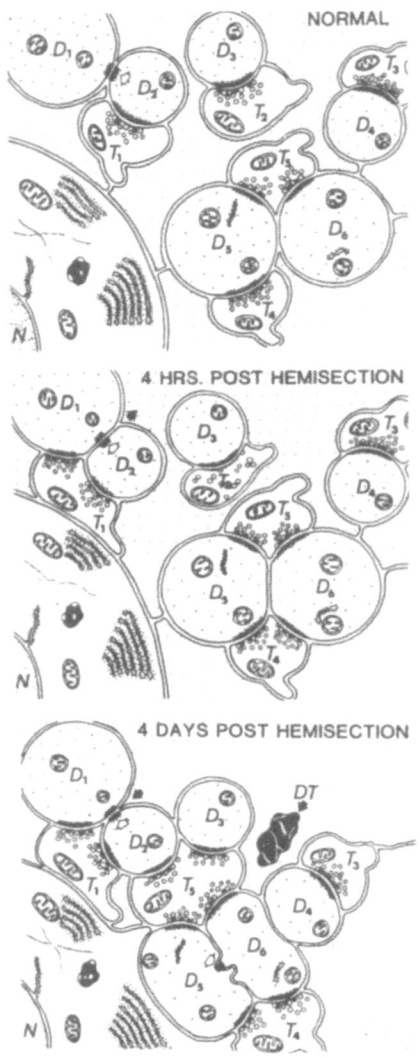

NORMAL

4 HRS. POST HEMISECTION

4 DAYS POST HEMISECTION

Fig. 2. A series of schematic drawings illustrating the ultrastructural features of the normal rat phrenic nucleus (top drawing) and the main changes of the phrenic neuropil which occur 4 hours (center drawing) and 4 days (bottom drawing) following ipsilateral spinal cord hemisection at C_2 (for details see text).

vity using a benzidine dihydrochloride technique (DeOlmos and Heimer, 1977). HRP-labelled phrenic motoneurons and their processes were identified in the 100 um sections under a light microscope. The labelled neurons were trimmed from their surrounding tissue, osmicated, dehydrated and embedded in Araldite. Adjacent sections were never processed for electron microscopy to insure that ultrastructural quantitative analysis (see below) did not involve overlapping areas of the phrenic nucleus.

Ultrathin sections stained with lead citrate and uranyl acetate were examined with the electron microscope until HRP-labelled phrenic motoneuron profiles were identified according to previous criteria established by our laboratory (Goshgarian and Rafols, 1984). After a labelled profile was found, 3-4 non-overlapping photomicrographs of the adjacent phrenic neuropil (defined as the area immediately adjacent to phrenic cell body profiles) were taken at 7000X. This process was continued until 25 photomicrographs were taken from several blocks in each animal. The photomicrographs were printed for analysis at a final magnification of 17500X.

Several morphological parameters of the phrenic neuropil were analyzed qualitatively and then quantitated with the aid of a Bioquant IV morphometric system integrated to an IBM-XT computer. For statistical purposes each photomicrograph was treated as a sample. The quantitative data from the 25 photo-micrograph samples in each of the 4 animals in a group were analyzed for variance (F-test, ANOVA) and then pooled when the test indicated that there was no significant difference between animals in each group. Thus, each group was ulti-mately represented by 100 random photomicrograph samples through the phrenic neuropil taken from 4 animals. Signifi-cant differences in specific morphological parameters of the phrenic neuropil between groups were determined by a two-tailed Student's t-test. Significance was established at $p < 0.05$.

RESULTS

The ultrastructural characteristics of the normal phrenic nucleus in the rat have already been described in detail by our laboratory (Goshgarian and Rafols, 1981, 1984) and are summarized in Fig. 2, top drawing. Normally, the longitudi-nally oriented phrenic dendrites (D_1-D_6) are not tightly fas-ciculated and, for the most part, are isolated from each other by astroglial processes (dotted areas). Occasionally, short dendrodendritic membrane appositions (e.g., between D_1 and D_2; and between D_5 and D_6) with punctum adhaerens (open block arrow) are seen. The majority of the synaptic terminals in the normal phrenic neuropil (T_1-T_4) form single synapses with postsynaptic dendritic profiles, but an occasional double syn-apse (between T_5 and D_5 and D_6) is also observed. Double synapses are defined as presynaptic boutons establishing active synaptic zones with more that one postsynaptic profile in the same plane of section.

By 4 hours after hemisection (Fig. 2, center drawing) the mean length of previously existing dendrodendritic appositions increases (Graph 1 and Fig. 2, between D_1 and D_2, and D_5 and D_6) significantly ($p < 0.005$) with no increase in the percentage of appositions per total dendrite number. The mean normal length of 1.42 ± 0.09 um inreased to 1.89 ± 0.12 um at 4 hours and further increased to 2.20 ± 0.20 um by 1 day post hemisection. Such an increase in mean length of the appositions may result from active retraction of glial processes (exemplified by the direction indicated by the solid block arrow). In addition, the number of double synapses increases significantly ($p < 0.001$) from the normal value of 71 ± 8 to 110 ± 8 by this time (Graph 2 and Fig. 2, between T_1 and D_1, D_2; and between T_4 and D_5, D_6).

At four days post hemisection (Fig. 2, bottom drawing) the mean length of a dendrodendritic apposition reverts back to normal levels (Graph 1 and Fig. 2, between D_1 and D_2) but

DENDRODENDRITIC APPOSITIONS

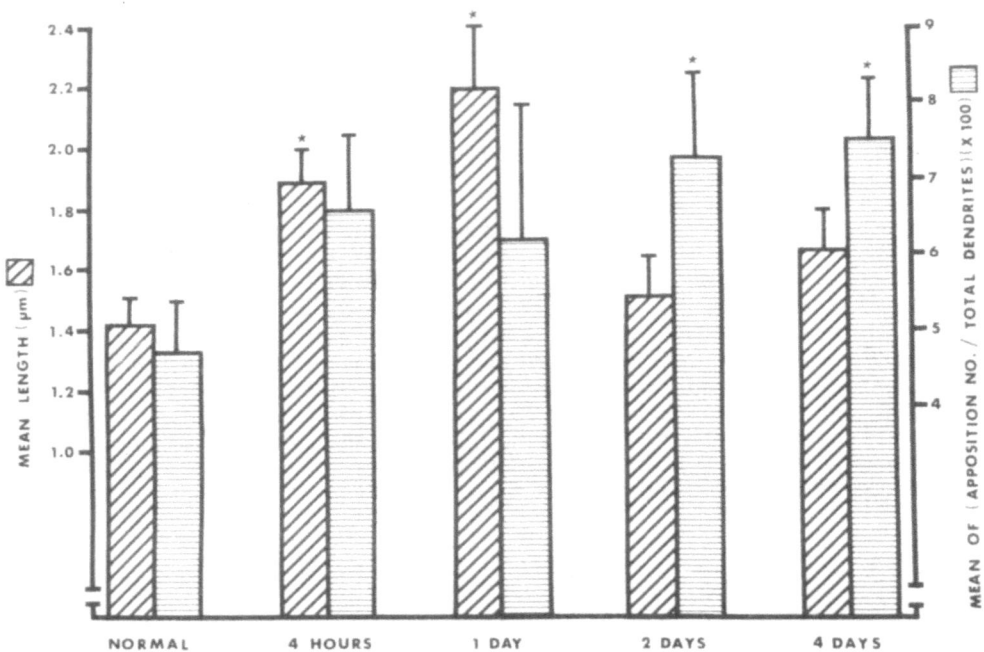

DOUBLE ☐ AND MULTIPLE ▦ SYNAPSES

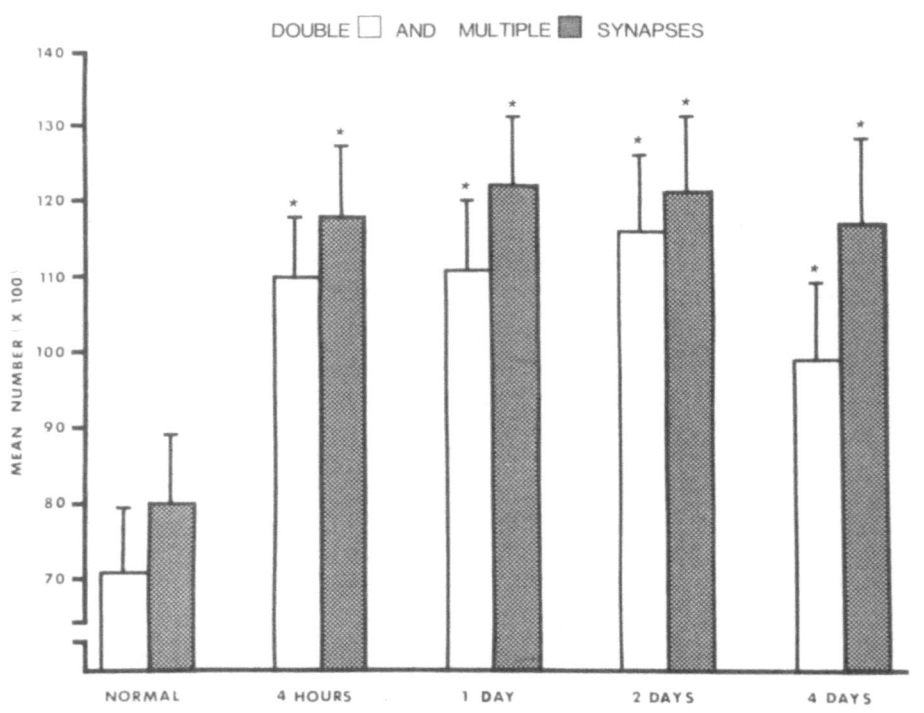

268

Figs. 3 and 4.   All quantitative data were derived from
                 100 electron micrographs taken from 4
                 animals in each groups.   Asterisks   in
                 each figure indicate values which were
                 significantly different than normal lev-
                 els.   All statistical analyses were per-
                 formed by the Student's t-test.

the percentage of appositions increases significantly (p<0.005) over normal values (Graph 1 and Fig. 2, between D_2 and D_3; D_4 and D_6). Some dendrites have flattened outlines (Fig. 2, D_4-D_5) and display finger-like processes with <u>puncta adhaerentia</u> (e.g. Fig. 2, open block arrow between D_5 and D_6) In addition, a slight decrease in the number of double synapses is accompanied by a corresponding increase in the number of triple and quadruple synapses (Graph 2 and Fig. 2, between D_5 and D_6). At this time, also some degenerated terminals (Fig. 2, DT which represented T_2 in the previous drawings) has been incorporated into glial processes.

DISCUSSION

Previously, injury-induced synaptogenesis has been described as an event requiring several days to develop because degenerating terminals must be first removed before new synapses are formed (Cotman and Nieto-Sampedro, 1984). Similar morphological changes as those described here have been observed in the hypothalamus and pituitary, not as a response to injury, but rather as a response to various physiological stimuli (i.e., dehydration, pregnancy, parturition, and lactation) which demand an increased production of the hormones ocytocin and vasopressin from these centers (see Hatton et al., 1984; Hatton, 1985, 1986 for reviews). We agree with Hatton's (1985) interpretation that glial retraction allows single synapses to further differentiate into double synapses. The time course of the present results, however, suggests that our type of spinal cord injury-induced synaptogenesis requires only a few hours for the morphological specialization of synaptic membranes. Similarly, we conclude that the increases in phrenic dendrodendritic membrane apposition are most likely due to an active retraction rather than passive displacement of the intervening glial processes, because measurements of the area and perimeter of phrenic dendrites did not change significantly from normal values at any of the post hemisection survival periods.

It is possible that the rapid morphological alterations which occur in the phrenic nucleus may be related to the unmasking of functionally ineffective synapses in our spinal cord injury model involving the crossed phrenic phenomenon. It has been suggested that glial process retraction from in between adjacent neurons would result in a rise in the extracellular potassium concentration because the glial processes would no longer be available to absorb the potassium which is released from the physiologically active cells (Hatton et al., 1984; Hatton, 1985, 1986). Elevated extracellular potassium would increase neuronal membrane excitability by partial depolarization (Hatton, 1985, 1986). In addition, the increase in neural membrane apposition resulting from glial retraction may create the development of electrical field effects around groups of neurons. The result of these field effects would be to synchronize the firing of cells, especially if electrical excitability were already elevated by depolarizing extracellular ion concentrations (Hatton, 1985).

Enhanced phrenic motoneuron excitability, augmentation of crossed phrenic presynaptic input, and synchronization of phrenic motoneuron output are all manifestations which would predispose to the unmasking of functionally ineffective synapses in our spinal cord injury model. It is possible that functionally ineffective synaptic terminations from the crossed phrenic pathway (Fig. 1) may be physiologically normal, but quantitatively insufficient to fully depolarize phrenic motoneurons in spinal hemisected young adult rats. Thus, when the animals are subjected to respiratory stress immediately after spinal hemisection, the summated action potentials of the crossed phrenic input may not be able to fire the cells. Within hours after spinal cord injury, however, glial retraction and increases in phrenic membrane apposition could result in enhanced excitability of phrenic motoneurons. The postsynaptic changes resulting in enhanced phrenic neuron excitability coupled with the rapid presynaptic augmentation of input mediated by double synapse formation, could be the mechanism enabling previously ineffective synapses of the crossed phrenic pathway to activate the phrenic motoneurons.

The rapid movement of glial processes coupled with the appearance of specialized synaptic profiles represents an aspect of placticity in the injured mammalian spinal cord which has never before been described. Moreover, the synaptic and glial changes which we observed within 4 hours after spinal cord injury may represent injury-induced, physiologically meaningful adaptations of spinal motoneurons. These adaptations may be correlated with differences in the expression of resumed motor activity in muscle that had been paralyzed behind the unmasking of a functionally latent motor pathway after spinal cord injury is unquestionably important. Such information will make a significant new contribution to our understanding of how the spinal cord reacts morphologically to injury and what the functional implications of this reaction may be.

ACKNOWLEDGEMENT

This work was supported by U.S. Public Health Service grant NS-14705.

REFERENCES

Aserinsky, E., 1961, Effects of usage of a dormant respiratory, Exp.Neurol., 3:467-475.
Basbaum, A.I., and Wall, P.D., 1976, Chronic changes in the response of cells in adult cat dorsal horn following partial deafferentation. The appearance of responding cells in a previously non-responsive region, Brain Res., 116:181-204.
Berman, N., and Sterling, P., 1976, Cortical suppression of retinocollicular pathway in the monocularly deprived cat, J. Physiol. (Lond.), 225:263-274.
Chatfield, P.O., and Mead, S., 1948, Role of the vagi in the crossed phrenic phenomenon, Am. J. Physiol., 54:417-422.
Cotman, C.S., and Nieto-Sampedro, M., 1984, Cell biology of synaptic plasticity, Science, 225:1287-1294.

Cragg, B., and McLachlan, E., 1978, A mechanism for the observed recovery from ineffectiveness of synapses in the central nervous system, J. Theor. Biol., 71:433-440.

DeOlmos, J.S. and Heimer, L., 1977, Mapping of collateral projections with the HRP-method, Neurosci. Lett., 6:107-114.

Devor, M., 1983, Plasticity of spinal cord somatotopy in adult mammals: Involvement of relatively ineffective synapses. Birth Defects: Original Article series, 19:287-314.

Devor, M., Basbaum, A.I., and Seltzer, Z., 1986, Spinal somatotopic plasticity: Possible anatomical basis for somatotopically inappropriate connections. in: "Development and Plasticity of the Mammalian Spinal Cord", M.E. Goldberger, A. Gorio and M. Murray, eds., Liviana, Padova, Italy.

Devor, M., Merrill, E.G. and Wall, P.D., 1977, Dorsal horn cells that respond to stimulation of distant dorsal roots, J. Physiol. (Lond.), 270:519-531.

Devor, M. and Wall, P.D., 1978, Reorganization of spinal cord sensory map after peripheral nerve injury, Nature, 275:75-76.

Devor, M. and Wall, P.D., 1981a , Effect of peripheral nerve injury on receptive fields of cells in the cat spinal cord, J. Comp. Neurol., 199(2):277-291.

Devor, M. and Wall, P.D., 1981b, Plasticity in the spinal cord sensory map following peripheral nerve injury in rats, J. Neurosci., 1:679-684.

Dostrovsky, J.O., Ball, G.J., Hu, J.N. and Sessle, B.J., 1982, Functional changes associated with partial tooth pulp removal in neurons of the trigeminal spinal tract nucleus, and their clinical implications. in: "Anatomical, Physiological and Pharmacological Aspects of Trigeminal Pain", R.G. Hill and B. Matthews, eds., Excerpta Medica, Amsterdam.

Dostrovsky, J.L., Millar, J., and Wall, P.D., 1976, The immediate shift of afferent drive of dorsal column nucleus cells following deafferentation in gracile nucleus and spinal cord, Exp. Neurol., 52:480-495.

Frank, J.I., 1980, Functional reorganization of cat somatic sensory-motor cortex (SMI) after selective dorsal root rhizotomies, Brain Res., 186:458-462.

Goshgarian, H.G., 1979, Development plasticity in the respiratory pathway of the adult rat, Exp. Neurol., 66:547-555.

Goshgarian, H.G., 1981, The role of cervical afferent nerve fiber inhibition of the crossed phrenic phenomenon, Exp. Neurol., 72:211-225.

Goshgarian, H.G., and Guth, L., 1977, Demonstration of functionally ineffective synapses in the guinea pig spinal cord, Exp. Neurol., 57:613-621.

Goshgarian, H.G., and Rafols, J.A., 1981, The phrenic nucleus of the albino rat: A correlative HRP and Golgi study, J. Comp. Neurol., 201:441-456.

Goshgarian, H.G., and Rafols, J.A., 1984, The ultrastructure and synaptic architecture of phrenic motor neurons in the spinal cord of the adult rat, J. Neurocytol., 13:85-109.

Guth, L., 1976, Functional plasticity in the respiratory pathway of the mammalian spinal cord, Exp. Neurol., 51:414-420.

Hatton, G.I., 1985, Reversible synapse formation and modulation of cellular relationships in the adult hypothalamus under physiological conditions, in: "Synaptic Plasticity", C.W. Cotman, ed., Guilford Press, New York.

Hatton, G.I., 1986, Plasticity in the hypothalamic magnocellular neurosecretory system, Fed. Proc., 45:2328-2333.

Hatton, G.I., Perlmutter, L.S., Salm, A.K., and Tweedle, C.D., 1984, Dynamic neuronal-glial interactions in hypothalamus and pituitary: Implications for control of hormone synthesis and release, Peptides., 5 (suppl. 1): 121-138.

Merril, E.G., and Wall, P.D., 1972, Factors forming the edge of a receptive field. The presence of relatively ineffective afferents, J. Physiol. (Lond.), 226:825-846.

Millar, J., Basbaum, A.I., and Wall, P.D., 1976, Restructing the somatotopic map and appearance of abnormal neuronal activity in the gracile nucleus after partial deafferentation, Exp. Neurol., 50:658-672.

Porter, W.T., 1895, The path of the respiratory impulse from the bulb to the phrenic nuclei, J. Physiol. (Lond.), 17:455-485.

Rhoades, R.W., Belford, G.R., and Killackey, H.P., 1987, Receptive-field properties of rat ventral posterior medial neurons before and after selective kainic acid lesions of the trigeminal brain stem complex, J. Neurophysiol., 57:1557-1600.

Seltzer, Z., and Devor, M., 1984, Effect of nerve section on the spinal distribution of neighboring nerves, Brain Res., 306:31-37.

Wall, P.D., 1977, The presence of ineffective synapses and the circumstances of neighboring nerves, Brain Res., 306:31-37.

Wall, P.D., and Egger, M.D., 1971a, Formation of new connections in adult rat brains after partial deafferentation, Nature, 232:542-545.

Wall, P.D., and Egger, M.D., 1971b, The plantar cushion reflex circuit: An oligosynaptic reflex, J. Physiol. (Lond.), 216:483-501.

MECHANISMS OF TRANSDIFFERENTIATION OF PIGMENT EPITHELIAL CELLS

INTO NEURAL RETINA: A HYPOTHESIS

Victor Mitashov

Institute of Developmental Biology
Academy of Sciences of the USSR
Ulitsa Vavilova, D.26
117334 Moscow, USSR

The adult amphibian eye is a very convenient model for studying the events involved in the process that has been called transdifferentiation. In this system, which involves regeneration of the lens and/or retina in the adult newt or certain other amphibia, fully differentiated cells of adult somatic tissue are released from the control of the differentiated state and are channeled into a new pathway of cell type differentiation.

There are several advantages in using eye tissues for investigations on transdifferentiation. The tissues under study - the iris, lens, pigmented epithelium and retina - are easily distinguished from one another. During regeneration, cells of the lens are formed from the two-layered iris. Regenerating retinal cells are formed from cells of the pigmented epithelium and the growth area located at the periphery of the eye (Stroeva and Mitashov, 1983).

A major goal of our investigations in this field is to study the changes of some patterns of macromolecular synthesis during the process of transdifferentiation. Our group has obtained much information through autoradiographic studies of DNA, RNA and protein synthesis, cell cycles and macrophage reaction. This paper will discuss one aspect of the phenomenon and examine the hypothesis involving proposed mechanisms underlying the formation of the retina from pigment epithelial cells. Understanding the mechanisms of transdifferentiation should reveal to us the general principle that operates in determination.

We have studied the regeneration of the neural retina in three different species of amphibians: the newt, the axolotl and the African frog, _Xenopus_ _laevis_. After removal of the neural retina in the adult newt, cells of the pigment epithelium provide the cellular origins of the regenerating retina (Stroeva and Mitashov, 1983). In the process of dedifferentiation they become depigmented, and the cells lose part of their cytoplasm (cytoplasmic shedding). Active proliferation

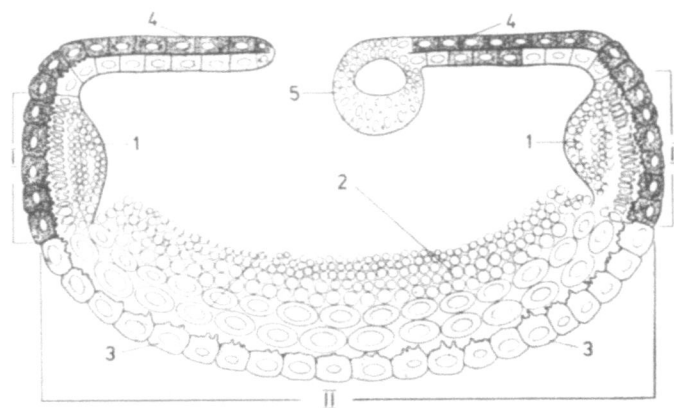

Fig. 1. Scheme of the regenerating eye. I. Peripheral
region; II. Central part of the fundus.
1. Retinal rudiment formed from the anterior
complex of the eye; 2. Retinal rudiment formed
from the retinal pigment epithelium; 3. Retinal pigment epithelial layer; 4. Iris; 5. Regenerating lens. All retinal pigment epithelial
cells in the periphery of the eye incorporate
[3]H-DOPA during early stages of neural retina
regeneration, wheres the ^{3}H-thymidine labelling index is equal to 4 to 6%. Incorporation
of ^{3}H-DOPA into pigment epithelial cells is
absent in the central part of the fundus of
the eye during formation of the retinal rudiment, but the ^{3}H-thymidine labelling index of
central pigment epithelial cells is 40 to 50%.

of the cells occurs after they cease synthesizing pigment granules. In adult newts all cells of the pigment epithelium
have the potential to transdifferentiate into neural retina.
In axolotls, only a population of cells located in the choroid
fissure (embryonic slit) can transdifferentiate into neural
retina. In albino Xenopus isolated patches of cells on the
surface of the pigment epithelium are capable of transformation into structures that have the appearance of a rudimentary
retina.

A summary diagram illustrating the regenerative processes
in the eye of the newt is shown in Fig. 1. After removal of
the neural retina, there is an activation of pigment epithelial cells in the central part of the eye. An early change is
a 90° degree rotation of the long axes of the pigment epithelial cells from parallel to the choroid layer (their normal
orientation) to perpendicular to the surface of the choroid
layer. The cells divide, and after the completion of their
divisions those cells move into the eye cavity and will become
cells of the retinal rudiment. Cells of the two-layered retinal rudiment contain pigment granules because these cells were
derived from the pigment epithelium. Later individual cells

276

Table 1.   Cell Cycle Parameters During Development and Regeneration of the Neural Retina in the Newt

Days After Operation	(in hours)			
	S	G_2	G_1+M	Total
8-10 (pigment epithelium)	20.0	2.5	20.5	43.0
14-16 (retinal anlage)	13.0-16.0	2.5	5.5	21.0-23.5
18-20 (retinal anlage)	21.5	2.5	4.5	28.5
Embryonic development of retina	24.0	2.5	4.0	30.5

emerge from this layer, forming the depigmented retinal rudiment.

All cells of the pigment epithelium are proliferative after removal of the retina. The total cell cycle in the anlage of the regenerating retina becomes 1.5 to 2.2 times shorter than it is in the pigment epithelial layer (Table 1).

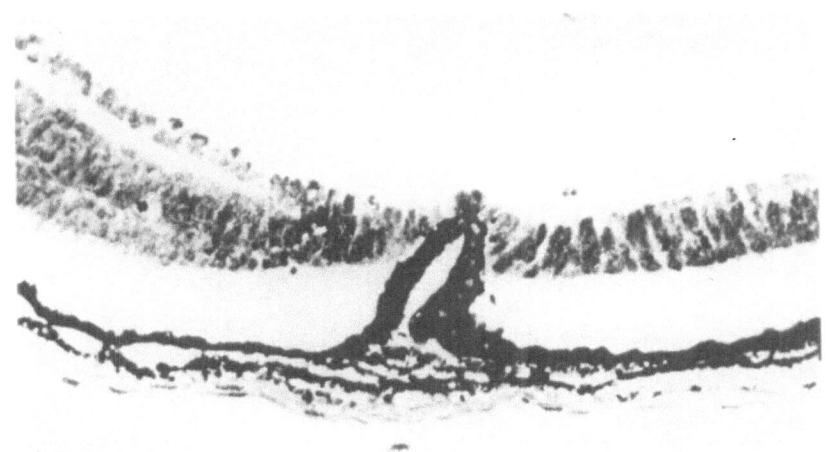

Fig. 2.   The choroid fissure (embryonic slit) in the ventral part of the eye in a control axolotl. This is a growth area of the retina and pigment epithelium. Many cells are labelled by ^{3}H-thymidine.

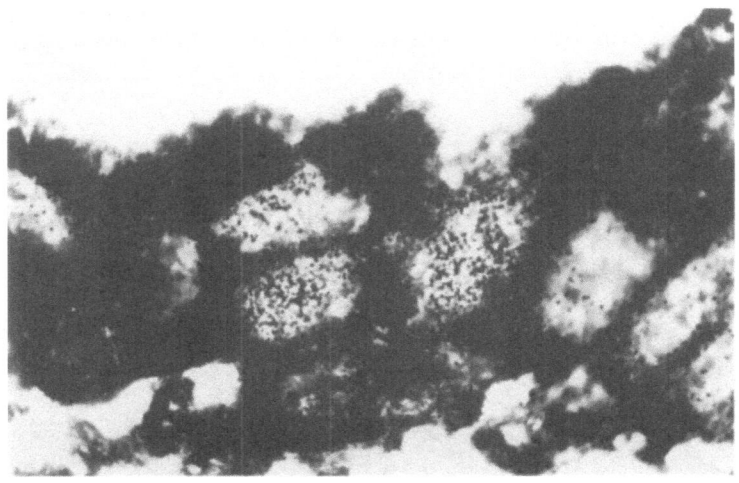

Fig. 3. Pigment epithelial cells in the choroid fissure
in a control axolotl. Approximately 20% of the
cells are labelled by ^{3}H-thymidine.

The proliferative activity of cells of the retinal rudiment
increases sharply after their formation from the pigment
epithelium. Another characteristic of the proliferative activity of cells of the pigment epithelium and regenerating retinal anlage is that the parameters of the cell cycles of these
cells resemble parameters of the cell cycles of corresponding
cells during embryonic development of the eye (Table 1).

The sharp changes in proliferative activity of the cells
of the retinal rudiment occur on the background of the depigmentation of these cells. This means that changes in the synthesis of certain specific proteins occur at fixed stages of
retinal regeneration.

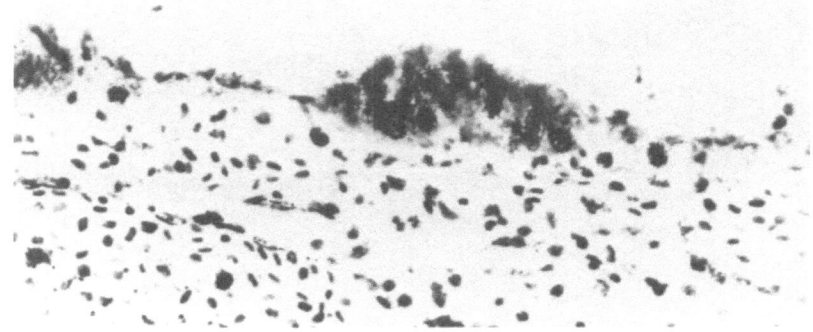

Fig. 4. A local zone of proliferation in the pigment
epithelium of albino Xenopus 20 days after
removal of the neural retina and iris. From
43-59% of the cells in areas like this are
labelled by ^{3}H-thymidine.

An autoradiographic study of the incorporation of ^{3}H-di-hydroxyphenylalanine (^{3}H-DOPA) by pigment epithelial cells illustrates this point. Cells of the normal pigment epithelium take up the isotope, but during the formation of the regenerating retinal rudiment, we did not observe the incorporation of ^{3}H-DOPA into the pigment epithelial cells in the central part of the eye. This is the area in which new retinal cells are formed from the pigment epithelium. In this situation, the transdifferentiation situation, the transdifferentiation of pigment epithelial cells into another pathway of cellular differentiation takes place upon the termination of synthesis of a specific product characteristic of the initial cell type.

We have also studied the role of cells of the pigment epithelium in regeneration of the neural retina in the axolotl and in Xenopus laevis. In axolotls, the neural retina regenerates from cells of the growth area, which is located along the peripheral part of the retina. The normal eye grows for a large proportion of the animal's lifespan. For our experimental work we used animals 6-9 months after metamorphosis. In these animals we examined the choroid fissure, which is an additional retinal growth area (Fig. 2). Many cells in this area were labelled by ^{3}H-thymidine, indicative for a subpopulation of pigment epithelial cells which is different from the remainder of the pigment epithelium. The main difference was that about 20% of the cells in this area incorporated ^{3}H-thymidine (Fig. 3), whereas only 1% of the cells in other regions of the pigment epithelial layer incorporated ^{3}H-thymidine.

We also grafted pigment epithelial cells into the eye cavity of the axolotl. In this experiment neural retina formed from pigment epithelial cells located in the choroid fissure. The same result was obtained in an experiment in which the retina, iris and cells of the growth area were removed from the eye. These results demonstrate that the eye

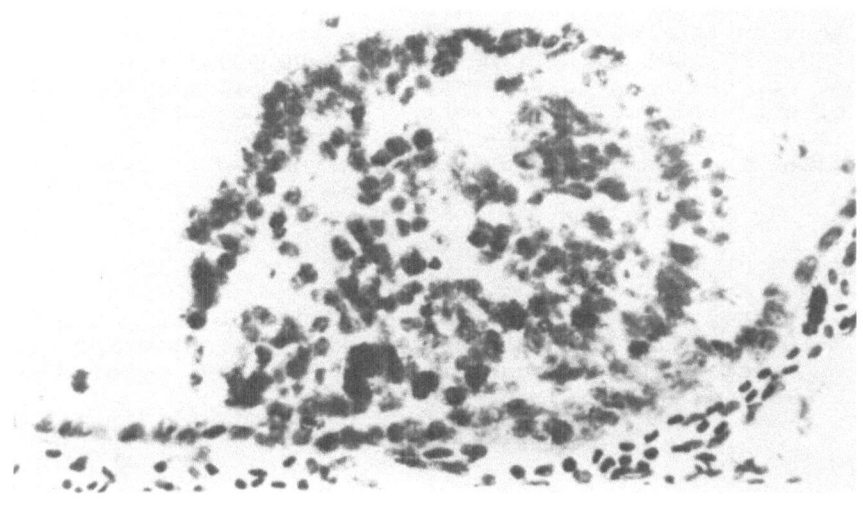

Fig. 5. A typical area of regenerating retina in albino Xenopus 75 days after removal of the neural retina and iris

of the axolotl has a subpopulation of pigment epithelial cells which have the ability to regenerate the retina.

Retinal regeneration has also been examined in the pigmented and albino strains of <u>Xenopus</u> <u>laevis</u>. In both strains the neural retina regenerates from cells of the growth area, like in the axolotl. However, when the retina and iris along with the growth area were removed in the albino animals, some local zones of proliferation were found in the remaining pigment epithelial layer 20 days after removal of the lens, iris and retina (Fig. 4). We found 10-15 local zones of proliferation per eye. From 43%-59% of the cells in these zones were labelled. Later, after the proliferative phase, structures like that illustrated in Figure 5 were found. In these structures, as well as in the neighboring pigment epithelium, cells incorporating ^{3}H-thymidine were found.

We have thus observed changes in certain properties of pigment epithelial cells during formation of the retinal rudiment in regeneration. According to some parameters, cells in the process of transdifferentiation are changed step by step into cells with embryonic characteristics. For example, proliferation of the cells of the retinal rudiment during regeneration follows the pattern of proliferation of cells of the embryonic retina. We also observed the termination of melanin synthesis.

For an explanation of the transdifferentiation phenomenon, the behavior of pigment epithelial cells of early stages of eye development and transdifferentiation were compared. At the stage of the optic cup in fishes, birds and mammals, the undifferentiated pigment epithelial cells can transform into neural retinal cells and vice versa. In embryological language, it can be said that these cells have regulative capacities. With this, the next question arises. During the process of dedifferentiation after removal of the retina, are the transdifferentiating cells changed in such a manner that they acquire regulative capacities similar to these in cells of the eye during the early stages of embryonic development? We propose that as the result of dedifferentiation (in newts and axolotls) or mutation (in albino <u>Xenopus</u>) the cells of the pigment epithelium acquire regulative properties characteristic of embryonic cells in fishes, birds and mammals, in which a new layer of neural retina forms from the wall of the optic cup after the neural retina has been removed or detached from it.

REFERENCES

Stroeva, O.G., and Mitashov, V.I., 1983, Retinal pigment
    epithelium: Proliferation and differentiation during
    development and regeneration, <u>Int. Rev. Cytol</u>.,
    83:221-293.

REGENERATION OF THE CENTRAL NERVOUS SYSTEM

OF THE ASCIDIAN <u>CIONA INTESTINALIS</u> (*)

Tomas Bollner

Department of Zoology, University of Stockholm
S-106 91, Stockholm, Sweden

SUMMARY

Regeneration of the neural complex of the protochordate <u>Ciona intestinalis</u> is morphologically described. The time needed to regenerate the central ganglion was found to be approximately 10-14 days at 18°C. Differentiated cells were characterized by staining for argyrophilia.

INTRODUCTION

The regenerative capacity of the adult ascidian is well known, ranging from replacement of minor parts to the complete development from a few cells only, especially in those genera that partly develop by budding (Berrill, 1950).

The central nervous system of the adult ascidian consists of two main parts, the cerebral ganglion and the neural gland. The ganglion is a compact structure, approximately 2 mm long in <u>Ciona intestinalis</u> with neurons arranged in layers around a fibrous medulla. The gland possesses a cavity which communicates with the pharynx via a ciliated duct and with the ovary by the so called dorsal strand (Fig. 1). The complete neural complex is developed from the left part of the embryonal tube (Grave, 1921; Elwyn, 1937). There is, however, still an open case whether the ciliated part of the duct is of neural origin or if it is derived from the pharyngeal epithelium.

It has been shown that ablation of the ganglion have little effect on animal response to external stimuli (Florey, 1951; Hoyle, 1952). Furthermore some reports state that a new ganglion forms shortly after the ablation (Bone, 1982). However, the pattern of the regenerative process has never been described before.

MATERIALS AND METHODS

Mature specimens of <u>Ciona intestinalis</u> were collected during scuba diving at the Tjarno Marine Biological Labora-

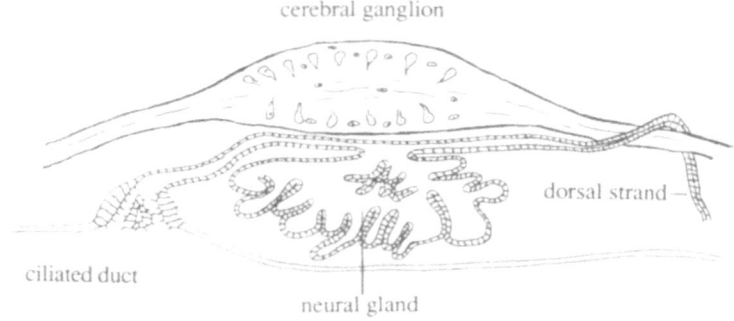

cerebral ganglion

dorsal strand

ciliated duct

neural gland

Fig. 1.   Schematic drawing of the neural complex
(sagittal view).

tory, Sweden. The animals were kept in tanks with through
flowing water at 12° or 18°C and allowed to recover overnight. They were then anaesthetized with MS222 (Sandoz) and
the complete neural complex was dissected out.

Following dissection the animals were again placed in the
tanks. After 2,4,6,12,18 and 24 days (12°) or 10, 12, 14, 17,
21, 28, or 35 days (18°) animals were anaesthetized and the
area around the ablated ganglion was dissected out and fixed
in cold Bouin's fluid for light-microscopy. From paraffin wax
embedded tissue 5 μm-10 μm sections were cut and stained with
either Harris haematoxylin-eosin or for argyrophilia according
to the Grimelius technique (1968) for the demonstration of
possible endocrine cells.

RESULTS

There was a clear difference between animals that were
kept at 12° and those kept at 18°, in that regeneration were
much quicker in the latter group. Since no animals from the
18° tanks were analyzed earlier than after 10 days post-opera-
tive time, the regeneration will be described as different
stages rather than at a time scale.

First stage

Animals kept at 12° and dissected 2-4 days after operation were characterized by high density of blood cells and
other small cells, sometimes elongated, building up a mesenchyme (Figs. 2,3).   After 6 days the mesenchymatous network
contained aggregations of slender cells. The epidermis was
still open during this period though covered by the tunic and
the density of tunic cells were much higher in these areas
than in any other.

Second stage

The second stage is characterized by the growth of the
ciliated duct and the dorsal strand. The ciliated duct was

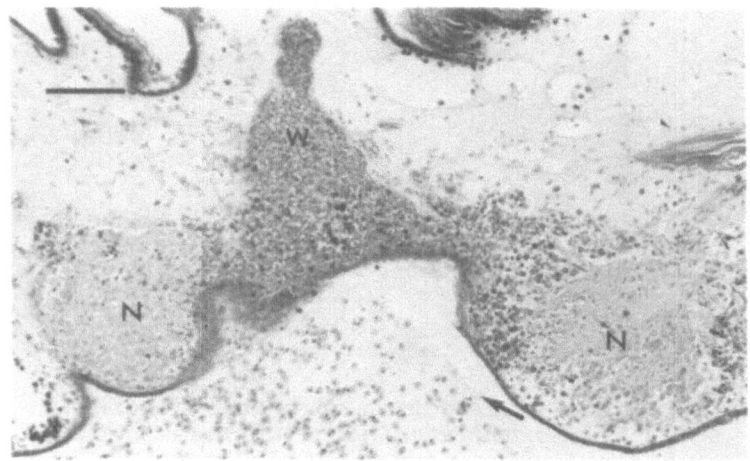

Fig. 2. Transverse section close to the cut showing
the mesenchyme (M) in the area between two cut
nerve trunks (N) and the tunic close to the
epidermis (arrow). Bar=50 μm.

often more developed than in non-operated animals. During the
growth these structures seemed to evaginate to form a new
neural gland. In close association to, as well as in these
new formations, clusters of small cells, thought to be proper
gland cells, were often observed (Fig. 4). This could be
observed after 6-12 days in the cold water group. Near the
end of the second stage, cells looking like normal neurons
appear for the first time, often near the duct or the strand
(Fig. 5). This happened after 12-18 days at 12° and only in
the least developed specimens kept at 18° for 10-14 days. Only
very few of these cells were stained by the Grimelius method
even though some cells of the ciliated duct showed staining.

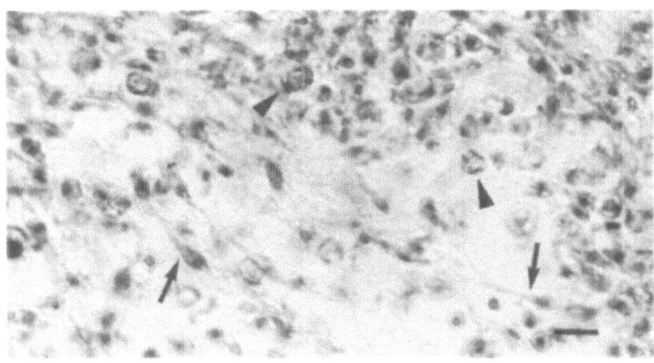

Fig. 3. High power micrograph showing elongated cells
forming a network (arrows) and also some blood
cells (arrow heads) in the mesenchyme. Bar=5
μm.

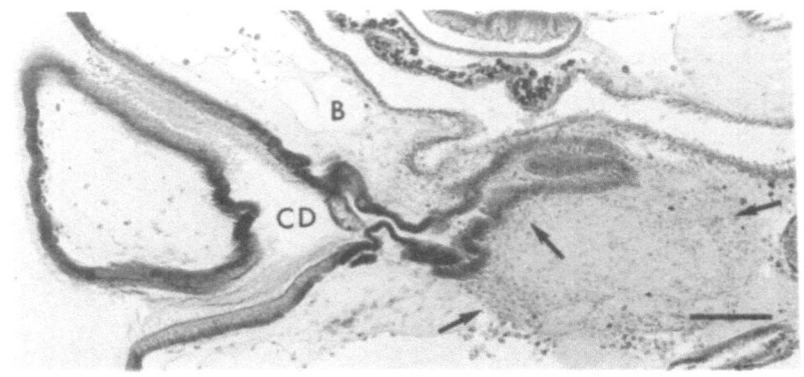

Fig. 4. Longitudinal section through regeneration area
with the ciliated duct (CD) mass of new gland
cells (arrows) and a blood sinus (B). Bar=50
μm.

Third stage

At this stage new cells of different sizes and shapes
were observed and they were arranged in small formations similar to a "normal" ganglion. These groups of new cells were
separated by foldings of the complex formed by the ciliated
duct, the gland and the dorsal strand. In the dissection
microscope several bundles were observed between the regenerate and each of the original nerve trunks as indicated in Fig.
6. As the regeneration continued the new ganglia expanded and
fused to one ganglion resembling the original one with cells
differentiated into neurons and arranged in layers. Cells
positive to the Grimelius stain were found in the glandular
structures as well as in the ganglion. The argyrophilic cells
of the ganglion were found both among the large and the
smaller cortical cells but rarely in the medulla (Fig. 7).

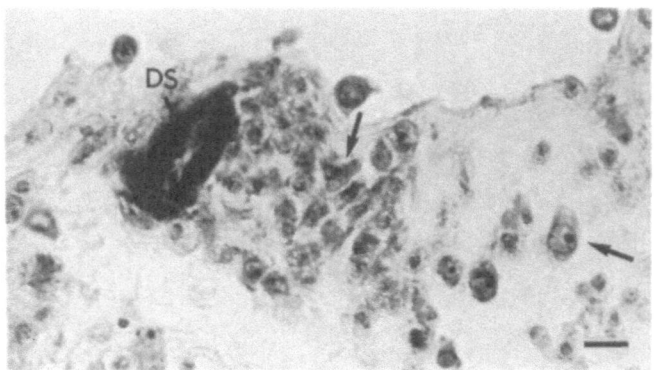

Fig. 5. High power micrograph of the area close to the
dorsal strand (DS) showing cluster of differentiating cells, the larger thought to be
neurons (arrows). Bar=5 μm.

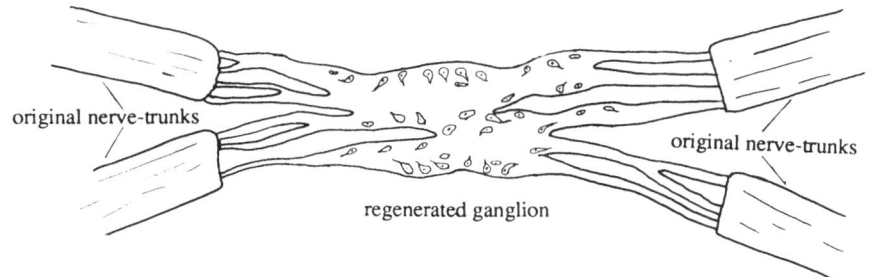

Fig. 6. Drawing illustrating how the neurons of the new ganglion form bundles that contact the old nerve trunks.

DISCUSSION

This investigation shows that even though all neuronal perikarya are ablated from the mature animal, <u>C. intestinalis</u> has the capacity to regenerate the complete central ganglion as well as the other components of the neural complex.

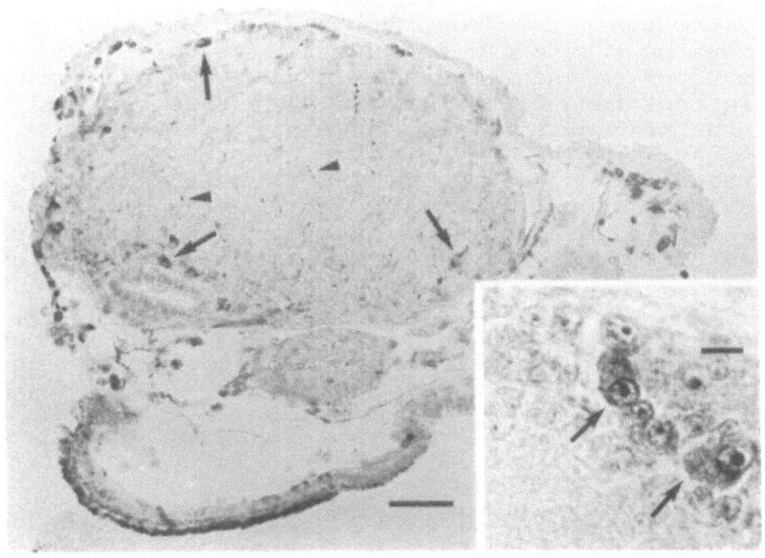

Fig. 7. Transverse section of regenerated ganglion stained with Grimellius' technique showing cells in the cortex (arrows) as well as fibers in the medulla (arrow heads). Inset: high magnification of Grimelius stained cells in the cortex (arrows). Bar=5 µm.

After the short period of wound healing during the first days following operation there seem to be a period when cells proliferate and partly occupy the empty space between the nerve stumps. Differentiation seem to start with the formation and growth of the neural gland and the ciliated duct and continue with the growth of larger cells interpreted as not yet fully differentiated neurons, as indicated by the lack of staining with the Grimelius' technique.

When the cells of the regenerated ganglion differentiate they send out axonal processes that grow towards the cut nerve stumps. It is possible that the neural gland is essential for axonal growth. And it is tempting to compare such a mechanism in the lower chordates with the importance of neuroepithelial cells in the guidance of axons in the newt proposed by Singer et al. (1979). This would be consistent with their theory that the neuroepithelium of the embryo play a role similar to that in the regenerate, since the gland is developed from the embryonal neural tube.

The regenerating neural gland and ciliated duct may come from growth of the dorsal strand. The origin of the new ganglion cells is still unknown but the mass of blood-cells and the location of the new ganglion in the close vicinity of the dorsal strand-gland duct tissue might indicate the importance of these structures in the regenerative process.

In the normal ganglion the cells containing argyrophilic granules, as demonstrated by the Grimelius stain, are to some degree identical with calcitonin as well as with prolactin and pancreatic polypeptide immuno-like cells. (Fritsch et al 1979; 1982). The distribution of these cells in different parts of the medulla is consistent with the findings in the normal ganglion. Therefore the argyrophilic cells might represent neurons that are differentiated at this stage. Further investigations on the occurrence of putative neurotransmitters and neuromodulators might indicate the state of differentiation of the regenerating ganglion.

Nothing is known about the substances active in the regenerating process. There is some evidence in Platyhelminthes for a role for substance P in regeneration phenomena (Salo and Baguna, 1986), and since that as well as other tachykinins are present also in the tunicate nervous system (O'Neil et al.,1987), they might be of importance. Several other peptides have been found and localized in nervous tissue of Ciona and other Tunicates by immunocytochemical techniques (Van Noorden, 1984, Thorndyke and Georges, 1987).

REFERENCES

Berill, N.J., 1950, "The Tunicata", Ray Society, London.
Bone, Q., 1982, Urochordata, in: "Electrical conduction and behaviour in "simple" invertebrates", A.B. Shelton ed., Oxford University Press, Oxford, pp. 473-535.
Elwyn, A., 1937, Some stages in development of the neural complex in Ecteinascidia turbinata, Bull neurol. Inst., N.Y., 6:163-177.

Florey, E.,1951, Reizphysiologische Untersuchungen ad der Ascidie _Ciona intestinalis L._, _Biol. Zentralbl._, 70:523-570.

Fritsch, H.A.R., Van Noorden, S., and Pearse A.G.E., 1979, Localization of Somatostatin, Substance P- and Calcitonin-like Immunoreactivity in the Neural Ganglion of _Ciona intestinalis L._ (Ascidiaceae), _Cell Tissue Res._, 202:263-274.

Fritsch, H.A.R., Van Noorden, S., and Pearse A.G.E., 1982, Gastro-intestinal and neurohormonal peptides in the alimentary tract and cerebral complex of _Ciona intestinalis_ (Ascidiaceae), _Cell Tissue Res._, 223:369-402.

Grave, C., 1921, _Amaroucium constellatum_ (Verril) II The structure and organization of the tadpole larva. _J. Morphol._, 36:71-101.

Grimelius, L., 1968, A silver nitrate stain for alpha cells in human pancreatic islets, _Acta Soc. Med._ Uppsala, 73:243-270.

Hoyle, G., 1952, The response mechanism in ascidians, _J. Mar. Biol. Ass. U.K._, 31:287-305.

O'Neil, G.S., Conlon, J.M., Deacon, C.F., and Thorndyke, M.C., 1987, Tachykinins in the central and peripheral nervous system of the ascidian _Ciona intestinalis_, _Gen. Comp. Endocrinol._, 66:314-322.

Salo, E. and Baguna, J., 1986, Stimulation of cellular proliferation and differentiation in the intact and regenerating Planarian _Dugesia (G) tigrina_ by the neuropeptide substance P., _J. Exp. Zool._ 237:129-135.

Singer, M., Nordlander, R.H., and Egar, M., 1979, Axonal guidance during embryogenesis and regeneration in the spinal cord of the newt: The blueprint hypothesis of neuronal pathway pattering, _J. Comp. Neurol.)_, 185:1-21.

Thorndyke, M.C., and Georges, G., 1987, Functional aspects of peptide neurohormones in protochordates, _in_: M.C. Thorndyke, G.J. Goldsworthy, eds., "Invertebrate Peptides and Amines", Cambridge University Press, Cambridge.

Van Noorden, S., 1984, The neuroendocrine system in protostomian and deuterostomian invertebrates and lower vertebrates, _in_: "Evolution and Tumour Pathology of the Neuroendocrine System", S. Falkmer, R. Hakanson, F. Sundler, eds., Elsevier, Amsterdam, New York, Oxford, pp. 7-38.

NEUROPEPTIDES CONTROL REGENERATION IN PLANARIANS (*)

Ch. P. Tiras and I. M. Sheiman

Institute of Biological Physics
USSR Academy of Sciences
Pushchino, Moscow Region, I42292, U.S.S.R.

Among existing regulators of regeneration neuropeptides attract particular attention as active modulators of various physiological functions in animals. It seems to be important to systematize the effects of neuropeptides on morphogenesis. Head regeneration of flatworms planarians may serve as a convenient model. The planarians used in the present study belong to the asexual race <u>Dugesia tigrina</u>. Regeneration was estimated using the method of vital computer morphometry based on registration of photocontrast between old (pigmented) and novel (without pigment) body parts of regenerates. Then images of planarians were put into computer memory by digitizer and relative criteria of regeneration were calculated. The criteria were based upon restoration of the initial relations between the lengths of the head and the total body and between the head area and the area of the whole body.

Normal Regeneration of Planarians

After cutting, all dimensional parameters change: the total area and length of the body decrease as well as the profile area of planarians, i.e. the animals become shorter and flatter while the area and length of blastema increase. The width of regenerates increases sharply just after cutting but then decreases to the initial state by the 20th day of regeneration.

During the first five days regeneration proceeds very rapidly, then the process slows down. By the 10th day after cutting, the initial relation between the head length and the total body length is restored completely. By the 30th day the relation between the areas of the head and the whole body reaches 100%. According to the data obtained one may indicate three stages of planarian regeneration: rapid growth (i), restoration of the relations between lengths (ii) and areas (iii) of the head and body.

The Effect of Neuropeptides on Regeneration

One-week starved animals were decapitated and placed (usually five) in 50 ml of peptide-water solution. Control group of animals was kept in pure water. 30 animals were used

289

in each series. Regeneration was observed during 10 days. The experiments were repeated no less than three times. Bioassay of peptides was carried out in the range of concentrations from 10^{-7} to 10^{-13}M; 10^{-9}M was a standard concentration. The following neuropeptides were used: arg-vasopressin, enkephalin analog-dalargin, hydra head activator (synthesized in All-Union Cardiologic Scientific Center, USSR Academy of Medical Sciences); luteinizing hormone - releasing hormone (LH-RH) (Institute of Bioorganic Chemistry, Byelorussian SSR Academy of Sciences); delta-sleep-inducing peptide (Institute of Bioorganic Chemistry, USSR Academy of Sciences) and somatostatin ("Diamalt"). The characteristics mentioned above were compared in control and experimental groups of animals using Student's t-criterion. Effective estimation of a peptide was obtained as an arithmetical mean of t-criteria in the whole course of regeneration.

It has been revealed that neuropeptides may stimulate, inhibit and even produce no effect on morphogenesis. Arg-vasopressin, somatostatin and dalargin were classified as stimulators. Hydra head activator being the most similar phylogenetically to planarians, also stimulated regeneration and asexual reproduction. Moreover as it turned out the latter neuropeptide is more effective in stimulating the regeneration of a head than a tail part. Sleep-inducing peptide had no effect on planarian regeneration and was classified as neutral one. LH-RH produced twofold inhibition of morphogenesis as compared to the control.

It should be noted that the neuropeptides differed in their effects depending on the stage of regeneration. For instance, arg-vasopressin stimulates regeneration from the very beginning, dalargin is effective only at the second stage of regeneration, i.e. in five days after cutting, and LH-RH inhibits morphogenesis in the whole course of regeneration.

Thus, neuropeptides of both vertebrates and invertebrates may regulate regeneration in planarians. The model described can be used to reveal morphogenetic functions of neuropeptides and other biologically active regulators for their subsequent classification.

Commentary

by Margaret W. Egar, on

NEURAL ASPECTS OF REGENERATION

The role of nerves in regeneration is a major area of concern and interest, with the papers presented at this workshop touching on broad aspects of neuronal involvement. Emphasis was placed on the multiplicity and complexity of the roles played by the nervous system and the differences that have been found between developing limbs (limb buds) and regenerating limbs (blastemata), as well as the differences that exist with age and between species.

The motoneuron response to regeneration after supernumerary limb formation, produced by grafting or by treatment with retinoic acid, differs slightly; possibly due to the age difference in these two paradigms (Lheureux; Holder). Even in axolotl adults, new neurons are being formed and thus regeneration may represent continued neurogenesis in this species (Holder). Ganglion formation during urodele tail regeneration is more irregular and probably is derived from a different cell of origin than ganglion formation during development (Geraudie). Tadpoles regenerate well, but froglets respond with only some tissue regrowth; although this response is still dependent upon nerves (Liversage). Age and species differences are evident in retinal regeneration (Mitashov).

After denervation, the accumulation of collagen type IV was characteristic of cone stage newt blastemata (Lassalle) while the opposite effect for hyaluronic acid was described in axolotls (Mescher). As the nerves grow back into the axolotl amputation stump, the dedifferentiated cells increase their production of this extremely large extracellular matrix molecule at the same time that the mitotic index is rising. The presence of nerves thus selectively alters the dedifferentiated cell's production of specific proteins, which in turn alters the extracellular environment, the cell density, cell migration and the general appearance of the mesenchymal tissue deep to the wound epithelium. Nerves also play a role in the production of a heavy basement membrane that may form a barrier between mesenchymal interaction and the overlying wound epidermis that is essential to successful regeneration (Lassalle).

The importance of wound epithelial interaction with the nerve and mesenchymal accumulation was stressed by several speakers. There may be an indirect action of the nerve in preventing cellular production of collagen IV (Lassalle). The interaction of wound epithelium and the nerve is essential in the formation of accessory limbs (Egar). In the deficient regeneration characteristic of froglets (Liversage), denervation results in the compaction, increase in extracellular matrix and finally, resorption of the regenerated tissue.

Even in these systems with limited tissue regeneration, the presence of neural tissue is necessary for this minimal response to amputation (Liversage). Here too, there is evidence of a quantitative requirement of the neurotrophic factor. Total denervation resulted in resorption of the regenerated tissue, partial denervation resulted in decreased response, and two implanted ganglia supported the regenerative response while one alone was insufficient to replace the missing nerves. This quantitative requirement of the nervous tissue was also stressed in accessory limb production where the required large amount of neurotrophic factor available may also require the large numbers of accompanying Schwann cells (Egar).

The outstanding fidelity with which central neurons reestablish contact with regenerated and with additional muscles, in the cases of reduplicated and supernumerary limbs, was explored in the newt (Lheureux) and in the axolotl (Holder) where species differences and development/regeneration differences were pointed out. Neuronal plasticity was evident in the younger experimental animals and to a lesser extent in older animals. In amphibian retinal regeneration, where levels of proliferative activity were reported to be species specific and where even a single mutation (albinism in Xenopus) resulted in a very different pattern of regeneration, striking species differences were seen, as well as age differences and differences dependent upon the method of retinal removal (Mitashov).

Glial influences were emphasized in the phrenic phenomenon (Goshgarian) and in thoracic spinal cord regeneration (Holder). The rapid activation (2-4 hours) of latent pathways in the phrenic nucleus involved glial-process withdrawal from between neuronal cell bodies with the formation of neuronal-neuronal contacts, and more frequent multiple synapses. These morphological changes in a motor pathway underline what physiologists have known for some time about sensory pathways, that ineffective synapses can become effective under some conditions of injury (and in several different areas of the central nervous system). In the case of the phrenic nucleus, rapid morphological changes were documented for the first time.

The influence of radial glia as cues in axonal regeneration differs in different areas of the central nervous system (Holder). Axons regenerate into a reoriented thoracic cord segment but never travel for very long in a wrong area. They rapidly grow to a corrected position but the information for this correction is not coming from long distances such as a target area, rather from the nearby radial glia. This sort of local glial sign-posting was predicted by the blueprint hypothesis of pathway formation (Singer).

An all important aspect of these presentations was that the multiple functions of the nervous system should not be ignored in studies of regeneration. Professor Singer pointed out that the Nobel laureate, T.H. Morgan (who made no mention in his book on regeneration, of Todds' work with the nerves) has given up on studying the complexities of regeneration in favor of the "simpler" area of genetics. The "genetics of regeneration" would be the avenue for future research, predicted Singer, and he implied that neurons and their supportive cells would be a major part of this avenue.

THE RETINOIDS

IN

REGENERATION

THE USE OF RETINOIDS TO EXPLORE THE CELLULAR AND MOLECULAR BASIS OF POSITIONAL MEMORY IN REGENERATING AMPHIBIAN LIMBS

David L. Stocum

University of Illinois
Department of Cell and Structural Biology
506 Morrill Hall, 505 S. Goodwin Ave.
Urbana, IL 61801, USA

THE CONCEPT OF POSITIONAL MEMORY

Epimorphic limb regeneration in amphibians, by definition, must include replacement of those parts that are amputated from the whole. But to make a functional regenerate, a stringent constraint is that the regeneration blastema not duplicate any structure proximal to its level of origin. Redifferentiation of only those structures distal to the amputation plane by the blastema is known as the "rule of distal transformation" (Rose, 1970). The mechanism underlying the rule of distal transformation has been investigated since the end of the 19th century. At a first level, we can ask whether the mechanism is intrinsic or extrinsic to the blastema. If the former, the blastema is an independently differentiating tissue; if the latter, it is a nullipotent or pluripotent tissue requiring an inductive signal from the adjacent differentiated limb tissues to specify its pattern of redifferentiation.

The results of transplantation experiments in the 1920s and 30s, in which heterotopically grafted blastemata appeared to develop according to their new location, supported the inductive hypothesis. These studies, however, were inconclusive, due to a lack of cell markers in the grafted tissue, and later experiments showed that the factors determining regenerate morphology and polarity are intrinsic to the blastema (see Wallace, 1981; Stocum, 1984, for reviews). Of particular interest are the results of experiments in which blastemata are grafted to more proximal or distal levels of the limb. In either case, the blastema develops strictly according to its level of origin The level disparity created by a proximal to distal shift is not recognized (Stocum and Melton, 1977), but cell marking experiments show that the pattern discontinuity created by a distal to proximal shift is eliminated by intercalary regeneration of the missing intermediate structures from the host limb stump (Pescitelli and Stocum, 1980). The autonomy of blastema development has led to the concept that its cells inherit a memory of their PD (proximodistal) level of origin. This <u>positional memory</u> (Carlson, 1983) constitutes a "lock" that prevents the blastema from forming patterns proximal to the amputation plane, thus defining the proximal boundary of the regenerate (Stocum, 1983). The PD axis of the limb can thus be viewed as a graded series of level-specific positional memories. Positional memory is thought to be impressed on limb cells by signals or cues that specify the original pattern of the limb during its embryogenesis (Wolpert, 1971; Carlson, 1983).

Grafting experiments in which the anteroposterior (AP) or dorsoventral (DV) axes of the blastema are reversed with respect to the stump, indicate that cells have level-specific positional memories along these axes as well. Reversal of the AP or DV axis triggers intercalary regeneration between the anterior and posterior or dorsal and ventral poles of blastema and stump, creating extra blastemata that develop into supernumerary limbs at the sites of intercalation (Bryant and Muneoka, 1986, for review). Cell marking experiments indicate that these limbs are derived from the edges of both host and graft (Stocum, 1982; Muneoka and Bryant, 1984). The structure of supernumeraries resulting from reversal of both the AP and DV axes of the blastema are more complex (Maden and Mustafa, 1984), but can also be interpreted in terms of positional memories.

Supernumerary regenerates are also obtained by rotating skin cuffs around the PD axis, or by the cross-transplantation of flexor and extensor muscles, provided that these manipulations are followed by amputation and the formation of blastema cells (Carlson, 1974, 1975b). The requirement for blastema formation indicates that positional memory is expressed only when limb cells dedifferentiate. Axolotl limbs with cross-transplanted muscles have been maintained up to two years before amputation, and still formed supernumerary regenerates, indicating that positioned memory is extremely stable (Carlson, 1975a).

Positional memory probably resides exclusively in the nonskeletal connective tissues of the limb, and the muscle (Carlson, 1983). Limbs whose skin has been replaced by tail epidermis regenerate normally after amputation, but skinned limbs receiving tail dermis regenerate tail-limb chimeras, even though the blastema becomes covered with limb epidermis (Glade, 1963). Axial rotation of bones or cuffs of epidermis has no effect on regeneration, but rotation of dermal cuffs results in supernumerary regenerates following amputation (Carlson, 1975b). It has been possible to test the connective tissue and myogenic components of muscles separately to determine which one or both are endowed with the positional memory carried by muscle. Freeze-thawing or x-irradiation of dermis abolishes its ability to elicit supernumerary formation after axial displacement (Tank, 1981). Provided these treatments do not alter molecular components of the matrix in a way that compromises their function, this result suggests that positional memory is a property of living cells, not extracellular matrix.

AN APPROACH TO ANALYZING THE CELLULAR BASIS OF
POSITIONAL MEMORY

Clearly, positional memory is an important component of
the mechanism that restores amputated limb parts, and a number
of formal models have been proposed to account for regeneration
in terms of intercalation between boundaries (see Tank and
Holder, 1981,for review; Mittenthal,1981; Meinhardt, 1983). A
deeper understanding of this mechanism requires knowledge of
the cellular and molecular basis of positional memory. One
approach to this problem is to identify cellular properties
that exhibit position-related differences along a limb axis
and which therefore might bear a direct relationship to positional memory. To demonstrate a direct relationship between
these properties and positional memory then requires showing
that the experimental modification of regenerate pattern is
paralleled by modification of the cellular property.

Two assays indicate that cell affinity is one property
that exhibits level-specific differences along the PD axis in
regenerating limbs. In one assay, blastemal mesenchymes from
the wrist or tarsus, elbow or knee, and upper arm or leg were
cultured _in_ _vitro_ in the nine possible binary combinations.
Pairs of mesenchymes from the same level (wrist/wrist,
etc.) fused to make a straight interface, whereas in pairs of
blastemata derived from different levels, the proximal
blastema attempted to surround the distal one. The engulfment
behavior was hierarchical, elbow and knee blastemata surrounding wrist and ankle blastemata and in turn being surrounded by
upper arm and leg blastemata (Nardi and Stocum, 1983). These
results suggest the existence of position-related differences
in cell adhesiveness along the PD axis (cf. Steinberg, 1970).

Unfortunately, the blastemal mesenchyme do not undergo
patterned redifferentiation _in_ _vitro_, so this assay is unable
to reveal whether experimentally-induced changes in regenerate
pattern are accompanied by coordinate changes in blastema cell
affinity. Therefore, we devised an _in_ _vivo_ assay that would
permit both blastemal morphogenesis and the detection of level-specific differences in blastema cell affinity (Crawford
and Stocum, 1988a). As diagrammed in Fig. 1, this assay consists of auto-or homografting forelimb blastemata derived from
three different levels (wrist, elbow, mid-upper arm) to the
mid-thigh level of regenerating hindlimbs. The grafts were
placed with their basal surface on a wound bed made on the
dorsal surface of the host blastema-stump junction, and
allowed to develop.

The results of this experiment were quite striking (Fig.
1). Each blastema developed with forelimb morphology and
according to its level of origin. At the same time, the
blastema was displaced distally to its corresponding level on
the PD axis of the host regenerate, thus eliminating the level
discontinuity between host and graft. Wrist regenerates
articulated with the host ankle, and elbow regenerates with
the host knee, while mid-upper arm regenerates remained at the
graft site. Like the _in_ _vitro_ assay, these results indicate
the existence of level-specific differences in blastema cell
affinity along the PD axis. Accordingly, we have termed the
blastemal displacement, or sorting behavior, "affinophoresis".
The mechanism of affinophoresis has not yet been analyzed.

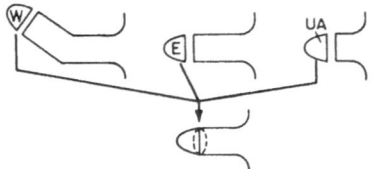

DONOR FORELIMBS REGENERATING FROM
WRIST, ELBOW OR MID-UPPER ARM

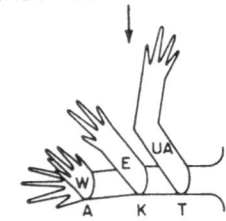

HOST HINDLIMBS REGENERATING FROM MID-THIGH

DISPLACEMENT OF GRAFTS
TO THEIR CORRESPONDING HOST LEVELS

Fig. 1. Affinophoresis assay. Wrist (W), elbow (E) or
mid-upper arm (UA) blastemata are grafted to
the blastema-stump junction of a host hindlimb
regenerating from the mid-thigh. The grafts
develop according to their level of origin,
and displace distally to their corresponding
levels on the host regenerate (A, ankle; K,
knee; T, mid-thigh).

Another conclusion that can be drawn from these results is
that very similar or identical sets of level-specific affinity
properties are shared by forelimb and hindlimb cells.

RETINOIDS MODIFY POSITIONAL MEMORY

To explore the relationship between level-specific
blastema cell affinity and positional memory, a pattern modifying agent is required. Such agents are the retinoids, natural and synthetic compounds related to retinol (vitamin A). A
number of retinoids have been assayed for their efficacy in
modifying regenerate pattern. The most effective of these is
the synthetic retinoid, arotinoid, followed by the natural
retinoids retinoic acid (RA), retinol, retinol palmitate and
retinyl acetate (Maden, 1982; Kim and Stocum, 1986c). Because
of its effectiveness and availability, RA is the most widely
used for pattern studies.

In the PD axis, positional memory of blastema cells is
__proximalized__ by RA, resulting in the formation by the regenerate of copies of limb structures proximal to the amputation
plane (Niazi and Saxena, 1978; Maden, 1982, Thoms and Stocum,
1984). The extent of proximalization is dose-dependent, the
effect of the RA is on limb mesodermal tissues, not epidermis

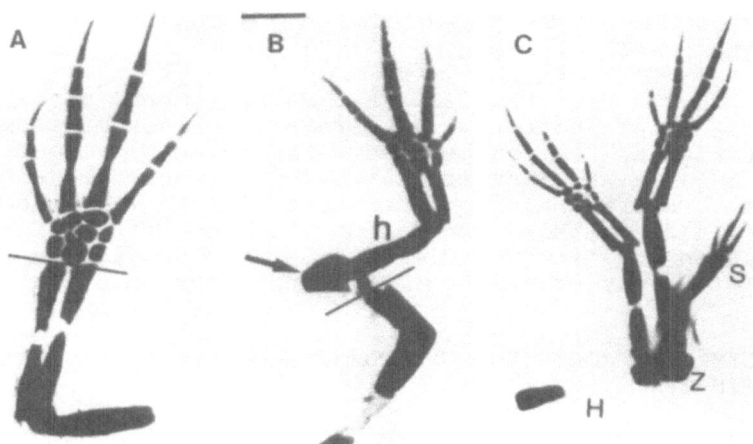

Fig. 2. Effect of intraperitoneal injection of RA [dissolved in dimethylsulfoxide (DMSO)] during the dedifferentiation phase of axolotl limbs regenerating from the wrist joint; methylene blue-stained whole mounts. (A) Normal regenerate of control animal injected with DMSO alone. Line indicates plane of amputation. (B) Regenerate of animal injected with 100 µg RA/gbw (gram of body weight). Line indicates regenerate/stump junction; extensive regression of the stump took place. The blastema duplicated a complete limb in the PD axis. Arrow, partial shoulder girdle; h, humerus. (C) Regenerates derived from double anterior lower arm; animal injected with 100 µg RA/gbw. Each anterior half completed the AP pattern to give a left and right limb, which was also duplicated in the PD axis. H=stump humerus; Z=stump radius and ulna (which regressed considerably). S=supernumerary formed by juxtaposition of anterior graft and posterior host tissue. Note that the supernumerary is duplicated in the PX axis. Bar=2 mm.

(Maden, 1984), and it is the blastemal mesenchyme cells that are actually proximalized, as opposed to say, the induction by RA of the distal migration of proximal stump cells to the amputation plane (Thoms and Stocum, 1984; Wallace and Maden, 1984). RA causes the production of supernumerary limb structures from the anterior side of regenerating frog tadpole hindlimbs (Niazi and Saxena, 1978; Maden, 1983a) and developing chick limbs (Summerbell, 1983). In amputated anterior half or double half limbs of axolotls, RA induces completion of the AP pattern, in addition to proximalizing the PD pattern, but inhibits the regeneration of posterior half or double half limbs (Kim and Stocum, 1986a). Similarly, RA induces completion of the DV pattern of amputated double dorsal half limbs, but inhibits the regeneration of double ventral half limbs (our unpublished results). These observations suggest that RA <u>posteriorizes</u> positional memory in the AP axis, and <u>ventralizes</u> it in the DV axis. Figure 2 illustrates examples of the effects of RA on the PD and AP axes of the regenerating axolotl limb.

It should be emphasized that RA treatment, at doses that modify regenerate pattern, first retards blastema formation. The retardation is associated with an inhibition of mitosis (Maden, 1983b), and lasts for a period of time (7-10 days) that is proportional to the dose of RA (Kim and Stocum, 1986b). In fact, this retardation is the first indication that pattern modification will take place, suggesting that the effects of RA may be exerted during this period.

RA COORDINATELY MODIFIES POSITIONAL MEMORY AND BLASTEMA CELL AFFINITY

We used RA-treated blastemata in the affinophoresis assay to demonstrate a direct relationship between blastema cell affinity and positional memory (Crawford and Stocum, 1988a). In the first experiment (Fig. 3), wrist, elbow and mid-upper arm blastemata of RA-treated animals were homografted to the blastema-stump junction of a host hindlimb regenerating from the mid-thigh. In the second experiment (Fig. 4), host animals were amputated through the ankle, and treated with RA. Then, untreated wrist, elbow and mid-upper arm blastemata were homografted to the blastema-stump junction of the RA-treated host hindlimb. RA-treated blastemata grafted to a normal host limb were proximalized to the level of the proximal stylopodium or girdle, but failed to displace distally, regardless of their level of origin (Fig. 3). RA-treated host hindlimbs were proximalized to the level of the proximal stylopodium or girdle, and the untreated donor blastemata, displaced distally to their corresponding levels on the host regenerate (Fig. 4). The results of both experiments show the same thing: that RA coordinately proximalizes blastema cell affinity and positional memory, leading to displacement and restoration of a normal neighbor pattern.

As a further test of the relationship between positional memory and blastema cell affinity, we compared the ability of control and RA-treated wrist or tarsus and elbow or knee blastemata to evoke intercalary regeneration from the mid-stylopodial level of forelimb or hindlimb stumps (Crawford and Stocum, 1988b). Control blastemata evoked normal intercalary regeneration. By contrast, RA-treated blastemata were proximalized to the level of the mid-stylopodium or beyond, and failed to evoke intercalary regeneration (Fig. 5). Thus, these results also support a direct relationship between positional memory and a blastema cell affinity property involved in recognizing pattern discontinuity.

Retinoids have been shown to modify the adhesion, growth, morphology and phenotypic differentiation of a variety of cultured normal and cancer cells (Roberts and Sporn, 1984). These retinoid-induced changes are correlated with changes in the expression and organization of cytoskeletal components, and the expression of various enzymes, receptors, oncogene proteins, and a variety of other proteins, as well as with qualitative and/or quantitative changes in cell surface glyco-conjugate composition and extracellular matrix composition (Roberts and Sporn, 1984; Chytil, 1986; Sporn et al., 1986).

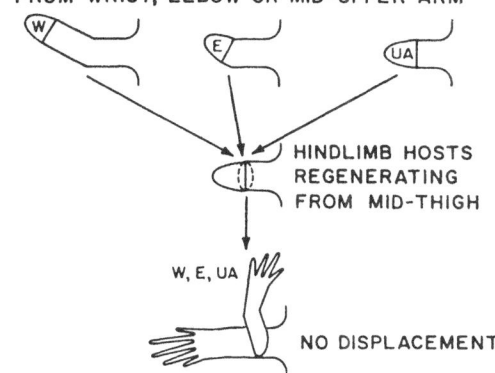

RA-TREATED FORELIMB DONORS REGENERATING
FROM WRIST, ELBOW OR MID-UPPER ARM

HINDLIMB HOSTS
REGENERATING
FROM MID-THIGH

W, E, UA

NO DISPLACEMENT

Fig. 3. Proximalization of donor regenerates, and abo-
lition of their displacement in the affinophoresis assay after RA treatment. Abbreviations
as in Fig. 1.

The adhesivity of blastema cells in regenerating limbs also
seems to be affected by RA treatment, because when the
blastema forms after the period of inhibition, its cells exhibit closer packing (Maden, 1983; Kim and Stocum, 1986b) that
is associated with the abnormal retention of fibronectin on
the cell surfaces (Maden and Keeble, 1987).

The effects of retinoids on cell adhesion fit well with
our data (Nardi and Stocum, 1983; Crawford and Stocum,
1988a,b) indicating that the positional memory of blastema
cells is related to level-specific differences in cell adhesiveness. Positional memory is thus likely to be based in the
composition and organization of cell surface molecules. There
is a wealth of indirect evidence indicating that differential
cell adhesiveness is one of the major forces driving morphogenesis (Townes and Holtfreter, 1955; Steinberg, 1963,
1978), and that cell surface glycoconjugates mediate cell recognition and adhesivity relationships that are important for
cell-cell and cell-substrate interactions during embryogenesis
(Edelman, 1983, 1985; Thiery et al., 1985; Gallin et al.,
1986).

ENDOGENOUS RA AND PATTERNS SPECIFICATION

The fact that exogenous RA has effects on regenerate pattern makes it reasonable to expect that it might be involved
in the patterning of normal regenerates. This has not been
demonstrated for the regenerating limb, but a gradient of
endogenous RA, with its high point posterior, has been measured across the AP axis of the embryonic chick limb bud (Thaller and Eichele, 1987). The mechanism whereby the gradient is
produced appears to be the metabolism of retinol to RA by
posterior limb cells (Thaller and Eichele, 1988). Further
evidence that RA is involved in specifying AP pattern is that

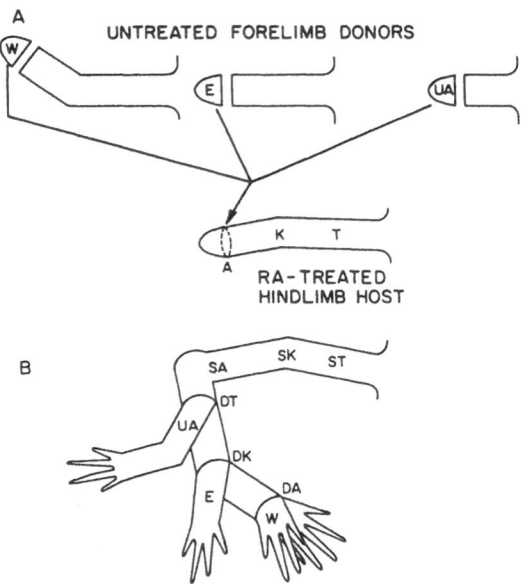

Fig. 4. Distal displacement of forelimb blastemata to
the matching level of a host ankle regenerate
that has been proximalized to the level of the
girdle by RA treatment. SA=stump ankle;
SK=stump knee; ST=stump thigh; DT, DK,
DA=thigh, knee and ankle of proximalized
regenerate. Other abbreviations as in Fig. 1.

ion exchange beads, soaked in RA and implanted under the
ectoderm at the anterior edge of the wing bud, create a second
RA gradient that is associated with the formation of a mirror-image set of digits in the AP axis. No such duplication
is observed when the beads are implanted into the posterior
edge of the wing (Tickle et al., 1985). RA-impregnated beads
thus mimic the action of the zone of polarizing activity (ZPA)
(Fallon and Crosby, 1977), which is postulated to specify AP
axial pattern by serving as the source of a gradient of morphogen to which limb cells respond (Tickle et al., 1985).
Although morphogen gradients have been invoked as pattern-
specifiers since the beginning of the century (Morgan, 1901),
RA appears to be the first one to be identified. Amphibian
limb regeneration blastemata respond somewhat differently than
chick limb buds to certain experimental manipulations, and to
exogenous RA, so it will be quite interesting to see if there
are gradients of RA along any of their axes, and whether one
of the effects of RA on positional memory might be to change
the rate which retinol is metabolized by blastema cells.

MOLECULAR ANALYSIS OF POSITIONAL MEMORY

Given its effects on pattern formation, the next question
is, what is the molecular mechanism of action of RA on cells?
This is extremely important to know, because it will give us
insights into the molecular basis of positional memory, and

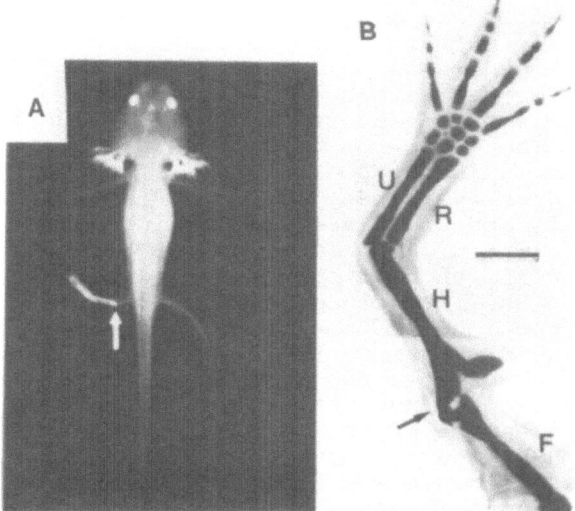

Fig. 5. (A) RA-treated wrist regenerate of a dark
axolotl grafted to the mid-thigh of a white
axolotl (2/3 life size). The blastema was proximalized to the level of the proximal upper
arm, and no intercalary regeneration took
place. Arrow, regenerate/stump junction.
(B) Whole mount showing skeletal structures of
a regenerate similar to (A). F=stump femur;
arrow=regenerate/stump junction; H,R,U=hum-
erus, radius and ulna of regenerate.

other aspects of pattern regulation. It is now generally
agreed that retinoids affect phenotypic changes by influencing
the genome to alter patterns of gene transcription. Retinoid
treatment induces changes in the sizes of keratin transcripts
and keratins synthesized by human keratinocytes (Fuchs and
Green, 1981; Eckert and Green, 1984). Similarly, the enhanced
synthesis of laminin and fibronectin induced in RA-treated F9
embryonal carcinoma cell is under transcriptional regulation
(Wang et al.,1985). Furthermore, treatment of cultured 3T3
cells with retinol increases the sensitivity of their chromatin to DNAase I digestion, suggesting that retinol causes
alterations in chromatin structure that allow greater accessibility of the DNA to RNA polymerase II (Ferrari and Vadali,
1985).

By what mechanism do retinoids alter patterns of gene
transcription? Two components have been identified that may
mediate retinoid effects on transcription. First, two cytosolic binding proteins for retinol and RA, with a molecular
weight of 15500 and a high degree of structural homology, are
found in most retinoid-responsive avian and mammalian cells
(Chytil and Ong, 1987). Retinol is the specific ligand for
the cellular retinol binding protein (CRBP), and RA is the
specific ligand for the cellular retinoic acid binding protein
(CRABP). Both proteins are found in all fetal tissues of the
rat, but CRABP is less ubiquitous in adult tissues and the

distribution of CRBP and CRABP is different in adult tissues. CRBP is detected in all adult tissues except heart and skeletal muscles; CRABP is also absent from these two tissues and a number of others as well (Chytil and Ong, 1987). CRBP and CRABP cDNA clones have been isolated from bovine and mouse libraries and sequenced (Shubeita et al., 1987; Sherman et al., 1987). The genes for both proteins map to chromosome 9 or 10 in the mouse (Wei, et al., 1987). Immunologically distinct second forms (designated II) of CRBP and CRABP have been isolated (Chytil and Ong, 1987; MacDonald and Ong, 198; Bailey and Siu, 1988). The binding sites of CRBP I and II are different. Only CRBP II can bind trans-retinaldehyde, and it is concentrated in the intestine where it may have a retinaldehyde-absorbing function (MacDonald and Ong, 1987). CRABP II binds RA less avidly than CRBP I (Bailey and Siu, 1988). Its distribution and possible function are as yet unknown (it was isolated from whole neonatal rat pups). There is also evidence for the presence of the binding proteins in the nucleus. By following ^{3}H-RA bound to cytosolic CRABP, it has been shown that the RA:CRABP complex is translocated from the cytoplasm to the nucleus (Russell et al., 1980).

The second component mediating retinoid activity is a nuclear receptor called the retinoic acid receptor (RAR), although it can bind both retinol and RA. The existence of a nuclear retinoid receptor was first suggested by Cope et al., (1984), and it was detected by Daly and Redfern (1987) in a fraction of F-9 cell nuclear extract. The receptor has been partially purified and characterized from human promyelocytic leukemia cells (HL-60) using the tritiated retinobenzoic acid, Am80, as ligand to bind the receptor, and then isolating the complex by HPLC (High Pressure Liquid Chromatography) (Hashimoto et al., 1988). The apparent MW of the receptor is 95000, and it is present in a 4:1 nuclear:cytoplasmic ratio. cDNAs for two different forms of the receptor have been isolated from human breast carcinoma and kidney libraries (RAR alpha; Petkovich et al., 1987; Giguere et al., 1987), and from human hepatocellular carcinoma and placenta libraries (RAR beta; Brand et al., 1988; Bendrook et al., 1988). The predicted MW of the alpha receptor is calculated from the 462 amino acid-coding reading frame of the cDNA to be about 48000, half that of the HL-60 receptor isolated by Hashimoto et al. (1988). Hence the receptor may exist as a dimer.

The alpha and beta RARs belong to the steroid/thyroid/ vitamin D3 family of nuclear receptors, and show the greatest structural homology with the thyroid receptor (Petkovich et al., 1987; Giguere et al., 1987). Such receptors are characterized by two major domains, a ligand binding domain, and a DNA binding domain with the finger motif. The DNA binding domain binds to enhancer cis regulatory sequences, usually upstream from promotors, to effect changes in the rate of transcription of structural genes (Yamamoto, 1985). Although the DNA-binding region binds to enhancer sequences in the absence of the ligand binding region, it appears that actual transcriptional enhancement requires binding of the ligand to the ligand-binding domain (Godowski et al., 1988).

Proof that the alpha and beta RARs bind RA and retinol was obtained by the "finger swap" technique, in which the DNA-binding region of an estrogen or glucocorticoid receptor cDNA was substituted for the DNA-binding region of the RAR, making a chimeric receptor cDNA (Petkovich et al., 1987; Giguere et al., 1987). The chimeric receptor cDNA was then cotransfected into cells along with a reporter construct consisting of the CAT gene plus steroid-responsive regulatory elements, and the cells challenged with a battery of ligands, of which only RA and retinol induced reporter activity. Retinol is bound with less affinity than RA by the alpha receptor, paralleling the biological efficacy of these retinoids (Petkovich et al., 1987). Beta RAR binds RA with 10-fold greater affinity than RAR alpha, and is found predominantly in epithelial tissues (Benbrock et al., 1988); the distribution of RAR alpha has not yet been reported.

The retinoids thus appear to function, through the mediation of a nuclear receptor, via a steroid-like mechanism to alter patterns of gene transcription. The functions of CRBP and CRABP in this process, however, are not as clear. Most retinoid-sensitive cells have detectable CRBP and CRABP (Chytil, 1984). Furthermore, the binding proteins leave their ligands in the nucleus and are recycled to the cytoplasm (Takase et al., 1986). These observations suggest that CRBP and CRABP mediate the transfer of retinol and RA to the nuclear receptor. But several normal cell types, and a number of mutant F-9 cell lines are sensitive to retinoids, yet have no detectable CRBP and/or CRABP, or else their sensitivity to retinoids bears no relationship to the binding affinity of retinoids for CRBP or CRABP (Trown et al., 1980; Matthei et al., 1983; Sani et al., 1984; Jetten et al., 1987). Hence another idea would be that the cytosolic binding proteins are not essential for transfer of retinoid to the RAR, or even into the nucleus, but serve only to buffer the rate of retinoid entry into the nucleus, where the free retinoid can bind to the RAR. This wound explain the lack of correlation described above between responsiveness of cells to retinoids and binding proteins in some cells.

It should be noted, however, that neither RAR alpha or beta synthesized from their cDNAs in vitro is able to bind RA in vitro, though they can do so when synthesized from their cDNAs in cells that have CRABP (Petkovich et al., 1987; Benbrook et al., 1988). This observation argues that CRABP is essential for the binding of RA to the RAR in these cells, although essential factors other than CRABP are not ruled out. The lack of correlation between cellular sensitivity to retinoids and their binding proteins might mean that their RARs are mutated in a way allowing them to bind free retinoid in the absence of CRABP.

With this background, we can construct the following working hypothesis to explain how RA might modify positional memory in regenerating amphibian limbs, or specify this memory during limb development. Cytosolic CRABP would bind RA and transfer it to a nuclear receptor. The ligand activates the receptor's DNA binding domain so that binding of this domain to cis enhancer sequences elevates the accessibility of RNA polymerase II to the promotor regions of structural genes, thus enhancing the production of transcripts whose proteins specify a "positional memory".

Some progress has been made toward analyzing the molecular mechanism of action of retinoids on positional memory in regenerating limbs. One obvious prediction of the hypothesis outlined above is that a nuclear receptor and cytosolic binding proteins for retinol and RA are present in regenerate cells. CRABP and CRBP have been detected in axolotl regenerates; the concentration of apo (unbound) CRABP is highest at the dedifferentiation-early blastema stages, which are the most sensitive to proximalization by retinoids (Keeble and Maden, 1986; McCormick et al., 1988). These binding proteins have also been detected in early stages of chick and mouse limb buds (Maden and Summerbell, 1986; Kwarta et al., 1987).

Using a lambda gt10 cDNA library constructed from poly A+RNA isolated from axolotl limb regeneration blastemata, we have cloned candidate cDNAs for both CRABP (CRABAX) and the nuclear receptor (RARAX). At this writing, these clones have been partially sequenced, and their structure will be compared with the bovine CRABP and the human RAR cDNAs to determine degree of homology. Once verified, these clones can be used to construct probes to analyze the spatial and temporal expression of the binding protein and receptor genes along limb and regenerate axes. Since the effect of RA on positional memory and on embryonic patterning is concentration-dependent, the axial concentrations of nuclear receptor and perhaps of CRABP are likely to be of importance to cells for making differential responses to different concentrations of endogenous or exogenous RA.

Beyond this is the question of how many and what genes are involved in specifying positional memory. Since our data (Nardi and Stocum, 1983; Crawford and Stocum, 1988a,b) suggest that positional memory resides at the cell surface, the gene products influenced by RA are likely to be structural proteins and/or enzymes associated with the cell surface, and perhaps the cytoskeleton. Level-specific differences in positional memory could be due to different gene products, or to quantitative variations in the same gene products.

Quantitative differences might seem more likely, in view of the fact that retinoids do not alter (as far as we know) cell phenotype in regenerating limbs, but only axial position, and in a graded, concentration-dependent manner. In addition, qualitative differences imply the involvement of a large and possibly uneconomical number of different genes, since each level of the limb presumably has a different positional memory. It should be pointed out, however, that different concentrations of RA induce different phenotypes (parietal and visceral endoderm, cardiac and skeletal muscle, astroglia) in F-9 cells (Edwards and McBurney, 1983). Since different gene products characterize these phenotypes, different concentrations of RA can therefore cause qualitative differences in gene expression, presumably at the level of transcription. Finally, retinoids might affect only the first gene in a cascade of regulatory/synthetic events, or affect all the genes or a fraction of them simultaneously. Whatever the case, concentration dependence might be mediated through the binding of different numbers of receptors to multiple enhancer sequences.

Clearly, the study of pattern formation and regulation has entered an exciting molecular phase. There will be a great feeling of satisfaction if we are able to relate molecular findings to the cellular interactions that have been deduced from tissue grafting experiments, thus gaining understanding of pattern regulation that spans several levels of biological organization.

REFERENCES

Bailey, J.S. and Siu, C.H., 1988, Purification and partial characterization of a novel binding protein for retinoic acid from neonatal rat, J. Biol. Chem., 263:9326-9332.

Benbrook, D., Lernhardt, E. and Pfahl, M., 1988, A new retinoic acid receptor identified from a hepatocellular carcinoma, Nature, 333:669-672.

Brand, N., Petkovich, M., Krust, A., Chambon, P., deThe, H., Marchio, A., Tiollais, P., and Dejean, A., 1988, Identification of a second human retinoic acid receptor, Nature, 332:850-853.

Bryant, S.V., and Muneoka, K., 1986, Views of limb development and regeneration, Trends in Genet., 2:153-159.

Carlson, B.M., 1974, Morphogenetic interactions between rotated skin cuffs and underlying stump tissues in regenerating axolotl forelimbs, Devel. Biol., 39:263-285.

Carlson, B.M., 1975a, Multiple regeneration from axolotl limb stumps bearing cross-transplanted minced muscle regenerates, Devel. Biol., 45:203-208.

Carlson, B.M., 1975b, The effects of rotation and positional change of stump tissues upon morphogenesis of the regenerating axolotl limb, Devel. Biol., 47:269-291.

Carlson, B.M., 1983, Positional memory in vertebrate limb development and regeneration, in: "Limb Development and Regeneration", J.F. Fallon and A.I. Caplan, eds., Alan R. Liss, Inc., New York.

Chytil, F., 1986, Retinoic acid: Biochemistry and metabolism, J. Am. Acad. Dermatol., 15:741-747.

Chytil, F. and Ong, D.E., 1984, Cellular retinoid-binding proteins, in: "The Retinoids", Vol. 2, M.B. Sporn, A.B. Roberts and D.S. Goodman, eds., Academic Press, Inc., Orlando.

Chytil, F. and Ong, D.E., 1987, Intracellular vitamin A binding proteins, Ann. Rev. Nutr., 7:321-325.

Cope, F.O., Knox, K.L., and Hall, R.L., 1984, Retinoid binding to nuclei and microsomes of rat testes interstitial cells: I-mediation of retinoid binding by cellular retinoid-binding proteins, Nutr. Res., 4:289-304.

Crawford, K. and Stocum, D.L., 1988a, Retinoic acid coordinately proximalizes regenerate pattern and blastema differential affinity in axolotl limbs, Development, 102:687-698.

Crawford, K. and Stocum, D.L., 1988b, Retinoic acid proximalizes level-specific properties responsible for intercalary regeneration in axolotl limbs, Development, in press.

Daly, A.K., and Redfern, C.P.F., 1987, Characterization of a retinoic-acid-binding component from F9 embryonal-carcinoma cell nuclei, Eur. J. Biochem., 168:133-139.

Eckert, R.L., and Green, H., 1984, Cloning of RDNAs specifying vitamin A-responsive human keratins, Proc. Natl. Acad. Sci., USA, 81:4321-4325.

Edelman, G.M., 1983, Cell adhesion molecules, Science, 219:450-457.

Edwards, M.K. and McBurney, M.W., 1983, The concentration of retinoic acid determines the differentiated cell types formed by a teratocarcinoma cell line, Devel. Biol., 98:187-191.

Fallon, J.F., and Crosby, G., 1977, Polarizing zone activity in limb buds of amniotes, in: "Vertebrate Limb and Somite Morphogenesis", D.A. Ede, J.R. Hinchliffe and M. Balls, eds., Cambridge University Press, Cambridge.

Ferrari, N. and Vidali, G., 1985, Effects of retinol on chromatin structure, Eur. J. Biochem., 151:305-310.

Fuchs, E. and Green, H., 1981, Regulation of terminal differentiation of cultured human keratinocytes by vitamin A, Cell, 25:617-625.

Gallin, W.J., Chuong, C.M., Finkel, L.H. and Edelman, G.M., 1986, Antibodies to liver cell adhesion molecule perturb inductive interactions and alter feather pattern and structure, Proc. Natl. Acad. Sci., USA, 83:8235-8239.

Glade, R.W., 1963, Effects of tail skin, epidermis and dermis on limb regeneration in Triturus viridescens and Siredon mexicanum, J. Exp. Zool., 152:169-193.

Giguere, V., Ong, E.S., Segui, P. and Evans, R.M., 1987, Identification of a receptor for the morphogen retinoic acid, Nature, 330:624-629.

Godowski, P.J., Picard, D. and Yamamoto, K.R., 1988, Signal transduction and transcriptional regulation by glucerticoid receptor-lex A fusion proteins, Science, 241:812-816.

Hashimoto, Y., Kagechika, H., Kawachi, E. and Shudo, K., 1988, Specific uptake of retinoids into human promyelocytic leukemia cells HL-60 by retinoid-specific binding protein: possibly the true retinoid receptor, Jpn. J. Cancer Res., (Gann), 79:473-483.

Jetten, A.M., Anderson, K., Deas, M.A., Kagechika, H., Lotan, R., Rearick, J.I., and Shudo, K., 1987, New benzoic acid derivatives with retinoid activity: Lack of direct correlation between biological activity and binding to cellular retinoid acid binding proteins, Cancer Res., 47:3523-3527.

Kim, W.S., and Stocum, D.L., 1986a, Retinoic acid modifies positional memory in the anteroposterior axis of regenerating axolotl limbs, Devel. Biol., 114:170-179.

Kim, W.S., and Stocum, D.L., 1986b, Effects of retinoic acid on regenerating normal and double half limbs of axolotls, Roux's Arch. Dev. Biol., 195:243-251.

Kim, W.S., and Stocum, D.L., 1986c, Effects of retinoids on regenerating limbs: comparison of retinoic acid and arotinoid at different amputation levels, Roux's Arch. Dev. Biol., 195:455-463.

Keeble, S. and Maden, M., 1986, Retinoic acid-binding protein in the axolotl: distribution in mature tissues and time of appearance during limb regeneration, Devel. Biol., 117:435-441.

Kwarta, R.F., Kimmel, C.A., Kimmel, G.L. and Slikker, W., Jr., 1985, Identification of the cellular retinoic acid binding protein (cRABP) within the embryonic mouse (CD-1) limb bud, Teratol., 32:103-111.

McDonald, P.N. and Ong, D.E., 1987, Binding specificities of cellular retinol-binding protein and cellular retinol-binding protein, type II, J. Biol. Chem., 262:10550-10556.

Maden, M., 1982, Vitamin A and pattern formation in the regenerating limb, Nature, 295:672-675.

Maden, M., 1983a, The effect of vitamin A on limb regeneration in Rana temporaria, Devel. Biol., 98:409-416.

Maden, M., 1983b, The effect of vitamin A on the regenerating axolotl limb; J. Embryol. exp. Morph., 77:273-295.

Maden, M., 1984, Does vitamin A act on pattern formation via the epidermis or the mesoderm? J. Exp. Zool., 230:387-392.

Maden, M. and Keeble, S., 1987, The role of cartilage and fibronectin during respecification of pattern induced in the regenerating amphibian limb by retinoic acid, Differentiation, 36:175-184.

Maden, M. and Mustafa, K., 1984, The cellular contributions of blastema and stump to 180° supernumerary limbs in the axolotl, J. Embryol. exp. Morph., 84:233-253.

Maden, M. and Turner, N., 1978, Supernumerary limbs in the axolotl, Nature, 273:232-235.

Maden, M. and Summerbell, D., 1986, Retinoic acid-binding protein in the chick limb bud: identification at developmental stages and binding affinities of various retinoids, J. Embryol. exp. Morph., 97:239-250.

Matthaei, K.I., McCue, P.A. and Shermam, M.I., 1983, Retinoid binding protein activities in murine embryonal carcinoma cells and their differentiated derivatives, Cancer Res., 43:2862-2867.

McCormick, A.M., Shubeita, H.E. and Stocum, D.L., 1988, Cellular retinoic acid binding protein: detection and quantitation in regenerating axolotl limbs, J. Exp. Zool., 245:270-276.

Meinhardt, H., 1983, A boundary model for pattern formation in vertebrate limbs, J. Embryol. exp. Morph.,76:115-137.

Mittenthal, J.E., 1981, The rule of normal neighbors: A hypothesis for morphogenetic pattern regulation, Devel. Biol., 88:15-26.

Morgan, T.H., 1901, "Regeneration", the McMillan Co., New York.

Muneoka, K. and Bryant, S.V., 1984, Cellular contribution to supernumerary limbs in the axolotl, Devel. Biol., 105:166-178.

Nardi, J.B. and Stocum, D.L., 1983, Surface properties of regenerating limb cells: evidence for gradation along the proximodistal axis. Differentiation, 27:13-28.

Niazi, I.A., and Saxena,1978, Abnormal hindlimb regeneration in tadpole of the toad, Bufo andersonii, exposed to excess vitamin A, Folia Biol., (Krakow), 26:3-11.

Pescitelli, M.J., Jr., and Stocum, D.L., 1980, The origin of skeletal structures during intercalary regeneration of larval Ambystoma limbs, Devel. Biol., 79:255-275.

Petkovich,M., Brand, N.J., Krust, A., and Chambon, P., 1987, A human retinoic acid receptor which belongs to the family of nuclear receptors, Nature, 330:444-450.

Roberts, A.B. and Sporn, M.B., 1984, Cellular biology and biochemistry of the retinoids, in: "The Retinoids", vol. 2, M.B. Sporn, A.B. Roberts and D.S. Goodman, eds., Academic Press, Orlando.

Rose, S.M., 1970, "Regeneration" , Appleton-Century-Crofts, New York.

Russell, P., Wiggert, B., Derr, J., Albert, D., Craft, J. and Chader, G., 1980, Nuclear uptake of retinoids: Autoradiographic evidence in retinoblastoma cells in vitro, J. Neurochem., 34:1557-1560.

Sani, B.P., Dawson, M.I., Hobbs, P.D., Chan, R.L.S., and Schiff, L.J., 1984, Relationship between binding affinities to cellular retinoid-acid-binding protein and biological potency of a new series of retinoids, Cancer Res., 44:190-195.

Sherman, D.R., Lloyd, R.S., and Chytil, F. 1987, Rat cellular retinol-binding protein: cDNA sequence and rapid retinol-dependent accumulation of mRNAs, Proc. Natl. Acad. Sci., USA, 84:3209-3213.

Shubeita, H., Sambrook, J.F. and McCormick, A.M., 1987, Molecular cloning and analysis of functional cDNA and genomic clones encoding bovine cellular retinoic acid-binding protein, Proc. Natl. Acad. Sci., USA, 84:5645-5649.

Sporn, M.B., Roberts, A.B., Rodie, N.S., Kagachika, H. and Shudo, K., 1986, Mechanism of action of retinoids, J. Am. Acad. Dermatol., 15:756-764.

Steinberg, M.S., 1963, Reconstruction of tissues by dissociated cells, Science, 141:401-408.

Steinberg, M.S., 1970, Does differential adhesion govern self-assembly processes in histogenesis? Equilibrium configurations and the emergence of a hierarchy among populations of embryonic cells, J. Exp. Zool., 173:395-434.

Steinberg, M.S., 1978, Cell-cell recognition in multicellular assembly: Levels of specificity, in:"Cell-cell Recognition", A.S.C. Curtis, ed., Cambridge University Press, Cambridge.

Stocum, D.L., 1982, Determination of axial polarity in the urodele limb regeneration blastema, J. Embryol. exp. Morph., 71:193-214.

Stocum, D.L., 1983, Morphogenesis of the amphibian limb regeneration blastema, in: "Nerve, Organ and Tissue Regeneration", F.J. Seil, ed., Academic Press, New York.

Stocum, D.L., 1984, The urodele limb regeneration blastema: Determination and organization of the morphogenetic field, Differentiation, 27:13-28.

Stocum, D.L., and Melton, D.A., 1977, Self-organizational capacity of distally transplanted limb regeneration blastemata in larval salamanders, J. Exp. Zool., 201:451-461.

Summerbell, D., 1983, The effect of local application of retinoic acid to the anterior margin of the developing chick limb, J. Embryol. exp. Morph., 78:269-289.

Takase, S., Ong, D.E. and Chytil, F., 1986, Transfer of retinoic acid from its complex with cellular retinoic acid-binding protein to the nucleus, Arch. Biochem. Biophys., 247:328-334.

Tank, P.W., 1981, The ability of localized implants of whole or minced dermis to disrupt pattern formation in the regenerating forelimb of the axolotl, Am. J. Anat., 162:315-326.

Thaller, C. and Eichele, G., 1987, Identification and spatial distribution of retinoids in the developing chick limb bud, Nature, 327:625-628.

Thaller, C. and Eichele, G., 1988, Characterization of retinoid metabolism in the developing chick limb bud, _Development_, 103:473-484.

Thiery, J.P., DuBand, J.-L. and Delouvee, A., 1985, The role of cell adhesion in morphogenetic movements during early embryogenesis, _in_: "The Cell in Contact: Adhesions and Junctions as Morphogenetic Determinants", G.M. Edelman and J.P. Thiery, eds., John Wiley and Sons, New York.

Thoms, S.D. and Stocum, D.L., 1984, Retinoic acid-induced pattern duplication in regenerating urodele limbs, _Devel. Biol._, 103:319-328.

Tickle, C., Lee, J. and Eichele, G., 1985, A quantitative analysis of the effect of all _trans-retinoic_ acid on the pattern of chick wing development, _Devel. Biol._, 109:82-95.

Townes, P. and Holtfreter, J. 1955, Directed movements and selective adhesion of embryonic amphibian cells, _J. Exp. Zool._, 128:53-120.

Trown, P.W., Palleroni, A.V., Bohoslawec, O., Richelo, B.N., Halpern, J.M., Gizzi, N., Geiger, R., Lewinski, C., Machlin, L.J., Jetten, A., and Jetten, M.E.R., 1980, Relationship between binding affinities to cellular retinoic acid-binding protein and _in vivo_ and _in vitro_ properties for 18 retinoids, _Cancer Res._, 40:212-220.

Wallace, H., 1981, "Vertebrate Limb Regeneration", John Wiley and Sons, New York.

Wallace, H. and Maden, M., 1984, Local action of vitamin A on amphibian limb regeneration, _Experientia_, 40:985-986.

Wang, S.Y., LeRosa, J. and Gudas, L.J., 1985, Molecular cloning of gene sequence transcriptionally regulated by retinoic acid and dibutyryl cyclic AMP in cultured mouse teratocarcinoma cells, _Devel.Biol._, 107:75-86.

Wei, L.N., Mertz, J.R., Goodman, D.S. and Nguyen-Huu, M.L., 1987, Cellular retinoic acid and cellular retinol-binding proteins: complementary deoxyribonucleic acid cloning, chromosomal assignment, and tissue specific expressions, _Mol. Endocrin._, 1:526-534.

Wolpert, L., 1971, Positional information and pattern formation, _Curr. Top. Devl. Biol._, 6:183-224.

Yamamoto, K.R., 1985, Steroid receptor regulated transcription of specific genes and gene networks, _Ann. Rev. Genet._, 19:209-252.

BIOCHEMICAL PATHWAYS INVOLVED IN THE RESPECIFICATION OF

PATTERN BY RETINOIC ACID

Malcolm Maden and Denis Summerbell

Limb Development Group
National Institute for Medical Research
Mill Hill, London, NW7 1AA, U.K.

SUMMARY

We describe here experiments designed to further our understanding of the ways in which cells of the developing chick limb bud and the regenerating axolotl limb detect and respond to retinoids during the process of pattern respecification. The binding proteins for retinoic acid (CRABP) and retinol (CRBP) have been identified and quantified in the chick and axolotl. In an attempt to determine the role of CRABP the potency (ability to respecify pattern) of a range of retinoid analogues was correlated with their binding affinities to CRABP. Analogues that were inactive did not bind to CRABP and with the exception of one analogue, arotinoid, the converse was also true. Immunolocalisation of CRABP in the chick limb bud showed that it was concentrated in the progress zone at the tip of the limb and distributed in a graded form across the anteroposterior axis with the high point on the anterior side. Immunolocalisation of CRBP showed that it was restricted solely to the posterior side of the limb bud, the converse of CRABP. The relationship between these distributions and the endogenous gradient of retinoic acid is discussed. Finally, experiments on retinoic acid-induced changes in proteins, particularly protein phosphorylation are described and it is suggested that these may be casual in the respecification of pattern.

INTRODUCTION

Although not identical in detail, the effects of retinoids on the developing limb buds of chicks and the regenerating limbs of amphibians are to respecify positional information in any axis which is undergoing regulation at the time of administration (Niazi and Saxena, 1978; Maden, 1982; Tickle et al., 1982; Maden, 1983a; Summerbell and Harvey, 1983; Kim and Stocum, 1986). In our search for the biochemical mechanisms of action of retinoids it therefore seems reasonable to assume that the same principles operate in chicks and amphibians. The limited amount of data so far available has confirmed this assumption as the experiments described below demonstrate.

We have been investigating the mode of action of retinoids in developing and regenerating limbs in the belief that if the process of pattern respecification can be fully understood this will lead to our comprehension of how the initial patterning process operated. Of course it is entirely possible that retinoic acid (RA) is itself a morphogen which organises pattern formation in the limb (Thaller and Eichele, 1987). If so then we are actually obtaining a molecular description of how cells detect and respond to the positional information inherent in a morphogenetic gradient of RA. These two aspects of the patterning process, detection and response, are the subject of this paper.

Detection of retinoids by cells

Retinoids are thought to act by attaching to specific binding proteins which are present in the cytoplasm of many cell types (Chytil and Ong, 1984). Retinol binds to cellular retinol-binding protein (CRBP) and retinoic acid binds to cellular retinoic acid-binding protein (CRABP). The retinoid-binding protein complex may then transfer to the nucleus to alter in some manner the pattern of gene activity. Specific binding sites for RA have been detected in the nuclei of rat testis, for example (Takase et al.,1986). Most recently, a class of high affinity receptors with both RA and DNA binding properties analogous to steroid receptors has been identified (Petkovich et al.,1987; Giguere et al.,1987), suggesting that the binding proteins may act as shuttle proteins transferring retinoids from the cytoplasm to the nucleus. One obvious question to ask therefore is whether these binding proteins can be detected in developing and regenerating limbs, in locations and at times where patterning events are taking place. We describe below experiments on the quantitation and spatial distribution of CRABP and CRBP.

Response of cells to retinoids

At the histological level, the two tissues which are most profoundly affected by retinoids are skin and cartilage. Hypovitaminosis A causes mucous secreting epithelia to become keratinising and, conversely, cultured skin treated with excessive retinoids changes from keratinising to mucous secreting (Fell and Mellanby, 1953). Cartilage matrix breakdown is induced in cultured cartilage rudiments (Fell and Mellanby, 1952), releasing proteoglycans into the medium. Both of these well-known effects are seen in the regenerating axolotl limb during the period of retinoid administration (Maden, 1983b). We have tested whether either of them was responsible for pattern respecification. After grafting RA-treated epidermis onto untreated blastemal mesoderm, regenerates were normal whereas untreated epidermis on treated mesoderm resulted in pattern respecification suggesting that the epidermal effects play no part (Maden, 1984). As regards cartilage degradation we have demonstrated that pattern respecification can still occur in limbs from which all the cartilage had been removed (Maden and Keeble, 1987) so the excessive production of proteoglycans from cartilage matrix cannot be responsible.

Retinoids are also known for their ability to inhibit cell division in a wide variety of cell types (Lotan, 1980) and this is the basis for their anti-tumourigenic properties. In the regenerating limb this effect can also be seen in the precipitous decline in thymidine labelling and mitotic indices (Maden, 1983b). This is unlikely to be responsible for pattern respecification because denervation can reversibly inhibit cell division yet the limb which eventually regenerates is perfectly normal. Two other effects of retinoid administration on the regenerating limb, the significance of which has not yet been established, is to, firstly, cause variations in cell density in the blastema to produce areas of densely packed cells and areas almost devoid of cells (Maden, 1983b; Kim and Stocum, 1986) and secondly, cause the blastema to change its shape from the elliptical cross-section of the distal blastema to the circular cross-section of the proximal blastema (Holder, unpublished).

At the cellular level, in cell culture systems retinoid-induced changes in the synthesis of a wide variety of proteins and enzymes such as fibronectin, laminin, collagen, glycosaminoglycans, plasminogen activator, protein kinases, transglutaminase, ornithine decarboxylase etc. have been described (Roberts and Sporn, 1984). We might therefore expect such differences in retinoid treated blastemal cells to be readily apparent and the experiments described below were designed to test this hypothesis.

MATERIALS AND METHODS

Chemicals: [11,12-^{3}H] all-trans-retinoic acid (30 Ci/mmol), arotinoid (ethyl p-[(E)-2-(5,6,7,8-tetrahydro-5,5,8,8-tetramethyl-2-naphthyl)-1-propenyl] benzoate), etretinate (ethyl all-trans-9-(4-methoxy-2,3,6-trimethylphenyl)-3,7-dimethyl 1-2,4,6,8-nonatetranoate), TTNPB (p-[(E)-2-(5,6,7,8-tetrahydro-5,5,8,8-tetramethyl-2naphthyl)-1 propenyl]benzoic acid) and Ro 10-1670 (all-trans-9-(4-methoxy-2,3,6-trimethyl-phenyl)-3,7-dimethyl-2,4,6,8-nonatetraenoic acid) were supplied by Hoffman-La Roche, Basel. [11,12(n)-^{3}H] vitamin A, free alcohol (60 Ci/mmol), [^{35}S] methionine (1300 Ci/mmol) and [^{32}P] orthophosphate (200mCi/mmol) were purchased from Amersham International plc. All other retinoids and chemicals were purchased from Sigma.

Administration of retinoids: in axolotl experiments 50-60mm larvae had their limbs amputated through the mid zeugopodium and retinoids administered in silastin blocks on day 4 after amputation (see Maden, et al.,1985 for details). In chick experiments small pieces of newsprint were soaked in solutions of retinoids made up in dimethyl sulfoxide (DMSO) and the newsprint then inserted into a slit made in the anterior margin of stage 18-19 limb buds (see Summerbell, 1983 for details).

Sample preparation and binding assays: cytoplasmic proteins were prepared as described previously (Keeble and Maden, 1986) by homogenising tissue from 100mm axolotls in phosphate buffered saline (PBS) and spinning the homogenate for 1 h at 100,000 g. Samples usually consisted of 2-3 mg of protein (250 µl of a 10mg/ml preparation), were made up to 500 µl with PBS and 40 pmoles of [^{3}H] RA or [^{3}H] retinol added in 10 µl DMSO. Excess unlabelled retinoids were added in 10 µl DMSO.

Sucrose gradient centrifugation was performed as previously described (Keeble and Maden, 1986). Specific binding was quantified by determining the drop in cpm with increasing molar excesses of unlabelled retinoids and using these values to calculate êC$_{50}$ for each retinoid.

Immunocytochemisty: tissues were fixed in Perfix (Fisher Scientific, New Jersey) for 3 hours, dehydrated, cleared in xylene and embedded in wax. 7 µm sections were cut. CRABP was immunolocalised with an affinity-purified anti-rat CRABP rabbit IgG as previously described (Porter et al.,1985). Colour was developed by the avidinbiotinylated peroxidase complex method with a kit from Vector Laboratories, Burlingame, California.

Protein labelling: animals with unamputated limbs, normal blastemata or RA treated blastemata were injected intraperitoneally with 200 µCi [^{35}S] methionine and returned to water for 5 hours. Tissue was then removed, homogenised in PBS-A/2% SDS/0.1M DTT and samples run on 5%, 10% or 15% SDS-PAGE gels. Gels were then stained and fluorographed. For phosphorylation experiments tissue from animals was first removed into phosphate-free Eagle's medium and cultured in 100 µCi of [^{32}P] orthophosphate for 5 hours. Tissue was then homogenised as described above and gels run and fluorographed.

RESULTS

CRABP: In the developing chick we have detected high levels of CRABP in stage 20, 25, 30 and 35 limb buds following sucrose gradient analysis (Maden and Summerbell, 1986, Table 1) and we estimate that there are about 10^5 molecules per cell at each of these stages. We have repeated this observation as recorded in Fig. 1A. In the unamputated axolotl limb CRABP can be detected, but only at low levels. After amputation, at the cone blastema stage the levels of CRABP rise 3-4 fold and then decrease again as redifferentiation begins (Keeble and Maden, 1986). We have repeated these observations which are shown in Fig. 1B and C and in Table 1. The time of highest activity levels corresponds with the time of maximum sensitivity to retinoic acid administration (Maden et al., 1985) suggesting that CRABP may play a role in pattern respecification.

One way to more directly test whether CRABP is involved is to see if the binding of various retinoid analogues to CRABP correlates with their potency at respecifying pattern. To this end we studied the relative potency of nine different analogues in the chick limb bud and the sequence shown in Table 2 was the result. TTNPB is an aromatic derivative of RA with a carboxyl end group and has previously been shown to be more potent than RA (Eichele et al., 1985). Here we found it to be more than twice as potent as all-trans-RA. Arotinoid is the ester of TTNPB and we found it to be just less than twice as potent as all-trans-RA. RA and Ro 10-1670 both have carboxyl end groups, Ro 10-1670 being about half as potent as RA. Only these 4 analogues reliably caused pattern respecification out of a total of 9 tested: etretinate, retinol, retinal, retinyl palmitate and retinyl acetate being inactive.

The binding of these analogues to CRABP showed specific

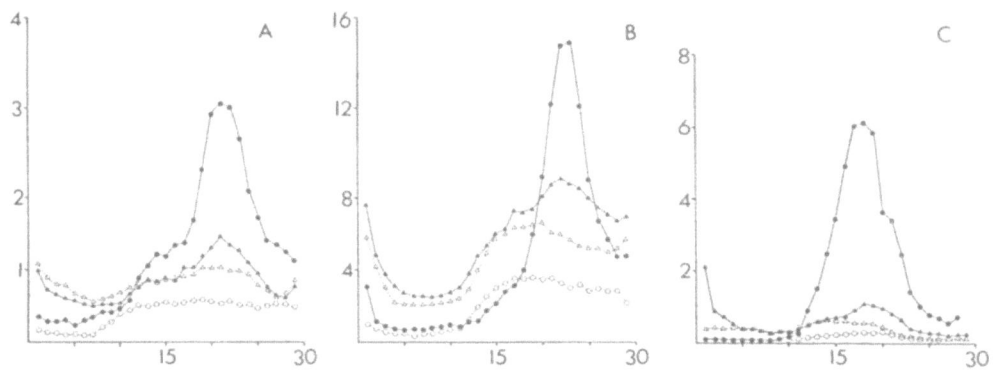

Fig. 1. Sucrose density centrifugation analysis of
cytoplasmic protein preparations from A) unamputated axolotl limbs, b) axolotl blastemata,
C) stage 25 chick limb buds. Ordinate: cpm x
10^{-3}. Abscissa: fraction number.
o——o = incubations with [^{3}H] RA + 100 fold
molar excess of cold RA.
▲——▲ = incubations with [^{3}H] retinol.
△---△ = incubations with [^{3}H] retinol + 100
fold molar excess of cold retinol. The specific peaks (established by the drop in peak
height with excess cold RA or retinol) are in
the 2S (approx. 17000 mol wt) region of the
gradient.
●——● = incubations with [^{3}H] RA

requirements for a carboxyl end group (Maden and Summberell,
1986), the esters did not bind and retinal bound only weakly
(Table 2). Significantly, TTNPB bound more effectively to
CRABP and was more potent than RA implying that there is a
casual link between binding to CRABP and pattern respecification. Ro 10-1670 bound with less affinity to CRABP and was
less potent than RA. However, arotinoid is the exception to

Table 1. Levels of CRABP and CRBP in Stage 25 Chick
Limb Buds and Cone Stage Axolotl Blastemata
Measured by Sucrose Density Centrifugation
Assays.

Tissue	Concentration (fmoles per mg cytosolic protein)			
	CRABP		CRBP	
chick limb	9100±1200	(n=24)	805±79	(n=8)
axolotl unampu-tated limb	700±30	(n=3)	162±44	(n=3)
axolotl blastema	2600±100	(n=3)	471±117	(n=4)

the generalization that the relative potencies of various analogues is the same as their relative binding affinities to CRABP. Arotinoid could be more potent than expected because it may be metabolised by esterases in the limb bud to the more potent analogue TTNPB and we are currently examining the metabolism of various retinoids to see whether this does in fact occur.

In the axolotl limb the same sequence of relative potencies as the chick was obtained (Table 2). TTNPB was at least 100x more potent at respecifying pattern than RA, arotinoid was about twice as potent as RA and Ro 10-1670 was less than 1/10th as potent as RA. The same analogues were found to be inactive: etretinate, retinol, retinal, retinyl palmitate and retinyl acecate.

The same sequence of binding affinities as in the chick was also obtained - only analogues with a carboxyl end group bound well with arotinoid and retinal binding weakly (Table 2). Again arotinoid did not conform to the precise relationship between binding to CRABP and potency at respecifying pattern. Arotinoid is more potent than expected from its binding affinity to CRABP and it may be metabolised by esterases to the far more potent analogue TTNPB. It is worth noting that these data on the relative potencies of retinoids in the axolotl were obtained by local administration of compounds in a silastin block. When applied in this manner retinol and retinyl palmitate were found to be inactive, yet when given to animals in their water then they are active (Maden, 1983b). This suggests that retinol and retinyl palmitate may be metabolised to an active product such as retinoic acid when administered systemically.

The data in Table 2 support the assumption presented earlier that the same principles of retinoid action operate in chicks and amphibians.

CRBP: We have determined the levels of the other retinoid-binding protein, specific for retinol, CRBP, in the chick and the axolotl. In both systems it is barely detectable on sucrose gradients (Fig. 1), being less than 1/10th of the level of activity of CRABP in the chick and about 1/6th of the level of activity of CRABP in the axolotl blastema (Table 2). Nevertheless, like CRABP, CRBP activity rises 3 fold from the level in the unamputated limb to its peak activity level in the cone stage regenerate (Table 2).

Since retinol has so far proved inactive in pattern respecification when administered to the chick or to the axolotl when administered locally and the levels of its binding protein are low, it may be assumed that CRBP is not a significant molecule in the mechanism of respecification. However, it is likely to be playing an important role in the establishment of pattern, at least in the chick, as described below.

Spatial distribution of CRABP and CRBP

Cytoplasmic preparations from whole limb buds or blastemata give no information about the cell types or spatial distribution of these binding proteins in individual limbs. To do this we need antibodies and to use them on section of the

Table 2. The Order of Relative Potencies (Ability to Respecify Pattern) and Binding Affinities for CRABP of
Various Retinoids in the Developing Chick Limb Bud
and Regenerating Axolotl Limb.

	Chick		Axolotl	
No.	Potencies	Binding affinities	Potencies	Binding affinities
1.	TTNPB*	TTNPB	TTNPB	TTNPB
2.	arotinoid	RA	arotinoid	RA
3.	RA	Ro 10-1670	RA	Ro 10-1670
4.	Ro 10-1670	retinal	Ro 10-1670	arotinoid
5.	-	-	-	retinal

*See Materials and Methods for chemical nomenclature.

relevant tissues. This has now been done with the chick limb
bud, but unfortunately the antibodies do not cross react with
amphibians so the same questions cannot be asked for limb
regeneration.

CRABP: the distribution of CRABP in a longitudinal section of a stage 24 chick limb bud shows a remarkable spatial
diversity (Fig. 2A). It is absent from the epidermis and apical ectodermal ridge (AER), present at high levels in the
progress zone at the tip of the limb bud and present at lower
levels in the undifferentiated tissues more proximally, it is
absent from differentiating cartilage and very high in differentiating muscle/dermis/connective tissue (Fig. 2A). In a
cross section through the progress zone to look at the distribution of CRABP across the anteroposterior axis we find that
there is a high level of staining at the anterior pole and it
is absent from the posterior pole (Fig. 2B). Interestingly,
this is the reciprocal of the distribution of endogenous retinoic acid (Thaller and Eichele, 1987).

CRBP: confirming the data from sucrose gradient analysis
(Table 2) CRBP is barely detectable in the stage 24 chick limb
bud. In longitudinal sections it is undetectable, however,
a more detailed analysis of serial cross sections showed that
it is apparent only in a strip of cells running along the
posterior edge of the limb (Fig. 2C). This is virtually the
same distribution as the cells of the zone of polarizing activity (ZPA) and so CRBP may well be a marker for ZPA cells. The
significance of the distribution of these two binding proteins
is discussed below.

Protein labelling

When chick chondrocyte cultures were treated with RA and
labelled with [^{35}S] methionine at least 10 differences in protein bands on SDS gels were revealed (Horton et al, 1987).
When we performed similar experiments no obvious differences
could be detected between control and RA treated blastema
(Fig. 3A, lanes 1 and 2) despite clearly observable differences between unamputated limbs and blastemata (Fig. 3A, lanes

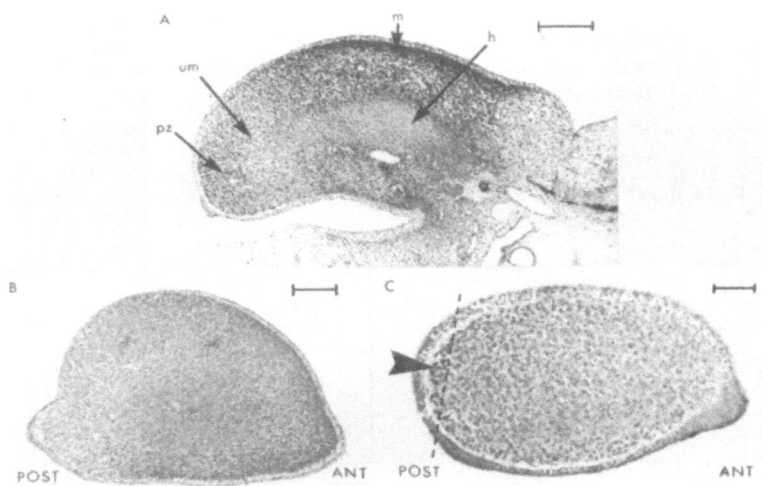

Fig. 2. Sections of chick limb buds treated with an
anti-CRABP antibody (A and B) or an anti-CRBP
antibody (C). A: longitudinal section of a
stage 24 bud. The progress zone (pz) stains
intensely whereas the undifferentiated mesenchyme behind it (um) does not. The differentiating humerus (h) has no immunoreactivity
whereas the muscle/connective tissue (m)
stains intensely. Bar = 150 µm. B: transverse section through the progress zone of a
stage 21 bud, anterior (ANT) and posterior
(POST) margins are marked. Note heavier
immunoreactivity with anti-CRABP on the anterior side. Bar = 100 µm. C: same type of
section as B treated with anti-CRBP. Here
reactivity is exclusively at the posterior
edge (arrowhead and dotted line). Bar=50 µm.

1 and 2). The same result was obtained in Coomassie blue
stained total proteins (Fig. 3A lanes 4-6). Here there were
at least 18 differences in proteins detectable between unamputated limbs and blastemata. More detailed studies of one protein, fibronectin, also failed to detect any differences
between control and RA treated blastemata by metabolic labelling, Western blotting or ELISA analysis (Maden and Keeble,
1987).

But when we looked at the phosphorylation of proteins
very clear differences were seen after RA treatment. On 5%
gels one band at a molecular weight of approximately 280.000
becomes phosphorylated (Fig. 3B lanes 1 and 2, upper arrowhead)
as well as two other proteins of molecular weight 20-35000
(lower arrowheads), more readily visible on 10% gels.

DISCUSSION

The experiments described above were designed to investi-

320

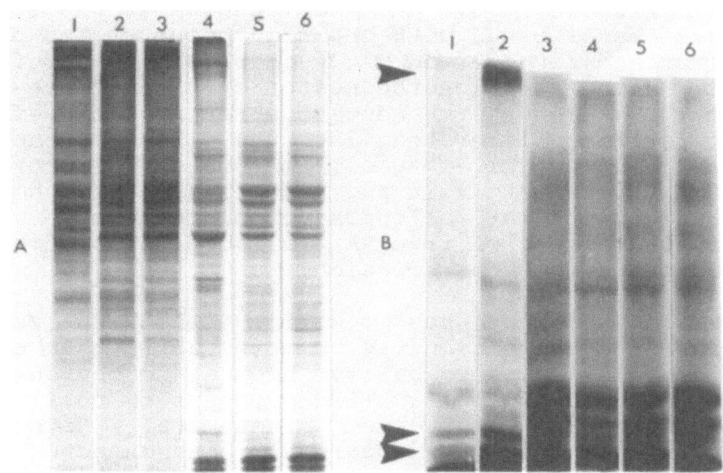

Fig. 3. A: 10% gels of [^{35}S] methionine labelled proteins from unamputated axolotl limbs (lane 1),
blastemata (lane 2), RA treated blastemata
(lane 3) and Coomassie blue stained total proteins from (lane 4) unamputated limbs, blastemata (lane 5) and RA treated blastemata (lane
6). B: Fluorographs of [^{32}P] labelled proteins on 5% gels of blastemata (lane 1), RA
treated blastemata (lane 2), skin (lane 3), RA
treated skin (lane 4), unamputated limbs (lane
5) and RA treated unamputated limbs (lane 6).
Arrowheads mark the position of 3 blastema
specific proteins which become phosphorylated
after RA treatment.

gate the ways in which cells of the developing chick limb bud
and the regenerating axolotl limb detect and respond to the
administration of retinoids, the ultimate effect of which is
to respecify positional information.

        With regard to the detection of retinoids, we believe
that CRABP plays an important role for several reasons.
Firstly, it is present at its highest levels in both the limb
bud and regenerating limb during the period when these tissues are most susceptible to RA. Secondly, the retinoids
which are most potent at respecifying pattern have the greatest binding affinity for CRABP and those retinoids which are
ineffective at respecifying pattern do not bind to CRABP.
Thirdly, CRABP is present at high levels in the progress zone
of the chick limb bud which is the area where pattern formation mechanisms are in operation during limb development.
Fourthly, CRABP is distributed in a gradient across the anteroposterior axis with the high point at the anterior margin of
the limb bud. Such an unusual distribution implies a role in
development.
        What this role may be can be hypothesised in conjunction
with the following findings: CRBP is only detected at the
posterior margin of the limb bud (Fig. 2C), endogenous retinol
is distributed at high levels uniformly across the limb bud

(Thaller and Eichele, 1987) and that endogenous RA is distributed across the limb bud with the high point at the posterior side (Thaller and Eichele, 1987). It is possible therefore that the CRBP in the cells on the posterior side of the limb bud (the ZPA cells?) sequesters endogenous retinol, converts it to RA which is then liberated into the milieu of the limb or passed from cell to cell along cell membranes. Either way RA will become distributed in a graded fashion across the limb and the gradient of CRABP of opposite polarity may serve to steepen the gradient of free RA available for binding to the nuclear receptors (see Introduction).

As regards the responses of limb cells to retinoids, changes in gene expression may be more subtle than cell culture studies have led us to believe because no obvious differences in $[^{35}S]$ methionine labelled proteins or total proteins could be detected in RA treated blastemata. Nevertheless, some protein changes have apparently been detected in other studies (Slack, 1983: Sharma and Anton, 1986). However, we did detect very clear changes in the phosphorylation of three proteins which only took place in RA treated blastemata, not in unamputated limbs, skin or tails. Since protein phosphorylation is kwown to be an important pathway for the control of gene regulation in cells (Cohen, 1985), we attribute some significance to these observations and are currently attempting to identify the proteins involved.

ACKNOWLEDGEMENTS

We thank Hoffman-La Roche for generous gifts of labelled and unlabelled retinoids, our collaborators in the immunocytochemistry, D.E. Ong and F. Chytil of the Department of Biochemistry, Vanderbilt University School of Medicine, Nashville, Tennessee and Katgriye Mustafa for excellent technical assistance.

REFERENCES

Cohen, P., 1985, The role of protein phosphorylation in the hormonal control of enzyme activity. Eur. J. Biochem., 151:439-448.

Chytil, F. and Ong, D.A., 1984, Cellular retinol-binding proteins in: "The Retinoids", M.B. Sporn, A.B. Roberts, S. Goodman, eds., Academic Press, Orlando.

Eichele, G., Tickle, C. and Alberts, B.W., 1985, Studies on the mechanism of retinoid-induced pattern duplications in the early chick limb bud: temporal and spatial aspects. J. Cell Biol., 101:1913-1920.

Fell, H.B. and Mellanby, E., 1952, The effect of hypervitaminosis A on embryonic limb-bones cultivated in vitro. J. Physiol., 116:320-349.

Fell, H.B. and Mellanby, E., 1953, Metaplasia produced in cultures of chick ectoderm by high Vitamin A. J. Physiol., 119:470-488.

Giguere, V., Ong, E.S., Segui, P. and Evans, R.M., 1987, Identification of a receptor for the morphogen retinoic acid. Nature, 330:624-629.

Horton, W.E., Yamada, Y. and Hassell, J.R., 1987, Retinoic
    acid rapidly reduces cartilage matrix synthesis by
    altering gene transcription in chondrocytes. Devel.
    Biol., 123:508-516.
Keeble, S. and Maden, M., 1986, Retinoic acid-binding protein
    in the axolotl: distribution in mature tissues and time
    of appearance during limb regeneration. Devel. Biol.,
    117:435-441.
Kim, W.S. and Stocum, D.L., 1986, Retinoic acid modifies posi-
    tional memory in the anteroposterior axis of regenerat-
    ing limbs. Devel. Biol., 114:170-179.
Lotan, R., 1980, Effects of Vitamin A and its analogues (reti-
    noids) on normal and neoplastic cells. Biochim.
    Biophys. Acta, 605:33-91.
Maden, M., 1982, Vitamin A and pattern formation in the regen-
    erating limb. Nature, 295:672-675.
Maden, M., 1983a, The effect of vitamin A on limb regeneration
    in Rana temporaria. Devel. Biol., 98:409-416.
Maden, M., 1983b, The effect of vitamin A on the regenerating
    axolotl limb. J. Embryol. exp. Morph., 77:273-295.
Maden, M., 1984, Does vitamin A act on pattern formation via
    the epidermis or the mesoderm? J. exp. Zool.,
    230:387-392.
Maden, M. and Keeble, S., 1987, The role of cartilage and
    fibronectin during respecification of pattern induced
    in the regenerating amphibian limb by retinoic acid.
    Differentiation, 36:175-184.
Maden, M. and Summerbell, D., 1986, Retinoic acid-binding pro-
    tein in the chick limb bud: identification at various
    developmental stages and binding affinities of various
    retinoids. J. Embryol. exp. Morph., 97:239-250.
Maden, M., Keeble, S. and Cox, R.A., 1985, The characteristics
    of local application of retinoic acid to the regenerat-
    ing axolotl limb. Roux's Arch. Dev. Biol., 194:228-235.
Niazi, I.A. and Saxena, S., 1978, Abnormal hind limb regener-
    ation in tadpoles of the toad, Bufo andersoni, exposed
    to excess vitamin A. Folia Biol. (Krakow), 26:3-8.
Petkovich, M., Brand, N.J., Krust, A. and Chambon, P., 1987, A
    human retinoic acid receptor which belongs to the fam-
    ily of nuclear receptors. Nature, 330:444-450.
Porter, S.B., Ong, D.E., Chytil, F. and Orgebin-Crist, M-C.,
    1985, Localization of cellular retinol-binding protein
    and cellular retinoic acid-binding protein in the rat
    testis and epididymis. J. Androl., 6:197-212.
Roberts, A.B., and Sporn, M., 1984, Cellular biology and
    biochemistry of the retinoids, in: "The Retinoids",
    M.B. Sporn, A.B., Roberts, S. Goodman, eds., Academic
    Press, Orlando.
Scadding, S.R., Vitamin A inhibits tail regeneration. Canad.
    J. Zool., 65:457-459.
Sharma, K.K. and Anton, H.J., 1986, Biochemical and ultra-
    structural studies of vitamin A induced proximalisation
    of limb regeneration in axolotl, in: "Progress in
    Developmental Biology", H.C. Slavkin ed., Alan R. Liss,
    New York.
Slack, J.M.W., 1983, Regional differences of protein synthesis
    in the limb regeneration blastema of the axolotl. Prog.
    in Clin. Biol. Res., 110A:557-563.
Summerbell, D., 1983, The effect of local application of reti-
    noic acid to the anterior margin of the developing
    chick limb bud. J. Embryol. exp. Morph., 78:269-289.

Summerbell, D. and Harvey, F., 1983, Vitamin A and the control of pattern in developing limbs. <u>Prog. Clin. Biol. Res.</u>, 110A:109-118.

Takase, S., Ong, D.E. and Chytil, F., 1986, Transfer of retinoic acid from its complex with cellular retinoic acid-binding protein to the nucleus. <u>Arch. Biochem. Biophys.</u>, 247:328-334.

Thaller, C. and Eichele, G., 1987, Identification and spatial distribution of retinoids in the developing chick limb bud. <u>Nature</u>, 327:625-628.

Tickle, C., Alberts, B., Wolpert, L. and Lee, J., 1982, Local application of retinoic acid to the limb bud mimics the action of the polarizing region. <u>Nature</u>, 296:564-566.

SURVEY OF RESEARCH WORK ON THE REGENERATION OF <u>TRITURUS</u> <u>ALPESTRIS</u> FORELIMBS TREATED WITH VITAMIN A PALMITATE

Stauros Koussoulakos

University of Athens - Zoological Laboratory
Panepistimiopolis 157 84, Athens, Greece

SUMMARY

These experiments were conducted to widen our understanding of the action of retinoids on urodele limb regeneration. Vitamin A palmitate was orally administered to terrestrial stages of adult <u>Triturus alpestris</u>, and the results were evaluated at both the morphological and cellular levels. It was observed, that the drug accelerated the speed of regeneration at the zeugopodial level, whereas, at the stylopod, growth rates lagged behind that of control animals. Proximalization was induced in 30% of the treated animals, but with a higher frequency on the right limb. In accord with the increased speeds of forearm regeneration, higher mitotic indices were scored in both mesenchymal and epidermal cells. On the contrary, treated animals displayed lower labelling indices than control ones. Microdensitometric scanning of the nuclear DNA in mesenchymal cells revealed a diminution in the duration of the S-phase, whereas, the time needed for cells to pass through G_2-phase was prolonged. The first days of vitamin A treatment were characterized by the appearance of some polyploid mesenchymal cells.

INTRODUCTION

The name "vitamin A" was originally given to characterize a water non-soluble compound, indispensable for the life of many vertebrates. Although it was at the beginning entitled "amine for life", it became widely known, later, by the discovery that high concentrations caused a series of toxic effects collectively termed "the hypervitaminosis A syndrome". Actually, the inteference of vitamin A and its derivatives (retinoids) with normal development and morphogenesis rendered these substances popular choices for experimental work. One of the most striking features of these works was the wide range of different and sometimes opposite effects. These effects are highly dependent on the retinoid tested, the biological system used, the developmental stage of the affected animals, the mode of administration and the duration of the treatment (for extensive reviews see Ganguly et al., 1980; Lotan, 1980; Sporn et al., 1984).

The activities of the retinoids include, among many others, reversal of keratinization (Fell, 1957), anencephaly and malformation of the eye (Marin-Padilla and Fern, 1965), phocomelia and micromelia (Kochhar, 1977), dedifferentiation of teratocarcinoma cells (Strickland and Mahdavi, 1978), promotion of mitotic activity, inhibition of normal and neoplastic growth (Lotan, 1980), promotion of growth factors receptors (Jetten, 1980), and modification of glycosaminoglycan and glycoprotein synthesis (Levin et al., 1983).

Niazi, at the University of Rajasthan, did not avoid the challenge to include retinoids in the study of anuran limb regeneration and surprised the whole world by the discovery that vitamin A palmitate was able to proximalize positional memory of the blastema cells (Niazi and Saxena, 1978), violating thus the rule of "distal transformation" (Rose, 1962). Similar investigations were further pursued in many other laboratories, and the main results can be summarized as follows:

    a)   Proximalization, posteriorization and ventralization of the regenerates (Maden, 1982; Kim and Stocum, 1986a; Stocum, this volume).

    b)   Retardation of blastema growth during the period of drug administration (Maden, 1983; Kim and Stocum, 1986b).

    c)   Increase in blastema cell adhesiveness (Maden, 1983; Kim and Stocum, 1986b).

    d)   Uniform action over the paired body appendages after systemic administration of retinoids (Maden, 1983).

These studies were performed mainly on water-living animals, whose physiology is in some aspects different from that of terrestrial forms. Moreover, vitamin A was administered by a variety of methods, except of the one, the animals themselves use for the uptake of this substance with their food. In order to fill in this gap and to complete these observations, vitamin A was given to terrestrial urodeles by gastric intubation.

MATERIALS AND METHODS

Animal Care

In the present study were used approx. 300 young, postmetamorphic terrestrial stage _Triturus alpestris_, which have received orally 250 IU of vitamin A palmitate/day/gram of body weight (gbw) for various time intervals. Throughout the course of the experiments, the animals were kept in individual plastic containers at a 14:10 hours light:darkness cycle of photoperiodism in a humidified atmosphere at 22±1°C, fed with Tubifex twice a week, and cleaned every other day.

Operations

Amputations were performed under MS 222 anesthesia. A sharp razor blade was used to transect both forelimbs at the middle of either the zeugopods or the stylopods. When necessary, the tissues at the wound place were trimmed to give a flat surface.

Observations on Growth Characteristics

Camera lucida drawings and Victoria blue staining were performed to assess growth rates and skeletal construction of the regenerates, respectively. For an accurate monitoring of limb elongation, measurements were performed in 2-day intervals for the first 20 days; thereafter growth was estimated every week. Since there are no sharp color demarcations between stump and blastema, the elongation was calculated by measuring the length of the existing appendage and subtracting the initial length of the stump (Tweedle, 1971).

Cell Proliferation and Differentiation

The ability of the quiescent limb cells to respond "pleiotypically" (Hershko et al., 1971) after amputation and vitamin A administration, was evaluated on the basis of DNA densitometric scanning and estimation of labelling and mitotic indices. To this end smears were prepared from mesenchymal cells, fixed in Lillie's fixative and stained using the Schiff reagent. DNA content was estimated in arbitrary units (A.U.) with a GN5 integrating microdensitometer. The mitotic index of mesenchymal cells was calculated by finding the number of cells in mitoses as a percentage of the total cell number. Semithin sections (1 μm) from animals injected with tritiated thymidine were used for histological examination, estimation of mitotic indices in epidermal cells, and calculation of labelling indices in both mesenchymal and epidermal cells (for details on the methods, see: Koussoulakos et al., 1988; Koussoulakos and Anton, 1988a; Koussoulakos and Anton, 1988b).

RESULTS

Morphological Observations

External size measurements: The animals used in these studies were raised ab ovo in the laboratory, and although they were of the same age (nine months old), and kept under identical conditions, body weight and body length (snout to tail tip), varied considerably. In addition, there was not a direct correlation between body size and limb length (shoulder to tip of the third digit), (see Table 1).

In a first experiment, the left and right forelimbs of each animal were transected at the middle of the stylopod and zeugopod, respectively. Vitamin A was given to the animals every day, at the same daily time and was continued in several animal groups for 4, 7, 11 and 14 days post amputation (dpa). During the first 20 days, the elongation of the treated stylopodial blastemata was slower than those of control animals, whereas, on the zeugopodial stumps, treated blastemata grew faster than control ones (Fig. 1). After this time period (that is shortly after the cessation of vitamin A administration), both zeugopodial and stylopodial treated blastemata displayed longer regenerates than their controls (Fig. 2).

Since the results concerning the growth promoting activity of vitamin A did not coincide well with the prevailing

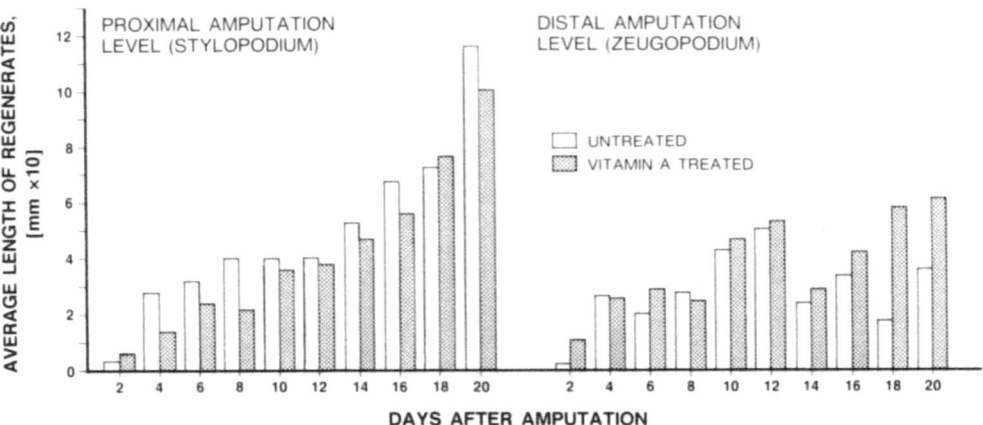

Fig. 1. During the first 20 days after limb amputation,
vitamin A palmitate seems to retard the elongation of the stylopodial blastemata; in contrast, treated zeugopodia grow faster than
their corresponding controls.

idea of retardation of limb growth during retinoid treatment,
a second experiment was conducted to discriminate between
eventual differences in the behavior of left and right limbs.
In this series, both forelimbs were amputated equidistantly
between the proximal and distal joints of the zeugopod.
Treated and control animals were kept under the same conditions, as in the first experiment. Again, vitamin A was found
to accelerate limb elongation in this series, too. The differences between the growth rates in control and treated animals were statistically significant (Fig. 3). A slight difference was also observed between left and right treated
limbs; right limbs grew a little faster than their left coun-
terpart and produced eventually longer appendages; however,
this difference was not statistically significant.

Examination of skeletal elements: Regenerated cartilage
was revealed by Victoria blue staining and clearing in methyl
benzoate. Limbs amputated at the stylopodial level, never
showed obvious proximalization (e.g. shoulder girdle was never
produced in tandem with the stump). However, at the zeugopodial level, the missing distal parts of radius and ulna were
shown to be regenerated, but at their "distal"? end, a second
functional elbow was formed, and then a second zeugopodium,
terminating in a complete autopodium was produced. Surprisingly, whereas some animals displayed bilateral proximalization of the regenerates (Fig. 4B), some others displayed proximalization only on the right limb, whereas, the left one was
never found simultaneously with a non proximalized right coun-
terpart. In addition, the phenomenon was shown to be dependent on the body weight of the affected animals (Table 1).

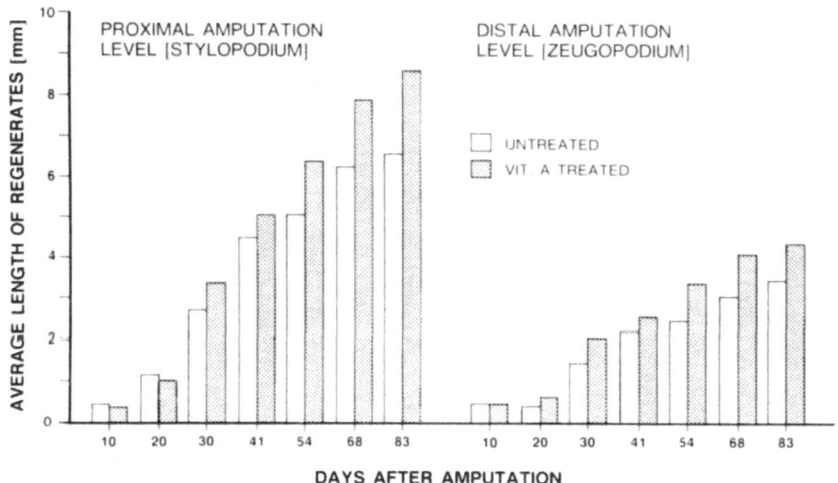

Fig. 2. Limb elongation measurements during a period of
83 dpa revealed that the vitamin A-treated animals produce longer appendages than the non-
treated ones over both the stylopodial and zeugopodial levels.

Histology: In another group of animals, vitamin A administration started 3 dpa, to allow wound epithelium to fully
cover the cut surface, and was discontinued on 13th dpa.

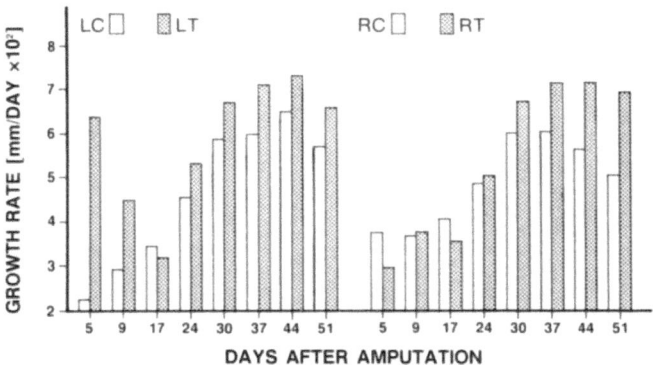

Fig. 3. Comparison of growth rates (mm/day) between
right (R) and left (L) regenerates of vitamin
A treated (T) and control (C) animals amputated bilaterally through the middle of the
zeugopod. The mean growth rates in control
limbs were 0.0637 and 0.0575 mm/day for the
left and right limb, respectively. The corresponding values for the vitamin A-treated animals were 0.0736 and 0.0784 mm/day.

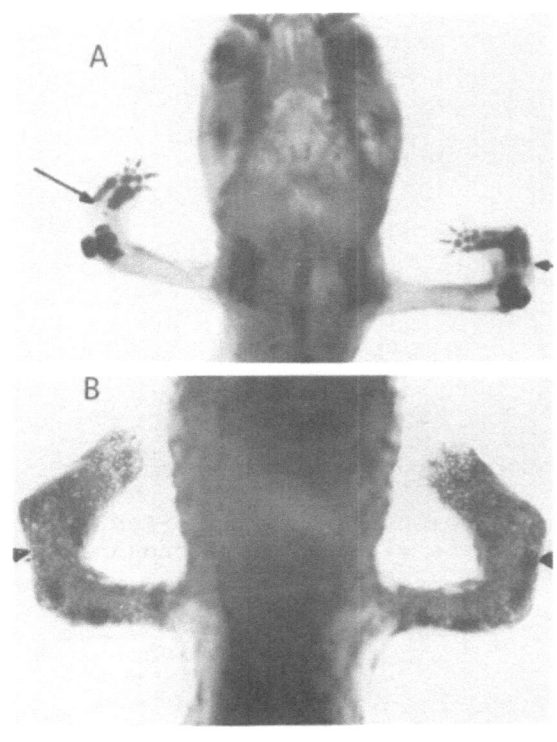

Fig. 4. <u>Triturus alpestris</u> had both forelimbs amputated at the middle of the zeugopod and were administered with vitamin A palmitate by gastric intubation. Approximately 30% of the treated animals (those ranging in body weight between 750 and 900 mgs) displayed the well known phenomenon of proximalization of the regenerates. Half of them had proximalized both forelimbs (B). Unexpectedly, the rest 15% showed duplication of zeugopodial elements only on the right stumps, whereas on the left stumps, distal, terminal regeneration has occurred (A). Arrows and the newly formed stained cartilage indicate the position of limb amputation. Photographs were taken before the full development of the regenerates.

Evaluation of the results obtained after 7 and 11 dpa (that is 4 and 8 days of exposure to the drug, respectively), confirmed the ability of retinoids to retard differentiation (keratinization) of the epithelial cells. Whereas 60% of the control epithelial cells were fully keratinized, this value was reduced to 40% for treated animals. In contrast to control blastemata, sections from vitamin A treated regenerates were characterized by a less homogeneous population of mesenchymal cells, regression of stump cartilage, thinner stratum corneum and enlargement of skin mucous glands. Externally, an elongation of the anteroposterior (AP) axis was noted (Fig. 5).

Cell Proliferation

In order to understand better the reasons for the growth activity of vitamin A on regenerating zeugopodia, several cellular parameters were studied.

Estimation of nuclear DNA content: Feulgen-stained nuclei were scanned at 560 nm and their color intensity was "translated" into A.U. of DNA content against the adjacent background. The sum of the A.U. found in treated stylopodia was slightly lower than the corresponding values from an equal number of mesenchymal cells from non-treated stylopodia; on the contrary, this value was found to be higher in the mesenchymal cells from zeugopodial blastemata of vitamin A-treated animals, with respect to their corresponding controls. The zeugopodial cells were classified according to their DNA-content, and their distribution into the individual cell-cycle

Table 1. List of Animals Displaying Proximalization on the
Left (L) and/or on the Right (R) Limbs in Relation
to their Body Weight, Body Length and Limb Length.

Animal	Body Weight (mg)	Body Length (mm)	Limb Length (mm)		Proximalization at	
			L	R	L	R
1	1240	59	10.26	10.05	−	−
2	1180	59	10.24	9.68	−	−
3	1110	57	10.56	9.60	−	−
4	1090	60	11.28	11.36	−	−
5	1080	55	11.86	10.08	−	−
6	1050	59	10.25	10.08	−	−
7	1040	54	10.08	10.72	−	−
8	1030	55	9.60	10.32	−	−
9	1010	57	10.72	10.16	−	−
10	980	56	0.00	10.18	−	−
11	975	56	9.92	9.92	−	−
12	970	54	10.64	9.92	−	−
13	960	53	9.84	10.00	−	−
14	940	56	10.64	10.32	−	−
15	930	57	10.32	10.56	−	−
16	920	58	10.64	10.48	−	−
17	920	53	10.72	10.72	−	−
18	910	51	9.70	9.60	−	−
19	900	52	9.84	10.08	−	+
20	890	58	10.64	10.48	−	+
21	890	54	10.56	10.56	+	+
22	870	56	9.76	10.24	+	+
23	870	50	10.08	9.84	+	+
24	830	56	10.04	10.00	+	+
25	830	54	10.88	10.88	−	+
26	820	54	9.92	10.16	+	+
27	820	50	9.60	9.44	−	+
28	800	53	9.68	9.44	+	+
29	780	50	8.40	8.72	−	+
30	770	53	9.68	9.68	−	+
31	750	50	8.80	9.12	−	−
32	720	52	9.68	9.52	−	−

phases is depicted in Table 2. It is obvious that vitamin A
facilitates the transition of cells from the G_1 to the S
phase, since the percentage of 2N(T) nuclei is lower than
2N(C) values, and the sum of S+G_2+polyploidic values are
nearly complementary to their corresponding G_1. The decrease
in the percentage of G_1 nuclei during the course of the experiment, displayed by the control and treated blastemata ought
to be accompanied by a corresponding increase in the number of
S-phase cells. However, 3N(T) values are lower than 3N(C)
ones. Two possible explanations could account for this phenomenon: either vitamin A inhibited DNA synthesis in some cells,
or it increased the speed of DNA replication with a concomitant decrease in the duration of S phase.

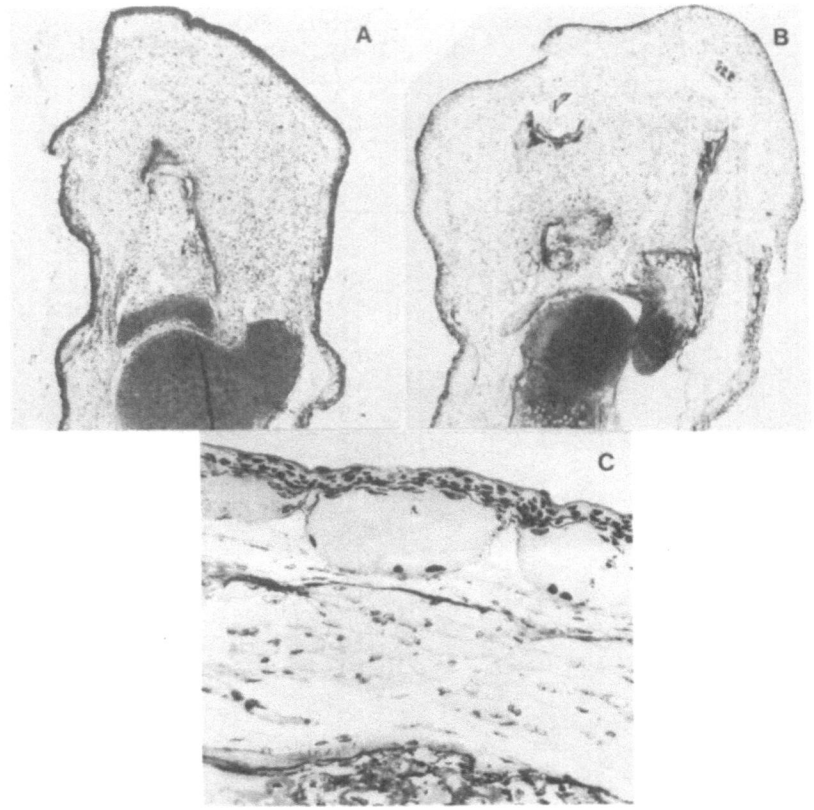

Fig. 5. 14 day control (A) and vitamin A treated
blastemata (B) and stump skin (C). Treated
blastemata are characterized by a flattening
along the AP axis, less homogeneous mesenchyme
and hypokeratinization of the wound epithelium.
Moreover, treatment results in enlargement of
the stump skin mucous glands.

Estimation of mitotic and labelling indices in mesenchyme
and epidermis: The number of mitotic figures scored on smears
from mesenchymal cells and on semithin sections from the wound
epithelium was divided by the total cells counted in each
case, and these values were used for calculation of mitotic
indices for mesenchymal and epidermal cells, respectively.
Labelling indices were determined on semithin sections
obtained from animals, injected for two hours with 0.01 µCi
tritiated thymidine per mg of body weight (sp.act. 1 mCi/ml).
Cummulative results are outlined in table 3. In concord with
the morphological observations and the results from DNA colo-
rimetry, vitamin A seems to increase the number of dividing
cells on both epidermis and mesenchyme. On the contrary, a
significant diminution in the labelling indices is observed
after the treatment of the animals with vitamin A.

Table 2. Percentages of Cells in G_0+G_1, S, and G_2 Cell Cycle Phases in Vitamin A-Treated (T) and Control (C) Blastemata at Various Times After Amputation.

DNA-content				Time (days)		
		O	4	7	11	14
$2N(G_0+G_1)$	C	97	85.53±5.14	84.37±6.91	81.29±4.30	80.76±4.22
	T		83.82±4.95	82.67±4.52	77.79±4.40	72.22±3.85
$3N(S)$	C		8.87±1.24	9.61±1.70	9.27±0.83	5.89±0.31
	T		6.92±0.73	7.51±0.36	7.61±0.61	11.48±0.55
$4N(G_2)$	C		4.13±0.16	3.52±0.57	6.05±0.30	10.83±0.33
	T		3.30±0.08	5.05±0.81	13.00±1.40	12.00±0.35
> 4N	C	0	0	0	0	
	T	5.94	2.75	0	0	
4NT:2NT		0.039	0.060	0.166	0.164	

DISCUSSION

Growth Rates of the Regenerates

In these experiments, vitamin A was found to retard the regeneration of the limbs developing on stylopodial stumps during the period of drug administration. After the cessation of the treatment, experimental limbs developed faster than controls. This effect was not repeated on the zeugopodial level; here, irrespectively of the time, treated regenerates grew faster than their appropriate controls. Thus, vitamin A behaves as a growth stimulant on the regeneration of the missing limb parts in Triturus alpestris. The different response of the early stylopodial regenerates can be accounted for by the fact, that proximal and distal blastemata of the same animal at the same time represent different developmental stages (Goss, 1969); in this respect, it is now well documented that the effects of retinoids are highly dependent on the developmental stage of the affected organ (Niazi et al., 1985). In addition, it was recently reported, that both the quantity and the distribution of positional values present in the stylopodium are different from those found in the zeugopodium (Bryant and Gardiner, this volume).

Proximalization of the Regenerates

The most impressive effect of retinoids on amphibian limb regeneration is the production of proximal limb elements from distal levels of amputation (Maden, 1982). It was conclusively demonstrated, that retinoids cause this effect by respecifying blastema cell positional memory and not by mobilizing proximal limb cells (Wallace and Maden, 1984).

Table 3. Labelling (L) and Mitotic (M) Indices of
          Treated (T) and Control (C) Mesenchymal (Me) and
          Epidermal (E) Cells

dpa	LI				MI			
	TMe	CMe	Te	CE	TMe	CMe	TE	Ce
4	10.75	21.10	–	–	1.33	1.98	–	–
7	10.37	20.21	–	–	2.16	1.08	–	–
11	5.25	8.13	3.4	5.1	0.12	0.16	1.5	0.25
14	10.85	13.25	–	–	0.16	0.12	–	–

The preceded experiments have indicated that even oral
administration of the drug was sufficient to produce on distal
stumps limb structures (elbow, proximal part of zeugopodium)
normally lying proximal to the amputation level. However,
although these structures were in perfect anatomical continuation with the stump (see Fig. 4), they were not in tandem
regarding level specific positional values. It seemed that,at
first, the distal parts of the missing zeugopodium were regenerated, and then, at the distal ends of radius and ulna,
instead of an autopodium, a second zeugopodium was developed
joint to the first one with a second functional elbow. Alternatively, it could be not ruled out, that the regenerated
parts on the stump zeugopodium were not the removed ones, but
mirror-image of the existing; in the case, a radual reverse
proximalization should have been accomplished. Whichever is
true, the proximalized values do not lie in tandem with the
distal values of the stump.

The dose of the drug which each animal received was proportional to its body weight. The smallest animal received a
total of 1800 IU of vitamin A palmitate, whereas, the largest
one received 3100 IU. On a specific basis, these values were
equivalent and equal to 250 IU/gbw/day. Nevertheless, animals
with body weight higher than 900 mgs and lower than 750 mgs
were regenerated only distally; the rest of the animals displayed unilateral or bilateral proximalizations. Since all
specimens originated from isochronous spawnings, dependence of
this effect on the age, can be ruled out. On the other hand,
it is well established that small animals have higher metabolic rates than large ones.Dose-response experiments, where
the quantity of vitamin A would be proportional not to the
body weight but to the metabolic weight of the animals, would
possibly clarify this phenomenon.

Left-Right Asymmetry

Anatomists and embryologists regard the paired appendages
of a "bilaterally symmetrical" animal as equivalent in respect
to growth characteristics. Although this seems true in a
first glance, detailed and convincing experiments especially
designed to test this preoccupation are not numerous. On the
contrary, there are some reports claiming differences between

the left and right limbs. Umanski et al. (1951), working on axolotls, found that the electric capacitance of the left arm skin was higher than that of the right one. Klein et al., (1981) observed that after oral administration of aspirin to rats, the degree of abnormalities was much higher on the developing right hindlimbs. The authors suspected that this handedness might be due to an asymmetric vascular supply. To my present knowledge there are no reports indicating a difference in the blood or nervous supply between left and right limbs in amphibians; nevertheless, growth differences are not infrequent (discussed by Kiortsis, in Trampusch and Harebomee, 1965). Choo et al., (1978) observed a slight difference in the rates of protein synthesis in cultured secondary upper arm blastemata of Notophthalmus viridescens, the incorporation in the rights being higher than in the left ones. Therefore it would be more logical to suggest that eventual differences observed between left and right limbs are intrinsic in nature. This hypothesis ought not surprise,since left-right asymmetries are frequently observed not only on external but on internal structures, too, and they are usually accompanied by a left dominance in a physiological left-right gradient (for revies see Wehrmaker, 1969; Oppenheimer, 1974; see also Koussoulakos et al., 1988).

In the present study vitamin A palmitate was found to induce, apart from bilateral, some cases of unilateral proximalizations, which were always restricted on the right forelimbs of the animals. It is the first time in the known literature, that retinoids show such an affection for the right side of the tested amphibia, and a search for the distribution of the cellular retinol binding protein (CRBP) and retinoic acid receptor (RAR) would be required for an understanding of this effect. However, three major differences of the present study with the others should be highlighted: (i) the mode of vitamin A administration might imply different pharmacological behavior of the drug. Gebhardt and Faber, (1966) found that aminopterin orally given to Ambystoma mexicanum causes polydactyly, whereas most other treatment inhibit regeneration; moreover it is reported, that aspirin given to rats by gastric intubation does not exert similar effects between fore- and hindlimbs on the one hand, and left and right hindlimbs on the other (Klein et al., 1981): (ii) Triturus alpestris is a terrestrial urodele with quite different physiological functions from the aquatic ones usually used in similar experiments, particularly in breathing and excretion; function which might be involved in the metabolism of the drug: (iii) Triturus alpestris is well known for its pecularity in displaying left-right asymmetries after various experimental interventions (Wehrmaker, 1969).

Cell Proliferation

The regeneration of an amphibian limb which has been accidentally or surgically removed, presupposes the active proliferation of a small number of mesenchymatous cells originating from dedifferentiation of adjacent tissues, mainly dermal fibrocytes. Cells undergoing division leave their G_0 or their prolonged G_1 phase and enter the S phase, which is characterized by DNA synthesis. Therefore, an increase in the mean DNA-content after experimental treatment in a population of cells, indicates that at least some of them have been stimulated to replicate their nuclear genome. On the contrary, if the treated cells have received a message to slow down the rate of DNA replication, then the control population should display higher DNA-values per cell.

In this respect, the corresponding results show that actually vitamin A instructs stylopodial blastemata to replicate their genome slower, whereas, zeugopodial blastemata respond to the stimulus by augmenting the mean value of DNA/cell. Although these findings were supported by those from morphological observations on external size measurements, the results obtained from the zeugopodial level did not coincide with the prevailing view of a retardation of regeneration after vitamin A administration. In order to settle this controversy, mitotic and labelling indices were calculated and the results were evaluated in relation to those from nuclear density scanning.

The mean DNA-content of the various mesodermal cells was at the time of amputation approx. 24 A.U., whereas the mitotic index was 0. As expected amputation triggers cells to enter the cycle so that at the 4th dpa the corresponding values were 26 and 0.71 respectively. In vitamin A treated animals, the same parameters were 29 and 1.33. Analogue results were obtained on the 7th dpa. In addition to that, this period was characterized by the presence of a certain number of polyploid cells, which disappear later. It was not possible to find a correlation between polyploidy and proximalization; nevertheless these cells were met only in treated blastemata. To my opinion, the increase in the average content of nuclear genome in the mesenchymatous population from zeugopodial blastemata, unequivocally indicates that actually vitamin A has, under the experimental conditions, a stimulatory effect on the replication of DNA. Although the results concerning an increase in the MI during the period of treatment do not coincide with those of Maden (1983), they are supported by the findings of Jangir and Niazi (1978) and Alam (1983), who observed intensification of blastema cell proliferation and increased mitosis in tadpoles of _Bufo melanostictus_. The MIs of the 11th and 14th dpa were very low, since the cells, after completing a specific cell-cycle (Sauer, 1981) redifferentiate to specific tissue cells (Brugal, 1971). Nevertheless, the number of dividing cells is still high enough to sustain blastema growth (Chalkley, 1954).

Counting of nuclei blackened by silver grains in control and vitamin A treated cells revealed that the number of labelled cells was much higher in control than in treated limbs. In this respect there is a satisfactory agreement with the findings of Maden (1983) and Sharma and Anton (1986) who also noticed a diminution in the labelling indices in axolotls. This results alone would possibly mean that vitamin A inhibits somehow the DNA synthesis. However, since it is not possible to have simultaneously inhibition of DNA synthesis on the one hand, and augmentation in the mean DNA-content and MIs on the other, another explanation should be required, and the most logical is, that vitamin A accelerates the speed of DNA replication with a concomitant decrease in the duration of the S phase. Actually, the microspectrophotometric data indicate an increase in the ratio of 4N to 2N cell percentages in vitamin A treated animals, thus favoring the view of a prolongation in the duration of G_2 phase to the detriment of S phase. The fact that, 3NT nuclei values are lower than 3NC ones, in combination with the higher 4N+polyploidic values met in the vitamin A treated series, advocates too, for a diminution in the duration of the S phase and a prolongation in the time needed for G_2 phase to be completed. In this respect it is well established that newts may decrease the duration of the S phase (Grillo, 1971).

Effects of Vitamin A on the Wound Epithelium

In various investigations at the cellular level, it was found that both the mesenchymal cells (Maden, 1983; Thoms and Stocum, 1984) and the extracellular matrix (Maden and Keeble, 1987) of vitamin A treated blastemata display a series of several modifications, when compared to control ones. We currently know little about the changes induced by vitamin A in epidermis of the blastema. This is a serious handicap in our problem since epidermis plays crucial roles in blastema formation and directional outgrowth, while its contribution to pattern specification is still a matter of debate (Stocum, 1985).

The notion that vitamin A exerts profound influences on epidermis is not new: low doses of the drug transform mucous secreting epithelia to keratinizing ones, whereas skin treated with excess of retinoids changes from keratinizing to mucous secreting (Fell and Mellanby, 1953). It is now generally agreed that such changes are due to alterations in the pattern of gene transcription (Eckert and Green, 1984).

In the present study the effects of retinoids on the hypokeratinization were confirmed, indicating that blastemal epidermis remained in a "Juvenile" state for relatively longer period of time. Thus, prolonged action on the apical epidermal cap, which is known for its mitogenic activity on blastema cells (Thornton, 1968) would possibly result in the attainment of the critical mass of mesenchymal cells needed for the escape of a regenerating limb from the distalizing influence of the epidermis (Faber, 1971).

ACKNOWLEDGMENTS

This work was performed at the University of Cologne, F.R. Germany. I would like to express my sincere gratitude to Prof. H.J. Anton for his hospitality and advice, and Dr. K.K. Sharma for his cooperation in some experiments.

REFERENCES

Alam, S., 1983, Studies on the morphogenetic influence of treatment of tadpoles of the anuran Bufo melanostictus (Schneider) with vitamin A palmitate on limb regeneration. Ph.D. Thesis, University of Rajasthan, Jaipur, India.

Brugal, G., 1971, Relation entre la proliferation et la differentiation cellulaires: etude autoradiographique chez les embryons et les jeunes larves de Pleurodeles waltlii (Michah, Amphibien Urodele), Devel. Biol., 24:301-321.

Chalkley, D.T., 1954, A quantitative histological analysis of forelimb regeneration in _Triturus viridescens_, J. Morphol., 94:21-70.

Choo, F.A., Logan, M.D., and Rathbone, P.M., 1978, Nerve trophic effects: an _in vitro_ assay for factors involved in regulation of protein synthesis in regenerating amphibian limbs. _J. Exp. Zool._, 206:347-354.

Eckert, R.L., and Green, H., 1984, Cloning of RDNAs specifying vitamin A responsive human keratins, _Proc. Natl. Acad. Sci._, USA, 81:4321-4325.

Faber, J., 1971, Vertebrate limb ontogeny and limb regeneration: Morphogenetic parallels _in_: "Advances in Morphogenesis", Abercrombie, M., Brachet, J., and King, T.J., eds., Vol. 9, pp. 127-147, Academic Press, New York.

Fell, H.B., 1957, The effect of excess of vitamin A on cultures of embryonic chicken skin explanted at different stages of differentiation. _Proc. Roy. Soc._, B146:242-256.

Fell, H.B., and Mellanby, E., 1953, Metaplasia produced in cultures of chick ectoderm by high vitamin A, _J. Physiol._, 119:470-488.

Ganguly, J., Rao, M.R.S., Murthy, S.K., and Sarda, K., 1980, Systemic mode of action of vitamin A, _Vitamins and Hormones_, 38:1-54.

Gebhardt, D.O.E., and Faber, J., 1966, The influence of aminopterin on limb regeneration in _Ambystoma mexicanum_. _J. Embryol. exp. Morphol._, 16:143-158.

Goss, J.R., 1969, "Principles of regeneration", Acad. Press, London.

Grillo, R.S., 1971, Changes in mitotic activity during limb regeneration in triturus, _Oncology_, 25:347-355.

Hershko, A., Mammont, R., Shields, R., and Tomkins, G.M., 1971, Pleiotypic response, _Nature (new biol.)_, 232:206-211.

Jangir, O.P., and Niazi, I.A., 1978, Stage-dependent effects of vitamin A excess on limbs during ontogenesis and regeneration in tadpoles of the toad _Bufo melanostictus_ (Schneider), _Indian J. Exp. Biol._, 16:438-445.

Jetten, A.M., 1980, Retinoids specifically enhance the number of epidermal growth factor receptors, _Nature_, 284:626-629.

Kim, W.S., and Stocum, D.L., 1986a, Retinoic acid modifies positional memory in the anteroposterior axis of regenerating axolotl limbs. _Devel. Biol._, 114:170-176.

Kim, W.S., and Stocum, D.L., 1986b, Effects of retinoids on regenerating normal and double half limbs of axolotls, _Roux's Arch. Dev. Biol._, 195:243-251.

Klein, L.K., Scott, J.W., and Wilson, G.J., 1981, Aspirin-induced teratogenesis: a unique pattern of cell deaths and subsequent polydactyly in the rat, _J. Exp. Zool._, 216:107-112.

Kochhar, D.M., 1977, Cellular basis of congenital limb deformity induced in mice by vitamin A, _in_: "Morphogenesis and Malformation of the Limb", Bergsma, D., and Lenz, W., eds, pp. 111-154, Liss, New York.

Koussoulakos, S., Sharma, K.K., and Anton, H.J., 1988, Vitamin A induced bilateral asymmetries in _Triturus_ forelimb regenerates, _Biol. Struct. Morphog._, 1:43-48.

Koussoulakos, S., and Anton, H.J., 1988a, Autoradiographic evaluation of vitamin A effects on _Triturus alpestris_ blastema cells, _Cell Biol. Intern. Reports_, 12:263-269.

Koussoulakos, S., and Anton, H.J., 1988b, Influence of vitamin A palmitate on the growth of regenerating <u>Triturus alpestris</u> forelimbs, <u>Biol. Struct. Morphog.</u>, 1:124-129.

Levin, L.V., Clark, J.N., Quill, H.R., Newberne, P.M., and Wolf, G., 1983, Effect of retinoic acid on the synthesis of glycoproteins of mouse skin tumors during progression from promoted skin through papillomas to carcinomas, <u>Cancer Res</u>., 43:1724-1732.

Lotan, R., 1980, Effects of vitamin A and its analogues (retinoids) on normal and neoplastic cells, <u>Biochem. Biophys. Acta</u>, 605:33-91.

Maden, M., 1982, Vitamin A and pattern formation in the regenerating limb <u>Nature</u>, 295:672-675.

Maden, M., 1983, The effect of vitamin A on the regenerating axolotl limb. <u>J. Embryol. exp. Morphol.</u>, 77:273-295.

Maden, M., and Keeble, S., 1987, The role of cartilage and fibronectin during respecification of pattern induced in the regenerating amphibian limb by retinoic acid. <u>Differentiation</u>, 36:175-184.

Marin-Padilla, M., and Fern, V.H., 1965, Somite necrosis and developmental malformations induced by vitamin A in the golden hamster. <u>J. Embryol. exp. Morphol.</u>, 13:1-8.

Niazi, I.A., Pescitteli, J.M., and Stocum, D.L., 1985, Stage-dependent effects of retinoic acid on regenerating urodele limbs, <u>Roux's Arch. Dev. Biol.</u>, 194:355-363.

Niazi, I.A., and Saxena, S., 1978, Abnormal hindlimb regeneration in tadpoles of the toad <u>Bufo</u> <u>andersonii</u>, exposed to excess vitamin A, <u>Folia Biol.</u>, (Krakov), 26:3-8.

Oppenheimer, M.J., 1974, Asymmetry revisited, <u>Amer. Zool.</u>, 14:867-879.

Rose, S.M., 1962, Tissue-arc control of regeneration in the amphibian limb, <u>in</u>: "Regeneration", Rudnick, D., ed., Ronald Press, New York.

Sauer, H., 1981, Developmental Biology today, an overview, <u>Fortschr. Zool.</u>, 26:7-13.

Sharma, K.K., and Anton, H.J., 1986, Biochemical and ultrastructural studies on vitamin A induced proximalization of limb regeneration in axolotl, <u>in</u>: "Progress in Developmental Biology", part A, Liss, New York.

Sporn, M., Roberts, A.B., and Goodman, D., 1984, "The Retinoids", Academic Press, New York.

Stocum, D.L., 1985, The role of the skin in urodele limb regeneration, <u>in</u>: "Regulation of Vertebrate Limb Regeneration", R.E. Sicard, ed., University Press, Oxford.

Strickland, S., and Mahdavi, V., 1978, The induction of differentiation in teratocarcinoma stem cells by retinoic acid, <u>Cell</u>, 15:393-403.

Thoms, S.D., and Stocum, D.L., 1984, Retinoic acid induced pattern duplication in regenerating urodele limbs, <u>Devel. Biol.</u>, 103:319-328.

Thornton, C.S., 1968, Amphibian limb regeneration, <u>in</u>: "Advances in Morphogenesis", Abercrombie, M., Brachet, J., and King, J.T., eds., vol. 7, pp. 205-249, Acad. Press, New York.

Trampusch, H.A.L., and Harrebomee, A.E., 1965, Differentiation, a prerequisite of regeneration, <u>in</u>: "Regeneration in Animals and Related Problems", Kiortsis, V., and Trampusch, H.A.L., eds., North Holland, Amsterdam.

Tweedle, C., 1971, Transneuronal effects on amphibian limb regeneration. <u>J. Exp. Zool.</u>, 177:13-30.

Umanski, E.E., Tkatsch, V.K., and Kondokotsev, V.P., 1951, Dielectric anisotropy of the skin in axolotls, <u>Doklady Akademiya Nauk SSSR</u>., 76:465-467.

Wallace, H., and Maden, M.,1984, Local action of vitamin A on amphibian limb regeneration. <u>Experientia</u>, 40:985-986.

Wehrmaker, A., 1969, Right-left asymmetry and situs inversus in <u>Triturus alpestris</u>, <u>Wilhelm Roux's Arch</u>., 163:1-32.

EFFECTS OF VITAMIN A ON LIMB REGENERATION IN THE LARVAL AXOLOTL, <u>AMBYSTOMA MEXICANUM</u>

Steven R. Scadding

Department of Zoology, University of Guelph
Guelph, Ontario, Canada, N1G 2W1

SUMMARY

Vitamin A is known to cause both proximodistal and anteroposterior duplication of parts of the limb during limb regeneration in amphibians. The objective of this study was to investigate the nature and location of the cellular changes induced by vitamin A when it causes these duplications in the larval axolotl, <u>Ambystoma mexicanum</u>. A variety of experimental techniques were employed to try and determine where and when vitamin A was acting when it induced changes in the pattern of the regenerate. Overall, the results were consistent with the hypotheses: that a) vitamin A acts at the gene level by switching on or increasing the expression of some gene or set of genes, that b) this leads to synthesis of some protein, likely a glycoprotein, which c) modifies certain as yet unknown cell activities leading to a change in the positional coding, such that proximodistal duplication occurs, as well as other characteristic histological changes, and that d) this change persists for some weeks after the cessation of the vitamin A treatment, but that eventually the cells revert to their normal positional coding, implying that the vitamin A induced gene expression is reversed.

INTRODUCTION

Vitamin A can induce proximodistal, anteroposterior, and dorsoventral duplications of parts of limbs during limb regeneration in several species of amphibians including the axolotl, <u>Ambystoma mexicanum</u> (Niazi and Saxena, 1978; Niazi et al, 1985; Maden, 1982, 1983a, 1984b; Scadding, 1983; Thoms and Stocum, 1984; Lheureux et al, 1986; Koussoulakos et al, 1986; Scadding and Maden, 1986a, 1986b, 1986c). Both how vitamin A induces these duplications, and the precise cellular target are unknown. Several hypotheses have been advanced to attempt to explain this action of retinoids, but the most convincing explanation may be that retinoids interact with specific binding proteins in the target cells which transport them to the nucleus, where they modify gene activity in a manner analogous to that proposed for the mechanism of steroid hormone action (O'Malley and Means, 1974; Roberts and Sporn, 1984). The identification of a cytoplasmic retinoic acid binding protein (CRABP) in the limbs of the axolotl lends support to this hypothesis (Keeble and Maden, 1986a, 1986b; McCormick et al, 1988). Taken together the evidence suggests a mechanism of action of vitamin A which involves a cytoplasmic retinoid binding protein, activation of gene transcription and subsequent protein synthesis (Scadding, 1988a) with a subsequent modification of the positional information (Stocum and Crawford, 1987; Scadding, 1988b) of the mesodermal cells of the limb (Maden, 1984).

The purpose of the present investigation was: 1) to seek further support for the hypothesis that vitamin A induces duplication by acting at the gene level, 2) to determine how persistent these vitamin A induced changes are, 3) to determine the location of the cells modified by vitamin A, and 4) to investigate some of the histological changes which are induced in regenerating limbs by retinoids. To meet these objectives a variety of experiments were carried out on axolotl larvae.

1) To seek further support for the hypothesis that vitamin A induces duplication by acting at the gene level, specific antibiotics were employed: actinomycin D, which blocks DNA-dependent RNA-synthesis; cycloheximide, which blocks protein synthesis; and tunicamycin, which interferes with the synthesis of glycoproteins by blocking glycosylation (Gale et al, 1981). All three of these drugs have been shown to antagonize some of the effects of vitamin A. Thus, axolotls were amputated and treated with retinol palmitate, and then a silastin block containing one of these antibiotics was implanted into the limb to ascertain if the duplication induced by the retinoid could be inhibited, i.e. to see if these drugs would antagonize the effects of vitamin A on the regenerating limb.

2) To determine how persistent these vitamin A induced changes are, two types of experiments were done. Series I: Limbs were treated with vitamin A and then amputated at the end of the 14 day retinol palmitate treatment period to determine if vitamin A induced changes in the cells, which persisted beyond the treatment period. Series II: To estimate the longevity of the induced changes, both forelimbs were amputated and treated with vitamin A. After 7 weeks, axolotls exhibiting proximodistal duplication were selected and limbs were reamputated as close as possible to the original amputation level. The subsequent regenerates were then examined to detect the occurrence of duplications.

3) To determine whether the cells modified by vitamin A were present in the blastema or in the stump or both, reciprocal blastema transplantations were carried out between black pigmented wild type axolotls which had been treated with vitamin A, and untreated white mutant axolotls. Regenerates were subsequently examined to evaluate the degree of duplication.

4) To investigate some of the histological changes which are induced in regenerating limbs by retinoids, limbs were amputated and treated in retinol palmitate, and fixed at intervals thereafter for examination in the light microscope or the scanning electron microscope (SEM).

342

The results of these studies were consistent with the hypotheses: that a) vitamin A acts at the gene level by switching on or increasing the expression of some gene or set of genes, that b) this leads to synthesis of some protein, likely a glycoprotein, which c) modifies certain as yet unknown cell activities leading to a change in the positional coding, such that proximodistal duplication occurs, as well as other characteristic histological changes, and that d) this change persists for some weeks after the cessation of the vitamin A treatment, but that eventually the cells revert to their normal positional coding, implying that the vitamin A induced gene expression is reversed.

MATERIALS AND METHODS

The axolotls used in these experiments were obtained as eggs or young larvae from the axolotl colony at the University of Ottawa. They were then raised in the laboratory on a diet of brine shrimp, chopped Tubifex worms, or intact Tubifex worms as their size increased. For several weeks prior to, and during each experiment, they were maintained in tap water in individual plastic dishes (D-8 cups, Canada Cup Co.) to preclude damage to the limbs as a result of biting by other axolotls. The axolotls were used in these experiments once they had reached a total length of between 4 and 6 cm. During surgical procedures, the axolotls were anaesthetized in 0.3 g/l tricaine methanesulfonate neutralized with sodium bicarbonate. Limbs were amputated so as to remove the distal third of the radius and ulna.

Vitamin A treatment of the animals was carried out by immersing the axolotls in 300 mg/l retinol palmitate (Sigma Type VII; 300 mg/l = 75 I.U./ml) and maintaining them in this solution for 10 or 14 days with changes every two or three days.

To analyze the skeletal patterns, resulting from the experimental manipulations, the limbs were amputated at the shoulder level, fixed in neutral buffered formalin and stained with Victoria Blue B for cartilage and cleared in methyl salicylate.

The effect of the retinol palmitate was scored using a strength of activity index (SAI) ranging from 0 to 5 to classify the duplications observed, similar to that used previously by Maden (1983a). This scale was based on the observation that as the dose of vitamin A was increased the degree of duplication increased as well. A normal forelimb regenerate was scored as 0. A forelimb regenerate with extra carpals scored 1; with partial extra radius and ulna scored 2; with complete extra radius and ulna scored 3; with partial extra humerus scored 4 (Fig. 1); and a complete extra humerus scored 5 (Fig. 3). Thus, the greater the degree of duplication, the higher the strength of activity score, implying a greater degree of proximalization.

Experiment 1. Silastin blocks were prepared as previously described (Scadding, 1988a) by mixing 10 mg of Actinomycin D, or 100 mg cycloheximide, or 100 mg of tunicamycin in 1.0 ml of silastin (Silastic 382 Medical Elastomer, Dow Corning), allowing them to set and then cutting blocks of appropriate sizes. Both forelimbs were amputated and the axolotls were treated with retinol palmitate for ten days commencing at the time of amputation. For implantation of the blocks, on day two post-amputation, a tunnel was made in the right forelimb from a site near the elbow to just proximal to the blastema using fine forceps. Then, the silastin block containing one of the antibiotics was implanted into this tunnel close to the blastema. The left limbs were used as controls and received either no implant or a control block of the silastin of the same size but without the antibiotic. On day six post-amputation, the initial blocks were removed (if still present), and a new block identical to the previous one was implanted. On day 49, the limbs were reamputated at shoulder level and fixed for examination of the skeletal structures.

Experiment 2. Series I: The left limb was amputated and the axolotls were treated with retinol palmitate for 14 days. On day 14, the right limb was similarly amputated and the retinol palmitate treatment was discontinued. Both limbs were then allowed to regenerate and were reamputated and fixed on day 56. Series II: Both forelimbs were amputated and axolotls treated with retinol palmitate for 14 days. On day 49, 14 axolotls were selected in which both forelimbs showed proximodistal duplication of the humerus. The left limb was then reamputated through the distal radius and ulna of the duplicated regenerate, and the right limb was reamputated through the original proximal radius and ulna as close as possible to the original amputation level. The limbs were again allowed to regenerate for a further six weeks, at which time the limbs were reamputated, fixed and stained for skeletal analysis.

Experiment 3. Both forelimbs were amputated from about 30 black wild type axolotls, which were then treated with retinol palmitate for 14 days. On day 14, 30 white mutant axolotls were similarly amputated and allowed to regenerate without vitamin A treatment. Because of the delayed regeneration in the vitamin A treated axolotls, cone stage regenerates were present on both sets of animals at about the same time 28 days after the initial amputation of the black axolotls (14 days after amputation of the white axolotls). Pairs of animals (one black and one white) were selected for reciprocal transplantation of a blastema. The blastema was removed from either the left or right limb of each pair, then the vitamin A treated blastema from the black axolotl was grafted onto the stump of the white axolotl, and vice versa, being careful to maintain the proper orientation of the blastemata and to ensure that they were not rotated. Then, the transplanted blastemata as well as those on the contralateral limbs which were not reamputated, were allowed to regenerate in the subsequent five weeks. The limbs were then reamputated, fixed and stained as described above.

Experiment 4. For the histological study of vitamin A treated regenerating axolotl limbs, about 135 axolotls had one or both forelimbs amputated. About half of these were then treated with retinol palmitate for up to 14 days, the other half were used as untreated controls. At intervals from days 1 through 49, limbs were reamputated and fixed for histo-

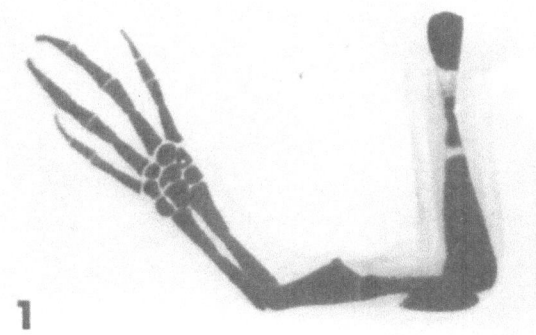

Fig. 1. Control left limb of an axolotl from Exp. 1
which was treated with retinol palmitate for
10 days. The regenerate includes a partial
duplication of the humerus (SAI 4).

logical analysis. Some were fixed in Bouin's fluid, embedded
in paraffin, sectioned longitudinally, and stained with haematoxylin and eosin for examination in the light microscope.
Other limbs were fixed in 2.5% glutaraldehyde in 0.1 M HEPES
buffer and then post-fixed in 2% osmium tetroxide in 0.1 M
phosphate buffer. Limbs so treated were then critical point
dried, sputter coated, and mounted for examination in the SEM.

RESULTS

Experiment 1. The implantation of the two silastin
blocks each containing 1.25, 0.625, 0.3125, or 0.156 µg of
actinomycin D, or 12.5 or 6.25 µg of cycloheximide completely
inhibited the proximodistal duplication induced in control

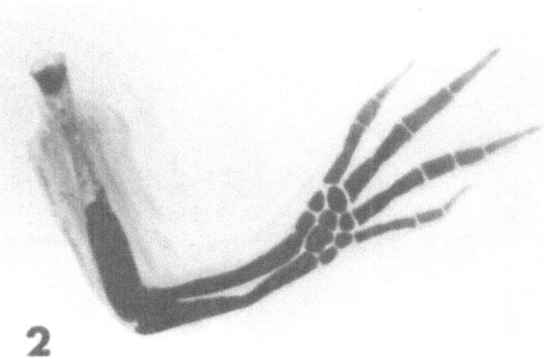

Fig. 2. Experimental right limb from the same axolotl
as Fig. 1. In this case, silastin blocks
containing 0.156 µg actinomycin D were
implanted on days 2 and 6 post-amputation.
This limb shows no evidence of antero-
posterior duplication and appears to be of
normal morphology.

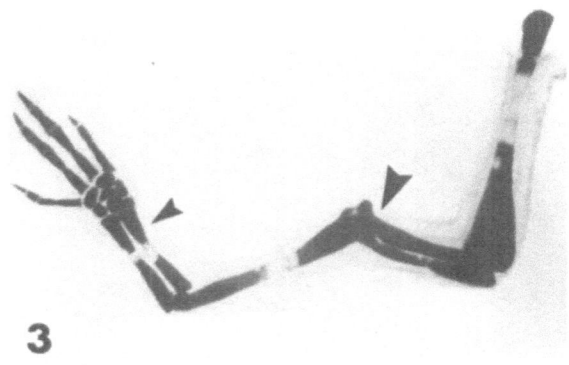

3

Fig. 3.   Left limb of an axolotl from Exp. 2 (Series
II) which was amputated (at the level indicated by the large arrowhead) and then allowed
to regenerate for 49 days at which time it
showed proximodistal duplication of the humerus.   This limb was then reamputated through
the distal radius and ulna (indicated by small
arrowhead).   The portion of the limb removed
by the reamputation regenerated.

limbs   by   the retinol palmitate treatment and allowed morphologically normal limb regeneration to occur   (Figs.   1   &   2).
Tunicamycin   loaded silastin blocks carrying 12.5 µg tunicamycin inhibited proximodistal duplication   (mean   SAI   =   0.38),
when   compared   to the contralateral control limbs (mean SAI =
1.75), but the effect was not complete and three out of   eight
cases   showed   the presence of extra carpals.   There were 8 to
10 axolotls in each experimental group.

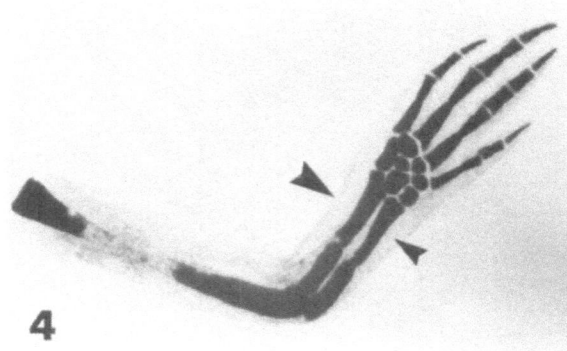

4

Fig. 4.   The right limb contralateral to that shown in
Fig. 3. In this case the reamputation (small
arrowhead) of the duplicated limb was through
the plane of the initial amputation (large
arrowhead). Only the portions normally present
distal to the amputation level regenerated.
A second duplicated limb was not produced.

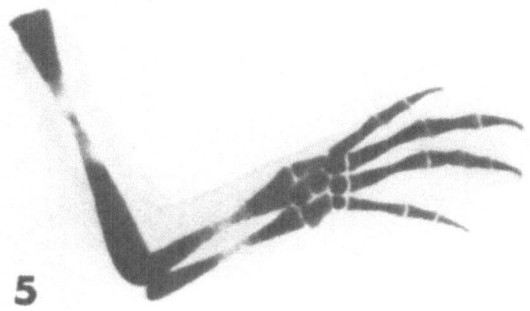

5

Fig. 5. A regenerated limb from a white axolotl of
Exp. 3 which was amputated and allowed to
regenerate without further treatment. The
regenerate appears normal in all respects.

Experiment 2. Series I: The right limbs treated with
retinol palmitate after amputation, produced regenerates with
a mean SAI of 5.0 (n=8; 2 out of 10 limbs did not regenerate).
The contralateral left limbs, which were amputated after the
14 day retinol palmitate treatment, produced regenerates with
a mean SAI of 1.6 (n=10). Series II: When a limb previously
having produced a regenerate with a proximodistal duplication
of the humerus was reamputated through the distal radius and
ulna of the duplicated regenerate, the result was always the
regeneration of the parts which had been reamputated (Fig. 3).
However, when the reamputation was at the original amputation
plane through the proximal radius and ulna, only 4 out of 14
cases regenerated the previously duplicated structures. In 9
of the 14 cases the regenerate was normal (Fig. 4), that is
the extra skeletal elements in the duplicated limb did not
regenerate, but regenerated only those elements which would

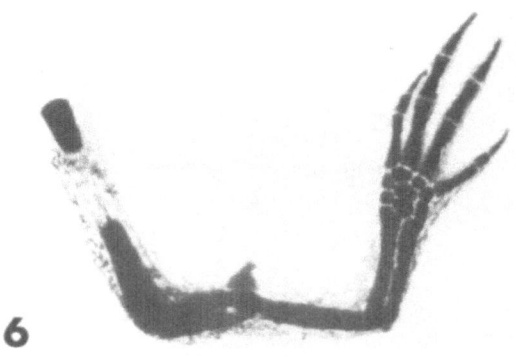

6

Fig. 6. A regenerated limb from a black vitamin A
treated axolotl of Exp. 3 which was amputated and then treated with retinol palmitate. The regenerate shows proximodistal
duplication of the entire humerus (SAI 5).

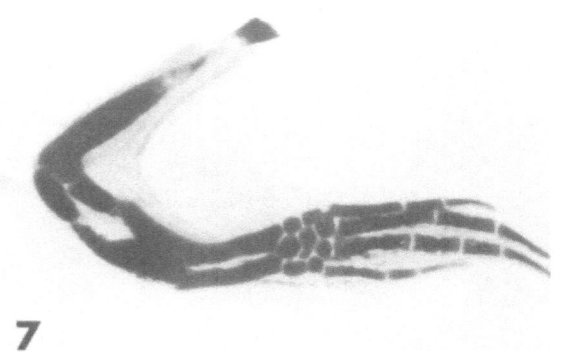

7

Fig. 7. This limb regenerate arose from the transplantation of a blastema from a black vitamin A
treated axolotl onto a white untreated axolotl. The regenerate exhibits proximodistal
duplication of part of the humerus (SAI 4).
Note that the regenerate is pigmented indicating its derivation from a black axolotl.

normally be present distal to the amputation plane.   One limb
out of 14 did not regenerate at all.

Experiment 3. Limbs of white untreated axolotls which
were not subjected to blastema transplantations, did not exhibit any proximodistal duplications (Fig. 5).   Limbs of black
vitamin A treated axolotls which were not subjected to

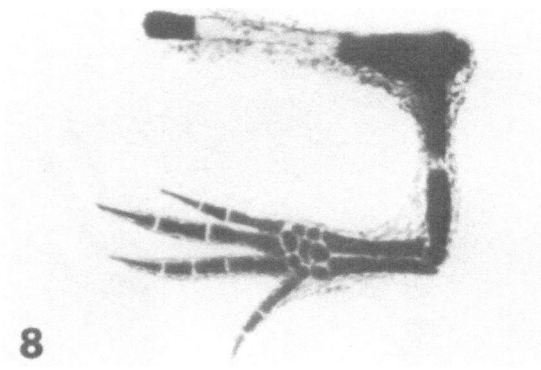

8

Fig. 8. This limb regenerate arose from the transplantation of a blastema from a white untreated
axolotl onto the limb of a black vitamin A
treated axolotl. This regenerate also exhibits proximodistal duplication of part of the
humerus (SAI 4). The entire limb is pigmented
due to the distal migration of melanocytes
into the regenerate.

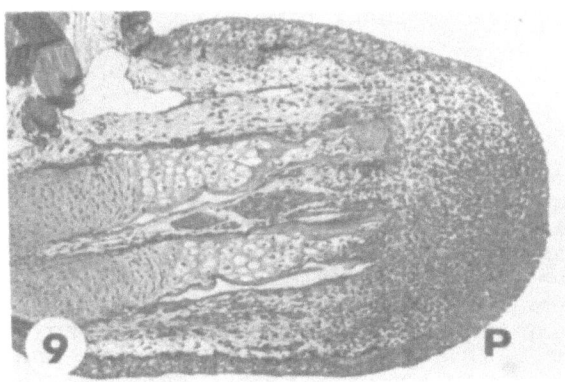

Fig. 9. This photomicrograph is a section of an axolotl
limb fixed 14 days after amputation during
which time it was treated with retinol palmitate. Note the eccentric thickening of the
epidermis, and the gradient in cell density
across the blastema. The region along the
posterior (P) edge of the limb beneath the
thickened epidermis is of greater cell density
than the more anterior part of the blastema.

blastema transplantation, showed proximodistal duplication
(Fig. 6) in 17 out of 20 cases (in 3 cases no regeneration
occurred). However, when a blastema from a black vitamin A
treated axolotl was transplanted onto the stump of a white
untreated axolotl, proximodistal duplications were observed in
9 out of 20 cases (Fig. 7). When a blastema from a white

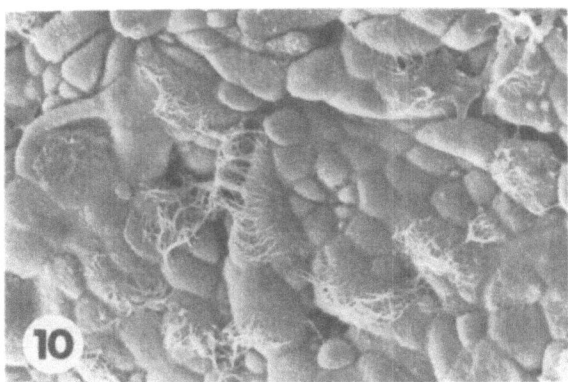

Fig. 10. Scanning electron micrograph of the epidermis
of a limb amputated and treated for 14 days in
retinol palmitate. The epidermis is characterized by irregularities in the cell sizes,
crevices in the epidermal surface, and the
development of cilia in some of the epidermal
cells.

untreated axolotl was transplanted onto the limb stump of a black vitamin A treated axolotl, proximodistal duplications were observed in 15 out of 22 cases (Fig. 8).

Experiment 4. The histology of normal axolotl limb regenerates has been described elsewhere and will not be repeated here (Tank et al., 1976, 1977; Stocum, 1979). Vitamin A induced a number of histological changes in the regenerating axolotl limb. These changes included a greater degree of initial resorption than the controls, and the development of an eccentrically located epidermal cap which was displaced towards the posterior side of the limb, unlike the centrally located epidermal cap of controls. Accompanying this was the development of a blastema in which the cell density showed a gradient across the blastema with the area of greatest cell density lying beneath the thickened epidermis on the posterior edge of the limb (Fig. 9). The more dense region of the blastema apparently gave rise to the distal parts of the limb while the less dense anterior portion of the blastema was associated with the more proximal portions of the limb regenerate (girdle and proximal region of the humerus). As well the retinol palmitate induced a marked thickening of the basement membrane.

Limb regeneration was generally inhibited by vitamin A such that at the end of the 14 day treatment period, the regenerates consisted only of a small flattened blastema (Fig. 9). Control limbs which had not been treated with vitamin A showed late cone to palette stage regenerates by day 14 which in some cases exhibited the beginning of digits. The epidermis was significantly disrupted by the retinol palmitate immersion treatment and showed a number of changes. The surface became much more irregular and was marked by crevices between groups of epidermal cells. The size of the cells on the epidermal surface also showed a great deal of variation. In addition, many of the epidermal cells were induced to produce cilia in response to the retinoid treatment (Fig. 10).

DISCUSSION

The results showed that both actinomycin D and cycloheximide completely inhibited the ability of vitamin A to cause proximodistal duplication during limb regeneration. This then provides support for the hypothesis that vitamin A brings about the proximodistal duplication of limbs during regeneration by acting at the gene level, initiating or augmenting the expression of some gene or set of genes. This interpretation of vitamin A action is also supported by the identification of a cellular retinoic acid binding protein in the axolotl limb, and the localization of labelled retinoic acid in the cell nucleus during limb regeneration (Keeble and Maden, 1986b).

The further observation that tunicamycin also antagonized the ability of retinoids to induce proximodistal duplication suggests that perhaps the end product of this gene activity is a glycoprotein. Glycoproteins have previously been suggested as possible candidates for the carriers of the positional information which underlies the process of pattern regulation (Slack, 1980; Maden, 1983a). Perhaps, tunicamycin might block the synthesis of new positional information, and hence inhibit the retinoid induced duplication.

The results of Exp. 2 (Series I) indicate that vitamin A can change cells in such a way that proximodistal duplications can be induced even when limb amputation is carried out after the cessation of the vitamin A treatment. However, when a duplicated limb is amputated through the same level after a period of seven weeks, it does not regenerate a second duplicated limb, but rather a normal limb regenerate (Series II). This suggests that some cellular change is induced by vitamin A, and that this change persists after the cessation of the treatment, but it is not permanent.

The reciprocal transplantations of Exp. 3 resulted in proximodistal duplications in a majority of cases. This suggests that both the cells of the blastema and of the stump have undergone some change as a result of the vitamin A treatment. Since this change is still expressed in the stump of vitamin A treated axolotls 14 days after the cessation of the vitamin A treatment, it suggests that the cellular change induced by vitamin A persists for at least this length of time in a majority of cases. Maden (1984) has shown that it is the mesodermal cells rather than the epidermal cells which mediate the vitamin A effects.

Some histological changes induced in regenerating limbs by immersion of the axolotl in retinol palmitate for 14 days have been identified in the present investigation. Kim and Stocum (1986) administered vitamin A to axolotls by a single injection of retinoic acid on day 4 post-amputation. They noted a number of histological features similar to those noted in this study including the unusual degree of limb resorption, the thickening of the wound epidermis on the posterior side of the limb, and the asymmetric cell density of the blastema with the region of greatest density underlying the thickened epidermis. However, a number of histological features not reported by Kim and Stocum (1986) were noted in the present investigation. These include the irregularity of the surface morphology of the epidermis, the great variation in cell size, and the induction of ciliated cells in the epidermis. It is quite possible that these epidermal changes result from the administration of the retinol palmitate by immersion which would presumably expose the epidermal cells to higher levels of the retinoids than they would receive when the retinoid was administered by injection.

The development of ciliated cells in retinol palmitate treated axolotls is another example of the cell metaplasia which retinoids can induce. Vitamin A is capable of inducing a variety of metaplastic changes in various cell types including the induction of feathers in the scale forming integument of the ckick embryo (Dhouailly and Hardy, 1978) and the induction of glandular structures from rodent hair follicles (Hardy, 1968).

Overall, the results of these experiments are interpreted as being consistent with the hypotheses: that a) vitamin A acts at the gene level by switching on or increasing the expression of some gene or set of genes, that b) this leads to synthesis of some protein, likely a glycoprotein, which c) modifies certain as yet unknown cell activities leading to a change in the positional coding, such that proximodistal duplication occurs, as well as other characteristic histological changes, and that d) this change persists for some weeks after the cessation of the vitamin A treatment, but that eventually the cells revert to their normal positional coding, implying that the vitamin A induced gene expression is reversed.

ACKNOWLEDGEMENTS

This research was supported by the Natural Sciences and Engineering Research Council of Canada. Thanks to Karin Davidson-Taylor for excellent technical assistance. Thanks to Dr. J.B. Armstrong and Mr. W. Fletcher, Department of Biology, University of Ottawa, for the gift of axolotl larvae.

REFERENCES

Dhouailly, D., and Hardy, M.H., 1978, Retinoic acid causes the development of feathers in the scale-forming integument of the chick embryo, Wilh. Roux's Arch. 185:195-200.

Gale, E.F., Cundliffe, E., Reynolds, P.E., Richmond, M.H., and Waring, M.J., 1981, "The molecular basis of antibiotic action", 2nd ed., John Wiley & Sons, New York.

Hardy, M.H., 1968, Glandular metaplasia of hair follicles and other responses to vitamin A excess in cultures of rodent skin, J. Embryol. Exp. Morph., 19:157-180.

Keeble, S., and Maden, M., 1986a, Retinoic acid binding protein in the axolotl: Distribution in mature tissues and time of appearance during limb regeneration, Dev. Biol., 117:435-441.

Keeble, S., and Maden, M., 1986b, The presence of cytoplasmic retinoic acid binding proteins in amphibian tissues and their possible role in limb regeneration, Prog. Clin. Biol. Res., 217A:309-314.

Kim, W.-S., and Stocum, D.L., 1986, Effects of retinoic acid on regenerating normal and double half limbs of axolotls. Histological studies, Roux's Arch. Dev. Biol., 195:243-251.

Koussoulakos, S., Kiortsis, V., and Anton, H.J., 1986, Vitamin A induces dorsoventral duplications in regenerating urodele limbs, IRCS Med. Sci., 14:1093-1094.

Lheureux, E., Thoms, S.D., and Carey, F., 1986, The effects of two retinoids on limb regeneration in Pleurodeles waltlii and Triturus vulgaris, J. Embryol. Exp. Morph., 92:165-182.

Maden, M., 1982, Vitamin A and pattern formation in the regenerating limb, Nature, 295:672-675.

Maden, M., 1983a, The effect of vitamin A on the regenerating axolotl limb, J. Embryol. Exp. Morph., 77:273-295.

Maden M., 1983b, The effect of vitamin A on limb regeneration in Rana temporaria, Dev. Biol., 98:409-416.

Maden, M., 1984, Does vitamin A act on pattern formation via the epidermis or the mesoderm?, J. Exp. Zool., 230:387-392.

McCormick, A.M., Shubeita, H.E., and Stocum, D.L., 1988, Cellular retinoic acid binding protein: Detection and quantitation in regenerating axolotl limbs, <u>J. Exp. Zool.</u>, 245:270-276.

Niazi, I.A., Pescitelli, M.J., and Stocum, D.L., 1985, Stage dependent effects of retinoic acid on regenerating urodele limbs, <u>Roux's Arch. Dev. Biol.</u>, 194:355-363.

Niazi, I.A., and Saxena, S., 1978, Abnormal hindlimb regeneration in tadpoles of the toad, <u>Bufo andersonii</u> exposed to excess vitamin A, <u>Folia Biologica (Krakow)</u>, 26:3-11.

O'Malley, B.W., and Means, A.R., 1974, Female steroid hormones and target cell nuclei, <u>Science</u>, 183:610-620.

Roberts, A.B., and Sporn, M.B., 1984, Cellular biology and biochemistry of the retinoids, pp. 209-286, <u>in</u>: "The Retinoids", Volume 2, M.B. Sporn, A.B. Roberts, and D.S. Goodman, eds., Academic Press, New York.

Scadding, S.R., 1983, The effect of retinoic acid on limb regeneration in <u>Xenopus laevis</u>, <u>Can. J. Zool.</u>, 61:2698-2702.

Scadding, S.R., 1988a, Actinomycin D, cycloheximide, and tunicamycin inhibit vitamin A induced proximodistal duplication during limb regeneration in the axolotl <u>Ambystoma mexicanum</u>, <u>Can. J. Zool</u>, 66:879-884.

Scadding, S.R., 1988b, Vitamin A modification of the positional information of blastema cells during limb regeneration in the axolotl, <u>Ambystoma mexicanum</u>, <u>Can. J. Zool.</u>, in press.

Scadding, S.R., and Maden, M., 1986a, Comparison of the effects of vitamin A on limb development and regeneration in the axolotl, <u>Ambystoma mexicanum</u>, <u>J. Embryol. Exp. Morph.</u>, 91:19-34.

Scadding, S.R., and Maden, M., 1986b, Comparison of the effects of vitamin A on limb development and regeneration in <u>Xenopus laevis</u>, tadpoles, <u>J. Embryol. Exp. Morph.</u>, 91:35-53.

Scadding, S.R., and Maden, M., 1986c, The effects of local application of retinoic acid on limb development and regeneration in tadpoles of <u>Xenopus laevis</u>, <u>J. Embryol. Exp. Morph.</u>, 91:55-63.

Slack, J.M.W., 1980, A serial threshold theory of regeneration, <u>J. Theor. Biol.</u>, 82:105-140.

Stocum, D.L., 1979, Stages of forelimb regeneration in <u>Ambystoma maculatum</u>, <u>J. Exp. Zool.</u>, 209:395-416.

Stocum, D.L., and Crawford, K., 1987, Use of retinoids to analyze the cellular basis of positional memory in regenerating amphibian limbs, <u>Biochem. Cell Biol.</u>, 65:750-761.

Tank, P.W., Carlson, B.M., and Connelly, T.G., 1976, A staging system for forelimb regeneration in the axolotl, <u>Ambystoma mexicanum</u>, <u>J. Morph.</u>, 150:117-128.

Tank, P.W., Carlson, B.M., and Connelly, T.G., 1977, A scanning electron microscopic comparison of the development of embryonic and regenerating limbs in the axolotl, <u>J. Exp. Zool.</u>, 201:417-430.

Thoms, S.D., and Stocum, D.L., 1984, Retinoic acid induced pattern duplication in regenerating urodele limbs, <u>Dev. Biol.</u>, 103:319-328.

VITAMIN A EFFECT ON LIMB REGENERATION: STUDIES ON THE

TADPOLES OF ANURAN AMPHIBIANS

I.A. Niazi, O.P. Jangir, S. Alam, K.K. Sharma and
C.S. Ratnassamy

Developmental Biology Laboratory, Department of
Zoology, University of Rajasthan, Jaipur 302004
India

INTRODUCTION

It has been known for long that excess of vitamin A
causes a variety of malformations and defects including those
of limbs in mammalian fetuses (Takekoshi, 1964). The initial
objective of the studies which begun at this laboratory about
20 years ago was to explore if the excess of this vitamin produced any teratogenic or inhibitory effects on the normal
development and on tail and limb regeneration in the anuran
amphibians. It was found that continuous immersion of frog and
toad tadpoles in aqueous suspension of vitamin A palmitate
inhibited regeneration of both tail (Niazi and Saxena, 1968)
and limbs (Saxena and Niazi, 1977). However, hindlimb regenerates of young tadpoles of Bufo andersonii, given a restricted
treatment with vitamin A postamputation revealed two unexpected features: in some cases more than one regenerates had
grown out from the same stump and some of them consisted of
all limb segments from thigh to foot arising from the shank
stump (Niazi and Saxena, 1978). The latter feature was particularly intriguing because it showed that vitamin A was instrumental in breaking the general rule of limb regeneration
according to which only the missing parts distal to the amputation level are restored.

The effect of vitamin A on limb regeneration observed by
Niazi and Saxena (1978) was soon confirmed in another toad
species, B. melanostictus (Jangir and Niazi, 1978; Jangir,
1979). They amputated one hindlimb in young tadpoles of foot-
spatula stage of development keeping the contralateral limb
intact. The tadpoles were then immersed in 15 IU/ml vitamin A
palmitate suspension for 3 days and thereafter reared in tap
water. Morphogenesis of the foot in the intact limbs was inhibited but good regenerates (often more than one per stump)
developed from the amputated shank stump. Most of these
regenerates were well formed complete limbs containing all
stump elements including a girdle in many cases (Figs. 1,2).

During the last six years the unique effect of vitamin A

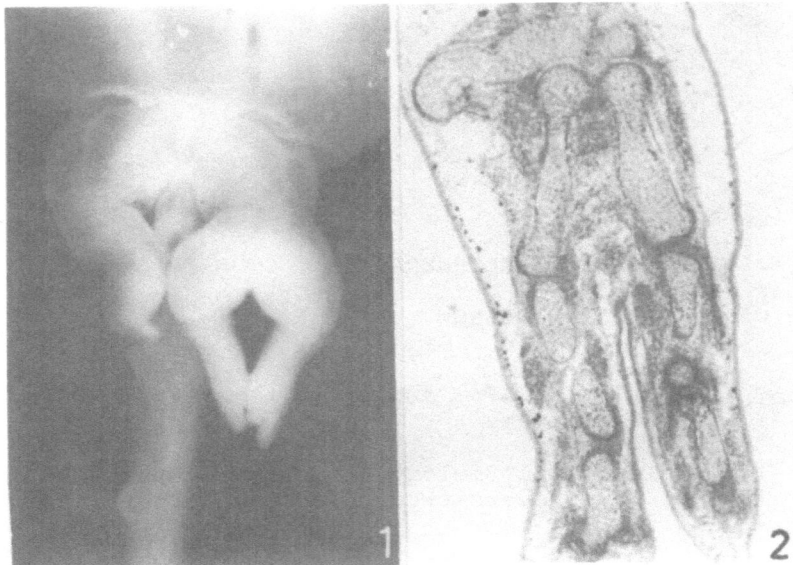

Fig. 1,2.   Two cases of twin proximalized regenerates
of young toad tadpoles treated with vitamin
A for 3 days after amputation through shank.
Girdle and femur were duplicated in the twin
regenerates (Fig. 2). Note inhibition of
foot   morphogenesis   in the intact limb
(Fig. 1).

and its derivatives (retinoids) on limb regeneration has been
confirmed and demonstrated in several amphibian species by a
number of workers from different laboratories. It is now
firmly established that suitable retinoid treatment of young
anuran tadpoles and urodele larvae as well as adults causes
serial duplication of some or all the stump elements in the
regenerates arising after amputation at any level of the limb.
This phenomenon is now referred to as PD duplication or proximalization of the regenerates (Maden, 1982, 1983; Thoms and
Stocum, 1984; Scadding and Maden, 1986; Lheureux et al., 1986;
Koussoulakos et al., 1988; Sharma and Niazi, 1988). The relationship of the proximalizing effect with the type and dose of
the retinoid, time of administration and duration of treatment
is well established. Retinoid acid (RA) is found to be the
most effective. The retinoids appear to cause proximalization
by their effect on the dedifferentiating cells accumulating to
form the blastema; but their administration immediately after
amputation during early wound healing phase is not essential.
Regeneration is adversely affected or inhibited if retinoid
treatment is commenced at late blastema/early redifferentiation stages or if it is continued beyond these stages (Jangir
and Niazi, 1978; Maden, 1982; Niazi and Ratnasamy, 1984, Niazi
et al., 1985).

With regard to the mechanism of action of vitamin A on
limb regeneration the information now available suggests the
possible cellular and molecular basis of its effect. It is
generally accepted that vitamin A influences cellular differ-

entiation and morphogenesis by way of regulation of transcriptional activity of the genes. Cytoplasmic binding proteins specific to retinoic acid (CRABP) and retinol (CRBP) have been identified in almost all cells which respond to these retinoids by phenotypic and/or biosynthetic changes. These proteins carry the retinoids into the nucleus and transfer them to receptors associated with chromatin (Takase et al., 1986; Daly and Redfern, 1987). In the mammalian tissues CRABP and CRBP levels are high during fetal and neonatal periods but drop significantly in the adult stage indicating the importance of these proteins in development (Chytil and Ong, 1984).

Among the amphibians CRABP has been detected in the skin, muscle and limb tissues of axolotls. It is found to increase severalfold early during limb regeneration and then drops to normal level as regeneration is completed. This is a normal feature of limb regeneration because exogenously administered RA did not raise the level of this protein in the cone stage blastema any higher (Keeble and Maden, 1986a). Most recently, McCormick et al. (1988) have found that limb amputation in axolotls is followed by a relative increase in the Apo-CRABP in the blastema which is increasingly converted to Holo-CRABP as it is saturated with endogenous RA reducing the Apo:Holo ratio. These findings strongly suggest that CRABP and also endogenous vitamin A may be essential for regeneration itself to occur. Sharma and Anton (1986) observed increased an number of pores in the nuclear membrane of the cells and changes in the protein profiles of the limb regeneration blastema of retinol palmitate treated axolotls. According to Keeble and Maden (1986b) exogenously given RA actually enters the nuclei of blastema cells. However, elucidation of the exact process by which retinoid mediated alteration of gene activity changes the pattern of limb regeneration must await results of future research.

In this report we present the results of some studies on anuran tadpoles carried out at our laboratory during the last few years which pose some further problems for research. They show that (i) the proximalizing effect of vitamin A is correlated with the presence of natural ability to regenerate and the sensitivity of tissues to this chemical; (ii) vitamin A influences regeneration in several other ways also and (iii) its influence persists for some time and the consequent morphogenetic effects are manifested even if regeneration is initiated some time after cessation of the treatment.

MATERIALS AND METHODS

The studies were made on the tadpoles of space-foot frog, Rana breviceps and the toad, Bufo melanostictus. The mode of treatment was immersion of the tadpoles in 15 IU/ml vitamin A palmitate suspension for the required period and subsequent transfer to tap water for the remaining period of the experiment. Previous experience had shown that this concentration of vitamin A was effective in producing proximalization of the regenerates. The drug employed was an oily solution of vitamin A palmitate marketed by Roche (India) under the trade name Arovit. A known quantity of the drug was dissolved in 1 ml ethanol and then diluted with tap water to obtain the required concentration of the vitamin. Hindlimbs were amputated through thigh, shank or ankle after narcotizing the tadpoles in 1:4000 solution of MS 222 (Sandoz). The regenerates were either stained with Victoria blue in toto or sectioned and stained with haematoxylin and eosin for histological study. Details of the methods adopted in various experiments are described at appropriates places in the text that follows.

RESULTS AND DISCUSSION

Effects of Vitamin A on Limb Regeneration in Anuran Tadpoles of Different Developmental Stages

In the anurans the ability to regenerate declines proximodistally along the limb in the growing tadpoles and disappears completely some time before metamorphosis. Consequently, at any developmental stage of the tadpoles the degree of this ability varies distoproximally along the PD axis of the limb. Therefore the experiments were designed to investigate the effects of vitamin A on limb regeneration in the tadpoles of different developmental stages after amputation through three PD levels of hindlimbs, viz., thigh, shank and ankle. The experiments were made on Bufo melanostictus tadpoles of stages 30/31, 34, 36 and 38 according to Khan (1965). The Rana breviceps tadpoles employed were approximately equivalent to Bufo stages 30/31 and 38. Following amputation the tadpoles were immersed in vitamin A suspension for 1, 2 or 3 days and thereafter reared in tap water. The similarly operated controls were not given vitamin A treatment.

The results are presented in Tables 1, 2 and 3. Four types of regenerates were obtained in different proportions at the three amputation levels in the tadpoles of different developmental stages; (i) Normal - In these the regenerate consisted of parts distal to amputation level. They were either perfect with 5 toes (pentadactylous) or hypomorphic and oligodactylous lacking one or more toes. (ii) Proximalized -In these, some or all the stump elements were duplicated. They included cases in which a single or more than one (multiple) such limbs had regenerated from the same stump. Very often the axes and orientation of these regenerates appeared to have no relationship with those of the stump. It seemed as if they had arisen from limb buds grafted on the amputation surface in a haphazard manner without caring to align the axes of the graft with those of the stump (Figs 3,4). Previously, such regenerates have been designated as whole limbs (Niazi and Alam, 1984; Niazi and Ratnasamy, 1984; Sharma and Niazi, 1988). We now consider that it is more appropriate to call them just proximalized because not all stump elements are duplicated in such regenerates although externally they may look like complete limbs consisting of all segments from stylopodium to autopodium. (iii) Inhibited blastemata - In these cases blastema was formed but it failed to redifferentiate. Microscopic examination of such cases revealed absence of mitotic figures or any histogenesis (Figs 5,6). Previously, they have been called persistent blastemata in publications referred to above. It seems more appropriate to call them inhibited blastemata. (iv) Spikes - These were small outgrowths containing some cartilage.

(1) Table 1 shows the results of 3 days vitamin A treat-

358

Table 1. Effect of 3 Days 15 IU/ml Vitamin A Palmitate Treatment of _Bufo mela-_
nostictus Tadpoles Following Amputation on Hindlimb Regeneration
(N: Normal; P: Proximalized; IB: Inhibited blastema)

Stage and amputation level	Group	Amputees regener- ating (%)	Types of regener- ates (%)			5-toe normal regener- ates (%)	Average number of toes per N regenerate
			N	P	IB		
Stage 30/31							
Thigh	Control	100	100	–	–	84	4.7
	Treated	100	0	90	10	–	–
Shank	Control	100	100	–	–	87	4.8
	Treated	100	0	95	5	–	–
Ankle	Control	100	100	–	–	85	4.8
	Treated	100	0	56	44	–	–
Stage 34							
Thigh	Control	100	100	–	–	40	3.9
	Treated	100	36	16	48	67	4.7
Shank	Control	100	100	–	–	47	4.1
	Treated	100	12	60	28	100	5.0
Ankle	Control	100	100	–	–	67	4.6
	Treated	100	8	76	16	100	5.0
Stage 36							
Thigh	Control	43	100	–	–	0	1.8
	Treated	52	53	0	47	14	3.8
Shank	Control	67	100	–	–	30	3.5
	Treated	72	100	0	0	50	4.8
Ankle	Control	73	100	–	–	27	4.0
	Treated	85	100	0	0	59	4.6

Stage 38 Control – Small spikes regenerated in 15% cases only at ankle
level.

Treated – Small spikes regenerated in 10, 20 and 50% cases at
thigh, shank and ankle levels, respectively.

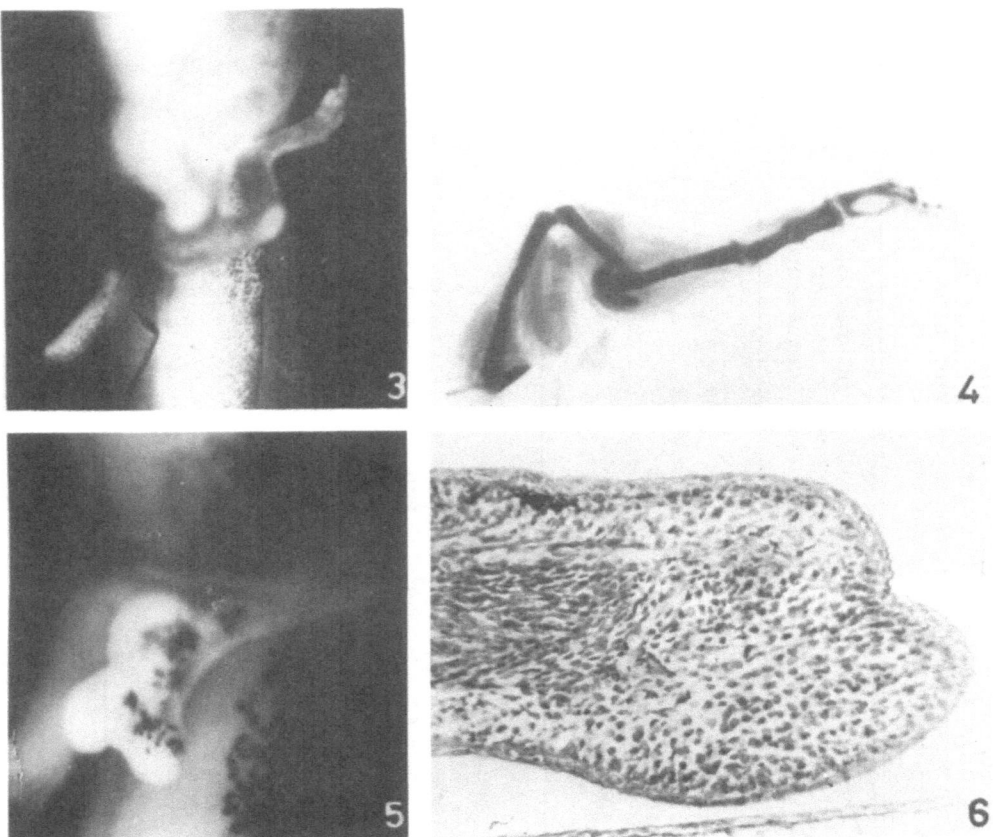

Figs. 3-6.   Four cases of regenerates of young <u>Bufo</u>
<u>melanostictus</u> tadpoles treated with vitamin
A for 3 days after shank level amputation
Figs. 3,4: Two single proximalized regenerates with atypical orientation and axes
with respect to stump. Figs. 5,6: Two cases
of inhibited blastemata in which morphogenesis and histogenesis did not occur.

ment of <u>B. melanostictus</u> tadpoles of 4 developmental stages
after amputation through thigh, shank and ankle. In the controls regenerative ability was very good throughout the limb
at stage 30/31; but it was reduced at all three levels at
stage 34 as indicated by decreased percentage of pentadactylous and increase in the number of oligodactylous regenerates.
This reduction was further enhanced at stage 36 at which the
percentage of amputees of all levels regenerating at all also
dropped significantly. By stage 38 there was very little
capacity to regenerate remaining even at ankle level where
only small spikes regenerated in just 15% cases.

In the treated groups proximalized regenerates were produced at stages 30/31 and 34 but not at older stages. At stage
30/31, when regeneration ability was very good, all the well
differentiated regenerates from the three amputation levels
were of the proximalized type; but at stage 34 they included

Table 2. Percentage of Normal (N), Proximalized (P) and Inhibited Blastema (IB) Type of Regenerates in Stage 30/31 Tadpoles of Toad and Frog Species with 15 IU/ml Vitamin A Palmitate Following Hindlimb Amputation

Species	Amputation level	Type of regenerates	Duration of treatment (days)		
			1	2	3
Bufo melanostictus	Thigh	N	60	0	0
		P	40	75	90
		IB	0	25	10
	Shank	N	50	0	0
		P	50	80	95
		IB	0	20	5
	Ankle	N	65	20	0
		P	30	45	56
		IB	5	35	44
Rana breviceps	Thigh	N	32	7	0
		P	53	77	31
		IB	15	16	69
	Shank	N	5	0	0
		P	95	40	27
		IB	0	60	73
	Ankle	N	97	2	0
		P	3	73	32
		IB	0	25	68

normal type as well. The frequency of perfect pentadactylous to regenerate along the length of the limb varied as thigh< shank<ankle. Significantly, the frequency of proximalized regenerates in treated tadpoles of this stage was also thigh< shank<ankle very clearly indicating the correlation of the proximalizing effect of vitamin A with the degree of the capacity to regenerate normally present at any level along the length of the limb. Similar results have been reported for Rana breviceps tadpoles (Sharma and Niazi, 1988). Earlier, Scadding and Maden (1986) had also observed that the reduction in the frequency of duplication of stump structures in the vitamin A treated limb regenerates of older Xenopus tadpoles was correlated with the general loss of the normal ability to regenerate with age. It may be the general rule for all amphibians. In the urodeles this effect has been demonstrated after amputation through any level in both fore and hindlimbs in the larvae as well as adults but these amphibians retain the ability to regenerate limbs throughout life.

Reference to Table 2 will show that there is a very definite difference between Bufo melanostictus and Rana breviceps tadpoles in their sensitivity and regenerative response to identical treatment with vitamin A. At each amputation level longer treatment was required to obtain the highest frequency of proximalized regenerates in the Bufo than in the Rana tadpoles. Three days treatment was actually inhibitory for the latter whereas it gave the best results in case of the former species. In stage 38 Bufo tadpoles this length of treatment could not induce regeneration of more than small spikes from any amputation level. In contrast, Sharma and Niazi (1988) found that in the advanced tadpoles of R. breviceps (equivalent to Bufo stage 38) the same treatment resulted in proximalized regenerates in fair number of cases at ankle and shank levels in this order; and even at thigh level, where regeneration ability was found to be completely absent in the controls, vitamin A induced good regeneration in at least 5 cases in which girdle was also duplicated. These examples reflect the differences among the two species with respect to when and at what speed they lose the ability to regenerate during larval growth. It is lost earlier and more rapidly in Bufo melanostictus than in Rana breviceps (Niazi, 1983).

We know from the findings of Keeble and Maden (1986a,b) and McCormick et al. (1988) on axolotls that CRABP and endogenous vitamin A are most probably essential for regeneration itself to occur. The unique morphogenetic effect of exogenous RA and other retinoids may also be mediated via specific cytoplasmic binding proteins in the blastema cells. So far there is no information about the presence or absence of such proteins in the tissues of intact or regenerating limbs of tadpoles of any anuran species. Could it be that the relative amounts of these proteins vary in the stylopodial, zeugopodial and autopodial tissues of the growing limbs of frog and toad tadpoles proximo-distally declining with age ? If this be so it may indicate one of the possible factors involved in the progressive loss of limb regeneration ability in anuran tadpoles with age, correlation of the proximalizing effect of vitamin A with the presence of this ability and the variations in the frequency of proximalized regenerates arising from different amputation levels. It may also explain the differences between frog and toad species in their response to vitamin A.

(2) Occurrence of inhibited blastemata was observed at all amputation levels in both Bufo and Rana tadpoles immersed in vitamin A suspension for 3 days or even less (Tables 1,2). Undoubtedly, this was also due to vitamin A effect for no such case was found in any untreated group. Vasan (1981) had observed that sternal chondrocytes of 13 days chick embryos exposed to RA in culture reverted to prechondrogenic morphology and synthetic activity but did not redifferentiate on transfer to normal medium. Probably, the high amount of the retinoid in the culture medium had irreversibly damaged the synthetic machinery of the cells necessary for differentiation. It is possible that in our experiments some tadpoles might have taken in too much vitamin A which stimulated the emergence of dedifferentiated cells from stump tissues but at the same time damaged the machinery for proliferation and redifferentiation. However, this alone does not explain the pattern of variations in the frequency of inhibited blastemata at the three amputation levels in tadpoles of different stages

Table 3. Percentage of Cases of Multiple Regeneration per
          Stump Among all Proximalized Regenerates in Young
          (Stage 30/31) Tadpoles of Bufo melanostictus (B.m.)
          and Rana breviceps (R.b.) Treated with 15 IU/ml
          Vitamin A Palmitate Following Hindlimb Amputation

Duration of Treatment	Thigh		Shank		Ankle	
	B.m.	R.b.	B.m.	R.b.	B.m.	R.b.
1 day	5	72	10	0	0	0
2 days	30	46	25	38	10	0
3 days	35	50	35	66	16	9

given identical treatment. It may be noted in Tables 1 and 2
that the pattern appears to have some relation not only with
the duration of treatment but also with the developmental
stage of limb tissues through which amputation was made.
Frequency of inhibited blastemata was high if the cut was made
through limbs whose tissues had attained a relatively higher
level of differentiation (older tadpoles) and also when it was
still too low (ankle level in young tadpoles).

(3) Frequent development of multiple regenerates from
the same stump in both frog and toad tadpoles was another
interesting effect of vitamin A. It occurred in as many as 1/3
to 1/2 of all proximalized regenerates. In the young tadpoles
its frequency was quite high at thigh and shank levels but it
was very rare at ankle level. Also, it occurred relatively
more often in Rana than in Bufo tadpoles. In both species its
frequency increased with length of exposure to vitamin A
(Table 3). In such cases usually 2 but sometimes 3 or even
more regenerates arose from the same stump. In most cases all
the individual members were definitely proximalized although
the degree of differentiation, morphogenesis and growth
attained by each was not always similar. In the case of twins
the two members were generally mirror images of each other. In
many, however, the individual members of a complex grew out in
all sorts of directions and some were quite bizarre (Figs. 7,
8). Multiple regeneration was found to occur when vitamin A
treatment was commenced immediately after amputation, one day
later (Niazi and Ratnasamy, 1984) and also when it preceded
amputation (Niazi and Alam, 1984; Tables 4 and 5 in this
article). Among the cases of inhibited blastema type of regenerates also there were many in which two blastemata were
formed (Figs 5,6).

Regeneration of multiple limbs from the same stump has
also been observed in the retinoid treated tadpoles of Rana
temporaria (Maden, 1983a), Xenopus laevis (Scadding and Maden,
1986), young larvae of the newt, Pleurodeles waltlii (Lheureux
et al., 1986), axolotls (Maden et al., 1985; Kim and Stocum,
1986) and also in the adult newt, Notophthalmus viridescens
(Stocum and Thoms, 1984).

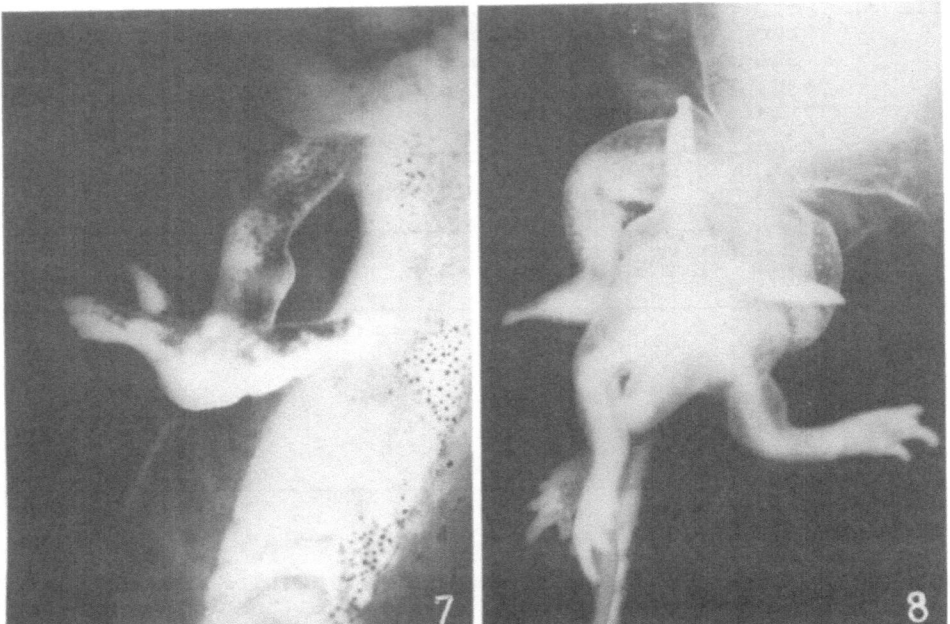

Fig. 7,8. Two cases of multiple regenerates arising
from the same shank stump of <u>B. melanostic-
tus</u> tadpoles treated with vitamin A for 3
days post-amputation. Note bizarre orientation of regenerates.

The factors responsible for the development of more than
one regenerates from the same stump are not known; but the
changes leading to this phenomenon must occur during the early

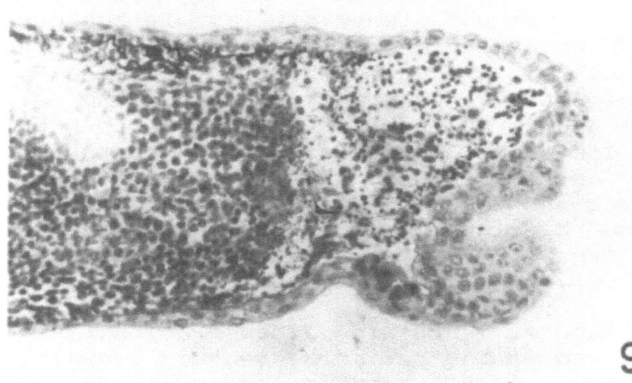

Fig. 9. A regenerating limb of a young <u>B. melanostictus</u>
tadpole sectioned longitudinally one day after
amputation and treatment with vitamin A. Note
extensive destruction of stump tissues.

stages of regeneration. We are of the opinion that the primary changes leading to multiple regeneration may lie in the distortion of wound healing pattern. Because of the very extennsive destruction and regression of stump tissues caused by vitamin A treatment (Fig. 9) it may be that the migrating epidermal cells, instead of forming a smooth cover over the cut end, might penetrate deep into the wound area creating 2 or more separated regions for blastema cell accumulation. Accordingly, the more intense the retinoid treatment and larger the wound area the greater will be the likelihood of multiple regenerates growing out from the same stump. This in fact has been found to be the case. Frequency of this type of regeneration rose with the increase in the dose of vitamin A given or the length of treatment (Sharma, 1982; Alam, 1983) and it occurred much more often at thigh and shank level which provided relatively larger wound area than at ankle level. It may also be noted that more such cases were obtained in the Rana than in the Bufo tadpoles which may be due to greater susceptibility of the tissues of the former to vitamin A action (Table 3). Examination of a large number of treated regenerating limbs sectioned during wound healing and blastema cell accumulation stages may provide the necessary information for or against this speculation.

(4) In many cases vitamin A treatment fails to result in the duplication of any stump element and production of proximalized regenerates (Tables 1,2). This may be due to relatively shorter treatment, developmental stage, susceptibility differences between species or between tissues of different limb levels. However, it was noted that the influence of vitamin A was expressed in certain other features: (a) the normal regenerates of treated tadpoles of both Rana and Bufo tadpoles were in general better formed than those of corresponding controls. As may be noted in Table 1, many of the good regenerates of stage 34 and all of stages 36 treated Bufo tadpoles were of the normal type but their morphological quality was superior than that of controls. The improvement was indicated by greater percentage of the normal regenerates of treated tadpoles possessing the complete set of 5 toes and a consistently higher average number of toes per regenerate from all amputation levels as compared to corresponding controls.
(b) At stage 36 greater percentage of limbs amputated at any level regenerated in the treated than in the control tadpoles.
(c) At stage 38 regeneration ability in the control tadpoles was completely absent at thigh and shank levels and even at ankle level small spikes regenerated only in 15% cases; but vitamin A treatment induced at least spike-like regenerative growth in as many as 10%, 20% and 50% cases at thigh, shank and ankle levels, respectively (Table 1).

Increasing oligodactyly of regenerates in anuran tadpoles is correlated with declining capacity of regeneration with age. Among other factors, insufficient number of blastema cells and/or earlier and more rapid redifferentiation induced by rising level of thyroid hormone may be responsible for increasing defects in limb regenerates (Wallace, 1981). Extensive dissolution of stump tissues caused by vitamin A may provide greater number of dedifferentiated cells for the blastema. Sharma (1982) noted much more intense acid phosphatase activity in the blastema of treated than that of untreated frog tadpoles. Increased activity of this enzyme

Table 4. Percentages of Normal, Proximalized and Inhibited Blastema Type of Regenerates of Stage 30/31 Tadpoles of <u>Bufo melanostictus</u> Treated with 15 IU/ml Vitamin A Palmitate for 3 Days Following Amputation Through Ankle and Thereafter Transferred to Water and Operated Limbs Reamputated Through Ankle, Shank or Thigh

Level of reamputation	Types of regenerates produced after reamputation		
	Normal	Proximalized	Inhibited blastemata
Ankle	8	62	30
Shank	20	40	40
Thigh	35	15	50

during the initial period of regeneration is considered to be associated with dedifferentiation and blastemal growth (Schmidt and Weary, 1963). Vitamin A reduces or suppresses thyroxine activity (Sharma and Niazi, 1983) and delays the onset of differentiation in regenerating limbs (Sharma and Anton, 1986; Kim and Stocum, 1986). It is likely that these effects of vitamin A may be instrumental in the structural improvement of the regenerates when its influence is not strong enough to cause proximalization. Sharma and Niazi (1988) have found this in frog tadpoles treated with vitamin A for 1-24 hours after ankle level amputation. The same factors may be involved in the induction of regenerative growth by vitamin A treatment in the otherwise non-regenerating limbs of older toad tadpoles (Table 1) and advanced tadpoles of frog (Sharma and Niazi, 1988).

Persistence of Vitamin A Influence

When the stumps of 3 days treated regenerating limbs were reamputated to remove the blastemata the subsequent regenerates were found to be proximalized in many cases although the tadpoles were not exposed to vitamin A after reamputation (Jangir, 1979). This indicated the persistence of vitamin A influence on stump tissues. Two questions were asked: (a) Does vitamin A treatment affect the tissues only in the immediate vicinity of the amputation level or those of the entire residual limb ? (b) How long does the influence persist after cessation of treatment ? Two series of experiments were performed on young stage 30/31 tadpoles of <u>Bufo melanostictus</u> to seek answers to these questions: in series I, hindlimbs were amputated through shank and the tadpoles immersed in 15 IU/ml vitamin A palmitate suspension for 3 days. Immediately after transfer of the tadpoles to water the operated limbs were reamputated, this time through ankle in one group, at a level slightly proximal to that of the first cut, through mid-shank in another and mid-thigh in the third group. In series II, hindlimb amputation through shank was followed by 3 days

Table 5. Various Types of Regenerates of <u>Bufo melanostictus</u> Tadpoles Transferred to Water After 3 Days 15 IU/ml Vitamin A Palmitate Treatment Following Amputation Through Shank and the Operated Limbs Reamputated Through the Same Level at Different Intervals After Cessation of Treatment

Day of reamputation after treatment	Group	Reamputated limbs regenerating (%)	Type of regenerates (%)		
			Normal	Proximalized	Inhibited blastemata
1 day	Control	100	100	0	0
	Treated	100	22	78	0
3 days	Control	88	100	0	0
	Treated	100	32	59	9
5 days	Control	52	100	0	0
	Treated	100	44	28	28
9 days	Control	21	100	0	0
	Treated	100	23	8	69

treatment as in series I but reamputation of operated limbs, this time again through shank stump, was delayed for 1, 3, 5 or 9 days after cessation of treatment and transfer of tadpoles to water. The controls of the two series were also amputated and reamputated similarly but were never exposed to vitamin A. The tadpoles were allowed to regenerate for 10-15 days after reamputation. The following results were obtained.

Series I - Subsequent to reamputation regeneration occurred in 100% cases of all control and treated groups. Regenerates of the former were all normal but in all the treated groups three types including normal, proximalized and inhibited blastemata (IB) were obtained. Among these the proportion of normal and IB types, especially of the latter, increased and that of the proximalized decreased as the level of reamputation became more proximal (Table 4). Nearly 1/3 of all proximalized and IB type regenerates were cases of multiple regeneration, most of them occurring when reamputation was through shank.

Series II - Vitamin A influence persisted even up to 9 days after cessation of treatment and was phenotypically manifested in the limb regenerates in one way or another (Table 5). As the regeneration ability declined with age the number of amputees regenerating in 3, 5 and 9-day groups of controls fell to 88%, 52% and 21%, respectively but it remained 100% in corresponding treated groups in which the usual three types of regenerates were produced but in different proportions according to delay in reamputation. In the treated groups the percentage of normal regenerates increased steadily up to 5 days but then fell quite steeply in the 9-day group which must be due to increasing age. Most such regenerates were oligodactylous but of lesser degree than those of respective control groups. The number of good proximalized regenerates was very high in the 1-day group (78%) steadily falling thereafter but even in the 9-day group at least one good proximalized generate (10%) was produced. Simultaneously, the proportion of inhibited blastema type regenerates rose from 0% in 1-day to 69% in the 9-day group.

The results clearly demonstrate that vitamin A affects the tissues of the entire residual limb and not merely those in the immediate neighbourhood of amputation level; and the influence persists, although in decreasing degree, for some time after the cessation of treatment. In fact, amputation before the commencement of treatment is not essential for the morphological effects of vitamin A to appear in the regenerates. Exposure of toad tadpoles to vitamin A only prior to amputation resulted in good proximalized regenerates in fair numbers (Niazi and Alam, 1984). One possibility can be that the exogenously given vitamin A ester persists in the blood or tissue fluids and when the limb is amputated before or after treatment it causes its characteristic morphological effects to appear in the regenerates. However, if this is the case it raises a very puzzling question. We have consistently found that treatment of frog and toad tadpoles with 15 IU/ml vitamin A palmitate beyond 3 days post-amputation or its commencement after this time inhibits redifferentiation of the blastema in all cases. How do we then explain that vitamin A even when given only during the initial 3 days but effectively persisting in the body beyond that period does not prevent post-blastemic development of normal or proximalized regenerate? At present we have no answer to this question.

Another hypothesis can be that vitamin A treatment perhaps induces some profound change(s) in the cells of the limb tissues along its entire length which destroy or modify the supposed positional information (or memory) encoded in them. Consequently, on amputation these changed tissues may contribute blastema cells with already altered or abolished positional memory and therefore the regenerates exhibit the morphogenetic effects of vitamin A. Slight changes may not affect the normal pattern of regeneration, those of a higher degree may result in proximalized regenerates and very severe ones may make the emerging blastema cells incapable of redifferentiation. The changes may be temporary and reversible, qualitative and/or quantitative affecting molecular configuration in the chromatin, cytoplasm and/or the cell surface. However, at present the necessary information is not available to support or contradict this proposition. One method of testing it can be to inject retinoic acid in urodele larvae and adults followed by amputation at different intervals. Retinoic acid is removed quickly and the use of urodeles will eliminate the difficulties due to reduction and loss of limb regeneration capacity with age in anuran tadpoles.

REFERENCES

Alam, S., 1983, Studies on the morphogenetic influence of treatment of tadpoles of the anuran <u>Bufo melanostictus</u> Schneider with vitamin A palmitate on limb regeneration, Ph.D. Thesis, University of Rajasthan, Jaipur, India.

Chytil, F. and Ong, D.E., 1984, Cellular retinoid-binding protein, <u>in:</u> "The Retinoids", M.B. Sporn, A.B. Roberts and D.S. Goodman, eds., Academic Press, New York.

Daly, A.K. and Redfern, C.P.F., 1987, Characterization of retinoid acid binding component from F9 embryonal carcinoma-cell nuclei, <u>Eur. J. Biochem.</u>, 168:133-139.

Jangir, O.P., 1979, Experimental studies on the ontogenesis and regeneration of limbs in the anuran <u>Bufo melanostictus</u> (Schneider), Ph.D. Thesis, University of Rajasthan, Jaipur, India.

Jangir, O.P. and Niazi, I.A., 1978, Stage dependent effects of vitamin A excess on limbs during ontogenesis and regeneration in tadpoles of the toad <u>Bufo melanostictus</u> Schneider, <u>Ind. J. Exp. Biol.</u>, 16:438-445.

Keeble, S., and Maden, M., 1986a, Retinoic acid binding protein in the axolotl; Distribution in mature tissues and time of appearance during limb regeneration, <u>Dev. Biol.</u>, 117:435-441.

Keeble, S., and Maden, M., 1986ab, The presence of cytoplasmic retinoic acid binding proteins in amphibian tissues and their possible role in limb regeneration, <u>in:</u> "Progress in Developmental Biology", Alan R. Liss, New York.

Khan, M.S., 1965, A normal table of <u>Bufo melanostictus</u>, (Schneider), <u>Biologia</u>, 11:1-39.

Kim, W.-S., and Stocum, D.L., 1986a, Effects of retinoic acid on regenerating normal and double half limbs of axolotls: histological studies, <u>Roux's Arch. Dev. Biol.</u>, 195:243-251.

Koussoulakos, S., Sharma, K.K. and Anton, H.J., 1988. Influence of vitamin A on proximal and distal blastemata of <u>Triturus alpestris</u>: a morphological and cytophotometric study, <u>Monogr. dev. Biol.</u>, 21:138-144.

Lheureux, E., Thoms, S.D. and Cary, F., 1986, The effects of two retinoids on limb regeneration in <u>Pleurodeles waltlii</u> and <u>Triturus vulgaris</u>, <u>J. Embryol. exp. Morph.</u>, 92:165-182.

Maden, M., 1982, Vitamin A and pattern formation in the regenerating limb, <u>Nature</u> (London), 295:672-675.

Maden, M., 1983, The effect of vitamin A on limb regeneration in <u>Rana temporaria</u>, <u>Dev. Biol.</u>, 98:409-416.

Maden, M., Keeble, S. and Cox, R.A., 1985. The characteristics of local application of retinoic acid to the regenerating axolotl limb, <u>Roux's Arch. Dev. Biol.</u>, 194:228-235.

McCormick, A.E., Shubeita, H.E. and Stocum, D.L., 1988, Cellular retinoic acid binding protein: Detection and quantitation in regenerating axolotl limb, <u>J. Exp. Zool.</u>, 245:270-276.

Niazi, I.A., 1983, Regeneration studies in India, in: "Developmental Biology. An Afro-Asian Perspective", S.C. Goel and R. Bellairs, eds. ISDB, Poona, India.

Niazi, I.A. and Alam, S., 1984, Regeneration of whole limbs from shank stumps in toad tadpoles treated with vitamin A, Roux's Arch. Dev. Biol., 193:111-116.

Niazi, I.A., Pescitelli, M.J. and Stocum, D.L., 1985, Stage dependent effects of retinoic acid on regenerating urodele limbs, Roux's Arch. Dev. Biol., 194:355-363.

Niazi, I.A. and Ratnasamy, C.S., 1984, Regeneration of whole limbs in toad tadpoles treated with retinol palmitate after the wound healing stage, J. Exp. Zool., 230:501-505.

Niazi, I.A. and Saxena, S., 1968, Inhibitory and modifying influence of excess of vitamin A on tail regeneration in Bufo tadpoles, Experientia, 24:852-853.

Niazi, I.A. and Saxena, S., 1978, Abnormal hindlimb regeneration in tadpoles of the toad, Bufo andersonii, exposed to excess vitamin A, Folia Biol. (Krakow), 26:3-11.

Saxena, S., and Niazi, I.A., 1977, Effect of vitamin A excess on hindlimb regeneration in tadpoles of the toad, Bufo andersonii, Boulenger, Ind. J. Exp. Biol., 15:435-439.

Scadding, S.R. and Maden, M., 1986, Comparison of the effects of vitamin A on limb development and regeneration in Xenopus laevis tadpoles, J. Embryol. exp. Morph., 91:35-53.

Schmidt, A.J. and Weary, M., 1963, The localization of acid phosphatase in the regenerating forelimb of adult newt, Diemyctilus viridescens, J. Exp. Zool., 152:101-114.

Sharma, K.K., 1982, Investigations on limb regeneration in tadpoles and froglets of the anuran Rana breviceps (Schneider), treated with vitamin A or electrically stimulated, Ph.D. Thesis, University of Rajasthan, Jaipur, India.

Sharma, K.K., and Anton, H.J., 1986, Biochemical and ultrastructural studies on vitamin A induced proximalization of limb regeneration in axolotl, in: "Progress in Developmental Biology", Alan R. Liss, Inc., New York.

Sharma, K.K. and Niazi, I.A., 1983, Effect of vitamin A on metamorphosis in tadpoles of Rana breviceps, Natl. Acad. Sci. Letters, 6:397-399.

Sharma, K.K. and Niazi, I.A., 1988, Variety of regenerative responses of different proximo-distal segments of young and advanced Rana breviceps tadpoles treated with vitamin A after amputation, Monogr. dev. Biol., 21:124-137.

Stocum, D.L. and Thoms, S.D., 1984, Retinoic acid induced pattern completion in regenerating double anterior limbs of urodeles, J. Exp. Zool., 232:207-215.

Takase, S., Ong, D.E., and Chytil, F., 1979, Cellular retinolbinding protein allows specific interaction of retinol with the nucleus in vitro, Proc. Natl. Acad. Sci. USA, 76:2204-2208.

Takekoshi, S., 1964, The mechanism of vitamin A induced teratogenesis, J. Embryol. exp. Morph., 12:263-271.

Thoms, S.D. and Stocum, D.L., 1984, Retinoic acid induced pattern duplication in regenerating urodele limbs, Dev. Biol., 103:319-328.

Vasan, N.S., 1981, Proteoglycan synthesis by sternal chondrocytes perturbed with vitamin A. J. Embryol. exp. Morph., 63:181-191.

Wallace, H., 1981, "Vertebrate Limb Regeneration", John Wiley, Chichester.

Commentary

by Malcolm Maden on

RETINOIDS IN REGENERATION

From the 1930's onwards an extensive knowledge had accumulated on the teratogenic effect of hypervitaminosis A and hypovitaminosis A on the development of embryos. Interestingly the effects of both these syndromes are remarkably similar: eye and retina defects, hydrocephalus, cardiac defects and limb defects which may tell us something about the embryonic requirement for retinoids. However, it was not until 1977 that Niazi reported that retinol palmitate administered to the regenerating limbs of _Bufo_ tadpoles, increases the amount of tissue regenerated: instead of just regenerating the foot after shank level amputations complete limb regenerates could be produced. Niazi's opening paper in this session summarised the results of his extensive work since that initial report and the principles that have emerged.

These principles, which must form the basis of all of the more biochemical approaches to the subject, are as follows. One, the effects are primarily on the proximodistal axis, but anteroposterior duplications are also commonly seen in the form of bifurcated regenerates (see below for further refinement). Two, the stimulatory effects are only seen during regeneration of the limb bud or fully developed limb. If vitamin A is applied during ontogenesis then the mammalian-style teratogenic effects are seen. Three, the effects are time dependent - the longer the time of administration, the greater the extent of duplication. Four, the effects are concentration dependent. Five, the effects are stage dependent - there is little effect if vitamin A is administered after the blastema stage. Six, the effects are local - when blastemata are grafted to non-limb sites after treatment, duplications still arise. Seven, there are a variety of different methods of administration of vitamin A, but all produce the same effect. Eight, the effects persist if the treatment is ended within a few days before amputation of the limb. Nine, vitamin A also induces regenerative growth in post metamorphic froglets.

This excellent summary provided the basis from which the remaining speakers in this session probed deeper into the mechanisms of action of vitamin A. But before discussing these let me draw two conclusions which are apparent from Niazi's summary. Firstly, the difference between administration during ontogenesis (teratogenic effects) and regeneration (stimulatory effects) imply that the state of the cells during these two processes must be completely different; otherwise they would respond in the same manner. This is in contrast to virtually all theories of pattern formation which assume identity of cell states and identity of pattern forming mechanisms. On the other hand, it must be said that other experimental results such as the exchange of rotated limb buds and blastemata producing supernumerary limbs do suggest identity; so there is clearly an unresolved issue here. Secondly, the phenomenology of vitamin A effects summarised above fits perfectly into the concept of positional information as elaborated by Wolpert. If a vitamin A compound such as retinoic acid is assumed to be the morphogen controlling the elaboration of pattern in the proximodistal axis then by increasing the amount of morphogen available more tissue would be expected to be produced. Similarly in the anteroposterior axis if there is a concentration gradient from a high point in the posterior side to the anterior side then adding extra morphogen could cause a new high point to arise. However, we can be certain that this type of description is far too simplistic, how, for example, does the same morphogen control each axis?

The cellular and biochemical mechanisms of action of retinoids were the subject of the subsequent papers. Stocum has been analysing this in a unique way by concentrating on the cell surface properties of blastemal cells. Obviously, blastemal cells must be able to "feel" their neighbours and make the necessary adjustments as new cells are produced by cell division. Stocum has shown that blastemata grafted to their wrong proximodistal location on another regenerating limb will relocate according to their original level specificity and that retinoid treated blastemata (which have been respecified, but which look exactly the same as control grafts) will relocate to their new position. So, whatever the mechanism of action of retinoids in blastemal cells, the ultimate phenotypic change must occur at the cell surface. Stocum also showed that the third cardinal axis, the dorsoventral axis, can also be respecified thereby adding further complexities to the simplistic interpretation mentioned above of retinoids being morphogens. A substance such as retinoic acid would have to act in each of the three axes !

The remaining talks concerned the initial effects of retinoids on cells. Maden and Summerbell demonstrated the involvement of the protein to which retinoic acid binds inside the cell, retinoic acid-binding protein (CRABP) in both the regenerating amphibian limb and the developing chick limb during the process of respecification. Unfortunately the antibodies to CRABP which are available and which would permit an examination of the spatial distribution of the binding protein do not cross-react with amphibians, only with chicks. In the chick they showed that in the anteroposterior axis there is a gradient of CRABP of opposite polarity to the endogenous gradient of retinoic acid. This may serve to lower the endogenous concentration of free retinoic acid, but the interactions of these two components with the newly discovered retinoic acid receptors make all these suggestions highly speculative. Both Stocum and Brockes are beginning to work with the retinoic acid receptors so we may soon have more information about this part of the mechanism of action of retinoic acid.

Koussoulakos presented some seemingly controversial data. Many other researchers have found that retinoids inhibit cell division and delay the progress of regeneration. It was a shock therefore to learn that his results showed that treated animals regenerate faster than controls and that the mitotic index is raised in treated animals. However, the number of cells in S phase of the cell cycle is diminished and polyploidy is induced, so it was a relief to learn that some of these effects could be explained by the effects on polyploidy.

Scadding added another piece of the jigsaw (and also enlarged it at the same time) by demonstrating that retinoid effects are abolished if limbs are treated with actinomycin D, cycloheximide and to a lesser degree tunicamycin at doses which permit normal regeneration of controls. He also showed that when untreated blastemata are grafted onto retinoid-treated stumps proximodistal duplications arise. These results provoked a lively discussion. Do the results with actinomycin D and cycloheximide mean that normal regeneration doesn't need RNA or protein synthesis? If not then the specific genes that are induced/repressed by retinoids must be selectively switched off/on which is very hard to accept. The grafting results argued against the dogma that only blastemal cells can change their positional specification because Scadding maintained that when animals are treated with retinoids the positional information of the cells of the unamputated limb are changed, but that we can never reveal that unless we amputate the limb, generate blastemal cells and get duplications. If the limb is not amputated then there is a gradual return to normal positional values over a period of several weeks. The other interpretation of these results is that there are some blastemal cells still in the stump when blastemata are grafted or grafting stimulates dedifferentiation of the stump to release new blastemal cells and that there is a sufficient amount of circulating retinoids released from storage in the liver to induce duplication in these new blastemal cells. This issue could be resolved in one of several ways: use retinoic acid instead of retinol palmitate to treat the animals because the former is not stored in the liver and has a rapid clearance rate from the body, measure circulating levels of retinoids during the course of the experiment or wait until we have a marker for individual positional values in the limb. If the latter course is taken we will still be waiting when NATO has long since gone.

It is clear that we are beginning to see a concerted attempt to understand at the biochemical and molecular level how the positional information of blastemal cells is changed by retinoids. I believe that once we know this then we will know how pattern is established in the first place and that is why it is such an important area of research. However, it will need a lot of new ideas and new techniques to reach this goal because we do not have available the powerful genetic approach which has proved invaluable for analysing Drosophila pattern formation.

PATTERN FORMATION

AND MORPHOGENESIS

OF

REGENERATES

POSITION-DEPENDENT GROWTH CONTROL AND PATTERN FORMATION IN LIMB REGENERATION

Susan V. Bryant and David M. Gardiner

Developmental Biology Center
University of California, Irvine
Irvine, California 92717, U.S.A.

Results from experimental studies in regenerating and developing amphibian limbs have led us to conclude that growth and pattern formation are coordinately controlled (French et al., 1976; Bryant et al., 1981; Bryant et al., 1987). This conclusion is based on two propositions: (1) cells possess information about their position in the limb (i.e. cells have positional values); and (2) whenever discontinuities exist in the array of positional values, interactions between adjacent cells with different positional values will result in the stimulation of growth and in the intercalation of appropriate intervening positional values. Consequently, growth will cease when all positional disparities have been resolved. As we discuss below, cellular interactions resulting in the stimulation of growth can occur during normal development as a result of cell rearrangements within the limb field, during regeneration as a result of cell migration associated with wound healing, or in experimental situations as a result of grafting to bring cells with disparate positional properties (e.g. anterior and posterior) into contact. Hence, this view can account for the initiation, maintenance and termination of limb growth and patterning during normal development, after amputation, as well as after a variety of experimental manipulations.

The view that growth and pattern formation are coordinately regulated by properties and behaviors intrinsic to cells is consistent with the idea that growth to final size of tissues and organs is, under normal conditions, controlled by mechanisms that are intrinsic to those tissues or organs. This idea is not new, and is supported by extensive experimental data in a variety of organisms (see Bryant and Simpson, 1984 for review). For example, developing amphibian limbs transplanted between small and large species develop to the size characteristic of the donor species, in disregard of the environment provided by the host (Twitty and Schwind, 1931). More recently, Sessions and Bryant (1988) demonstrated that limb regenerative ability in Xenopus, involving both growth and pattern formation, is also an intrinsic property of limb tissues. They observed that limb buds capable of normal regener-

377

ation can regenerate even when grafted to an older stage tad-pole whose limbs have lost that ability. Conversely, blaste-mata from later stages did not regain regenerative properties when grafted onto a host that can regenerate a normal limb pattern. These results indicate that in addition to whatever growth permissive factors are provided by the host (e.g. hor-mones, growth factors, nerves), limbs develop and regenerate under the control of mechanisms intrinsic to the limb tissues.

The intrinsic nature of growth and pattern regulation has also been demonstrated quantitatively in experimental studies of both developing and regenerating limbs. In such studies, the nature and degree of cellular interactions have been manipulated experimentally so as to either stimulate or inhi-bit growth. In both cases, the amount of growth and the extent of the pattern formed are coincident. An example of the experimental stimulation of excess growth and pattern is provided by the well-known ability of contralateral limb bud or blastema transplants to engender the formation of supernum-erary limbs (see Muneoka and Bryant, 1984). Additional growth and pattern are stimulated when cells from disparate positions are brought together; whereas identical procedures, but without the confrontation of cells from different limb posi-tions, do not lead to any additional growth or patterning. Other experimental procedures that also generate the formation of well developed supernumerary limbs (see Tank and Holder, 1981; Maden and Holder, 1984; Egar, 1988), also involve either the deliberate juxtaposition of cells with different positional values or extensive wound healing, which as we dis-cuss below, results in the confrontation of cells from oppo-site sides of the wound. Therefore, all situations in which excess growth is stimulated involve an increased probability of interactions between cells with disparate positional val-ues.

Position-dependent growth and patterning can be nega-tively regulated in experimental situations by the juxtaposi-tion of cells with similar positional values. A normal upper arm or leg stump is asymmetrical and possesses positional val-ues for anterior, posterior, dorsal and ventral positions. Symmetrical stumps can be created by surgery such that one set of positional values is removed (e.g. anterior or posterior) and one set is represented twice in a symmetical arrangement. If amputation of symmetrical upper arms or legs is performed close to the time of grafting, we have found that in the early regenerate each half of the limb is spatially isolated from the other (unpublished observations). Therefore immediate amputation of symmetrical limbs does not provide a test of the consequences of interactions of cells with similar positional values across the wound surface. However, when grafts are allowed to heal together for long periods of time (30-60 days), subsequent amputation leads to dramatically reduced growth and patterning, resulting in small spikes or no new structures at all (Bryant, 1976; Bryant and Baca, 1978; Sto-cum, 1978; Tank, 1978). Control experiments in which limbs have been grafted to make symmetrical limbs, and then regrafted to make normal but reversed limb stumps show that such limbs regenerate as expected, indicating that the proce-dures involved in creating symmetrical limbs and in allowing them to heal for long periods do not interfere with regenera-tive ability (Bryant and Baca, 1978). Furthermore, grafts of symmetrical blastemata to normal limb stumps and vice versa result in the formation of a supernumerary limb where disparate circumferential positional values of host and graft are confronted. Such grafts also stimulate proximal-distal intercalary regeneration when distal to proximal shifts in level are involved (Holder and Tank, 1979; Stocum, 1980,1981). Hence, the cells originating from symmetrical limbs clearly are capable of participating in limb outgrowth but do not do so in the absence of positional disparities. Since all the permissive requirements for limb outgrowth are present in these symmetrical limbs, we conclude that growth is curtailed because the positional disparities that normally drive outgrowth are severely reduced, and are replaced by non-growth promoting confrontations between cells with similar positional values. Tank (1987) has come to a similar conclusion to explain the inhibition of regeneration that occurs when progressively larger grafts of non-limb skin are made to limb stumps. In this case, inhibition is proposed to occur because non-limb cells which lack positional information interfere with the interactions between limb cells that are essential for growth and patterning.

In summary, we have proposed that limb outgrowth and patterning are coordinately controlled by a mechanism intrinsic to the limb that involves position-dependent interactions between cells. In addition, the amount of growth and pattenring achieved is dependent upon the degree of positional disparity - greater positional disparity than normal leads to excess growth and patterning, less positional disparity leads to reduced growth and patterning.

In the subsequent sections of this paper we will consider four questions that we have been able to investigate because of the existence of position-dependent growth in limbs: how is positional information organized in the limb; are the positional signals similar in different vertebrates; which cells possess positional information and finally, can we study position-dependent growth in vitro. We conclude with an overview of how position-dependent processes could operate to account for the initiation, maintenance and termination of limb outgrowth.

THE ORGANIZATION OF POSITIONAL INFORMATION

In order to understand how positional information is utilized during regeneration, it is necessary to first understand how this information is organized within the limb. Previous studies have shown that the positional information present in a full circumference of limb skin is sufficient for the formation of a normally patterned (but muscleless) limb (Dunis and Namenwirth, 1977; Lheureux, 1983). In addition, limb stumps lacking skin possess sufficient information to generate a complete regenerate (Tank, 1979). While the organization of positional information in skin has been well studied in various grafting experiments (Slack, 1980; Tank, 1981; Lheureux, 1975; Rollman-Dinsmore and Bryant, 1982), little is known about how positional information is arranged in the internal tissues. In a three-dimensional model for the upper arm, Stocum (1980) proposed that the center of the limb contains the same set of angular (circumferential) positional

values as the periphery, with additional codings for proximal-
distal and inside-outside specifications. Using position-
dependent growth as an assay, we have recently completed a
series of experiments to investigate the organization of circumferential positional values in the center of the upper and
lower arms. The central cartilaginous elements with adherent
connective tissue were removed and implanted beneath the skin
of the upper arm of the same animal. Limbs were amputated
through the region of the graft and regenerates were observed
for signs of supernumerary structures. By placing the
implants into different positions around the limb circumference we were able to determine what positional information was
present in the graft.

We have discovered that the circumferential positional
values present at the limb periphery are not projected in a
simple way to the limb center. In the center of the upper
arm, we could detect only anterior and ventral positional values. However in the center of the lower arm, all peripheral
positional values are present. Further investigation of the
lower arm revealed that only anterior and ventral positional
values are present around the radius, while in contrast, all
positional values are present around the ulna (Gardiner and
Bryant, in preparation).

Our results provide a clear explanation for the results
from earlier experiments analyzing the regenerative ability of
half limbs. Wigmore and Holder (1985; 1986) and Wigmore
(1986) reported on the regenerative abilities of half upper
versus half lower arms, and on upper and lower arms in which
half of the circumference of limb skin was replaced by head
skin, which lacks positional information. In the half limb
studies, the limbs showing the poorest regenerative abilities
were half anterior or half ventral, which we now know would
contain a low diversity of positional information. In contrast half posterior limbs or half dorsal limbs contain internal anterior and ventral positional information and regenerate
well. In the experiments where half of the circumference of
limb skin was replaced by head skin, all lower arms regenerated well, reflecting the fact that they contain all positional values internally. Upper arms with posterior skin
remaining regenerated normal limbs, whereas upper arms with
only anterior skin remaining regenerated very poorly, as
expected from the arrangement of positional information we
have described.

Symmetrical limb behavior can be more clearly interpreted
in light of these new results. Regardless of healing time,
symmetrical anterior upper arms never regenerate more than a
tapering structure, as would be expected from limbs containing
a symmetrical set of anterior positional values and no posterior positional values (see Tank and Holder, 1978). Conversely, each half of a posterior symmetrical upper arm, after
short healing times, is able to regenerate autonomously to
give rise to a broad symmetrical regenerate which can expand
further following amputation. Since each posterior half contains a full complement of positional values (posterior values
in the skin and internal anterior values) each half is able to
regenerate almost the complete limb pattern. Nevertheless,
internal positional values seem to play a subservient role in
outgrowth, as shown by the fact that when internal tissues are

normal and asymmetrical and skin is symmetrical (posterior) the regenerates follow the organization of the skin (Slack, 1980). Similarly, after long healing times symmetrical posterior upper arms either fail to regenerate or form truncated outgrowths (see Bryant et al., 1982), despite the fact that they contain internal anterior positional values.

Anatomists have for many years attempted to identify the metapterygial axis of the tetrapod limb - a theoretical line drawn through the limb to link the proximal and distal skeletal elements. The goal of this exercise has been to identify homologous structures and to define the phylogenetic relationships between different tetrapod limbs. A recent reevaluation of the pattern of early cartilage differentiation (Shubin and Alberch, 1986) reaffirms earlier interpretations that the metapterygial axis passes through the humerus and the ulna. The ulna then appears to branch and segment to form the carpals or tarsals and digits. This apparent anatomical and developmental connectedness between the ulna and the more distal elements has become known as Gregory's pyramid (see Westoll, 1943). The pattern of connections cannot be based on cell-lineage relationships between the elements because lineage studies of regenerates formed on half-diploid/half-triploid lower arms show no such cell-lineage relationship (Muneoka et al. 1985). In these studies, it was shown that descendents of cells in the anterior of the lower arm give rise to the anterior half of the regenerate, and vice versa for the posterior half. However, the anatomical branching and connections between elements could be based on the pattern of positional information and the way in which it is elaborated during outgrowth. Elements thus would appear to branch during development to give separate structures based on the continuity of positional information, rather than on cell lineage. Similarly secondary connections or fusions would occur, as a result of interactions of cells with similar positional values but not necessarily of a similar lineage. The symmetrical arrangement of positional values in the anterior of the lower arm would permit only segmentation and eventual termination (due to decay of the diversity of positional information) of anterior derivatives, whereas the complete information in the posterior would allow for repeated branching as well as segmentation to form the distal elements of the pattern, until they in turn become symmetrical and terminate (see Stock and Bryant, 1981 and discussed further below).

COMPATIBILITY OF POSITIONAL SIGNALLING MECHANISMS AMONG TETRAPODS

Our view that an understanding of the ontogeny and phylogeny of the tetrapod limb will necessarilly involve an understanding of pattern forming mechanisms, raises the issue as to what extent the similarities in pattern among tetrapod limbs reflect similarities in patterning mechanisms. Conversely, at what level do differences in morphology reflect differences in developmental mechanisms? Position-dependent growth can be used to investigate whether different vertebrates share similar mechanisms for limb outgrowth. When limb cells from different species are grafted together to create positional disparities, the formation of supernumerary growth and pattern indicates a compatibility of patterning mechanism and of the signals used to execute the mechanism. However, a failure to generate supernumerary structures when positional disparities exist cannot be interpreted. Since vertebrates differ widely in their physiological states, some combinations of cells are not feasible. Results from earlier studies of limb patterning mechanisms in urodeles and anurans provided indirect evidence that the limb patterning mechanisms of both are similar (see Muneoka and Murad, 1987). We have recently shown directly that the signalling systems and the basic patterning mechanisms underlying limb outgrowth in urodeles and anurans are in fact compatible (Sessions, et al., in press). Limb buds grafted between axolotls and Xenopus at equivalent stages of development lead to the formation of supernumerary limb structures in experimental (with positional disparities) but not control (without positional disparities) grafts. Previous studies have shown that grafts of posterior limb cells of a variety of amniotes (reptiles, birds, mammals) (see Fallon and Crosby, 1977) into the anterior of chick wings also lead to the formation of supernumerary structures. Taken together these results show similarity of signals and mechanisms between urodeles and anurans, and between reptiles, birds and mammals. Whether the same mechanism exists in amphibians and amniotes is difficult to test for technical reasons. However, the only major phylogenetic dichotomy that has been proposed to exist within the vertebrates falls between urodeles and anurans, and this proposed dichotomy is based in part on differences in limb structure and development (see Hanken, 1986 and Schubin and Alberch, 1986 for reviews). Our findings indicate that whether or not a phylogenetic dichotomy exists within the amphibia, there is not a difference in the basic mechanism of limb pattern formation. Hence, it is likely that all vertebrate limbs follow the same developmental rules and thus share similar mechanisms of growth control and pattern formation. Based on our present understanding of such regulatory mechanisms, it appears that positional information may be specified in an equivalent manner in all tetrapod limbs and the morphological similarities among limbs is likely to be a reflection on this commonality of mechanism. The basic mechanisms involved in receiving positional information appear also to have been conserved, thus, differences in details of limb patterns perhaps reflect differences in the details of how such information is finally interpreted, as suggested by Wolpert (1969).

IDENTIFICATION OF LIMB CELLS WITH POSITIONAL INFORMATION

It is important to an understanding of limb pattern formation to know whether all limb bud or blastema cells, regardless of their origin or eventual fate, possess the positional information required for position-dependent growth control and patterning, or whether this is the property of only a few or perhaps a single cell type. Mature limbs consist of a variety of different cell types organized in characteristic patterns. Developing limb buds contain precursors to these cell types, and regeneration blastemata which are derived from the mature tissues of the stump, also consist of cells with a variety of origins. The results of several types of experiments can be used to deduce which cell types in the limb possess positional information.

The epidermis of the limb forms a discrete lineage of cells at the limb surface. It is known that the apical epidermis is required for limb outgrowth and patterning because if it is removed and either fails to reform the apical specialization (e.g. chicks, Saunders, 1948) or is prevented from reforming the apical specialization (e.g. amphibians, Stocum and Dearlove, 1972) limbs develop that are truncated distally. In contrast, similar deletion experiments of the dorsal epidermis does not affect subsequent pattern formation and growth (Martin and Lewis, 1986). Despite the demonstrated necessity of apical epidermis for limb outgrowth and patterning, experimental evidence indicates that epidermis does not possess positional information. In chicks, when limb bud mesoderm is heterochronically recombined with ectodermal jackets, development proceeds normally and in accordance with the stage of the mesoderm (Rubin and Saunders, 1972). In addition, chick limb bud mesoderm grafted beneath flank epidermis in a variety of orientations develops autonomously and is not influenced by the different orientation of the epidermis (Saunders and Reuss, 1974). In regenerating limbs, when the epidermis is reoriented with respect to the mesodermal tissues, limbs develop that conform to the orientation of the mesoderm (Carlson, 1975). Furthermore, supernumerary structures do not develop in these experiments, indicating that there is no position-dependent growth between reoriented ectoderm and mesoderm. In summary, while the ectoderm fulfills an essential role in limb outgrowth, there is no evidence that it possesses positional information.

Limb muscle, like epidermis, forms a discrete lineage within the limb. During the development of the chick limb, muscle precursor cells migrate into the limb from adjacent somites, but they are never present in the distal tip of mesoderm where pattern-forming events are occuring (Newman et al., 1981). The most direct evidence that the myogenic lineage does not possess positional information comes from studies in chicks, in which it has been shown that muscle precursor cells from inappropriate regions of the body can nevertheless invade the limb and take up residence to form a pattern of muscles that is appropriate for the limb, rather than a pattern that corresponds to their normal fate (Chevalier et al., 1977). Similar evidence has been obtained concerning the muscles of the head (Noden, 1986). Analogous experiments have not been performed in regenerating limbs. In order to do so, it would be necessary to introduce myogenic cells without any other cell types. This is possible in development because grafts can be made prior to the mingling of myogenic cells and connective tissue to form muscle. Less direct, but comparable experiments have been performed in amphibian limbs. If muscle precursor cells are prevented from entering the regenerating blastema, limbs are formed that are well patterned and contain all tissues except muscle (Dunis and Namenwirth, 1977; Lheureux, 1983). Such results indicate that myogenic cells in amphibian limbs also represent a discrete lineage and that they are not necessary for pattern formation. Nevertheless, mature muscle tissue, which contains both myogenic cells and connective tissue, does stimulate position-dependent growth. When muscle tissue is transplanted to a new location in the limb stump prior to amputation, extra growth and supernumerary structures result, indicating that at least one of the components of muscle possesses positional information (Carlson,

1975a,b). By the reasoning outlined above it is unlikely that myogenic cells have positional information, and therefore, we assume it is the connective tissue component that is involved in growth stimulation.

Direct evidence that limb fibroblasts possess positional information and are capable of position-dependent growth comes from two types of studies. When dermis alone (fibroblasts plus matrix) is transplanted to a disparate position around the limb circumference, it stimulates the formation of extra growth and patterning at amputation (Tank, 1981). Furthermore, if the dermis is treated to kill the cells, the matrix alone does not stimulate a response (Tank, 1981). In addition, whole skin (dermis plus epidermis) transplanted to a new position around the circumference,also stimulates the formation of supernumerary structures (Rollman-Dinsmore and Bryant, 1982). Since epidermis does not possess positional information, we conclude that the effect is mediated by dermal fibroblasts. Similarly, when cartilage cleaned of its adherent connective tissue is transplanted either to different positions in the stump or in different orientations, it does not provoke the formation of supernumerary structures after amputation (Goss, 1956; Carlson, 1975b), whereas cartilage with its connective tissue sheath intact does (Gardiner and Bryant, in preparation). Finally, we (Muneoka et al., 1986) have looked at the relative numbers of cells derived from dermis in the young blastema relative to those present in the stump. This was accomplished by grafting triploid skin onto a diploid host prior to amputation, and calculating from direct cell counts the percentage of dermal cells at the amputation plane and in the blastema. We found that cells of dermal origin represent a larger percentage of the cells within the blastema than within the stump (a 2-3 fold increase). In contrast, the percentage of cartilage-derived cells and nerve sheath-derived population expands through cell proliferation as the blastema forms. Cells derived from cartilage and nerve sheath, which do not possess positional information based on the results of grafting tests, would not be expected to respond to or to generate position-dependent growth stimuli. Consistent with this expectation, cells from these tissues are underrepresented in the blastema relative to their presence in the stump.

In conclusion, the only cells that clearly show evidence of possessing positional information and of being able to participate in position-dependent growth reside in loose connective tissue. Although we do not yet understand the extent to which these cells represent a homogenous population, we presently refer to them collectively as fibroblasts. Hence, we have proposed (Bryant, et al., 1987) that fibroblasts have two major roles in limb patterning: (1) they interact with one another in a position-dependent fashion to construct the pattern in outline form and (2) they interact with other cell types of the limb to control their position and growth. The coordination of these activities would ensure the production of a limb of normal composition, pattern and size.

POSITION-DEPENDENT GROWTH IN VITRO

We are in the process of establishing a system for studying position-dependent growth of axolotl cells in vitro in order to investigate the molecular nature of the interactions of fibroblasts with other fibroblasts, and with other cell types. Because blastemata are composed of up to 80% cells of fibroblast origin (Muneoka et al., 1986) we have initiated this study by examining the growth responses of blastema cells from anterior and posterior positions, cultured either singly or in combination. Results of similar _in vitro_ experiments demonstrating position-dependent growth of chick limb bud cells have recently been reported (Aono and Ide, 1988). Under the reported culture conditions, explants from different regions of the limb bud produced soluble factor(s) that stimulated growth of limb bud cells, the most dramatic stimulation being that of anterior cells by posterior explants. The role of short range cell-cell interaction in position-dependent growth has not yet been investigated. In our studies to date, we have observed that AA co-cultures show the lowest labelling index (29%) at seven days after explantation, whereas PP and AP co-cultures show higher indices (PP: 39%; AP: 45%). Since we have shown that the posterior half of the limb contains both anterior and posterior positional information, this result is expected since PP co-cultures, like AP co-cultures will contain cells with both anterior and posterior positional information. An _in vitro_ assay system for position-dependent growth will allow us to investigate position-dependent effects of known growth factors on growth stimulation, as well as to identify cell fractions active in position-dependent growth stimulation.

OVERVIEW OF LIMB OUTGROWTH

We propose that limb outgrowth is initiated as a result of interactions between fibroblasts with different positional values. The most direct demonstration of this is achieved by contralateral grafting of blastemata or limb buds to bring cells with different positional information into contact. Supernumerary limbs only develop at the sites of these contacts. We have also shown that cells with different positional values come into contact after limb amputation (Gardiner, et al., 1986). To do this we grafted a full circumference of limb skin to create limbs consisting of triploid skin on a diploid stump or _vice-versa_. We than amputated these limbs through the graft region and analyzed the distribution of dermal fibroblasts in whole mount preparations of the wound area during the first two weeks of regeneration. Rather than remaining in their original positions at the wound edge, dermal fibroblasts begin to migrate towards the center of the wound at about 5 days after amputation, and by ten days have reached the wound center. In subsequent studies, we have observed that while the majority of the cells from a particular circumferential quadrant remain within the sector that projects from the limb periphery to the limb center, a minority migrate outside of this sector (Gardiner and Bryant, unpublished). The direction taken by the minority population corresponds to the previously described directional bias that has been inferred from cell marker studies after intercalation has occurred (Muneoka and Bryant, 1984; Muneoka, et al., 1986). Hence, prior to blastema formation dermal fibroblasts behave in a way that brings them into contact with fibroblasts with different positional information (i.e. from different limb positions). Studies from other laboratories are also consistent with this idea since they show that cell division is initiated at about 5 days after amputation (Kelly and Tassava, 1973; Tassava et al., 1974; Maden, 1978), a time that corresponds to the time at which fibroblasts begin their migration and thus begin making contacts with cells with different positional information.

Few data are available concerning the initiation of outgrowth during limb development. A prediction of our hypothesis is that contacts between previously non-adjacent cells will be a necessary prelude to outgrowth. Borgens (1984) has proposed that endogenous electrical currents resulting from local "leaks" in the epithelium set up voltage gradients that could direct the aggregation of mesenchymal cells. The existence of ionic currents in the limb forming regions has been observed in both axolotls and Xenopus (Borgens et al., 1983; Robinson, 1983). Electron microscopic observations of the ectoderm and mesenchyme in the pre-limb bud flank have shown that in Xenopus the basement lamella becomes disorganized in localized regions, and previously scattered mesenchyme cells aggregate in these regions (Kelley and Bluemink, 1974). Similar observations have been made by Borgens et al (1987) in axolotls. These authors have also noted that ectodermal cells of the limb forming region have fewer hemidesmosomes and that some appear to be dying and desquamating. They have proposed that it is these changes in the integrity of the ectoderm that lead to the generation of currents, which in turn cause the aggregation of previously separated mesenchymal cells. Thus confrontations between previously separated cells apparently occur coincident with the onset of outgrowth, as they also do during the initiation of regeneration. Once outgrowth has been initiated, position dependent growth between newly generated cells representing all circumferial positional values at the tip of the blastema or limb bud will be maintained as described previously (Bryant et al., 1981; Bryant et al., 1987).

Just as the initiation and maintenance of outgrowth are processes that are intrinsic to the pattern-forming cells, so too is the process of growth termination. We have proposed that growth termination occurs as a consequence of fragmentation of the circumference of the limb into digits (Stock and Bryant, 1981; see Bryant et al., 1987). Evidence for such a fragmentation event comes from the finding that the level of the digit bases is the only position along the proximal-distal axis of the limb at which amputation leads to abnormal regeneration. The abnormalities are those that would be expected if the normal process of fragmentation of the circumference is disrupted by cell movements occuring during wound healing, namely, additional digits, branched or fused digits, and increased or decreased numbers of phalanges. The subsets of positional values that we propose are formed by positional fragmentation will necessarily become symmetrical as a result of position-dependent growth. Experiments with symmetrical limbs (limbs in which a partial set of circumferential positional values are represented twice in mirror symmetry) have shown that symmetry can lead to growth termination (see Bryant et al., 1982). In comparison to asymmetrical structures, when cells from symmetrical structures come into contact, they have a higher probability of having identical positional values. Therefore instead of stimulating growth, such interactions bring about its termination. The fact that digits are bilaterally symmetrical, and that peripheral digits are shorter than central digits strongly support this view (see Stock and Bryant, 1981 for a detailed discussion of this issue).

In conclusion, the view that limb fibroblasts possess positional information and interact in a manner that controls both growth and patterning, provides a conceptual framework within which many features of limb outgrowth can be understood. With this knowledge, we can now turn to the question of how positional information is encoded in cells, how cells signal this information to one another and how cells respond to and interpret the information they receive.

ACKNOWLEDGEMENTS

Authors' research supported by PHS grants HD06082; HD20662; NSF grant DCB-8615513 and a gift from the Monsanto Company.

REFERENCES

Aono, H., and Ide, H., 1988, A gradient of responsiveness to the growth-promoting activity of ZPA (Zone of polarizing activity in the chick limb bud), Devel. Biol., 128:136-141.

Borgens, R.B., 1984, Are limb development and limb regeneration both initiated by an integumentary wounding? A hypothesis., Differentiation, 28:87-93.

Borgens, R.B., Callahan, L. and Rouleau, M.F., 1987, Anatomy of axolotl flank integument during limb bud development with special reference to a transcutaneous current predicting limb formation, J. Exp. Zool., 244:203-214.

Borgens, R.B., Rouleau, M.F. and DeLanney, L.E., 1983, A steady efflux of ionic current predicts hind limb development in the axolotl, J. Exp. Zoll., 228:491-503.

Bryant, P.J. and Simpson, P., 1984, Intrinsic and extrinsic control of growth in developing organs, Quart. Rev. Biol., 59:387-415.

Bryant, S.V., 1976, Regenerative failure of double half limbs in Notophthalmus viridescens, Nature, 263:676-679.

Bryant, S.V. and Baca, B., 1978, Regenerative ability of double-half and half upper arms in the newt, Notophthalmus viridescens, J. Exp. Zool., 204:307-324.

Bryant, S., French, V. and Bryant, P., 1981, Distal regeneration and symmetry, Science, 212:993-1002.

Bryant, S.V., Gardiner, D.M. and Muneoka, K., 1987, Limb development and regeneration, Amer. Zool., 27:675-696.

Bryant, S.V., Holder, N. and Tank, P., 1982, Cell-cell interactions and distal outgrowth in amphibian limbs, Amer. Zool., 22:143-151.

Carlson, B.M., 1975a, Multiple regeneration from axolotl limb stumps bearing cross-transplanted minced muscle regenerates, Devel. Biol., 45:203-208.

Carlson, B.M., 1975b, The effects of rotation and positional change of stump tissues upon morphogenesis of the regenerating axolotl limb, Devel. Biol., 47:269-291.

Chevallier, A., Kieny, M. and Mauger, A., 1977, Limb-somite relationship: Origin of the limb musculature, J. Embryol. exp. Morph., 41:245-258.

Dunis, D.A. and Namenwirth, M., 1977, The role of grafted skin in the regeneration of X-irradiated axolotl limbs, Devel. Biol., 56:97-109.

Egar, M.W., 1988, Accessory limb production by nerve-induced cell proliferation, Anat. Rec., 221:550-564.

Fallon, J.F. and Crosby, G.M., 1977, Polarising zone activity in limb buds of amniotes. in: "Vertebrate Limb and Somite Morphogenesis", D.A. Ede, J.R. Hinchliffe and M. Balls, eds., Cambridge University Press, Cambridge.

French, V., Bryant, P.J. and Bryant, S.V., 1976, Pattern regulation in epimorphic fields, Science, 193:969-981.

Gardiner, D.M., Muneoka, K. and Bryant, S.V., 1986, The migration of dermal cells during blastema formation in axolotls, Devel. Biol., 118:488-493.

Gardiner, D.M. and Bryant, S.V., 1988, The organization of positional information within the urodele limb (in preparation).

Goss, R.J., 1956, The relation of bone to the histogenesis of cartilage in regenerating forelimbs and tails of adult Triturus viridescens, J. Morph., 98:89-123.

Hanken, J., 1986, Developmental evidence for amphibian origins, Evol. Biol., 20:389-417.

Holder, N. and Tank, P.W., 1979, Morphogenetic interactions occurring between blastemas and stumps after exchanging blastemas between normal and double-half forelimbs in the axolotl, Ambystoma mexicanum, Devel. Biol., 68:271-279.

Kelley, R.O. and Bluemink, J.G., 1974, An ultrastructural analysis of cell and matrix differentiation during early limb development in Xenopus laevis, Devel. Biol., 37:1-17.

Kelly, D.J. and Tassava, R.A., 1973, Cell division and ribonucleic acid synthesis during the initiation of limb regeneration in larval axolotls (Ambystoma mexicanum), J. Exp. Zool., 185:45-54.

Lheureux, E., 1975, Nouvelles donnees sur les rôles de la peau et des tissus internes dans la regeneration du membre du triton Pleurodeles waltlii Michah (Amphibien Urodele), Wilhelm Roux's Arch., 176:285-301.

Lheureux, E., 1983, The origin of tissues in the x-irradiated regenerating limb of the newt Pleurodeles waltlii, in: "Limb Development and Regeneration, Part A", J.F. Fallon and A.I. Caplan, eds., Alan R. Liss, Inc., New York.

Maden, M., 1978, Neurotrophic control of the cell cycle during amphibian limb regeneration, J. Embryol. exp. Morph., 48:169-175.

Maden, M. and Holder, N., 1984, Axial characteristics of nerve induced supernumerary limbs in the axolotl, Roux's Arch. Dev. Biol., 193:394-401.

Martin, P. and Lewis, J., 1986, Normal development of the skeleton in chick limb buds devoid of dorsal ectoderm, Devel. Biol., 118:233-246.

Muneoka, K. and Bryant, S.V., 1984, Cellular contribution to supernumerary limbs in the axolotl, Ambystoma mexicanum, Devel. Biol., 105:166-178.

Muneoka, K., and Murad, E.H.B., 1987, Intercalation and the cellular origin of supernumerary limbs in Xenopus, Development, 99:521-526.

Muneoka, K., Fox, W. and Bryant, S.V., 1986, Cellular contri-
    bution from dermis and cartilage to the regenerating
    limb blastema in axolotls, Devel. Biol., 116:256-260.
Muneoka, K., Holler-Dinsmore, G.V. and Bryant, S.V., 1985, A
    quantitative analysis of regeneration from chimaeric
    limb stumps in the axolotl, J. Embryol. exp. Morph.,
    90:1-12.
Muneoka, K., Holler-Dinsmore, G. and Bryant, S.V., 1986, Pat-
    tern discontinuity, polarity and directional intercala-
    tion in axolotl limbs, J. Embryol. exp. Morph.,
    93:51-72.
Newman, S.A., Pautou, M.-P. and Kieny, M., 1981, The distal
    boundary of myogenic primordia in chimeric avian limb
    buds and its relation to an accessible population of
    cartilage progenitor cells, Devel. Biol., 84:440-448.
Noden, D.M., 1986, Patterning of avian craniofacial muscles,
    Devel. Biol., 116:347-356.
Robinson, K.R., 1983, Endogenous electrical current leaves the
    limb and prelimb region of the Xenopus embryo, Devel.
    Biol., 97:203-211.
Rollman-Dinsmore, C. and Bryant, S.V., 1982, Pattern regula-
    tion between hind- and forelimbs after blastema
    exchanges and skin grafts in Notophthalmus viridescens,
    J. Exp. Zool., 223:51-56.
Rubin, L. and Saunders, J.W., Jr., 1972, Ectodermal-mesodermal
    interactions in the growth of limb buds in the chick
    embryo: Constancy and temporal limits of the ectoder-
    mal induction, Devel. Biol., 28:94-112.
Saunders, J.W., Jr., 1948, The proximo-distal sequence of ori-
    gin of the parts of the chick wing and the role of
    ectoderm. J. Exp. Zool., 108:363-403.
Saunders, J.W., Jr. and Reuss, C., 1974, Inductive and axial
    properties of prospective wing-bud mesoderm in the
    chick embryo, Devel. Biol., 38:41-50.
Sessions, S.K. and Bryant, S.V., 1988, Regenerative ability is
    an intrinsic property of limb cells in Xenopus, J. Exp.
    Zool., (in press).
Sessions, S.K., Gardiner, D.M. and Bryant, S.V., 1989, Compa-
    tible limb patterning mechanisms in urodeles and anu-
    rans, Devel. Biol., (in press).
Shubin, N.H. and Alberch, P., 1986, A morphogenetic approach
    to the origin and basic organization of the tetrapod
    limb, Evol. Biol., 20:319-387.
Slack, J.M.W., 1980, Morphogenetic properties of the skin in
    axolotl limb regeneration, J. Embryol. exp. Morph.,
    58:265-288.
Stock, G.B. and Bryant, S.V., 1981, Studies of digit regener-
    ation and their implications for theories of develop-
    ment and evolution of vertebrate limbs, J. Exp. Zool.,
    216:423-433.
Stocum, D.L., 1978, Regeneration of symmetrical hindlimbs in
    larval salamanders, Science, 200:790-793.
Stocum, D.L., 1980, Intercalary regeneration of symmetrical
    thighs in the axolotl, Ambystoma mexicanum, Devel.
    Biol., 79:276-295.
Stocum, D.L., 1981, Distal transformation in regenerating
    double anterior axolotl limbs, J. Embryol. exp. Morph.,
    65 (supplement):3-18.

Stocum, D.L. and Dearlove, G.E., 1972, Epidermal-mesodermal
    interaction during morphogenesis of the limb regener-
    ation blastema in larval salamanders, J. Exp. Zool.,
    181:49-62.
Tank, P.W., 1978, The failure of double-half forelimbs to
    undergo distal transformation following amputation in
    the axolotl, Ambystoma mexicanum, J. Exp. Zool.,
    204:325-336.
Tank, P.W., 1979, Positional information in the forelimb of
    the axolotl: Experiments with double-half tissues,
    Devel. Biol., 73:11-24.
Tank, P.W., 1981, The ability of localized implants of whole
    or minced dermis to disrupt pattern formation in the
    regenerating forelimb of the axolotl, Amer. J. Anat.,
    162:315-326.
Tank, P.W., 1987, The effect of nonlimb tissues on forelimb
    regeneration in the axolotl, Ambystoma mexicanum, J.
    Exp. Zool., 244:409-423.
Tank, P.W. and Holder, N., 1978, The effect of healing time on
    the proximo-distal organization of double-half forelimb
    regenerates in the axolotl, Ambystoma mexicanum, Devel.
    Biol., 66:72-85.
Tank, P.W. and Holder, N., 1981, Pattern regulation in the
    regenerating limbs of urodele amphibians, Quart. Rev.
    Biol., 56:113-142.
Tassava, R.A., Bennett, L.L. and Zitnik, G.D., 1974, DNA syn-
    thesis without mitosis in amputated denervated fore-
    limbs of larval axolotls. J. Exp. Zool., 190:111-116.
Twitty, V.C. and Schwind, J.L., 1931, The growth of eyes and
    limbs transplanted heteroplastically between two
    species of Amblystoma, J. Exp. Zool., 59:61-86.
Westoll, T., 1943, The origin of the primitive tetrapod limb,
    Proc. R. Soc. Lond. B, 131:373-393.
Wigmore, P., 1986, Regeneration from half lower arms in the
    axolotl, J. Embryol. exp. Morph., 95:247-260.
Wigmore, P. and Holder, N., 1985, Regeneration from isolated
    half limbs in the upper arm of the axolotl, J. Embryol.
    exp. Morph., 89:333-347.
Wigmore, P. and Holder, N., 1986, The effect of replacing dif-
    ferent regions of limb skin with head skin on regener-
    ation in the axolotl, J. Embryol. exp. Morph.,
    98:237-249.
Wolpert, L., 1969, Positional information and the spatial pat-
    tern of cellular differentiation, J. Theoret. Biol.,
    25:1-47.

SUPERNUMERARY LIMBS DEVELOPING AFTER SMALL ANGLE DISLOCATIONS OF THE LIMB BUD OF THE NEWT

S. Papageorgiou*, D. Venieratos** and
E. Vamvassakis**

*NRC "Demokritos" Aghia Paraskevi, Athens, Greece
**Department of Anatomy, University of Athens
Greece

SUMMARY

We have performed the following experiments on the regenerating forelimbs of the newt _Triturus cristatus_: 1) Limb bud blastemata were ipsilaterally rotated at angles 90° or 270°. 2) Blastemata were contralaterally transplanted so that their anteroposterior axis coincided with the dorsoventral or ventrodorsal axis of the stump. The morphology and histological structure of the emerging supernumerary limbs for both experimental setups were examined. Different models for regeneration provide diverging predictions for the above experiments. Our results are in agreement with an extended version of the Polar Coordinate Model.

INTRODUCTION

Limb regeneration in Amphibians has been studied for a long time. However, the last decade or so this phenomenon has been intensively scrutinized and better understood for mainly 3 reasons: a) the development of new experimental techniques like exact cell marking; b) the observation of unexpected results as the regeneration after treatment with retinoic acid (Maden, 1982); and c) the introduction of novel ideas like the polar coordinate model (PCM) of French et al. (1976).

With the ambition to describe the existing data of distal outgrowth after contralateral and ipsilateral limb bud graftings an extended version of PCM has been proposed (Papageorgiou, 1984). This model, designated as hierarchical polar coordinate model (HPCM), is based on two topological notions: continuity and congruence. When a normal sequence of positional values is disrupted, intercalation takes place and restores continuity through the shortest path. In the case of an amphibian limb, two positional values (variables) are necessary for the process of pattern formation. One is along the proximodistal (PD) axis and the other fixes the position around the ring of the limb section transverse to the PD axis.

In experiments where the circular variables are disrupted (contralateral or ipsilateral transplantations), intercalation might take the form of circular paths which then trigger the outgrowth of supernumerary structures. As it turns out, these intercalating contours are congruent to both stump and graft only in the case of contralateral limb bud graftings (Papageorgiou, 1984). Under congruence we mean that the intercalating sequence of cells possesses positional values in the same direction as determined by the adjacent cells of the stump (or graft). In ipsilateral graft rotations it is impossible to form congruent intercalating contours to stump and graft unless one is allowed to deform the contour by twisting it. According to the HPCM, a rule for the probabilistic classification of the intercalating contours is based on plausibility arguments: congruent paths are favored versus non-congruent ones and between simple and twisted contours simple ones are prefered. The HPCM can then describe the existing data on contralateral and ipsilateral transplantations. Furthermore, this model predicts the formation of abnormal supernumerary structures in contralateral grafts but with very low probability compared to the normal supernumerary limbs of the stump handedness. This prediction has been recently confirmed on tadpoles of the Anuran Bufo bufo (Costaridis et al., 1988).

The ipsilateral or contralateral grafting at a small angle of dislocation of the anteroposterior (AP) axis between stump and graft presents considerable interest: for the ipsilateral case the HPCM predicts: a) lower probability of supernumerary limb formation as compared to a 180° ipsilateral limb bud rotation. b) the supernumerary limbs should be preferentially incomplete (fewer digits than a normal limb). c) the muscular structure of these limbs should be mainly abnormal (double-dorsal or double-ventral).

For the case of contralateral grafts where the dorsoventral (DV) axis of the graft coincides with the AP axis of the stump, the PCM (together with the HPCM) clearly predicts the formation of normal supernumerary limbs of the stump handedness located at the positions of maximal positional disparity. No detailed histological analysis has been performed as yet [see however the experiment of Wallace and Watson (1979) which records many macroscopic data]. For these graftings, other models based on the cardinal AP and DV axes (Cartesian models) provide quite different predictions. In the relevant contralateral experiments one of the AP/DV axes is in coincidence between stump and graft whereas in the present setup neither axis of the stump coincides with the corresponding axis of the graft. In particular, the boundary model of Meinhardt (1983) predicts abnormal supernumerary structures for this grafting, in contrast to the expectations of the PCM. This divergence of different model predictions motivated the present experiment to determine the muscular structures of the outgrowths.

MATERIALS AND METHODS

All operations were performed on the newt Triturus cristatus. The animals were anaesthetized with MS 222 and the forelimbs were amputated through the mid-humerus level and then allowed to regenerate. The blastemata were cut off at the stage of palette or early digits and treated according to

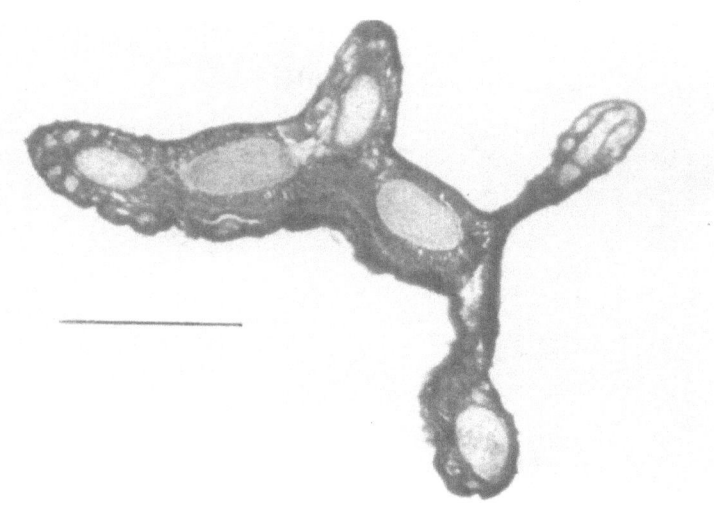

Fig. 1. A transverse section of a developed graft at
the metacarpal level. Two independent supernumerary digits have appeared whose muscular
structure is not defined. Bar = 0.5 mm.

either of the following processes: 1) the right limb bud was
rotated clockwise so that the anterior pole coincided with the
dorsal pole of the right stump; the left limb was turned
anticlockwise. 2) the right blastema was transplanted on the
left stump so that the anterior pole of the graft coincided
with the dorsal pole of the stump; the left blastema was
transplanted contralaterally by inverting the sense of the
rotation so that again the anterior graft pole contacted the
stump dorsal pole. Control grafts were amputated and placed
back on their position without rotation. The blastemata and
supernumerary limbs developed in room temperature and observed
twice a week.

The multiple limbs were harvested and fixed in Bouin's
fluid then they were decalcified in EDTA, dehydrated, stained
with Victoria-blue and cleared with methyl salicylate. The
skeletal structure was recorded and subsequently the limbs
were prepared for serial transverse sectioning on a microtome.
The 10 μm thick sections were stained with haematoxylin and
eosin so that their muscular structure could be observed.

RESULTS

From the 10 control limbs 8 regenerated normally while
one developed with 3 digits all normal in skeletal and muscular structure. One control limb developed normally but an
extra digit (spike) grew on the ventral side.

1)   Ipsilateral 90° or 270° Rotations

The operated limbs were 61 while the collected supernumerary outgrowths were 23. The features of these multiple limbs
are presented in Table 1. The skeletal structures are mainly
incomplete. One was complete and furthermore was normal in its

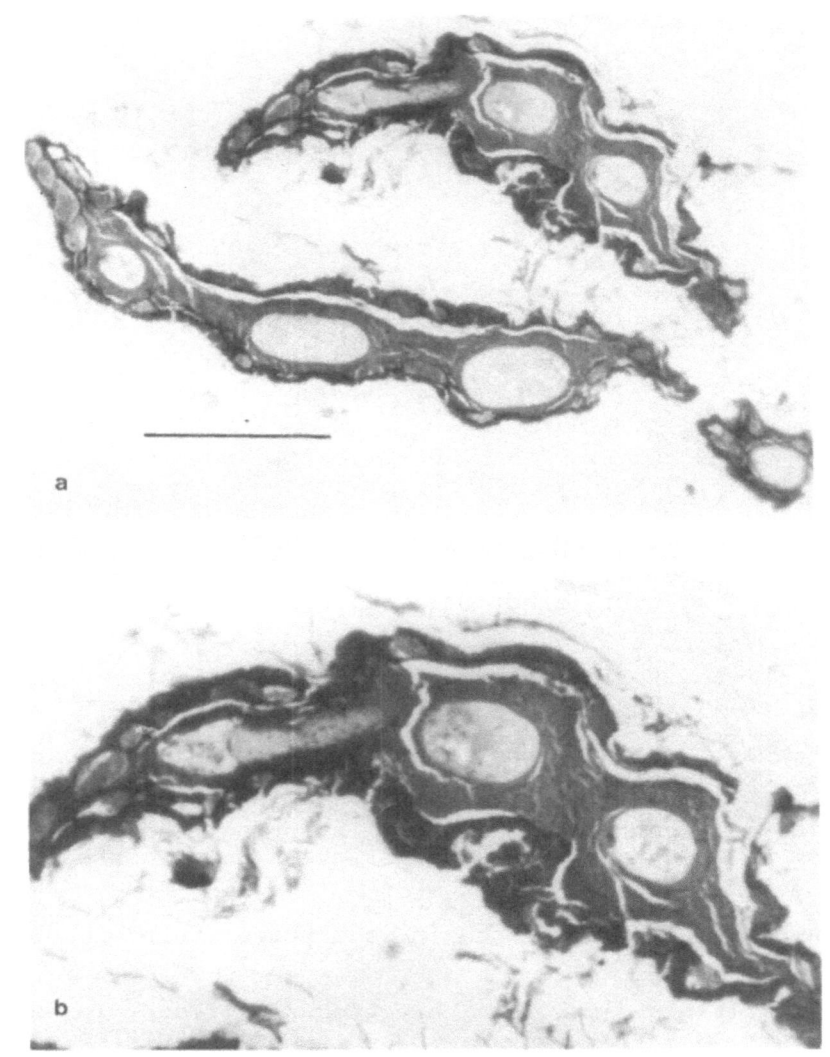

Fig. 2. a. The section below belongs to the graft while
above lies the section of the double-ventral
3-digit supernumerary limb. Bar = 0.5 mm.
b.   Higher magnification of the double-ventral
regenerate.

muscular pattern but it was derotated from its initial transplantation position. All other multiple limbs were abnormal
(symmetrical) either double-dorsal or   double-ventral.

In Figs. 1 and 2 two characteristic sections are shown
from the above supernumerary limbs.

2) Contralateral Graftings

Neither of the cardinal axes AP/DV is in coincidence between stump and host. 24 animals were operated and we collected 13 supernumerary limbs which appeared off the A, D, P and V-poles. The data are tabulated in Table 2.

Table 1. Supernumerary Outgrowths in Ipsilateral 90° or 270° Rotations

	N	DD	VV	un.	Tot.
4 dig.	1	1	2	4	8
3 dig.	3	1	-	2	6
2 dig.	2	2	2	2	8
1 dig.	-	-	-	1	1
Tot.	6	4	4	9	23

N: normal, DD: double-dorsal VV:
VV: double-ventral, un: undefined

Table 2. Supernumerary Outgrowths in Contralateral Dorsal to Anterior Graftings

	N	DD	VV	un.	Tot.
4 dig.	2	-	-	1	3
3 dig.	4	-	-	1	5
2 dig.	3	-	-	-	3
1 dig.	-	-	-	2	2
Tot.	9	-	-	4	13

The notation is as in Table 1.

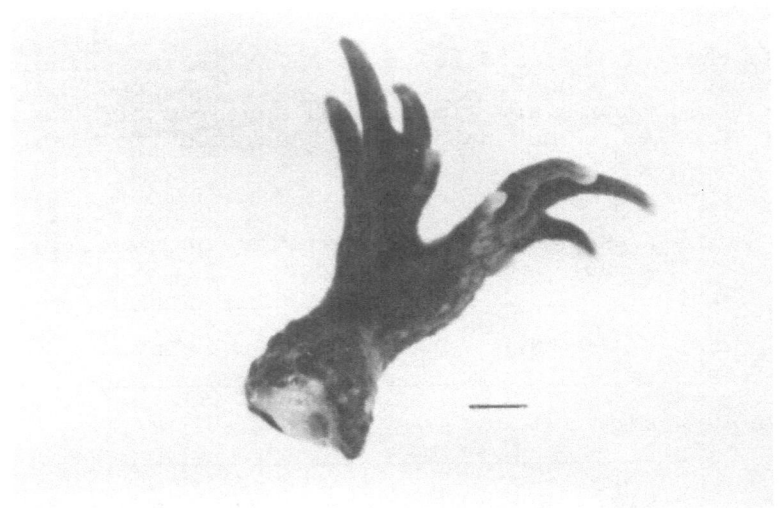

Fig. 3. A right blastema (left side of the figure) was
grafted on a left stump so that its anterior
pole coincided with the dorsal side of the
host. A normal supernumerary limb of the stump
handedness (left) developed in the PD-quadrant
of the stump. Bar = 1 mm.

In all defined cases the supernumerary limbs were normal
and of the stump handedness. Some macroscopic and microscopic
views from these supernumerary limbs are shown in Figs. 3 and
4.

DISCUSSION

In the 90° or 270° ipsilateral limb bud rotations performed on urodeles (e.g. Wallace and Watson, 1979) the analysis was limited to the macroscopic features of the supernumerary outgrowths. The overall conclusion was that multiple limbs
outgrow with quite variable number of digits. The present
analysis is not contradictory to this result but it stresses
more the fact that incompleteness extends to both the AP and
DV axes. Most of the supernumerary limbs have less than normal digits and at the same time are symmetrical double-dorsal
or double-ventral (Figs. 1,2). These results comply with the
HPCM predictions mentioned in the introduction. From the data
of Table 1 it is clear that the frequency of their appearance
is lower than in a 180° ipsilateral rotation (Stock et al.,
1980). This comparison agrees with the HPCM which, for small
angle ipsilateral rotations, allows fewer contours to be
formed. Therefore, the probability for a supernumerary outgrowth formation is lower than in a 180° ipsilateral grafting.

In the contralateral limb grafts most of the supernumerary limbs were normal of the stump handedness as shown in
Table 2 and Figs. 3,4. Their location (near the middle of the
quadrants AD, DP, PV or VA), corresponds to the points of max-

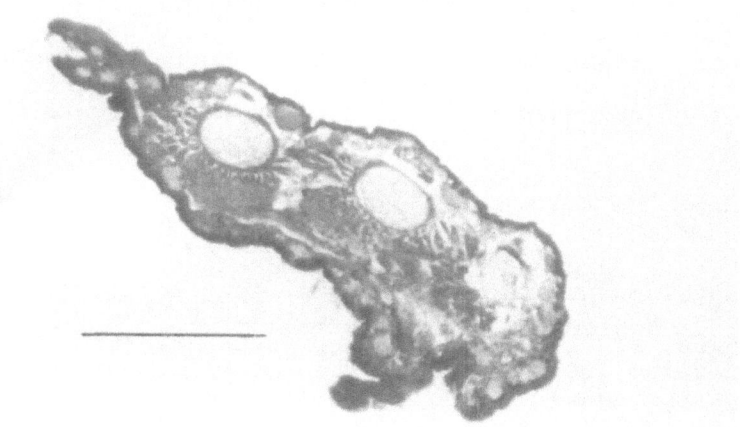

Fig. 4. A transverse section of another supernumerary
         limb with 3 digits and normal musculature-
         dorsal and ventral sides lie above and below
         respectively. Bar = 0.5 mm.

ximal circumferential disparity between stump and graft.
According to the PCM, this is where the supernumerary
limbs should appear. So the above results agree completely
with the predictions of the PCM (and the HPCM) and at the same
time strongly support the basic polar coordinate hypothesis:
in the limb pattern formation no predominant axes are involved
but the positional values are more or less equipotential and
lie on a peripheral ring. In contrast, the models based on a
cartesian system of axes like the AP and DV reproduce all data
of contralateral graftings as long as one of these predominant
axes is in coincidence between stump and graft. This is not
the case in the present experiment and it is interesting to
see what these models would imply. Consider for example the
boundary model (Meinhardt, 1983) which reproduces many data of
supernumerary limb formation. It is based on the assumption
that gradients are formed along the AP and DV axes and distal
outgrowth occurs if a DV border is established in an anterior
environment flanked by polarizing posterior cells. This condition is met in the present experiment but the expected supernumerary outgrowths are abnormal: it turns out that the model
cannot create any of the cardinal axes. The present experiment
indicates that PCM is more suitable to describe amphibian limb
regeneration than the boundary model which is based on a
cartesian system of coordinates.

     Another point of interest is the inclusion of anatomical
discontinuities in the framework of the HPCM. The twisting of
a contour does not disrupt the continuity of positional values
around the limb circumference but, inside the contour discontinuities are unavoidable. Such discontinuities appear usually
along the AP axis when ventral muscles lie next to dorsal
ones. If now we adopt the hypothesis that dermal fibroblasts

are the carriers of circumferential positional information
(Bryant et al., 1987), (e.g. mixed handed supernumeraries in
180° ipsilateral rotations), the density of dermal fibroblasts
in the middle of a transverse section should be higher than in
all other classes of supernumeraries corresponding to simple
contours. The present experiment is still in progress and we
search for supernumerary outgrowths with features not yet
observed.

REFERENCES

Bryant, S.V., Gardiner, D.M. and Muneoka, K., 1987, Limb
        development and regeneration, Amer.Zool., 27:675-696.
Costaridis, P., Zafiratos, C., Kiortsis, V. and Papageor-
        giou, S., 1989, Diverse supernumerary structures
        develop after inverting the antero-posterior limb axis
        of the Anura, Devel.Biol., (in press).
French, V., Bryant, P.J. and Bryant, S.V., 1976, Pattern for-
        mation in epimorphic fields, Science, 193:969-981.
Maden, M. 1982, Vitamin A and pattern formation in the regen-
        erating limb, Nature, 295:672-675.
Meinhardt, H., 1983, A boundary model for pattern formation in
        vertebrate limbs, J.Embryol. exp. Morph., 76:115-137.
Papageorgiou S., 1984, A hierarchical polar coordinate model
        for epimorphic regeneration. J. Theor.Biol.,
        109:533-554.
Stock, G.B., Krasner, G.N., Holder, N. and Bryant, S.V., 1980,
        Frequency of supernumerary limbs following blastemal
        rotations in the Newt, J. Exp. Zool., 214:123-126.
Wallace, H. and Watson, A., 1979, Duplicated axolotl regener-
        ates, J. Embryol. exp. Morph., 49:243-258.

EXPERIMENTAL ANALYSIS OF INTERCALARY LIMB REGENERATION

CAPACITY IN URODELES

Hermann Josef Anton

Zoological Institute, I. Chair
University of Cologne
Weyertal 119, D 5000 Koln 41, F.R. Germany

SUMMARY

Intercalary regeneration has been investigated after the removal of the distal part of the stylopodium, the whole zeugopodium, and the proximal region of the basipodium and the immediate dorso-dorsal, ipsilateral transplantation of an autopodium onto the amputation surface of the stylopodial stump in larval <u>Triturus alpestris</u> (Glucksohn stage 53-59) (i), in postmetamorphic young <u>Triturus alpestris</u> (ii), and in <u>Salamandra salamandra</u> as well as in the chimeric limbs of <u>S. salamandra</u> and <u>Ambystoma mexicanum</u> (iii). From all series it is concluded that stylopodial repair and regeneration of the completely missing zeugopodium are different types of developmental processes: Repair of the injured structures takes place by activation of the affected tissues without the loss of their tissue-specific basic determination. In the case of a completely missing segment (zeugopodium) the formation of a mesodermal blastema is required, recruited by dedifferentiated cells released from their tissue bindings. (i) The intercalary regeneration capacity of larval hindlimbs has a negative linear correlation, whereas the deformities of the regenerates have a positive linear correlation, to the larval developmental stages. (ii) In postmetamorphic limbs, intercalation of zeugopodial structures is provoked by digit amputation to induce dedifferentiation within the transplanted autopodium. (iii) In larval <u>S. salamandra</u>, the intercalation of zeugopodial structures is not suppressed by transplantation of mature autopodia onto a stylopodial stump. Chimeric limbs composed of a <u>Salamandra</u> stylopodial stump and an <u>Ambystoma</u> autopodium regenerate intercalary zeugopodia in a higher percentage of cases than the <u>Salamandra</u> autoplastics. In the chimeras no dramatic incompatibility reactions have been observed even after metamorphosis. The reciprocal composition in contrast showed only stylopodial repair. A low continuous resorption of all chimeric <u>Salamandra</u> autopodia took place within a year. Restoration of basipodial elements takes place by repair in proximal direction. Cellular participation of the graft in the intercalating regenerative processes is indicated by the presence of graft pigment cells within the intercalated area.

INTRODUCTION

Among vertebrates the repair or restitution of injured bones, tendons and muscles is a more widespread phenomenon than terminal regeneration after the loss of distal parts of appendages. The intercalary capacity of amphibians in developing as well as in regenerating limbs is remarkable and has been investigated for some years (see Wallace, 1981; Sicard, 1985; Anton, 1988). In discussions of the results of these investigations, particular attention was focused on the sources of the cells contributing to blastema formation. The rule of distal transformation during terminal regeneration (Wolpert, 1971) has been repeatedly doubted and corroborated by different authors (see Wallace, 1981; Sicard, 1985). A majority of investigators of intercalary regeneration tends to accept the rule of distal transformation though the results of vitamin A treatment during the early regeneration phase contradict this rule because proximalization of the terminal regeneration occurs (see Niazi and Saxena, 1978; Maden, 1982; Sharma and Niazi, 1988).

To obtain insight into the developmental events during intercalary regeneration, three groups of experiments have been carried out in recent years in our laboratory: 1) the analysis of the intercalary regeneration capacity of the developing larval hindlimb of <u>Triturus alpestris</u> (Anton et al., 1988); 2) the intercalary regeneration capacity of fore-and hindlimbs in young post-metamorphic <u>T. alpestris</u> (Anton and Sobanski, 1988); 3) the intercalary regeneration capacity of <u>Salamandra salamandra</u> larvae and of xenoplastic chimeras between <u>S. salamandra</u> larvae and <u>Ambystoma mexicanum</u> (Anton and Winter, 1988).

MATERIALS AND METHODS

In all experiments intercalary regeneration was initiated by removal of the zeugopodium and the distal part of the stylopodium by cross-cutting the stylopodium through the distal third of the diaphysis immediately followed by autoplastic (<u>T. alpestris</u> (1, 2) and <u>S. salamandra</u> (3)) or xenoplastic (<u>S. salamandra</u>, <u>A. mexicanum</u> (3)) transplantation of an autopodium, obtained by amputating across the proximal basipodium, onto the amputation surface of the stylopodial stump in dorso-dorsal orientation.

In the autoplastic experiments on <u>T. alpestris</u> (1) and <u>S. salamandra</u> (3) larvae, the autopodia were reimplanted onto the amputation surface of the stylopodial stump of the same limb. The autoplastic experiments in postmetamorphic <u>T. alpestris</u> (2) were performed on juvenile terrestrial animals two to ten months after metamorphosis. In addition, all or only two digits had been amputated on different days before the autopodium was grafted. The xenoplastic experiments (3) were carried out using mature <u>S. salamandra</u> larvae and corresponding <u>A. mexicanum</u> larvae as host and donor respectively. All animals (except <u>A. mexicanum</u>), in which the transplantation was carried out during the larval stage, were reared until metamorphosis had finished.

The experiments had been evaluated after the regenerates

Table 1. Intercalary Zeugopodial Regeneration During
         Hindlimb Development in T. alpestris (Staging
         after Glücksohn, 1931)

| Devl. stage No. | Cases n | Intercalated zeugopodia | | | | | Stylopodial repair only | |
| | | complete | | incomplete | | deform. | | |
		n	%	n	%	%	n	%
53	5	5	100	0	0	0	0	0
54	12	11	91.7	1	8.3	8.3	0	0
55	33	19	57.6	5	15.2	20.8	9	27.3
56	17	7	41.2	4	23.5	36.4	6	35.3
57	9	2	22.2	2	22.2	50	5	55.6
58	6	0	0	2	33.3	100	4	66.7
59	12	0	0	1	8.3	100	11	01.7

were stained in toto with Victoria blue (see Bryant and Iten,
1974) (1).   Alizarin red S/ Alcian blue 8GX (2) and Alcian
blue 8GX (3). (For more details see: Anton et al., 1988;
Anton and Sobanski, 1988; Anton and Winter, 1988).

RESULTS AND DISCUSSION

Intercalary Regeneration Capacity of the Hindlimb in
Triturus alpestris Larvae

      The stage by stage investigation of the developing hind-
limb in T. alpestris (Anton et al., 1988) has shown that the
capacity to intercalate a totally removed zeugopodium
decreases gradually from 100% perfect regenerates in stage 53
(knee-joint and two digits recognizable) to only imperfect and
very rare intercalary regenerates in stages 58/59 (fourth and
fifth digits differentiated, hindlimb with definitive proportions) (see Fig. 1; Table 1). Simultaneously, the number of
incompletely regenerated zeugopodia increases. At the same
time the number of deformities (which were not only restricted
to the regenerated area but occurred also within the autopodium) also increases from 0% in stage 53 to 100% in stage
58/59.

      In contrast to this age-dependent, decreasing regenera-
tive tendency, the potencies to restore the stylopodial structures remain constant in all stages. Both the decreasing
capacity to intercalate zeugopodia and the increasing quantity
of deformations are linearly correlated to the larval stage as
well as to the age. This is expressed by correlation coefficients (r) of r=-.98 when the percentage of zeugopodial
intercalation is plotted against the developmental stages
(st.), and r=-.97 when it is plotted against the age of the
larvae (age). The percentage of deformations shows correspondingly high but reciprocal correlation coefficients r=.96
(st.) and r=.95 (age) (see Fig. 2). When the percentage of
the complete zeugopodial intercalary regenerates is compared
with the incomplete, the correlation coefficients are r=-.99
(st.) and r=-.9 (age) respectively for complete and r=.98
(st.) and r=.95 (age) respectively for incomplete regeneration
(see Fig. 3).

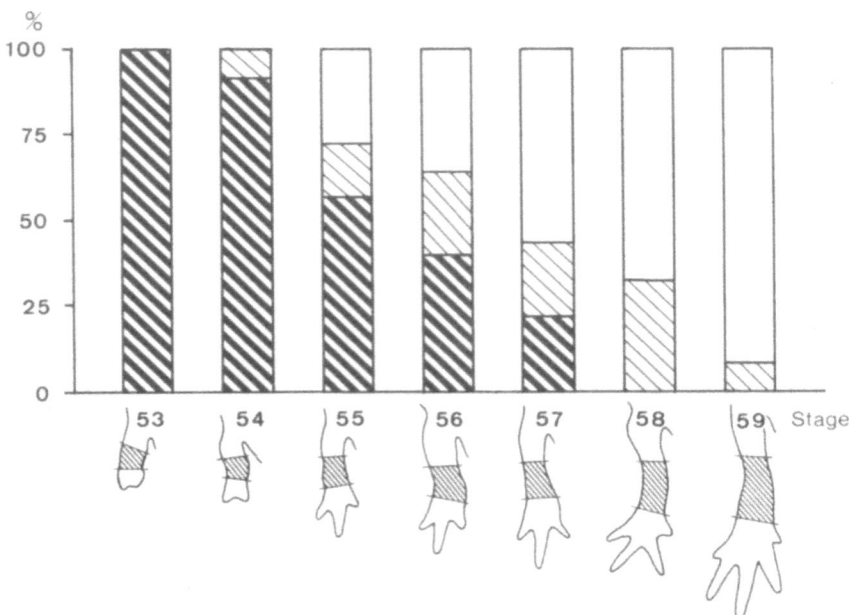

Fig. 1. Intercalary regeneration of hindlimb zeugopodia in <u>T.</u> <u>alpestris</u> larvae (complete = dark shading; incomplete = light shading, no zeugopodial intercalation = white). Drawings: hindlimbs of the corresponding stages (removed area = hatched).

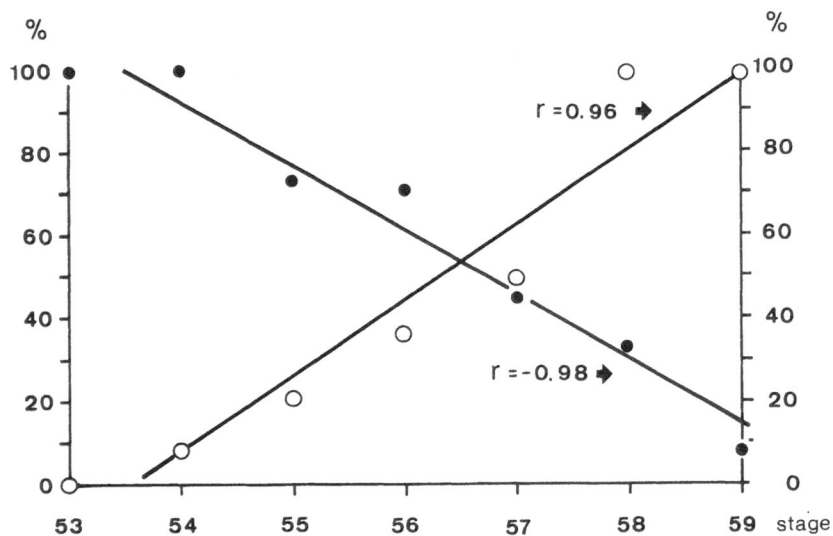

Fig. 2. Relationship between the percentage of intercalated zeugopodia and deformities in <u>T.alpestris</u> larvae (black circles = percentage of intercalated zeugopodial structures, white circles = deformed regenerates; r = coefficient of correlation).

Hyper-regeneration was never found within the intercalated zeugopodial area. Regulatory tendencies occurred not only within the newly formed area, but also within the basipodial region through which the autopodium had been severed from the zeugopodium, and sometimes within the acropodium. This is obviously a consequence of interactions among the stylopodial stump, the intercalating blastema, and grafted autopodium. The abnormalities within the intercalated zeugopodia ranged from the hypomorphic development of the tibia and fibula to an abnormal cartilaginous bulk being either only one element or the two elements fused. Abnormal differentiations within the autopodia occurred by fusion or reduplication of tarsalia as well as by the formation of additional digits. It may be that slightly incorrect alignments of the axes of the stump and graft cause such irregularities (Stocum, 1975). It is in any case the expression of the regulatory potencies within and between the intercalating area and the graft. The more the limb development progressed and the less the zeugopodial intercalation took place, the less deflections of the normal autopodial pattern occurred. The time- and stage-dependent linear decrease of the intercalary regeneration capability together with the decreasing ability to verify the genetic developmental pattern is in sharp contrast with the terminal regeneration capability, which occurs in urodeles in all developmental stages even in sexually mature adults, i.e. independently of the grade of respective tissue differentiation. The capability to form a mesodermal blastema which is able to redifferentiate the missing distal parts exists in all limb regions, and each level of the limb is capable of informing such a blastema of what has to be built. From this point of view there is no reason to suppose that the intercalation capacity is genetically limited. Consequently there is no reason to believe that the mesodermal tissues of an adult limb cannot dedifferentiate to mesodermal blastema cells to form a starting population for the intercalation of a missing segment.

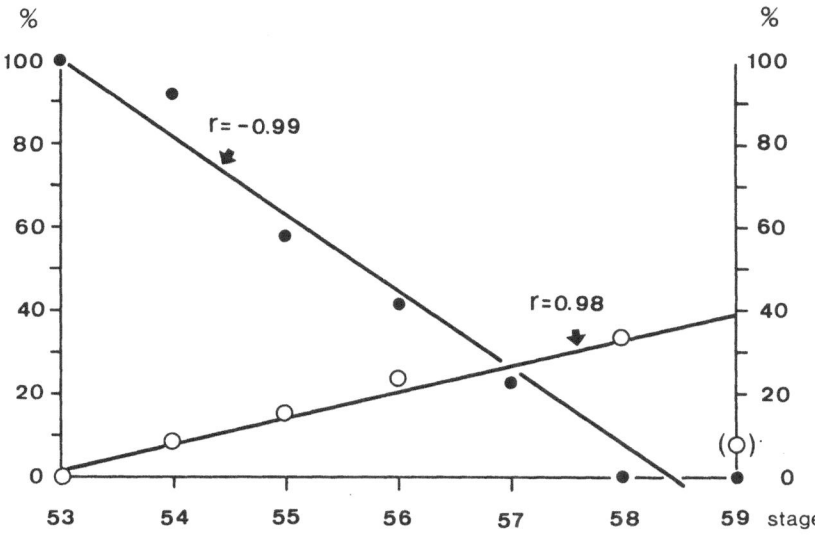

Fig. 3. Relationship between the percentage of complete and incomplete intercalated zeugopodia in T. alpestris larvae (symbols see Fig. 2).

Intercalary Limb Regeneration in Postmetamorphic Triturus alpestris

The transplantations carried out in the experiments on the young post-metamorphic _Tr. alpestris_ correspond to those using larvae. As expected from previous experiments (Bryant and Iten, 1977; Pescitelli and Stocum, 1980) no other results were obtained when the same experimental conditions were maintained: Only stylopodial repair and no zeugopodial intercalation took place (see Fig. 4a,b). After all digits had been amputated and a sham operation was carried out 10 days later (regression phase) by amputating such a digit-less autopodium and immediately disto-distally replanting it onto the original place, the fingers regenerated normally and no intercalary activity was observed even when a hindlimb autopodium was replaced by a forelimb autopodium. In such a case the proximal basipodial elements can be recognized as tibiale, intermedium, and fibulare (foot basipodium) and the middle and distal elements as centrale 1,2 and carpale d1-2,3,4 (hand basipodium). The fingers have regenerated to normal (see Fig. 4c,d). The combination of both the amputation of all or some digits and the disto-proximal transplantation onto the mid-stylopodial level resulted in stylopodial restitution plus intercalary regeneration of a zeugopodium in a more or less perfect shape. However, such results have only been obtained when the autopodia were

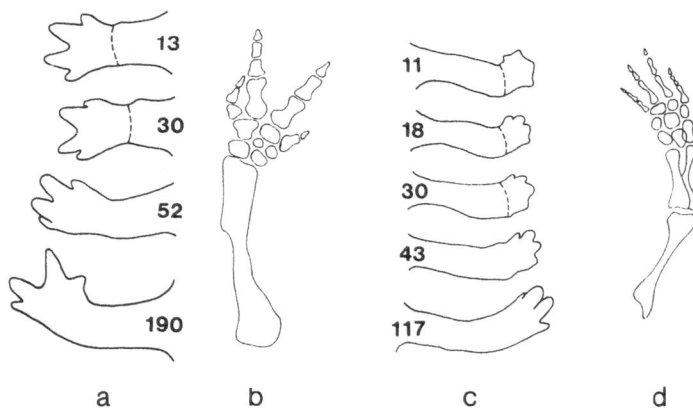

a   b   c   d

Fig. 4.   a,b) Disto-proximal transplantation of a fore-
          limb autopodium with intact digits onto the
          midstylopodial level hindlimb stump. a)
          regenerate at different days a.t., b) skeleton
          6 months a.t.: The distal stylopodium has been
          restored to normal. No zeugopodial intercala-
          tion took place. c,d) Disto-distal transplan-
          tation of a forelimb autopodium without any
          rotation from which all digits had been ampu-
          tated 10 days before onto the corresponding
          level of the hindlimb from which the autopo-
          dium was previously amputated, c) regenerate
          at different days a.t., d) skeleton 4 months
          a.t.: Fingers have been regenerated to normal.

transplanted during wound healing or the regression phase. When all digits had been amputated, the intercalated zeugopodium was nearly perfectly formed (see Fig. 5a,b). The amputation of only two digits resulted in less perfect zeugopodial structures (see Fig. 5c,d). Autopodia which had been transplanted during the digital redifferentiation phase suppressed intercalary zeugopodial regeneration but allowed stylopodial restoration (see Fig. 5e,f).

From these results it is evident that mesodermal blastema formation between two different proximo-distal levels of mature mesodermal tissues depends on the grade of differentiation of the graft. Mature tissue normally prevents tissue dedifferentiation and suppresses intercalary mesodermal blastema formation. The periosteal cells in contrast are activated when bone is injured. This weaker dedifferentiation does not seem to be suppressible by the transplanted distal mature material. Consequently, in various experiments including our own, the stylopodial bone had been restored by a periosteal blastema which was eventually supported by dedifferentiating chondrocytes (Anton, 1965, 1968). In a similar way the muscles within the stump regrew together with the connective tissues and tendons by partial dedifferentiation, multiplication, and subsequent redifferentiation to form adequate or inadequate functional units. The steady genetic control on all levels of organization, responsible for the morphological and functional status of the whole body, acts continuously during the regenerative processes by positional information. Morphogens may be the mediators. Cellular information and cell recognition mechanisms act to realize the species specific patterns. Externally conditioned disturbances of these patterns result in the activation of the respective genes of those cells which have been affected directly or indirectly by the disturbance and which react suitably within their differentiation limits. If the activation is triggered only by a simple segmental injury, fibrocytes and/or periosteal cells are activated to restore the original pattern or a state as near as possible to the genetically scheduled pattern. In case of missing segments (e.g. the whole zeugopodium), genetically intact mesodermal blastema cells are able to recognize the developmental information of their surroundings (morphogenetic field information). This causes transformation and new-formation of the totally missing parts by following the genetic blueprint within the limits of the developmental stage, i.e. an immediate verification of the stage specific structure and not a kind of repetition of embryonic development.

Intecalary Regeneration in Salamandra salamandra, Ambystoma mexicanum and in Chimeric Limbs

In the search for species suitable for xenoplastic transplantation experiments the intercalary regeneration capacity of S. salamandra larvae was investigated and compared with A. mexicanum. In contrast to the latter, which showed only stylopodial repair (see also Pescitelli and Stocum, 1981), S. salamandra additionally intercalated zeugopodia with well recognizable tibia and fibula in 8 (17%) of the 48 operated animals. Three of them (6%) also regenerated missing tarsalia. In another 5 cases (10%) zeugopodia with only reduced skeletal elements regenerated. Among urodeles high

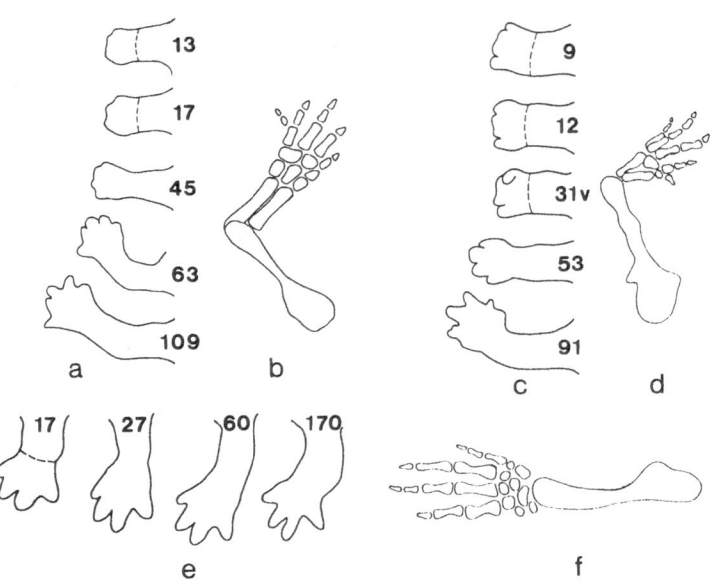

Fig. 5. Disto-proximal transplantation of the forelimb
autopodium onto the stylopodial stump after
removal of the total zeugopodium and the distal
stylopodium: a,b) All digits had been removed 10
days b.t. a) regenerate at different days a.t.,
b) skeleton 4 months a.t.: The stylopodium has
been restored and a zeugopodium with radius and
ulna has intercalated. Within the basipodium,
carpalia 3-4 are fused and centrale 2 is fused
with the intermedium. The third digit consists
only of three phalangae. An additional digit has
formed behind the fourth. c,d) The digits two and
three had been removed immediately b. t. : c)
regenerate at different days a.t., d) skeleton 6
months a.t.: The regenerated limb consists of
the well restored stylopodium, the intercalated
zeugopodium with separate cartilaginous radius and
ulna which are small in relation to the humerus.
Within the basipodium, consisting only of the carpale d1-2,3,4 and centrale 1, the centrale 2,
radiale, intermedium, and ulnare have not been
intercalated. In addition to the 4 digits (2 and
3 are well regenerated), an additional phalan-
gial(?) cartilage has developed. e,f) all digits
had been removed 22 days (redifferentiation phase)
b.t. e) regenerate at different days a.t.,
f) skeleton 6 months a.t.: The regenerated
limb consists of the well restored stylopodium
and the intercalated zeugopodium in which a
relatively small cartilaginous radius and ulna
are visible. Within the basipodium with only
the carpale d1-2,3,4 and centrale 1, the centrale 2, radiale intermedium and ulnare have
not been intercalated. In addition to the 4
digits (2 and 3 are well regenerated), an
additional (phalangial?) cartilage has developed.

differences exist in the limb regeneration capacity (see Scadding, 1977; Wallace, 1981), and this obviously also applied to intercalary regeneration potencies. It is surprising that our results indicate that S. salamandra, which is born as a quadruped larva with fully differentiated limbs which undergo metamorphosis within a relatively short time after birth, has a much higher intercalary regeneration capacity than T. alpestris larvae.

From our xenoplastic experiments we know that with Ambystoma as the stylopodial base and Salamandra autopodia as the graft only stylopodial repair occurred. However, in the reciprocal combination with Salamandra as stylopodial base, zeugopodial intercalations regenerated under the influence of the grafted Ambystoma autopodium in 8 (62%) of 13 cases. While the Salamandra/Ambystoma chimeras outlast metamorphosis without any signs of immunological complications, the Ambystoma/Salamandra chimeras underwent incompatibility reactions with a very slow but continuous progress in resorption of the Salamandra graft. Within a year the graft was reduced to a small remnant only consisting of skin and connective tissue.

In all zeugopodial intercalation cases we have observed both the stump and the graft participate in supplying cells to the intercalated area: pigment cells deriving from A. mexicanum graft were found everywhere in the intercalated region, and the muscles also had an appearance suggesting origination from the graft material. The repair of injured basipodial bones was affected by periosteal cells of the particular bone. In some cases in which no zeugopodial structures had been intercalated, a proximally directed outgrowth of a basipodial bone had taken place.

The fact that the intercalation of a zeugopodium was significantly higher in the xenoplastic chimeras composed of a stylopodial stump of S. salamandra with an A. mexicanum autopodium, in comparison to the results in similar autoplastic combination in S. salamandra is, in terms of our hypothesis, the consequence of a temporarily limited weak incompatibility reaction within the graft tissues. The resulting dedifferentiation allows the formation of a mesodermal blastema and the subsequent regeneration of the zeugopodium. Tissues of a certain degree of dedifferentiation are not able to inhibit these genetic potencies which allow mesodermal cells to become meodermal blastema cells.

The restoration of the stylopodial structures in all series seem to me to be a proof that repair of injured structures and the new formation of completely removed segments are two entirely different developmental processes. Repair results from activated (blastematic) cells of the particular tissue (i.e., epidermis, dermis, muscle, connective tissues, periosteum and/or chondro- and osteocytes, pigment- and Schwann-cells). In contrast, new-formation results from a mesodermal blastema.

REFERENCES

Anton, H.J., 1965, The origin of blastema cells and protein synthesis during forelimb regeneration in *Triturus*. in: "Regeneration in Animals and Related Problems" (Ed.: V. Kiortsis, H.A.L. Trampusch) North-Holland Publ. Co., Amsterdam: 377-395.

Anton, H.J., 1968, Autoradiographische Untersuchungen uber den Eiweisstoffwechsel bei der Extremitatenregeneration der Urodelen, *Roux' Arch. Entw. Mech. Org.*, 161:49-88.

Anton, H.J., 1988, "Control of Cell Proliferation and Differentiation During Regeneration," *Monogr. devl. Biol.*, 21, Karger, Basel.

Anton, H.J., Michael, M.I. and Sayed-Ahmed, A.A., 1988, Intercalary regeneration in the different hindlimb stages Triturus alpestris larvae, *Monogr. devl. Biol.*, 21:156-168.

Anton, H.J. and Sobanski, H., 1988, Intercalary regeneration in postmetamorphic forelimbs of *Triturus alpestris*, *Monogr. devl. Biol.*, 21:169-180.

Anton, H.J. and Winter, H.G., 1988, Intercalary limb regeneration after auto- and xenoplastic grafting of mature autopodia at the stylopodial stump in larval *Salamandra salamandra* and *Ambystoma mexicanum*, *Monogr. devl. Biol.*, 21:181-195.

Bryant, S.V., and Iten, L.E., 1974, The regulative ability of the limb regeneration blastema of *Notophthalmus viridescens*, *Roux' Arch. Entw. Mech. Org.*, 174:90-101.

Bryant, S.V., and Iten, L.E., 1977, Intercalary and supernumerary regeneration in regenerating and matur elimbs of *Notophthalmus viridescens*, *J. exp. Zool.*, 202:1-16.

Glucksohn, S., 1931, Aussere Entwicklung der Extremitaten and Stadieneinteilung der Larvenperiode von *Triton taenitus* Leyd. und *Triton cristatus* Laur. *Roux Arch. Entw. Mech. Org.*, 125:341-405.

Maden, M., 1982, Vitamin A and pattern formation in the regenerating limbs, *Nature (London)*, 295:672-675.

Niazi, I.A. and Saxena, S., 1978, Abnormal hindlimb regeneration in tadpoles of the toad, *Bufo andersonii*, exposed to excess vitamin A, *Folia Biol.*, (Krakow) 26:3-11.

Pescitelli, M.J. and Stocum, D.L., 1980, The origin of skeletal structures during intercalary regeneration of larval *Ambystoma* limbs, *Dev. Biol.*, 79:255-275.

Scadding, S.R., 1977, Phylogenetic distribution of limb regeneration capacity in adult *Amphibia*, *J. exp. Zool.*, 202:57-67.

Sharma, K.K. and Niazi, I.A., 1988, Variety of regenerative responses of different proximo-distal limb segments of young and advanced *Rana breviceps* tadpoles treated with vitamin A after amputation. *Monogr. devl. Biol.*, 21:124-137.

Sicard, R.E., 1985, "Regulation of Vertebrate Limb Regeneration", Oxford University Press, New York, Oxford.

Wallace, H., 1981, "Vertebrate Limb Regeneration", John Wiley and Sons, Chichester, New York, Brisbane, Toronto.

Wolpert, L., 1971, Positional information and pattern formation, *Curr. Top. Develop. Biol.*, 6:183-224.

FORELIMB REGENERATION IN REFERENCE WITH HINDLIMB REGENERATION

IN STAGES OF THE EGYPTIAN TOAD, BUFO REGULARIS REUSS

M.I. Michael[1] and M.H. El-Tawil[2]

[1]Department of Zoology, Faculty of Science
Alexandria University, Egypt
[2]Department of Zoology, Faculty of Science
Menufia University, Egypt

SUMMARY

1.   Forelimb regeneration was studied in seven larval and metamorphic stages of the toad, Bufo regularis Reuss, after transection at the antebrachium level.

2.   The power of forelimb regeneration at the level of zeugopodium matches with that of the hindlimb at the level of stylopodium within the same stages of the same species.

3.   Fading of the power of regeneration of the extremities of tadpoles in the concerned stages, takes place in a cranio-caudal order along the body axis.

4.   Five-fingered cases appeared only when the early forelimbs were amputated before the hand plate had differentiated its normal four digits.

5.   Syndactyly, when present, took place between fingers 5 and 4 or between 3 and 2 but not in a single case, of the present study, between fingers numbers 4 and 3.

INTRODUCTION

Urodeles, unlike anurans, among amphibians, are able to regenerate the missing parts of their amputated limbs throughout life (see Goss, 1969). Work on limb regeneration in anurans was somewhat limited and most of the available data were on the hindlimbs which normally show conspicuous decline in their regenerative abilities shortly before or around the period of metamorphosis (see Wallace, 1981).

During the last two decades, studies were performed in the laboratories of the Faculty of Science, Alexandria University, to investigate aspects of hindlimb regeneration in stages of the Egyptian common toad, Bufo regularis Reuss. Firstly, the normal power of distal regeneration of these limbs, after amputation at various levels were studied (see Michael and El-Malkh, 1969; Michael and El-Mekkawy, 1977a). Thereafter, series of investigations were carried out to analyse the causal factors for the enhancement of the powers of limb regeneration in late metamorphic stages by the use of some mechanical means (Michael and Aziz, 1975) or by means of chemicals as sodium perchlorate (Michael and Aziz, 1976), sodium chloride, glucose and acetic acid (Michael and Aziz, 1978). Further studies to test the abilities of these limbs, in stages of the same species, for intercalary regeneration after removal of the thigh or the shank regions, were performed by autografting the ipsilateral autopodium without rotation of the limb axes (Michael and Hassona, 1982). Quite recently, Anton et al., (1986-1987) reported on the abilities of these limbs for intercalary regeneration after removal of the shank region, but the grafted ipsilateral autopodium was rotated clockwise by 90° or 180°. There are evidences that both the stump and graft tissues share in the intercalated limb part. Intercalation was found to fade in a proximo-distal order in the hindlimbs and nearly disappeared before reaching the metamorphic climax.

Based on the previous studies on the morphogenesis and histogenesis of the hindlimb regeneration in various larval and metamorphic stages, a survey was provided on the sequence of incidence of malformations in regenerates with special emphasis on their skeletal configurations (Michael and El-Mekkawy, 1977b).

The forelimbs, unlike the hindlimbs, normally begin to develop underneath the operculum, in the peribranchial chamber, just embracing the hyobranchial apparatus (see Sedra and Michael, 1958, 1961). The concealed forelimbs during early developmental stages, hindered experimenters to get easy reach of them. In the literature, data so far available on forelimb regeneration in anurans, were mostly on postmetamorphic and even adult stages in which the limbs were already exposed (see Goode, 1967; Michael and Sammak, 1970; Kurabuchi and Inoue, 1988). Our knowledge on the regenerative abilities of the forelimbs of early larval stages in anuran tadpoles is quite abbreviated. Limited information, so far known, was provided by Newth (1948) on Xenopus laevis and Lecamp (1964) on Alytes obstetricans.

The present study aimed to test the power of the fore-limbs in stages of the toad, Bufo regularis for regeneration, when amputated, and to compare this power with that of the hindlimbs in comparable stages of the same species. The question posed here: Do both fore- and hindlimbs have the same regenerative power?

MATERIALS AND METHODS

Couples of adult toads of Bufo regularis, collected from the fields in the vicinity of the Faculty of Science, Moharram Bey, Alexandria University Campus, were kept in special aquaria till spawning. Fertilized eggs were raised in the laboratory at a temperature of approximately 26°C±2 and 73%±2 relative humidity (RH). Larvae were fed on freshly boiled leaves of lettuce but tadpoles which passed the metamorphic period were supplied with woodlice as <u>Leptotrichus panzeri</u> and <u>Philoscia africana</u>.

Seven larval and metamorphic stages were selected and these are stage numbers 52, 53, 54, 55, 56, 57 and 66, according to the Normal Table of the same species (Sedra and Michael, 1961). The most distinctive external criteria for these stages were based on the morphogenesis of the hindlimbs, till the exposure of the forelimbs. These are summarized in the following. <u>Stage 52</u>: Dorsal part of hindlimb, paddle-shaped; first indication of knee joint construction. <u>Stage 53</u>: Foot paddle shows fourth toe protuberance. <u>Stage 54</u>: Early appearance of ankle constriction with shank; foot has toe protuberance 5-2. <u>Stage 55</u>: Foot plate indented between all five toe protuberances. <u>Stage 56</u>: Hindlimb 1.9 mm long, three major limb parts are delimited, five toes with a depigmented area on tip of each. <u>Stage 57</u>: A preclimactic stage; hindlimb 3 mm long; first appearance of prehallux, fourth toe distinctly longer than other toes 5-1. Forelimbs still concealed under the opercular skin. <u>Stage 66</u>: A newly metamorphosed toadlet; hindlimb 17 mm long, exposed forelimb 7.3 mm long with distinct elbow and wrist joints; hand with four fingers 5-2.

As anaesthetic a solution of chloretone in tap water (1:2000) was used. Operations were performed by following two steps: (a) incision through the opercular skin and lining of the peribranchial chamber, and (b) exposure of the forelimb through the cut window in the operculum (see Fig.1). The exposed limb was amputated through the mid-zeugopodial level (antebrachium). Precautions were taken to avoid contact of the stump skin with the cut edge of the opercular skin, otherwise the transected limb will be anchored.

Experimentals were kept for some time in half concentration anaesthetic solution before they were transferred to normal dechlorinated tap water. Cases were reared till they passed the metamorphic period and in later stages even for 3-9 weeks postmetamorphosis. Among fixatives used were Bouins's fluid, Zenker-acetic and formalin. Alizarin and methylene blue transparencies were prepared to study the skeletal elements regenerated. For sectioned material some stains used as Mallory, Masson or Azan (Mahoney, 1973). A total of 360 cases were studied in these stages.

Fig. 1. A diagrammatic drawing of a right side view of a tadpole (stage 52) showing the cut through the opercular skin to expose the forelimb which was transected across the zeugopodium (see arrow).

RESULTS

Normal Ontogenesis of the Forelimbs

Detailed description of the forelimb ontogenesis in stages of <u>Bufo regularis</u> will appear in a separate publication (Michael and El-Tawil, under publication). It seems worthy to refer here to some essential landmarks for these limbs based on the earlier observations of Sedra and Michael (1961).

By stage 54 all four fingers (numbers 5-2) are differentiating. By stage 55 free parts of fingers about equally long as broad. By stage 57 fingers are stretched out in forelimb atrium, concealed by operculum skin; second finger about 2/3 of the third one, third and fifth fingers are more or less equal to each other, and each of them about 2/3 of the fourth finger. The first finger is missing. Around stage 58 the forelimb becomes exposed. By stage 66 the forelimb attains the normal limb proportions met with in post-metamorphic stages.

Regeneration of the Forelimbs

(a) <u>In early stages</u> (52 and 53); four fingers on hand plate not yet differentiated. (see Table).

By stage 52; out of 57 operated cases, four had five fingers each (polydactylia; see Fig. 2), 48 cases had four fingers each, two with three fingers each and three with two fingers each.

By stage 53; out of 59 operated cases, two had five fingers each, 36 cases with four fingers each, 13 with three fingers each, four with two fingers each, two with one finger each and the remaining two cases could not restore any finger.

(b) <u>In later stages</u> (54-66); all four fingers appeared on hand. (see Table).

By stage 54; out of 37 operated cases, 16 restored all four fingers each, 9 cases with three fingers each, three with two fingers each, four with one finger each and five cases partially restored the missing parts but had no fingers.

By stage 55; out of 70 operated cases, four with four fingers each, 10 with three fingers each, 12 with two fingers each, 21 with one finger each, and 23 cases partially restored the antebrachium with/without wrist elements but had no fingers.

By stage 56; out of 62 operated cases, two had three fingers each, one case only with two fingers, 6 with one finger each, while the remaining 53 cases were negative.

By stage 57; out of 37 operated cases, one case only had two fingers, 6 with one finger each, and the other 30 cases were negative.

By stage 66; out of 38 operated cases, 37 were negative but in one case only the regenerate extended slightly into two atrophied digit-like protuberances, covered with normal skin.

412

Table. Regeneration of Forelimb in Stages of *Bufo regularis*, Reuss,
After Transection at the Antebrachium Level

Stage	Number of regenerates with corresponding fingers						A total of 360 cases	Post-operative period (days)
	0	1	2	3	4	5		
52	–	–	3	2	48	4	57	24 – 33
	–	–	5%	4%	84%	7%		
53	2	2	4	13	36	2	59	24 – 45
	3%	3%	7%	22%	61%	3%		
54	5	4	3	9	16	–	37	22 – 47
	14%	11%	8%	24%	43%	–		
55	23	21	12	10	4	–	70	20 – 49
	33%	30%	17%	14%	6%	–		
56	53	6	1	2	–	–	62	27 – 40
	85%	10%	2%	3%	–	–		
57	30	6	1	–	–	–	37	25 – 45
	81%	16%	3%	–	–	–		
66	37	–	1	–	–	–	38	21 – 60
	97%	–	3%	–	–	–		

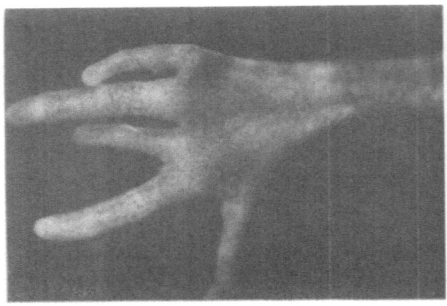
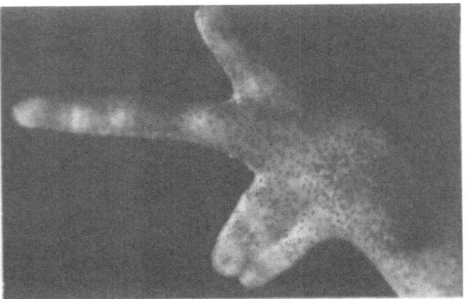

Fig. 2. A five-fingered
regenerate of the forelimb
amputated at stage 52,
fixed after 33 days. x12.

Fig. 3. A deformed four-
fingered regenerate of
the forelimb amputated at
stage 54, fixed after
40 days. x18.

(c) Malformations in Finger Arrangement

Among those cases operated by stages 54 and 55, which
restored all four fingers in each, some showed deformity in
finger arrangement. Superficial syndactyly due to synderma-
tosis between two adjacent fingers was met with. When present, it took place between fingers 5 and 4 or between fingers 3 and 2 but not in a single case between fingers 4 and
3, (Fig. 3).

DISCUSSION

Forelimb regeneration showed higher abilities in early
larval stages (numbers 52 and 53) where 84% of the cases in
the former and 61% of the cases in the latter, could restore
the missing parts of the transected limbs and developed four
fingers each. In later larval and metamorphic stages, the
regenerative abilities of the forelimbs began to decline gradually since 43% of the cases of stage 54 and 6% of those of
stage 55 could restore the full number of fingers in each. In
those cases of stage 55, the one-fingered type became prevalent among others with fingers. However, nearly one third of
the total number of cases of that stage were negative.

The power of regeneration showed conspicuous decline in
cases of stage 56 and stage 57 and nearly disappeared in stage
66. More than 80% of cases of the first two stages and about
97% of those of the last stage could not restore the missing
limb parts or developed well defined fingers.

It is worth mentioning to discuss the present results on
forelimb regeneration with those published earlier on hindlimbs in stages of the same species (Michael and El-Malkh,
1969; Michael and El-Mekkawy, 1977a). By taking number and
structure of digit formations besides distal regeneration of
the missing limb parts as baseline for assessment of the limb
regenerative power, one might simply notice a lower grade of
that power in the forelimbs than in the hindlimbs. Actually,
the present results on forelimb regeneration, after transec-

tion across the zeugopodium, broadly match with those on hind-limbs after transection across stylopodium, a more proximal level (Michael and El-Malkh, 1969). This conclusion is concordant with that of Lecamp (1964) on the regeneration of the fore and hindlimbs in stages of Alytes obstetricans. The difference between the grades of regeneration in both limbs might be related to difference in the chronology of their ontogenesis. Balinsky (1972) pointed out such a difference at the ultra-structural level in limb buds in stages of Bufo regularis. Such a difference in the staging and ontogenesis of both limbs is not uncommon in other groups of vertebrates such as the newborn opossum (Mizell, 1968).

Thus, the power of regeneration of the extremities in tadpoles of Bufo regularis declines in a cranio-caudal order along the body axis. Needless to say, the tail in larvae of anurans and in different stages of urodeles are ready to regenerate after amputation (Goss, 1969; Wallace, 1981).

Cases of polydactylia reaching 7% in stage 52 and 3% in stage 53 out of the total number but none among cases of later stages (54-66) deserve reference. The cells of the blastema that stemmed from such early forelimb stumps were evoked to differentiate beyond their normal limits and developed the five fingers, recalling the ancestral pattern of the pentadactyl limb.

REFERENCES

Anton, H.J., Michael, M.I., and Sayed Ahmed, A.A., 1986-1987, Intercalary regeneration in developmental stages of hindlimbs of the Urodele Triturus alpestris and the Anuran Bufo regularis, Arch. Anat. mic. Morphol. exper., 75:304-305.
Balinsky, B., 1972, The fine structure of the amphibian limb bud, Acta Embryol. exp. Morphol., 22:455-470.
Goode, R., 1967, The regeneration of limbs in adult anurans, J. Embryol. exp. Morphol., 18:259-267.
Goss, R.J., 1969, "Principles of Regeneration", Acad. Press, New York.
Kurabuchi, S., and Inoue, S., 1988, Limb regenerative capacity in anuran amphibia, in: "Regeneration and Development", S. Inoue et al., eds., Okada, Maebashi, Japan.
Lecamp, M., 1964, Régénérations comparées des ébauches des membres postérieurs et antérieurs chez le têtard du crapaud accoucheur Alytes obstetricans, Laur, C.R. Acad. Sci. Paris, 295:4160-4162.
Mahoney, R., 1973, "Laboratory techniques in zoology", Butterworth and Co., London.
Michael, M.I., and Aziz, F.K., 1975, Effect of mechanical means on the restoration of the limb regenerative ability in metamorphic stages of Bufo regularis Reuss. Acta Biol. Acad. Sci. Hung., 26:15-21.
Michael, M.I., and Aziz, F.K., 1976, Effect of sodium perchlorate on the restoration of the limb regenerative ability in a metamorphic stage of Bufo regularis, Reuss, Folia Biol.(Krakow), 24:309-315.

Michael, M.I., and Aziz, F.K., 1978, Effect of chemical means on the restoration of the limb regeneration in stages of <u>Bufo regularis</u> Reuss <u>XIXth Morphol. Congr. Sympos. Charles Univ. Prague</u>: 143-147.

Michael, M.I., and El-Malkh, N.M., 1969, Hindlimb histogenesis and regeneration in larvae and metamorphic stages of the Egyptian toad, <u>Bufo regularis</u> Reuss, I. Transection at the thigh level, <u>Arch. Biol.</u>, (Liege), 80:299-326.

Michael, M.I., and El-Mekkawy, D.A., 1977a, Hindlimb regeneration in larvae and metamorphic sages of the Egyptian toaad, "Bufo regularis" Reuss, II. Transection at the shank and ankle levels, <u>Arch. Biol.</u>, (Bruxelles) 88:373-391.

Michael, M.I., and El-Mekkawy, D.A., 1977b, Progress of malformations during limb regeneration in stages of the Egyptian toad, <u>Bufo regularis</u> Reuss, after transection at various proximodistal levels, <u>Mem. Soc. Zool. France, Symp. L. Gallien</u>, 41:203-208.

Michael, M.I., and El-Tawil, M.H., Ontogenesis of the forelimbs in the Egyptian toad, <u>Bufo regularis</u>, Reuss, (under publication).

Michael, M.I., and Hassona, A.A., 1982, Intercalary and supernumerary regeneration in limbs of stages of the Egyptian toad, <u>Bufo regularis</u> Reuss. <u>J. Exp. Zool.</u>, 220:207-218.

Michael, M.I., and Sammak, A.J., 1970, Regeneration of limbs in adult <u>Rana ridibunda ridibunda</u> Pallas, <u>Experientia</u>, 26:920-921.

Mizell, M., 1968, Limb regeneration: Induction in the newborn Opossum, <u>Science</u>, 161:383-385.

Newth, D.R., 1948, The early development of the forelimbs in <u>Xenopus laevis</u>, <u>Proc. Zool. Soc. Lond.</u>, 118:559ff.

Sedra, S.N., and Michael, M.I., 1958, The metamorphosis and growth of the hyobranchial apparatus of the Egyptian toad, <u>Bufo regularis</u> Reuss, <u>J. Morph.</u>, 103:1-30.

Sedra, S.N., and Michael, M.I., 1961, Normal table of the Egyptian toad, <u>Bufo regularis</u> Reuss, with an addendum on the standardization of the stages considered in previous publications, <u>Cs. Morfol.</u>, 9:333-351.

Wallace, H., 1981, "Vertebrate limb regeneration", John Wiley Sons, Chichester.

EXPERIMENTAL ANALYSIS OF

MOUSE LIMB DEVELOPMENT IN SITU

Ken Muneoka
Department of Biology, Tulane University
New Orleans, Louisiana 70118, USA

SUMMARY

Using exo utero surgical techniques we have fate mapped
various limb stages and have used these fate maps to analyze
the regulative ability of the early mouse limb. Fate mapping
studies demonstrate that the anterior-posterior extent of the
limb bud develops asymmetrically with the posterior half giving rise to 3 digits while the anterior half forms only 2
digits. In addition, we find that individual digits are formed
by the distal addition of phalangeal elements as opposed to
the segmentation of an initial cartilagenous rod that forms in
the footplate. The regulative ability of early limb buds was
investigated following amputation through the footplate and by
grafting tissues to confront anterior and posterior cells. In
both instances we find that partial regulation occurs: partial
regeneration of peripheral digits following amputation and the
formation of supernumerary digits tips following heterochronic
grafts of anterior wedges into posterior positions.

REGENERATION AND MORPHOGENESIS

Rosine Chandebois

Laboratory of Animal Morphogenetics
University of Provence
Marseille, France

INTRODUCTION

The problem raised by traumatic regeneration in multicellular animals cannot be tackled without a sound conception of embryonic development. Actually the adult organism is the issue of ontogenesis during which cells have acquired their respective properties and structural features have emerged. We have to know the intimate mechanisms of these phenomena if we intend to understand how the specific structural patterns of an eliminated part of the adult body are reproduced during its regeneration.

The present concepts of adult structural regulation -more particularly Wolpert's concept of positional information (1969; 1971) - are focused on the narrow problem of position effect in cell differentiation. The solutions they propound are based on the same general idea: any visible structural pattern is created and further maintained by a pre-existing invisible organization, the "prepattern" which is generally viewed as a molecular gradient. In order to avoid all possible confusion we use the term "gradient-prepattern" for referring to these "blueprints" of visible structures. How are the prepatterns unfolded and how they specify cell individualities are the fundamental questions which have to be solved. Thus the complicated problem of morphogenesis seemed clarified but its experimental approach became impossible. Consequently theoreticians were - and are still - reduced to propound speculative models.

Actually the basic idea of these concepts is an old hypothesis, a conviction which forced itself upon embryologists more than half-a-century ago, when most of their experimental results were still uninterpretable in the state of their knowledge of development. These concepts, because they do not speak of the real embryo are extremely simplistic and take advantage of this. They remain in vogue although they are at variance with the important findings of experimental embryologists. We must say that the latter, for their part, have neglected to conceptualize their findings. It is high time to realize that the notion of gradient - prepattern dramatically hampers the progress of experimentation and that we

419

must search after new theoretical clues in the synthesis of a large range of experimental data rather than in a speculative approach.

In this paper I shall first recall why the notion of gradient-prepattern has been propounded and why it has finally been accepted as a certainty. I shall explain why it has evolved without reference to the progress of experimental embryology. Then I shall present a general outline of my concept of Cell Sociology which helps to grasp the real nature of gradients and their role in development and regeneration. Finally, I shall discuss experimental data obtained with planarians which show the role of tissue differentiation in the maintenance of gradients in the uninjured organism and in regulation of structures during traumatic regeneration.

ORIGIN AND EVOLUTION OF THE NOTION OF GRADIENT - PREPATTERN

The idea that visible patterns are underlaid with prepatterns was a concensus reached by workers in the field of embryonic development (see Waddington, 1956) and regeneration (Guyénot, 1927). Some years before, Spemann had laid the foundation of modern embryology with the discovery of embryonic induction. Nevertheless at the same time, he had put a riddle that it was tempting to solve by means of gratuitous suppositions. Induction does not only trigger new specific activity in cells; it is followed by the sudden and considerable increase in structural complexity of the induced anlage. The most puzzling case was the regionalization of the neural plate in amphibians. We know now that the complicated organization which is settled early in the neural tube is determined by factors interferring with the progress of neural differentiation induced by the archenteron roof - the notochord anlage. Firstly the influence of the non-neuralized ectoderm creates prosencephalic structures. Later the same inducer exerts a second induction called "transformation". Since its transforming capacity gradually declines from front to back the neural ectoderm acquires an antero-posterior organization (prosencephalon, rhombencephalon, spinal cord) (references in Nieuwkoop et al., 1985). Half-a-century ago these data were unknown. A solution was sought in the organization of the archenteron roof, only, but the latter is too simple and does not account for the complexity of the neural tube. Moreover, typical neural structures are obtained when ectoderm is neuralized with heterogeneous inducers (e.g. liver or kidney). The only solution which was borne in upon minds was that some invisible organization is already present in the ectoderm before induction. From that time forth, morphogenesis appeared as a phenomenon causally distinct from tissue differentiation. The idea was clearly expressed by Waddington (1956) when he coined the terms "evocation" which calls the change of tissue identity in induced cells and "individuation" which calls the emergence of specific structures in the induced anlage.

From the beginning embryologists have contended with another difficulty: the regulation of structures. They supress parts of embryos or graft them into other places with the object of disturbing cell interactions and impairing development. Generally the grown embryo is normal or partly duplicated. Then it seemed that the invisible organization which manifests itself through the emergence of visible structural features is self-regulating. It must be maintained by some control in the normal embryo and is remodelled in the experimental embryo. During the process of regulation cells are switched on other differentiation pathways. This led to suppose that invisible organization and cells are physically independent of each other. At the same time another aspect of regulation,traumatic regeneration, was extensively investigated. Since cell communication and tissue interactions in the adult animal were still unknown the possible conversion of one part of the organism into another seemed at variance with the stability of adult organization. Again one had no choice but to allot the maintenance and remodelling of visible patterns to controls at the supra-cellular level.

The development of embryonic anlagen after partial removal or transplantation and the behaviour of the adult organism after amputation or grafting reveal properties which are curiously reminiscent of those of physical fields. A particular part acts as a "centre" from which the "intensity" of the field declines gradually. The structure of the field is re-established after an amputation. It is shifted if the centre is displaced or reduplicated if a second centre is added. The term "morphogenetic field" was coined in order to help the formulation of experimental results. Unfortunately the analogy was pushed too far. Just as physical fields are studied independently of matter, morphogenetic fields were conceived without taking cell properties into consideration. Finally P. Weiss (1939) stipulated that "the field concept is not only a circumlocution but an expression of physical reality". This assertion was in good agreement with the discovery of metabolic gradients which were found wherever one looked for them thanks to the archaic technique of vital staining under anaerobic conditions. The relation between the fate of a cell and its position within the morphogenetic field seemed evident in the case of the sea-urchin blastula. On the one hand two opposite metabolic gradients (the animal and the vegetal ones) were shown. On the other case, once more, experimenters have been deluded by preconceived ideas. We presently know that the organization of the blastula is settled by the progressive induction exercised by micromeres, and that this results in the unfolding of the vegetal gradient. The animal gradient just creates the metabolic conditions required for the determination of micromeres at the vegetal pole (references in Chandebois and Faber, 1983).

The interest taken in fields has culminated after the publication of Child's concept of metabolic gradients (1941). Nevertheless this concept raised great scepticism because one cannot imagine how the variation of O_2 consumption along a polar axis may suffice to determine the whole diversity of cells in the organism. Towards the sixties it seemed that the notions of field and gradient were falling into disuse. Unexpectedly they revived when Wolpert reformualted Child's ideas in his famous concept of positional information and reduced the mechanisms of morphogenesis to over-simplified principles. A gradient is assumed to be set up by substances called "morphogens" which are produced at one extremity (the source) and destroyed at the other (the sink) at different rates. Consequently both extremities are the "reference points" of the system: they have different "boundary values". They release "positional information" which is interpreted by the cells depending on the number of cells separating each of them from the reference points. After an amputation the boundary value of the end of the gradient which has been removed is re-established at the level of the cut. Consequently the gradient and the corresponding visible pattern are restored.

This concept raises scepticism from among those experimental embryologists who study the intricate cell interactions involved in the emergence of structural features. For example Nieuwkoop (1973) wrote "... describing biological phenomena in terms of morphogenetic fields or positional information does not necessarily have much explanatory value and may, in some cases, simply be a cloak of our ignorance". Nevertheless, this concept is a universal success which is rather surprising when one realizes the number of fundamental questions that are left unanswered. Morphogens have been imagined for the sake of the theory. Some authors claimed that they have identified such substances but we still cannot explain how cells measure their amount and respond appropriately. Morphogens cannot be produced and destroyed otherwise than by specialized cells. How are these cells determined as source or sink during development? In what respect do they differ from cells which interpret positional information? How is a reference point re-established and its boundary value specified when one extremity of the gradient has been removed? If we take into account the wealth of data provided by the rational analysis of cell interactions in the growing animal and in the adult one, we can easily discern the error made by theoreticians. Instead of simplifying the problem they have shifted our interest from the actual visible organization to intangible imaginary prepatterns.

WHAT IS A STRUCTURAL PATTERN?

The reality of a structural pattern is difficult to grasp. Moreover we hardly feel necessary to give it an unambiguous definition, so familiar is animal form to us. Nevertheless, we must attempt to fill this gap if we intend to put the problems of morphogenesis on the right track (references in Chandebois and Faber, 1983).

Embryologists, anatomists and histologists teach us that in a given part of a given animal we can distinguish a number of intermingled structural features. Some of them are unique features which confer its unique aspect to the part in question (e.g. relative positions of tissues, specific morphology of organs, ornamentation of integuments, etc). Other features are "motifs" repeated in this part of the animal (e.g. somites, precartilage condensations, etc) or found in any animal group (e.g. acini, tubules, hexagonal patterns). Any structural pattern includes invisible motifs which I call "covert graded patterns" (cGP). A certain cell property (e.g. O_2 consumption in planarians, transforming property in the archenteron roof, production of maintenance factor in limb bud mesenchyme) gradually decreases along an axis of polarity. cGPs are evidenced by molecular analysis so that they appear to be molecular gradients if the reality of cells is not taken into account.

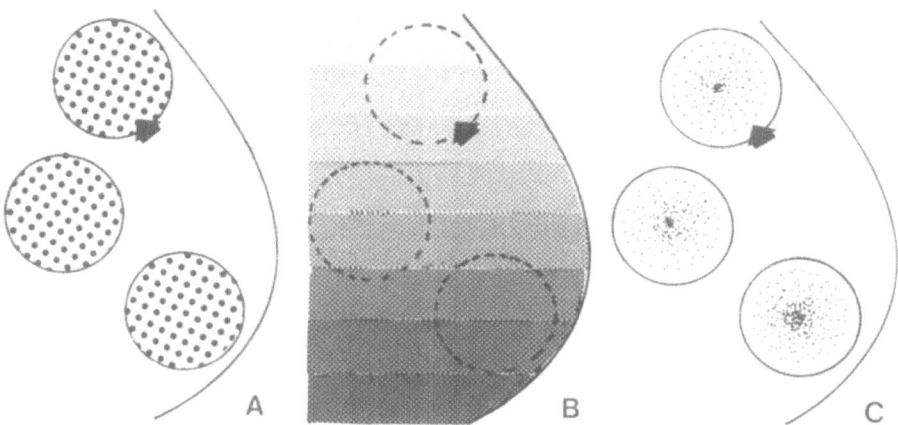

Fig. 1. The interference of motifs in the limb mesenchyme. One cell (arrow) is involved in a precartilage condensation since it synthesizes
chondroitin sulphate (A), in the antero-
posterior cGP of the bud with the value of its
quantitative metabolism (B), in the graded
pattern of the condensation with its degree of
cell adhesiveness (C).

    Each structural feature of the adult pattern emerges at a
particular stage of development. The salient features may be
viewed as indelible traces left by transient common cell activities. Cell displacement (e.g. the formation of the mesodermal layer, the closure of the neural plate) and cell rearrangement (e.g. the segmentation of the somitic mesoderm,
the formation of precartilage condensations) count among the
most important processes involved in organogenesis. Certain
features are established by differential growth at a given
stage of development and are irreversibly fixed even though
further changes may occur in mitotic rate. For example the
brain of higher vertebrates acquires its specific architecture
as a result of considerable regional differences in mitotic
rate. This architecture is retained even though the neurons
later lose their capacity to divide. Cell death is another
very importance factor in morphogenesis since it leaves marks
in the part of the animal where it occurs (e.g. the separation
of the digits in tetrapod vertebrates).

    Each cell possesses a particular "individuality" which
comprizes various traits (e.g. nature of luxury proteins,
selective membrane adhesiveness, metabolic rate). Since any
structural feature or motif is established with a certain cell
property, any single cell is involved in several features or
motifs. For this reason the latter interfere with each other.
This point is of the utmost importance in the understanding of
experimental development. The pattern of a vertebrate limb
bud can be quoted as an example (Fig. 1). Three main motifs
are observed in the rudiment: (1) the even distribution of
precartilage condensations within the mesenchyme, (2) the
antero-posterior gradient of maintenance factor in the mesenchyme which specifies the antero-posterior polarity of the

bud, (3) a graded pattern in each condensation (towards the periphery the cells are flatter). A given cell of a precartilage condensation is involved in motif (1) by its tissular identity (it synthesizes chondroitin sulphate), in motif (2) by the peculiar value of its quantitative metabolism, in motif (3) by its degree of adhesiveness.

THE SOCIAL BEHAVIOUR OF CELLS AND THE REPRODUCTION OF STRUCTURAL PATTERNS DURING DEVELOPMENT

When Wolpert formulated his concept of positional information he expressed the same opinion as Waddington did: the problems of morphogenesis and those of tissue differentiation must be worked out separately. On the contrary experimental analysis of pattern formation unambiguously shows that the progressive emergence of structural features, with all their details, proceed from the progressive setting-up of tissue-specific activities (Chandebois and Faber, 1983, 1987).

Theoreticians were constrained to consider that controls at the supra-cellular level are required for pattern unfolding and pattern maintenance because they referred to an over-simplified conception of the functionning of cells. They still forget a fact repeatedly proved: the individuality of a given cell is not entirely specified by the item of extra-cellular information it registers in the instant. Cells are endowed with cytoplasmic memory. This means that when a cell has undertaken a certain activity because it registers a certain item of extra-cellular information, a trace of this activity will persist in it and will condition the further behaviour of the cell and of its descent - namely the interpretation of the successive items of extra-cellular information which they will register subsequently. Cell individuality is not based on individual memory only. We have to do, to different degrees in different tissues, with a collective memory retained by way of continuous exchange of information which is generally effected through cell contact relations. For instance cells frequently loose their tissular identity when they are disaggregated.

When a population of embryonic cells engaged in the same differentiation pathway is isolated, it generally continues to evolve for some time - a phenomenon I called "autonomous progression of differentiation". When the progression stops the population assumes the structure of a tissue which normally emerges at a later stage of development. Since the properties of cell membranes change in the meanwhile the cells re-arrange themselves. This often leads to the emergence of motifs (e.g. condensation of precartilage cells in limb bud mesenchyme, tubules in kidney mesenchyme). The nature of the progression is fixed at the start by the cytoplasmic information. As soon as the cells possess this information the population is determined. Disaggregation reversibly stops the autonomous progression. This shows that the cells have a certain "elementary social behaviour" which is specified by the determinative event. While cells are propelled towards new tissular activities they divide at a certain rate and they re-arrange themselves spatially in a novel manner. The complete differentiation of any tissue entails numerous determinative events. Each of them requires information coming from another cell popula-

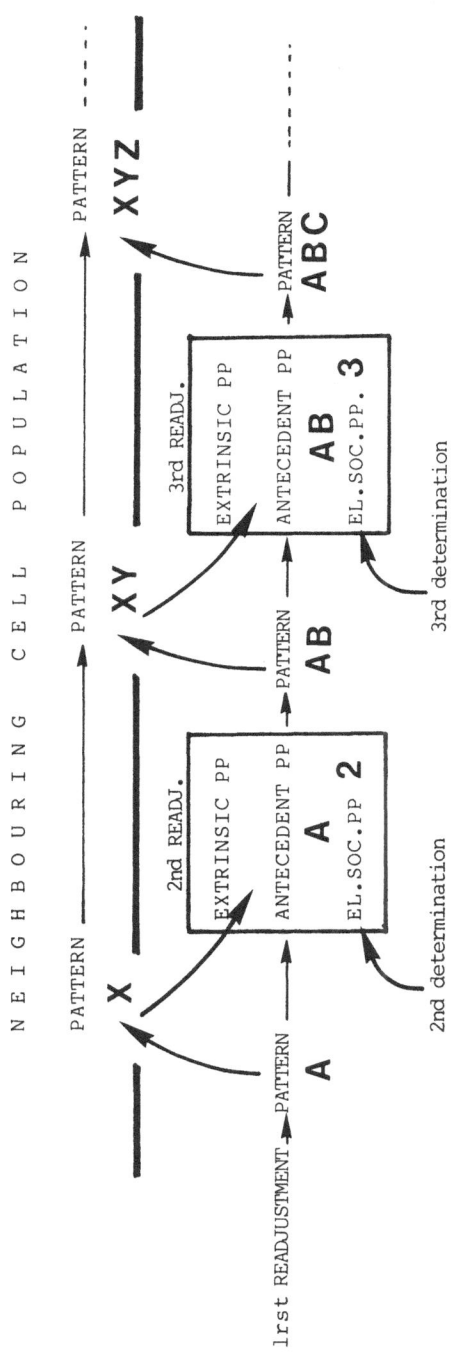

Fig. 2. Diagram showing the concatenation of readjustments in a cell population in normal development, and its effects on neighbouring cell populations. Explanation in text.

425

tion (induction, hormone) which either restarts the autonomous progression or deflects the course of the autonomous progression in which the cells are engaged and would have persisted in its absence.

Cell individualities can be almost infinitely modulated within one and the same tissue because cells engaged in the same autonomous progression can only remain identical if they register the same extra cellular information at the same time. Actually they interprete this information as a function of the properties of their own cytoplasm which are continuously modified. During an autonomous progression each cell of a population must continuously readjust its individuality to the changes occurring in its neighbours. Under the influence of its heterogeneous environment the population as a whole performs a readjustment during which cell individualities are extremely diversified. Consequently, structural features and motifs (interferring with those resulting from the elementary social behaviour of cells) emerge in the population because cells either rearrange themselves in various manners or divide at different rates - incidently because some of them undergo cytolysis.

The readjustment is an automatic process which represents the execution of a specific programme. The latter includes various components established with distinct sources of information which are experimentally dissociable. The components of the programme of a readjustment deserves the name of prepattern (even when the informing group of cells is not patterned) because each one imparts certain feature(s) or motif(s) in the ultimate pattern. One prepattern is established with the determination of the elementary social behaviour of cells (elementary social prepattern), the others with the properties, relative positions and structures of the neighbouring populations (extrinsic prepatterns).

The full differentiation of any tissue throughout development involves a concatenation of several autonomous progressions. For each readjustment the organization established during the preceding ones is kept by the collective memory of cells and serves as "antecedent prepattern". It will interfere with the new structural features and motifs which emerge. The structural amplification which results from the process in its turn leads to a greater complexity since patterns are involved in the programming of the readjustments performed by the neighbouring cell populations (Fig. 2).

Every part of the adult animal has a specific structural pattern because each readjustment involved in their formation is a unique event in the history of development. In their patterns traits and motifs may be identified. They emerged at various developmental stages, each with a peculiar prepattern established with peculiar temporary cell activities. They are maintained through permanent exchange of information between cells and tissues. It is clear that there is another reason why theoreticians still have recourse to the notion of gradient-prepattern: they still disregard autonomous progression which is the prime mover of development.

THE FORMATION OF cGPs AND THEIR ROLE IN THE UNFOLDING OF VISIBLE PATTERNS

Like any motif of another kind a cGP is created by cell interactions during a readjustment and kept by collective memory. Then it forms part of the antecedent and extrinsic prepatterns of further readjustments. The unfolding of a cGP (which will eventually interfere with other cGPs) is due to the exquisite diversification of the quantitative metabolism of cells through divisions or as a result of the interference of various processes (e.g. the propagation of homoiogenetic induction, cell aggregation) with the autonomous progression (see models in Chandebois and Faber, 1983). A cGP unfolded duirng a readjustment may serve for the creation of new visible features during the following ones. The cGPs which play the major role in morphogenesis are those which thus specify axes of polarity and levels in the organism as as whole and in its constituent organs. A cGP may also distort features which will be determined by other prepatterns. Frequently both extremities of the cGP - not every level - are sufficient for the correct visible pattern to appear. For instance, a complete and well proportionned neural axis is formed when ectoderm is neuralized by both liver (which induces prosencephalic structures) and bone marrow (which induces mesoderm) (Toivonen and Saxen, 1955).

In the adult animal regeneration cannot be normal unless every features and motifs which belong to the pattern of the eliminated part - including the cGPs which specify its axes of polarity and levels - are reformed in the regenerate. Nevertheless, the readjustments involved in regeneration are not completely identical with those which created the same pattern during embryonic development. Actually the cells which build the regenerate have already performed adult functions - and necessarily keep traces of them. Moreover, they are submitted to the influence of the differentiated cells of the stump. In consequence, the knowledge of cell interactions during development does not substitute for analyzing cell interactions during regeneration. This has been already shown in the case of the amphibian limb bud (Faber, 1971).

THE CGPs IN PLANARIAN REGENERATION

Planarians are most suitable models for studying the regulation of structural patterns. In this group regeneration is not only possible in the anterior and posterior directions (i.e. after transverse cuts) but also in the medio-lateral direction (i.e. after longitudinal cuts). An analysis of atypical regenerates which frequently develop in marine planarians has already revealed that any region of the animal is the intersection of one antero-posterior level (which belongs to the antero-posterior self-regulating system) and one medio-lateral level (which belongs to the medio-lateral system) (Chandebois, 1976). A similar proposal was made later for insect and amphibian appendages (French et al,1976). More recently (Chandebois, 1985b) it has been shown that both systems are located at the periphery of the worm. Terminal and

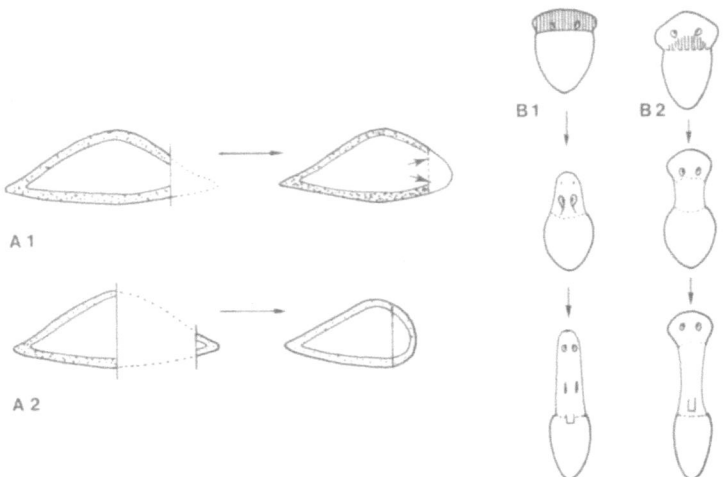

Fig. 3. Regulation of cGPs in planarians. A=Medio-
lateral cGP; Al=terminal regeneration;
A2=fusion of distant medio-lateral levels.
B=Anterio-posterior cGP. Remodelling of a
head piece cut just behind the eyes (B1) or
behind the auricular grooves (B2) joined to
the caudal extremity. Hatched=prospective
area of the intercalary regenerate. Further
explanation in text.

intercalary regeneration are possible on one face of the animal (dorsal or ventral) even when the other is uninjured
(Chandebois, 1988).

At first sight it seems that Wolpert's concept of positional information perfectly applies to planarian regeneration. Each point of the body has two positional values: one
is specified by both reference points of the antero-posterior
gradient (the extremity of the head and the one of the tail),
the other by the reference points of the medio-lateral system
(located in the sagittal plane and in the lateral edges). Nevertheless, this interpretation is at variance with current
observations. When a piece of a planarian is isolated by
transverse cuts both reference points of the antero-posterior
gradient are eliminated. Anyhow, when regeneration does not
occur, the organization of the piece remains unchanged. In
most planarian species it happens that the regenerate reforms
the extremity which is just the opposite of the one expected:
a head instead of a tail or a tail instead of a head, an
abnormality called heteromorphosis. In terms of Wolpert's
concept each sample is formed by identical opposite portions
of the antero-posterior gradient. It has two "sources" or two
"sinks" - a situation which should lead to the disappearance
of the gradient and consequently to the complete loss of antero-posterior organization. We must conclude that structural
patterns in planarians are not maintained by permanent control
at the supra-cellular level, whatever the mechanisms propounded.

If the concepts of positional information were right, the
gradient should be promptly re-established after an amputation

because of the even re-distribution of the morphogen between the two reference points - the one kept at the extremity of the stump, the other reformed at the cut level. Consequently, shortly after amputation, the whole stump should be thouroughly remodelled. Actually, the distal level (the reference point) of the eliminated part is the first to be reformed but the other levels of the regenerate are produced one by one, according to a basipetal sequence. Moreover, only the distal part of the regenerate (the blastema) is built up by dedifferentiated cells. When the blastema is developing, and under its influence, the structure of the adjacent part of the stump is gradually remodelled, a phenomenon called "morphallaxis". In the same way, when two originally widely separated body levels are joined together intercalary regeneration ensues. It is generally accomplished by morphallactic remodelling. This process concerns one of both pieces only (Fig. 3B). It starts at the margin of the suture and progressively extends through the piece which - with the exception of the head - is never entirely remodelled (Chandebois, 1984). It is obvious that the presence of differentiated tissues create a kind of vis inertiae which prevents the conversion of one level into another.

Now let us take into account the fact that an adult pattern is formed by interfering traits and motifs. Levels and axes of polarity are specified by antero-posterior and medio-lateral cGPs. In both cGPs each cell has a value which is the arithmetical mean of the vcalues of adjacent cells. This value has been irreversibly fixed when the progression of its differentiation has reached a certain critical level. There is no possible re-equilibration of a cGP after an injury unless the tissue in which it has been unfolded is repaired by undifferentiated or dedifferentiated cells. Each of them must take the value which is the arithmetical mean of the values of its neighbours. We assume that the latter is not irreversibly fixed before the critical stage of differentation.

We know that certain tissues are formed by permanent cells which are not renewed in the uninjured worm. When a part of such a tissue is eliminated its repair is possible through dedifferentiation and proliferation of cells of its own or through transdifferentiation of cells belonging to foreign tissues. We think that the medio-lateral cGP resides in a tissue of that kind, probably a peripheral muscle layer (Fig. 3A). When a blastema develops on one half (left or right) of a transverse section it forms the corresponding half of a head (or tail) since it is determined by the stump. Nevertheless, in normal conditions, the blastema promptly reacquires a complete transverse organization. This shows that the values of the medio-lateral system are not yet irreversibly fixed in the incompletely differentiated tissue and can be promptly redistributed. On the contrary, when two medio-lateral levels which are widely separated in the normal planarian are joined (e.g. the saggital plane and a level close to the lateral edge), homologous tissues fuse with each other. Tissue specific cell activities are thus maintained on either side of the suture, preventing the shift of medio-lateral levels. Consequently intercalary regeneration never ensues (Chandebois, 1985a). After longitudinal section morphallaxis never occurs. The border of the worm is entirely reformed by the epimorphic blastema. It seems that the tissue which comprizes the medio-lateral cGP does not take part in the formation of the regenerate because it is maintained in its completely differentiated state by a neighbouring tissue - in contrast with those which produce the differentiated cells of the blastema.

Other tissues are subjected to cell renewal. Each time a generative cell divides, one of the daughter cells remains as a generative cell. The other cell completes its differentiation and we assume that, meantimes, its quantitative metabolism acquires a certain value, according to the values of its neighbours. In the uninjured worm level specification is thus maintained by means of cell communication despite cell renewal. Consequently, in the experimental worm, a change in level specification is possible through cell renewal, without tissue dedifferentiation. When transverse levels belonging to cephalic and caudal region, respectively, are joined morphallactic regeneration is observed without fail. This suggests that the antero-posterior cGP resides in a tissue normally undergoing renewal - probably the subepidermal pigment layer -since the pattern of pigment is changed in the remodelled part from the start of intercalary regeneration (Fig. 3B).

In a previous work (Chandebois, 1976) I propounded a model accounting for the reformation of levels according to a basipetal sequence in epimorphic regenerates (Fig. 4A). To fix our thoughts let us suppose that a transection is made at level 20. During healing the latter reforms the distal level of the regenerate: level 1. The first cells that are made available in the stump receive information from level 1 and yield level 2 in the blastema. This level 2 will enable other cells to produce level 3 and so on. It now appears that this model is not satisfactory since, between the time when distalization is performed and the time when the complete sequence of missing levels is re-established, there should be a gap in the cGP. This is at variance with the postulate that level properties are rapidly re-established in an incomplete undifferentiated blastema. Presently, I think that every time a layer of cells enter the blastema it takes the mean value between the one of the level of the cut and the one of the last level specified, after which the cGP is immediately re-equilibrated. In the previous situation the first level would take the mean value 10. The second level, placed between 10 and 21 would take the mean value 15. Then the system would be immediately re-equilibrated so that levels 10 and 15 take the values 7.6 and 14.4 respectively. In the same manner, when the third level has been specified, the system would be re-equilibrated once more. There are now three levels 5, 10 and 15, and so forth. In short, each level produced would not take its definitive value immediately. Rather, it would be gradually promoted to more and more distal values.

I have also propounded a model for explaining why, during morphallaxis, the reformed levels which have not yet re-acquired their complete differentiation are still smaller in size (Chandebois, 1984). They are, so to speak, compressed in the remodelled part, as if a given level of the stump was converted into a certain number of other levels of reduced size. For instance (Fig. 4B) suppose that morphallaxis starts when level 12 is reformed in the blastema. According to the given conception level 21 is remodelled and yields levels 13 and 14

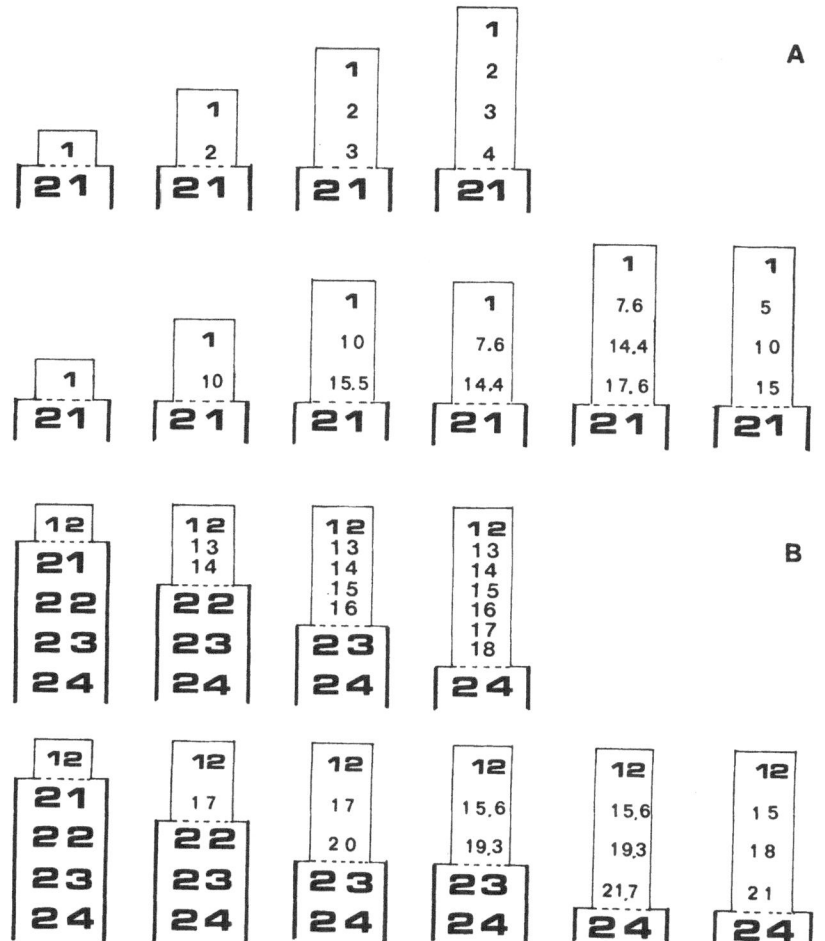

Fig. 4. Diagrams showing the reformation of levels during terminal regeneration. A=epimorphosis.
B=morphallaxis. Above=the given conception.
Below=the new conception. Further explanation
in text.

under the influence of level 12. Level 22 finds itself next
to level 14 and yields levels 15 and 16, and so on. Presently, I think that cells generated between levels 12 and 22
in the tissue forming the antero-posterior cGP take the value
17 instead of 21, those generated by level 22, the value 20.
Then the cGP is re-equilibrated and the reformed levels take
the values 15.6 and 19.3, respectively. As the remodelling
concerns more and more posterior levels, the difference
between the value of the worn-out cells and the value taken by
the cells which replace them is smaller and smaller. The
remodelling stops when the difference is reduced to zero.

Generally, during intercalary regeneration the sequence
of levels normally intervening between the two levels joined
are reformed by the partial remodelling of one of both associ-

ated pieces. This intercalary regeneration can be viewed as the manifestation of "relative dominance" (Child, 1941), a phenomenon which is still attributed to some basic property of the gradient (Wolpert, 1969). This rather appears to be the consequence of level differences in tissue properties. In fact the extent of dominance does not gradually vary along the antero-posterior axis. It may change drastically from one level to the other in the head whereas it seems approximately the same everywhere, from the anterior pharyngeal zone to the postpharyngeal zone. We can predict that the piece where the cellular turnover is faster must be remodelled under the influence of the other. This is in good agreeement with experimental data. A head cut at the level of eyes and joined to a piece from any other part of the worm is totally reshaped. On the other hand, after X-irradiation (which primarly affects mitotic cells), the head surface shrinks and completely disappears within a few day so that the eyes are shifted to the prepharyngeal zone (Chandebois, 1963). This corroborates the hypothesis that cell renewal proceeds more rapidly in the head than elsewhere in the worm. The conversion of one level into another may require the conversion of one tissue into another. For this reason, dominance may also be exerted by cells which cannot dedifferentiate (Chandebois, 1973).

The last question to be discussed is the reformation of the "reference points" of gradients - in other words, the extremities of the cGPs. In planarians, as in other animal groups (Chandebois, 1979), the meeting of the dorsal and ventral epidermis during healing is in itself a necessary and sufficient condition for the formation of a blastema. In planarians there are three different patterns of distalization, each reproducing the topographical characteristics of the extremity or the edge which has been eliminated (Chandebois, 1980). The junction between the dorsal and ventral epidermis is shifted ventrally, as in heads, in anterior sections and dorsally, as in tails, in posterior sections. In longitudinal sections the junction between the dorsal and the ventral epidermis lies at the bend between the dorsal and the ventral faces as in the lateral edges. Distalization cannot be attributed to the self-regulating properties of the cGPs. It rather seems that the new topographical relations between differentiated cells confer to undifferentiated (dedifferentiated) cells close to the section the value of the extremity of the cGP.

My previous work led me to the conclusion that any region arbitrarily defined in a planarian is characterized by a certain equilibrium between cells which depends upon the nature of neighbouring equilibriums (Chandebois, 1973). This idea is still valid and may be stated more precisely. The fundamental property of cGPs is the self-re-equilibration of metabolic rates of its consistuent cells, which is the prime mover of regeneration, i.e. the conversion of one region into another. The properties of all the tissues of the worm and all the controls which they exert on each other (which have been settled during development independently of cGPs) are of the utmost importance. They play a decisive role in the start of regeneration and determine the modalities of the reformation of visible patterns in the regenerated part.

REFERENCES

French, V., Bryant, P.J., and Bryant, S.V., 1976, Pattern regulation in epimorphic fields: (Cells make use of polar coordinate systems for assessing their positions in developing organs), <u>Science</u>, 193:969-981.

Chandebois, R., 1963, Action des rayons X sur la differenciation cellulaire de <u>Planaria subtentaculata</u>, <u>Bull. Soc. Zool. Fr.</u>, 86:632-644.

Chandebois, R., 1973, General mechanisms of regeneration as elucidated by experiments on planarians and amphibians and by a new formulation of the morphogenetic field concept, <u>Acta Biotheor.</u>, 22:2-33.

Chandebois, R., 1976, Histogenesis and morphogenesis in planarian regeneration, vol. 11, <u>Monogr. Dev. Biol.</u>, Karger, Basel.

Chandebois, R., 1979, The dynamics of wound closure and its role in the programming of planarian regeneration. I. Blastema emergence, <u>Dev. Growth Different.</u>, 21:195-204.

Chandebois, R., 1980, The dynamics of wound closure and its role in the programming of planarian regeneration. II. Distalization, <u>Dev. Growth Different.</u>, 22:693-704.

Chandebois, R., 1984, Intercalary regeneration and level interactions in the fresh-water planarian <u>Dugesia lugubris</u>, I. The anteroposterior system, <u>Wilhelm Roux's Arch., Dev. Biol.</u>, 193:149-157.

Chandebois, R., 1985a, Intercalary regeneration and level interactions in the fresh-water planarian <u>Dugesia lugubris</u>, II: Evidence for independent antero-posterior and mediolateral self-regulating systems, <u>Roux's Arch. Dev. Biol.</u>, 194:390-396.

Chandebois, R., 1985b, Intercalary regeneration and level interactions in the fresh-water planarian <u>Dugesia lugubris</u> III. The peripheral localization of the anteroposterior self-regulating system and the lack of interactions between the dorsal and ventral faces, <u>Roux's Arch. Dev. Biol.</u>, 194:397-403.

Chandebois, R., 1988, The handling of half-thickness pieces in planarians: preliminary results, <u>Monogr. Dev. Biol.</u>, vol. 21:235-241.

Chandebois, R., and Faber, J., 1983, Automation in animal development, vol. 16, <u>Monogr. Dev. Biol.</u>, Karger, Basel.

Chandebois, R., and Faber J., 1987, From DNA transcription to visible structure: what the development of mutlicellular animals teaches us, <u>Acta Biotheor.</u>, 36:61-119.

Child, C.M., 1941, "Patterns and Problems of Development", University Press, Chicago.

Faber, J., 1971, Vertebrate limb ontogeny and limb regeneration: morphogenetic parallels, <u>Adv. Morphog.</u>, 9:127-147.

Guyenot, E., 1927, Le probleme morphogenetique dans le regeneration des urodeles, <u>Revue Suisse Zool.</u>, 34:127-154.

Nieuwkoop, P.D., 1973, The organization center of the amphibian embryo: its origin, spatial organization, and morphogenetic action, <u>Adv. Morphog.</u>, 10:1-39.

Nieuwkoop, P.D., Johnen, A.G., and Albers, B., 1985, The epigenetic nature of early chordate development, <u>Development and Cell Biol. Monogr.</u>, vol. 16, Cambridge University Press, Cambridge.

Toivonen, S., and Saxen, L., 1955, The simultaneous inducing
        action of liver and bone-marrow of the guinea-pig in
        implantation and transplantation experiments with
        embryos of <u>Triturus</u>, <u>Expl. Cell. Res</u>., 3 (Suppl.):
        346-357.
Waddington, C.H., 1956, "Principles of Embryology". Allen
        and Unwin, London.
Weiss, P., 1939, "Principles of Development", Holt, New York.
Wolpert, L., 1969, Positional information and the spatial pat-
        tern of cellular differentiation, <u>J. Theor. Biol</u>.,
        25:1-47.
Wolpert, L., 1971, Positional information and pattern forma-
        tion, <u>Curr. Top. Dev. Biol.</u>, 6:183-224.

PATTERN REGENERATION IN AN INSECT SEGMENT

Katharina Nübler-Jung

Biologisches Institut I (Zoologie)
Alberstrasse 21
D-7800 Freiburg, F.R. Germany

SUMMARY

The integument of an insect segment displays a sequence of differentiated cells and a polarity pattern of uniformly oriented cuticular protrusions. It has been proposed that both pattern features are controlled by the concentration gradient of some diffusible morphogen. I report that an abnormal cell sequence is repaired by cell sorting and/or by intercalary regeneration of missing pattern elements. Patterning in the intercalary regenerate need not,though may, rely on a graded morphogen concentration. A disturbed polarity pattern, on the other hand, repairs itself by cell interactions that operate independently of a gradient.

INTRODUCTION

In a complex organism the cells are arranged in specific ways to form characteristic patterns. Upon disturbance such patterns are often repaired. The repair of a multicellular pattern as well as its generation during embryogenesis seem to call for some supracellular cue that informs individual cells where they are and how they should behave (e.g. Wolpert, 1971; Harrison and Tan, 1988). We want to understand how the pattern in an insect segment is being maintained during normal growth and how it is repaired after a disturbance.

The insect integument has long been a model system for the analysis of pattern repair mechanisms (e.g. Wigglesworth, 1959). Earlier authors concluded that in each body segment or leg segment a supracellular graded cue controls the spatial sequence of differentiated cells (Bohn, 1965; Locke, 1967). Later it has been proposed that the concentration gradient of some diffusible "morphogen" not only controls the sequence of differentiated cells but also provides orientation, in single cells as well as in the segment as a whole (Lawrence, 1987; Lawrence et al., 1972; Meinhardt and Gierer, 1980).

Here I report that different features of the pattern in an insect segment are repaired by different mechanisms. An abnormal cell sequence is repaired by cell sorting and/or by local cell divisions (intercalary regeneration). It will be discussed that patterning in the intercalary regenerate may rely on a graded morphogen concentration but could also result, like cell sorting, from local cell interactions.

The posterior orientation of cuticular denticles may also result from local cell interactions. Yet, while the sequence of differentiated cells always conforms to the gradient model, various abnormal but characteristic polarity patterns are not compatible with any gradient model. A supracellular graded cue may thus not be necessary to control the polarity pattern in an insect segment.

MATERIALS AND METHODS

Cotton bugs (<u>Dysdercus intermedius</u> Dist, Heteroptera) were reared as described in Nubler-Jung (1977). The experiments were done on 3rd or 5th instar larvae in the ventral part of the second or fifth abdominal segment. The results were recorded after subsequent moults. For further details see Nübler-Jung (1987a,b) and Nübler-Jung et al.,(1987).

RESULTS

The insect integument is subdivided into segments. It consists of a monolayer of epidermal cells and the overlying cuticle. A new cuticle is secreted before each moult.

Each abdominal segment of the cotton bug <u>Dysdercus</u> displays two distinct pattern features. Position dependent pigmentation constitutes a spatial sequence of distinctly differentiated cells (<u>differentiation pattern</u>) which can be seen through the transparent cuticle: anterior cells in a segment synthesize red pigment, posterior cells synthesize white pigment. In a colour mutant (Hollweg, 1972) the anterior cells lack the red pigment and appear grey instead.

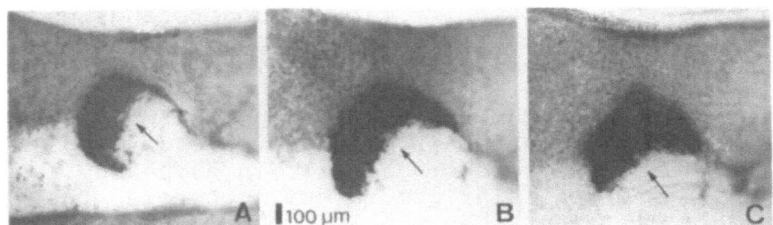

Fig. 1. Wild type graft in colour mutant rotated through 90° during 3rd larval instar.
(A) early 4th; (B) early fifth; (C) late 5th instar. The graft successively derotates into its original orientation when epidermis is detached from the old cuticle before a moult. Arrow: local peculiarities persist at red/white graft border.

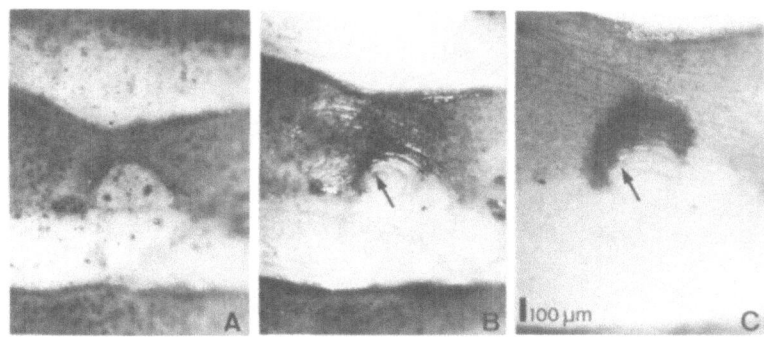

Fig. 2.   White graft from wild type posterior segment
           region transposed anteriorly into colour
           mutant during 3rd larval instar. (A) mid-4th;
           (B) early 5th; (C) late 5th instar. A red
           strip is intercalated at anterior host/graft
           border. Note folds in intercalary regenerate
           before and after moults, and persistence of
           local peculiarities at red/white graft border
           (arrow).

     The polarity pattern is seen in the adult cuticle: in
some regions (e.g. ventral part of 5th abdominal segment)
each cell forms a polarized cuticular structure, either a
bristle or a denticle. Their uniform posterior orientation
indicates a supracellular topological polarity.

Repair of the differentiation pattern

     Earlier grafting experiments indicate that in an insect
segment cell adhesiveness declines from anterior to posterior
(Nübler-Jung, 1977).   Cells from different anteroposterior
segment levels apposed by grafting tend to minimize contact by
sorting out according to their origin. For example, a graft
rotated by 90° re-rotates so that cells of similar origin
regain contact (Fig. 1, Bohn, 1974; Lawrence, 1974; Nübler-
Jung, 1974).

     When discrepancies in cell adhesiveness cannot be elimi-
nated by cell sorting then the positional values missing
between graft and host tissue are inserted by "intercalary
regeneration".   For example, a graft transposed from the posterior into the anterior segment region produces an intercalary
regenerate that grows more than segment or graft tissue (Fig.
2).   Intercalary growth and cell sorting may both result from
divergent cell adhesions along the host/graft border (Nübler-
Jung,1977; Mittenthal and Mazo, 1983) and therefore may occur
independently of some diffusible morphogen.

     Divergent cell adhesion, however, does not explain why
and how the intercalated cells form the missing pattern elements.   Possibly, some diffusing morphogen conveys positional
information to the intercalated cells (Lawrence, 1987; Harrison and Tan, 1988). It is another possibility, however, that
cells from divergent anteroposterior segment levels interact
locally and in a specific way so that, concomitant with cell

437

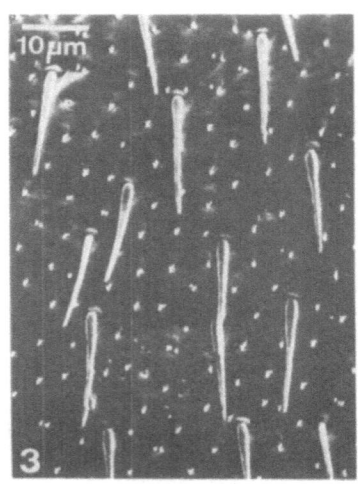

Fig. 3. Adult cuticle from 5th segment. Denticles (small structures) and bristles (large structures) point posteriorly. Scanning EM.

divisions, they produce the missing cell states. This will be discussed below.

Repair of the polarity pattern

In addition to the apico-basal polarity which characterises all epithelial cells, the insect epidermis expresses an asymmetry in the plane of the cell sheet. In <u>Dysdercus</u> this

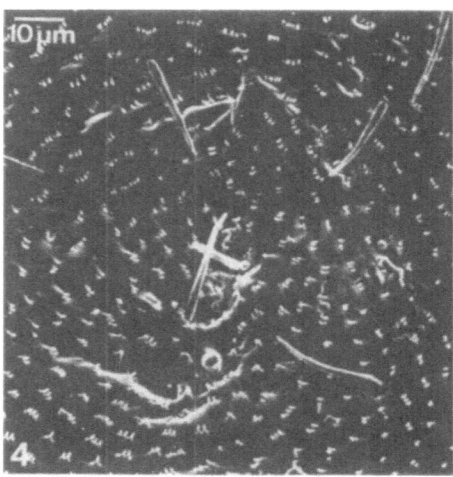

Fig. 4. Adult cuticle from animal where epidermal cells were killed in a circular spot during 5th instar. The wound was closed by inward migration of epidermal cells. Denticles and bristles tend to point towards former wound centre. Scanning EM; top = anterior.

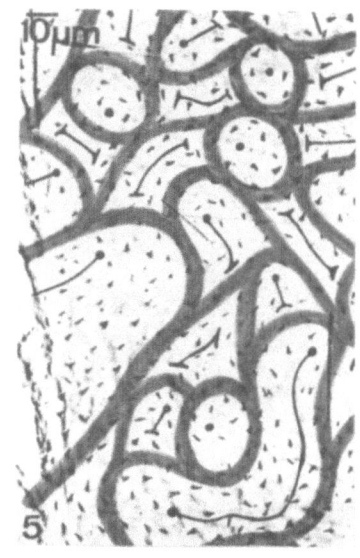

Fig. 5. Denticle pattern in an adult injected with colchicine 6 days before imaginal moult. Broad
bands that trace uniform denticle orientations
form essentially two pattern elements a loop
and a π (see Fig. 6A and 6B). Thin lines separate denticles with tangentially opposing
orientations. Phase contrast; top = anterior.

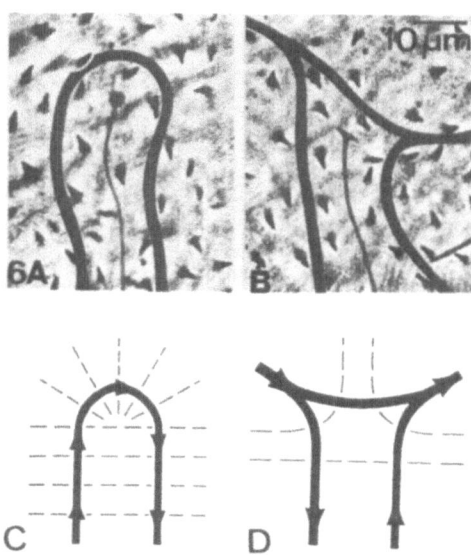

Fig. 6. Pattern elements in Dysdercus (A,B) can be formally derived from pattern elements formed by
confluent fibroblast cultures (thin lines in C
and D), by orthogonal transformation and introduction of polarity (broad bands with arrows in
C and D).

439

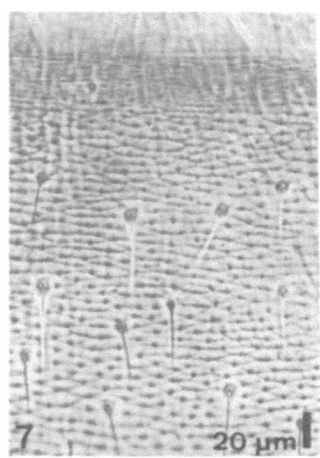

Fig. 7.   Cuticle from 5th instar larva near anterior
           boundary of 2nd segment with highly elongated
           cells.   Ridges delineate cell shape during
           cuticle secretion.   Denticles point to the
           posterior of animal. Phase contrast.

"planar" polarity is expressed in a uniform posterior orientation of cuticular bristles and denticles (Fig. 3).

    The differentiation pattern in an insect segment is main-
tained by an axial decline in cell adhesion (see above). However, a decline in cell adhesion does not easily explain how
planar polarity is established.   We therefore started to

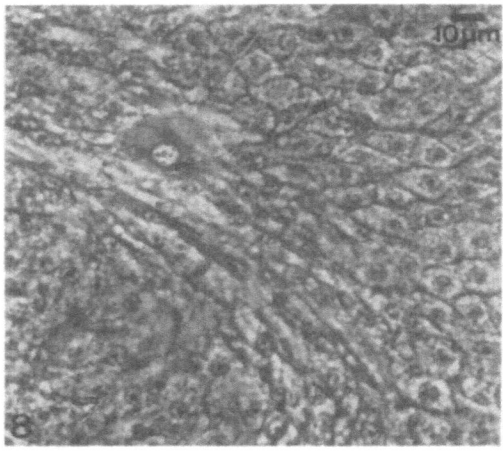

Fig. 8.   Elongated epidermal cells around upper right
           host/graft border during 5th instar.   Graft was
           rotated during 4th instar by 180°. Phase contrast
           of depigmented cells.

440

analyse the cellular basis of cell polarity and of supracellular tissue polarity with the following results:

(1) Colchicine can abolish the uniform posterior orientation of denticles (Nübler-Jung, 1978a). This suggests that individual cells in the sheet can establish planar polarity without reference to some covert supracellular cue. Moreover, this result indicates that the cytoskeleton could be one of the components in the intercellular communication system that orients individual cells.

(2) Migrating epidermal cells tend to orient denticles in the direction of cell movement (Fig. 4). This indicates that cell migration and denticle formation may share similar orienting mechanisms in a cell (Nübler-Jung et al., 1987).

We have now found that the centrosome in the epidermal cells of <u>Dysdercus</u> is located preferentially in the apico-posterior cell region <u>only</u> while the denticle is being formed (in prep.). This indicates that the centrosome, too, may be one of the components that determine the orientation of a nascent cuticular denticle.

(3) The abnormally oriented denticles seen after colchicine treatment tend to form small arrays with uniform orientation (Fig. 5). Adjacent arrays with divergent orientations realize a pattern composed of only two pattern elements (Fig. 6A,B) which are not compatible with a gradient model. Yet, we obtain these abnormal patterns (but also the normal uniform polarity pattern) from patterns that form spontaneously fibroblast cultures (Elsdale and Wasoff, 1976): After orthogonal transformation of the fibroblast pattern and introduction of planar cell polarity the fibroblast model would produce the denticle patterns (Fig. 6C,D).

In the insect epidermis tissues apposed with divergent polarity tend to adopt a uniform orientation (e.g. Piepho, 1955; Lawrence, 1974; Nübler-Jung and Grau, 1987). Therefore, abnormal polarity patterns may simplify, like fibroblast patterns, by mutual elimination of unlike pattern elements to form the final uniform pattern, with posteriorly oriented denticles (Nübler-Jung, 1987b). The cellular basis for this seems to be that during cuticle secretion insect epidermal cells, like fibroblasts, elongate along each other and along the parallel segment boundaries (Fig. 7). The cellular basis of orthogonal transformation of the fibroblast patterns into the denticle patterns seems to be that insect epidermal cells tend to orient their denticles perpendicular to their long axes and into the same direction.

(4) Lately we observed that in grafts transposed along the anteroposterior segment axis (i.e. up or down the putative segmental gradient) the denticles invariably point outwards (in prep.). Yet, the gradient model would predict inward orientation for anteriorly grafts.

The persistent outward orientation is consistent with the assumption that epidermal cells elongate parallel to a discontinuity in cell adhesion and therefore orient their denticles perpendicular to the host/graft border.

Fig. 8 shows that indeed epidermal cells elongate along the host/graft boundary when cells from divergent antero-posterior segment levels meet (see also Campbell, 1987). Control grafts replaced to their normal segment level show normal denticle orientation and normal cell shape. We are now testing whether cell shape can directly determine the axis of planar cell polarity in the insect epidermis.

DISCUSSION

The epidermis in an abdominal segment of Dysdercus displays two pattern features which are based on different properties of the constituent epidermal cells: The position-dependent cell pigmentation (differentiation pattern) superimposed on which is the polarity pattern of cuticular denticles which uniformly point to the posterior.

Upon disturbance this complex pattern may repair itself (i.e. undergo regeneration in a broader sense). Depending on the kind of disturbance, different repair mechanisms may become active. We distinguish between restoration of a pattern (i.e. rearrangement of disarranged pattern elements) and pattern regeneration s.str. (i.e. replacement or insertion of missing pattern elements).

Pattern restoration

Pattern restoration after 90°-rotation of a graft can be explained without a diffusible morphogen. For example, cell sorting during graft de-rotation (Fig. 1) seems to be elicited by a divergence in cell adhesion, i.e. by a position-dependent cell surface property which normally declines from anterior to posterior within a segment (Nübler-Jung, 1977). Under normal conditions the cells do not migrate. The axial decline in cell adhesiveness thus seems to maintain the normal sequence of differentiated cells by keeping epidermal cells where they are.

A similar decline in cell adhesion was observed along the proximodistal axis in an insect wing anlage (Nardi and Kafatos, 1976) and in the regeneration blastemata of an amphibian leg (e.g. Crawford and Stocum, 1988). Position dependent cell adhesion may thus be one cellular basis for the formally similar regeneration behaviour of the insect epidermis and of the amphibian leg (e.g. French et al., 1976).

Pattern restoration also occurs after disturbance of the uniform polarity pattern (Nübler-Jung, 1987b). It has been proposed that insect epidermal cells orient along a graded morphogen concentration (Lawrence et al., 1972; Meinhardt and Gierer, 1980). However, the abnormal but characteristic polarity patterns that occur after colchicine treatment (Figs. 5 and 6) or after transposition of cells along the putative gradient axis (in prep.) are not compatible with a gradient model. Rather, these abnormal, but also the normal uniform polarity patterns, are related to the patterns that form spontaneously in confluent fibroblast cultures (Elsdale and Wasoff, 1976). This indicates that uniform cell orientation in an insect segment is controlled independently of a gradient and may rely, like the supracellular pattern in fibroblasts, on local cell interactions.

We have observed that epidermal cells orient their denticles into the direction of cell migration, that colchicine disturbs the uniform orientation of denticles, and that during denticle formation the centrosome is located in the apico-posterior cell region. I, therefore, propose that, analogous to migrating cells in culture (e.g. Snyderman and Goetzel, 1981; Shields and Haston, 1985), planar polarity of insect epidermal cells correlate with an asymmetric distribution of intra-cellular components (e.g. cytoskeleton, centrosome) and with an asymmetric distribution of cell surface molecules. The mediolateral cell elongation during cuticle secretion (Fig. 7) thus may reflect an increased membrane contact between cells about to transmit orientation in anteroposterior direction (Sander and Nübler-Jung, 1981).

Orientation along the anteroposterior segment axis may arise because epidermal cells tend to elongate along the segment border, possibly because here cells with divergent adhesiveness meet. The specific posterior orientation may be determined by the opposite orienting properties of the anterior and of the posterior intersegmental regions (Nübler-Jung,1979).

We cannot exclude that the orienting properties of the intersegmental region are based on freely diffusing molecules. However, since uniform orientation <u>within</u> the segment seems to be achieved independently of a gradient, a model where the intersegmental membrane, too, transmits orientation via cell bound structures still appears attractive (Sander and Nübler-Jung, 1981).

Pattern regeneration (s.str.).

We have seen above that a disturbed cell sequence can restore itself by cell sorting and/or by intercalary regeneration. The local cell divisions for intercalary regeneration may be elicited by divergent cell adhesions (Nübler-Jung, 1977; Mittenthal and Mazo, 1983). It remains unexplained, however, how the intercalated cells are instructed to differentiate the pattern elements missing along the host/graft interface.

During segmentation in the <u>Drosophila</u> embryo different genes are becoming active in different (partly overlapping) stripes within a future segment. For example, the gene <u>engrailed</u> is active in the posterior compartment of each segment (Kornberg et al., 1985). On the other hand, we have seen that in the body segment of <u>Dysdercus</u> cell adhesion declines from anterior to posterior. This may indicate a continuum of cell states.

Unfortunately, we are still ignorant of whether cell adhesion is graded cell by cell and thus may be a consequence of some continuous prepattern such as a graded morphogen concentration between boundaries of specific stripes of gene expression (Lawrence, 1987). Alternatively, the consecutive stripes of discrete and qualitatively different cell states within a segment (Martinez-Arias et al.,1988) could constitute a series of discrete and (from anterior to posterior) successively <u>lower</u> cell adhesions. The segment boundary may be but one of these boundaries, where, however, cell adhesion <u>increases</u> (from anterior to posterior) from a very low to a very high level (Nübler-Jung.1979).

At present we cannot decide between these alternatives. The observation that the intersegmental region can exert <u>long range effects</u> on the differentiation state of a cell and on its polarity might indicate diffusion of some morphogen. For example, in <u>Dysdercus</u> anterior (red) cells transposed near the posterior segment border begin to produce white pigment many cell diameters away from the host/graft junction (Nubler-Jung, 1979; see also Marcus, 1962). In analogy, it has been proposed that in <u>Drosophila</u> the product of the <u>wingless</u> gene mediates some of the effects traditionally ascribed to gradients (Baker, 1987; 1988; Cabrera et al., 1987). Yet, the more local <u>intercalary regeneration</u> in an insect larva seems to favour local interactions between abnormally apposed cells. Interestingly, in early <u>Drosophila</u> embryos with a genetic deficiency for one of the four stripes, the abnormally apposed cells intercalate (concomitant with a single cell division) the third stripe that separates these cells. This indicates that in an insect segment positional information may be encoded not in a diffusible morphogen but in a set of discrete cell states (Martinez-Arias et al., 1988).

<u>Pattern control versus pattern generation</u>

We have seen that the differentiation pattern in an insect segment is controlled by cell adhesion that normally declines from anterior to posterior. An abnormal cell sequence is repaired (by cell sorting and/or by intercalary regeneration) so as to apparently conform to the gradient hypothesis.

Planar polarity of adjacent cells, on the other hand, can be oriented independently of the prevailing cell adhesion conditions (Nübler-Jung, 1987b; Nübler Jung and Grau, 1987), possibly by local cell interactions via an asymmetric distribution of cell bound structures (Sander and Nübler-Jung, 1981). Yet, this mechanism often does not predict the orientation in an intercalary regenerate, while the gradient model does (Bohn, 1965; French et al., 1976; Lawrence et al., 1972). This indicates that during <u>de novo</u> generation of the pattern (during embryogenesis or during intercalation) the differentiation pattern and the polarity pattern are being laid down in a coordinate manner, but that later these same pattern features may be controlled (or repaired) by different and independent mechanisms.

ACKNOWLDEGEMENTS

I thank Prof. Dr. K. Sander for comments on the manuscript.

REFERENCES

Baker, N.B., 1987, Molecular cloning of sequences from wing-less a segment polarity gene in Drosophila: the spatial distribution of a transcript in embryos, EMBO J., 6:1765-1773.

Baker, N.B., 1988, Embryonic and imaginal requirements for wingless, a segment polarity gene in Drosophila, Devl. Biol., 125:96-108.

Bohn, H., 1965, Analyse der Regenerationsfähigkeit der Insektenextremität durch Amputations - und Transplanta-tions- versuche an Larven der afrikanischen Schabe (Leucophaea maderae Fabr.) II. Mitteilung. Achsendet-ermination, Wilh. Roux' Arch. EntwMech. Org., 146:449-503.

Bohn, H., 1974, Pattern reconstitution in abdominal segment of Leucophaea maderae (Blattaria), Nature, 248:608.

Cabrera, C.V., Alonso, M.C., Johnston, P., Phillips, R.G. and Lawrence, P.A., 1987, Phenocopies induced with anti-sense RNA identify the wingless gene, Gene, 50:659-663.

Campbell, G.L., 1987, Cell behaviour during postembryonic pat-tern regulation in the insect abdomen (Oncopeltus fas-ciatus). II. Intrasegmental regulation, Development, 101:237-246.

Crawford, K. and Stocum, D.L., 1988, Retinoic acid coordi-nately proximalizes regenerate pattern and blastema differential affinity in axolotl limbs, Development, 102:687-698.

Elsdale, T. and Wasoff, F., 1976, Fibroblast cultures and der-matoglyphics: the topology of two planar patterns. Wilh. Roux' Arch. Dev. Biol., 180:121-147.

French, V., Bryant, P.J., and Bryant, S.V., 1976, Pattern regulation in epimorphic fields. Science, 193:969-981.

Harrison, L.G., and Tan, K.Y., 1988, Where may reaction-diffusion mechanisms be operating in metameric pattern-ing of Drosophila embryos? Bioessays, 8:118-124.

Hollweg, G., 1972, Eine neue Farbmuster-Mutante "white" der roten Baumwollwanze Dysdercus intermedius Dist. (Heter-optera, Pyrrhocoridae), Biol. Zentralbl., 91:545-556.

Kornberg, T., Siden I., O' Farrell, P. and Simon, M., 1985, The engrailed locus of Drosophila: in situ localization of transcripts reveals compartment specific expression. Cell, 40:45-63.

Lawrence, P.A., 1974, Cell movement during pattern regulation in Oncopeltus. Nature, 248:609-610.

Lawrence, P.A., 1987, Pair-rule genes: do they paint stripes or draw lines? Cell, 51:879-880.

Lawrence, P.A., Crick, F.H.C. and Munro, M., 1971, A gradient of positional information in an insect, Rhodnius, J. Cell Sci., 11:815-853.

Locke, M., 1967, The development of patterns in the integument of insects, Adv. Morphogen., 6:33-88.

Martinez-Arias, A., Baker, N.E. and Ingham, P.W., 1988, Role of segment polarity genes in the definition and mainte-nance of cell states in the Drosophila embryo, Develop-ment, 103:157-170.

Meinhardt, H. and Gierer, A., 1980, Generation and regener-ation of sequence of structures during morphogenesis, J. theor. Biol., 85:429-450.

Mittenthal, J.E. and Mazo, R.M., 1983, A model for shape generation by strain and cell-cell adhesion in the epithelium of an arthropod leg segment, J.Theor. Biol., 100:443-483.

Nardi, J.B. and Kafatos, F.C., 1976, Polarity and gradients in lepidopteran wing epidermis: II. The differential adhesiveness model: gradient of a non-diffusible cell surface parameter, J. Embryol. Exp. Morphol., 36:489-512.

Nübler-Jung, K., 1974, Cell migration during pattern reconstitution in the insect segment (Dysdercus intermedius Dist., Heteroptera), Nature, 248:610-611.

Nübler-Jung, K., 1977, Pattern stability in the insect segment. I. Pattern reconstitution by intercalary regeneration and cell sorting in Dusdercus intermedius Dist., Wilh. Roux's Arch., 183:17-40.

Nübler-Jung, K., 1979, Pattern stability in the insect segment: II. The intersegmental region. Wilh. Roux's Arch., 186:211-233.

Nübler-Jung, K., 1987a, Insect epidermis: disturbance of supracellular tissue polarity does not prevent the expression of cell polarity, Roux's Arch. Dev. Biol., 196:286-289.

Nübler-Jung, K., 1987b, Tissue polarity in an insect segment: denticle patterns resemble spontaneously forming fibroblast patterns, Development, 100:171-177.

Nübler-Jung, K., Bonitz R. and Sonnenschein M., 1987, Cell polarity during wound healing in an insect epidermis, Development, 100:163-170.

Nübler-Jung, K. and Grau, V., 1987, Pattern control in insect segments: superimposed features of the pattern may be subject to different control mechanisms. Roux's Arch. Dev. Biol., 196:290-294.

Piepho, H., 1955, Über die Ausrichtung der Schuppenbälge und Schuppen auf dem Schmetterlingsrumpf, Biol. Zbl., 42:22.

Sander, K. and Nübler-Jung, K., 1981, Polarity and gradients in insect development, in: "International Cell Biology 1980-1981", H.G. Schweiger ed., Springer Verlag, Berlin, pp. 497-506.

Shields, J.M. and Haston, W.S., 1985, Behaviour of neutrophil leucocytes in uniform concentrations of chemotactic factors: contraction waves, cell polarity and persistence, J. Cell Sci., 74:75-93.

Snyderman, R. and Goetzel, E.J., 1981, Molecular and cellular mechanisms of leukocyte chemotaxis, Science, 213:830-837.

Wolpert, L., 1971, Positional information and pattern formation, Curr. Top. devl. Biol., 6:183-224.

REGENERATION AND MORPHOGENESIS IN THE FEATHER STAR ARM

M. Daniela Candia-Carnevali, Ludovica Bruno
Suzanne Denis Donini and Giulio Melone

Dipartimento di Biologia, Universita di Milano
Via Celoria 26, 20133 Milano, Italy

SUMMARY

The regeneration of the arm of the crinoid <u>Antedon</u> <u>mediterranea</u> has been studied by means of LM, TEM, and SEM techniques as regards the development of both the external form and the internal anatomical organization. According to the timing and the modalities of development of the different structures and their reciprocal interactions, three main stages can be recognized in the morphogenesis of the whole arm: 1) a coelom-dependent stage, 2) a nervous system-dependent stage and 3) a skeleton-dependent stage. Moreover, using histofluorescence and immunohistochemical methods to visualize catecholamines and indolamines, it has been possible to reveal the presence of dopamine and serotonin and follow the fluctuations and the different involvements of these neuro-trasmitters during the regeneration process.

INTRODUCTION

Crinoid echinoderms are endowed with remarkable regenerative abilities. It is well known that they readily cast off arms, when accidentally injured or under unfavourable environmental conditions, then quickly regenerate all of them. The arms are characterized by a surprising adaptability that minimizes the consequences of possible amputations promoting a quick repair of the injury and the regeneration of the lacking part. Thanks to the ease and the speed at which they can replace one or several arms, either spontaneously or traumatically cut off, it is possible to obtain and follow experimentally all the steps of arm development.

The regenerative phenomena in crinoids have not been extensively investigated in the past (Minckert, 1985; Perrier, 1873; Przibram, 1901; Reichensperger, 1912) as well as recently (Holland and Greimmer, 1981; Mladenov, 1983), in respect of their power of regeneration and a complete analysis of the regenerative process is up to date lacking. In particular many fundamental questions need an answer, first of all the problem of the origin of the regenerative elements and that of the structures and/or the agents involved in morphogenesis of the highly organized structural pattern of the arm.

The microscopic and submicroscopic anatomy of the arm is remarkably complex (Candia-Carnevali and Saita, 1985 a,b) and is strictly related to the versatility of the movements performed, as well as to the evolutionary history of the animal group. In order to interpret this complexity we have examined the regenerative development of the whole arm in the comatulid <u>Antedon</u> with particular emphasis on the roles and the reciprocal relationships of the different elements involved in the movement of the arm. These are the <u>endoskeleton</u>, consisting of articulated brachial ossicles, the <u>hydroskeleton</u>, consisting of a multiple system of coelomic cavities, the <u>muscles</u>, consisting of two ventral flexor bundles, the <u>ligaments</u>, consisting of one dorsal and two lateral bundles, and the <u>nervous system</u>, mainly consisting of two, a ectoneural, mostly sensitive, component (subepithelial plexus) and an entoneural, mostly motor, component (central nervous trunk) (Cobb, 1986; Hyman, 1955).

Series of regenerating arms have been carried out in parallel and studied particularly for what concerns: 1) <u>development of form</u>, by employing light microscopy (LM) and scanning electron microscopy (SEM) techniques, in order to recognize the characteristic features of the growth and the main developing stages of the arm; 2) <u>development of an organized pattern</u> and 3) <u>cell differentiation</u>, by employing LM and TEM (transmission electron microscopy) techniques on serial sections from different regenerative stages, in order to understand the timing and the modalities of development of the different structures and their reciprocal interactions. In particular, suitable methods of histofluorescence and immunohistochemistry to visualize catecholamines and indolamines have been used to reveal the presence of endogenous amines and to check the possible control exerted by the nervous system on the mechanism of regeneration.

MATERIALS AND METHODS

The regeneration process was analyzed in many regenerating arms of different ages, which developed in superimposable conditions. The amputations were always carried out reproducing the spontaneous autotomies, i.e. in the distal-intermediate arm regions, in correspondence with rigid articulations (syzygies). Progressively growing regenerating arms were dissected every day during 4 weeks and prepared according to different methods.

Light Microscopy and Transmission Electron Microscopy

The regenerates were prefixed with 4% paraformaldehyde-25% glutaraldehyde mixture in 0.2 M cacodylate buffer (5 hours), then, after an overnight washing in the same buffer, postfixed with 1% osmic acid in 0.2 M cacodylate buffer. After a normal dehydration in the ethanol series, the samples were embedded in Epon-Araldite 812. The semithin and thin sections, cut with LKB ultramicrotomes IV and V and stained with standard methods, were observed under Leitz Dialux 20 light microscope and Jeol 100 SX electron microscope respectively.

Scanning Electron Microscopy

The samples fixed according to the previous procedures, were dehydrated in acetone, critical point dried in liquid CO_2 with a CPD 020 Balzers, then gold coated by means of a Sputtering coater (Nanotec) and observed under a Cambridge Stereoscan 250 MK2 electron microscope.

Monoamine Histofluorescence

Following the Faglu-Peg method (De Biasi et al., 1984; Scholer and Armstrong, 1982) the samples were fixed in a 4% paraformaldehyde, 0.5% glutaraldehyde mixture in 0.1 M phosphate buffer (pH 7.0) for 2 hours at 4°C, and then embedded in molten Peg (m.w.1500), after previous passages in 30%, 50%, 70%, 90% solutions of molten Peg dissolved in the same fixative. 10 μm sections, cut on a sliding microtome, after floating on warm fixative, were mounted on slides with paraffin oil and observed under a Zeiss III photomicroscope equipped with a HB 50 mercury lamp and a suitable set of filters: BP 405/8, FT 420, LP 418.

Immunohistochemistry

The samples were fixed in a 4% paraformaldehyde; 0.1% glutaraldehyde mixture in 0.1 M phosphate buffer (pH 7.2) for 2 hours, and then washed, dehydrated and embedded in Epon-Araldite 812 according to standard procedures. Serotonin was visualized in 1.5 μm sections (Lauder et al., 1988) by means of overnight incubation with antiserotonin antibody (Technogenetics) in dilution 1:1000. The sections were then washed thoroughly in PBS, incubated with the PAP complex and revealed with diaminobenzidine. Dopamine containing cells were visualized with an antibody against tyrosine hydroxylase (Eugene Tech) in dilution 1:60 following the same procedure.

RESULTS AND DISCUSSION

In order to give an overall idea of the morphogenesis of the arm, the different results obtained must be examined and discussed together in a very schematic form. The development of the whole structure, considered both in its functional external morphology and its anatomical internal organization, follows a typical apical growth (at a rate of about 5 pinnules per week) and is characterized by three main stages.

Stage I (up to 1 week after amputation)

This stage corresponds to a regenerative bud growing out from the stump and showing roughcast lateral prominences (pinnules) and ventral groove (Fig. 1a). The anatomical pattern is very simplified. The blastema mainly consists of a bulk of mesenchyme, where the coelom is the first element to differentiate and represents the pilot-structure (Figs. 1b, c). Its role seems to be fundamental in giving rise to or transporting pluripotential wandering coelomocytes at the lesion sites and in acting as a primary organizer. In fact, in parallel with the progressive formation and subdivision of the coelomic cavities, the arm assumes a more defined architecture, particularly for what concerns its feeding system, represented by a

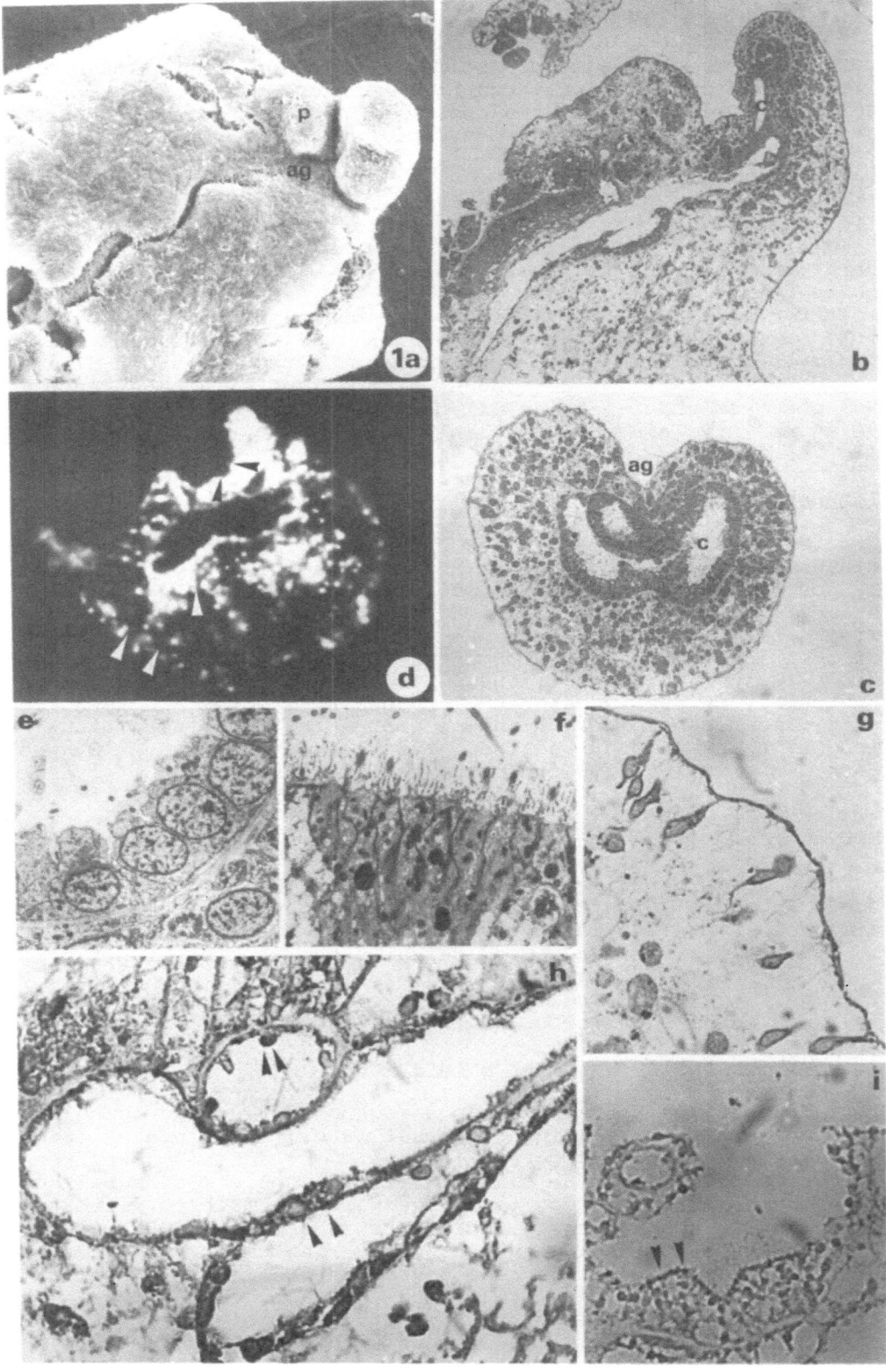

Fig. 1. Different aspects of the <u>coelom-dependent</u> stage.

a) 5th day of regeneration. SEM, ventral view of the regenerating bud. ag=ambulacral groove; p=pinnula (x120),

b) 6th day of regeneration. LM, longitudinal section of the bud showing the anatomical organization of both the new and the old arm; m=mesenchyme; c=coelom (x100).

c) 6th day of regeneration. The LM cross section shows the proliferation and splitting of the coelom (c) in its different compartments. The ambulacral groove (ag) is well outlined, (x200),

d) 5thday of regeneration. Monoamine histofluorescence. Cross section showing an intense fluorescence in the regenerating arm. The specific (blue) fluorescence of the catecholamine, localized in the ectoneural tissue (black arrows),and the widespread (yellow) fluorescence of the indolamines (white arrows) cannot be distinguished in the black/white print, (x200).

e,f,g) 7th day of regeneration. Differentiation of the epithelia.

e) TEM detail of the coelomic epithelium, (x7000).

f) TEM detail of the epithelium of the ambulacral groove, (x7000).

g) LM detail of the epithelium covering the arm, (x1200).

h) 7th day of regeneration. Immunohistochemistry. Intense 5-HT immunoreactivity of the coelomic epithelium (arrows) visualized using anti-5-HT antibodies, (x1200).

i) 7th day of regeneration. Immunohistochemistry. TH$^+$ cells of the ectoneural tissue (arrows) visualized using antibodies against tyrosine hydroxylase, (x500).

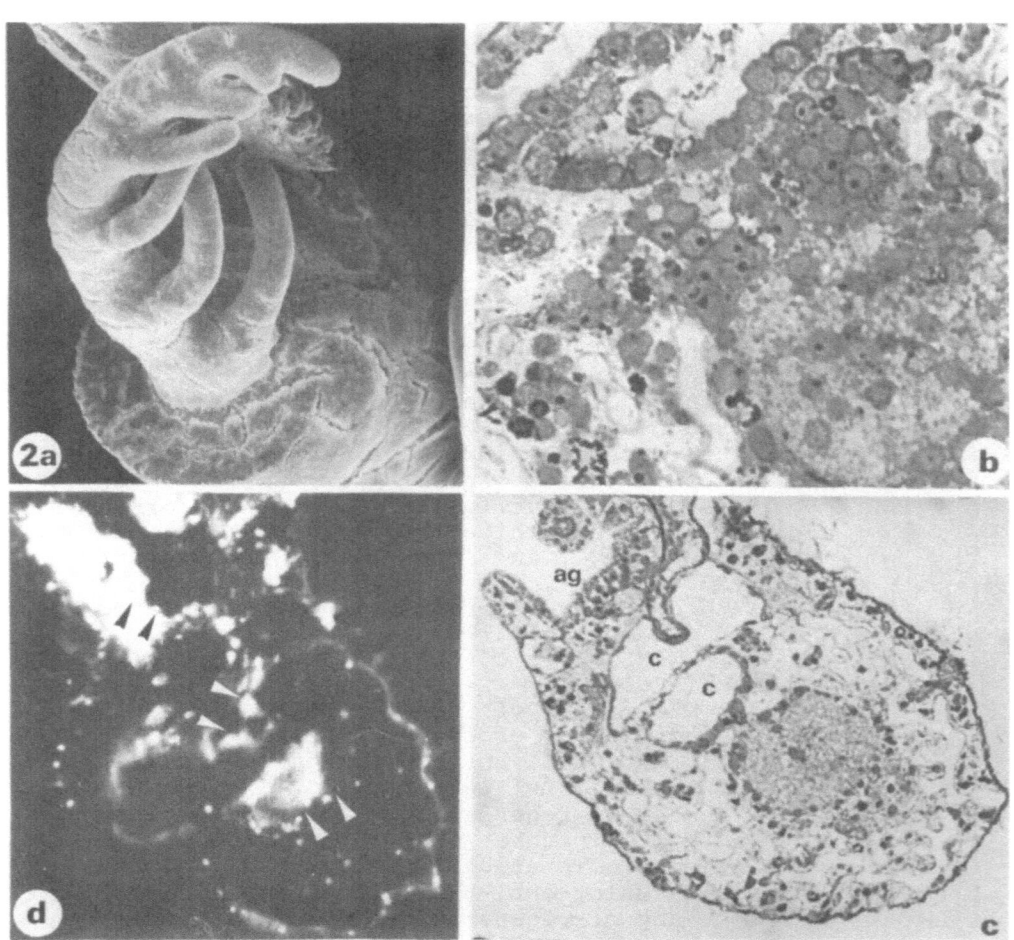

452

Fig. 2. Different aspects of the <u>nervous system-dependent</u> stage.
a) 15th day of regeneration. SEM lateral view showing a well developed regenerating arm with its lateral pinnulae. The growth is typically apical, (x50).
b) 8th day of regeneration. LM detail in cross section to show the proliferation of coelomic cells migrating in the region of the nervous trunk, (x1500).
c) 9th day of regeneration. LM cross section of a whole arm. The structural pattern is more organized. The nervous trunk partially maintains connections with the coelomic epithelium. c=coelomic cavities, n=nervous trunk, ag=ambulacral groove, (x250).
d) 9th day of regeneration. Monoamine histofluorescence. Cross section comparable to Fig. 2c. The distribution of fluorescence is particularly intense in the ectoneural tissue (catecholamine blue-fluorescence, marked by black arrows) and in both the coelom and the nervous trunk (serotonin yellow-fluorescence, marked by white arrows), (x200).

sufficiently differentiated ambulacral groove (Fig. 1c). During this first week, the development of the tissues seems to be limited to three different specialized epithelia characteristic of the normally developed arm: they are the monostratified-monociliated epithelium lining the coelom (Fig. 1e), the pseudostratified-multiciliated epithelium of the ambulacral groove (Fig. 1f), and the monostratified sunken epithelium covering the arm (Fig. 1g). The differentiation of the other tissues occurs later. This stage, therefore, can be considered as strictly <u>coelom-dependent</u>.

Moreover, histofluorescence studies of endogenous monoamines have revealed the presence of serotonin indicated by a typical yellow fluorescence widely diffused in the whole new structure and particularly concentrated at the level of the coelom (Fig. 1d). This has been confirmed through immunohistochemistry using anti-5-HT antibodies. In particular the distribution of 5-HT immunoreactivity is quite intense in some cells of the monociliated epithelium lining the coelom (Fig. 1h). Whether such cells synthesize 5-HT is still unknown. Serotonin could be synthesized and liberated from the nervous system and then specifically uptaken by some proliferating coelomocytes. In addition, the histofluorescence methods reveal the sensitive component of the nervous system, which is associated with the ambulacral groove and its accessory structures as a peculiar subepithelial plexus; this ectoneural component appears to be the only part of the nervous system already developed at this early stage and is characterized by an intense blue fluorescence (Fig. 1d) typically indicating the presence of dopamine in agreement with observations by Gobb (1969) in other echinoderms. These results have also been confirmed through immunohistochemistry using antibodies against tyrosine hydroxylase (TH), the first enzyme of the catecholamine biosynthetic pathway. These experiments indicate the presence of some TH^+ cells in both the ectoneural plexus and the strictly related external epithelia of the regenerative bud (Fig. 1i).

Stage II (up to 2 weeks)

The regenerating arm is more developed since it includes two lateral series of pinnulae and a completely differentiated ambulacral groove (Fig. 2a). The arm, however, is still unable to move or actively react. At this stage the anatomical pattern is characterized by the appearance of the main nervous trunk (the entoneural component of the nervous system), which originates both from migrating coelomic cells, tending to lie peripherally in the trunk (Figs. 2b, c), and from a fibrous central mass, perhaps proliferating from the old stump (Reichensperger, 1912). The nervous trunk successively form conspicuous lateral branches (Fig. 3a) which approach and innervate the presumptive regions where muscles and ligaments begin to differentiate in parallel (Fig. 3b), starting from other undifferentiated cells presumably migrating from the coelom itself. It seems that the nervous system behaves as an inducer/organizer for both developing muscles and ligaments, which acquire an increasingly defined macro- and microstructure and arrangement around the nervous endings, typically filled by a lot of different vesicles (Fig. 3e). The endoskeleton also begins to develop starting from single calcite elements produced and deposited all around in the mesenchyme by differentiating sclerocytes. This stage can be considered as <u>nervous system-dependent</u>. It is characterized by an intense discriminated fluorescence, localized in the two main components of the nervous system (Fig. 2d), indicating a circumscribed presence of dopamine at the level of the sensitive subepithelial plexus, and a more diffuse distribution of serotonin, which nevertheless is highly concentrated at the level of the nervous trunk and its lateral branches (Fig. 3c). Again these results were confirmed through immunohistochemistry, which however gives a more precise distribution of serotonin in the developing arm. Some cells of the nervous trunk are intensely labelled with the 5-HT antibodies, while others, which most probably represent the real motor component of the nervous system and are supposedly cholinergic, are negative (Fig. 3d). Moreover some punctate staining is also observed at the level of the muscles and of the ligaments.

<u>Stage III</u> (3-4 weeks)

At this stage the arm is well developed and is supported by a conspicuous endoskeleton (Fig. 4a), which now consists of a perfect sequence of articulated ossicles and performs a predominant role. The anatomical pattern is complete (Fig. 4b). Only at this step do the muscles form coherent and defined bundles and are attached to their respective skeletal levers. They reach their final functional morphology and the contractile apparatus can actively work (Fig. 4c). The ligaments too reach their complete arrangement, which is comparable to that of the "mutable collagenous tissues" (Wilkie, 1984) typical of many echinoderms. This stage which can be considered as <u>skeleton-dependent</u>, seems to represent an important moment for the assembly and coordination of all the structures involved in movement.

The histofluorescence and immunohistochemical methods applied at this stage do not reveal any qualitative change in the localization of the neurotransmitters if compared to stage II, where the nervous system has already reached its final configuration and contacted its potential targets. Only some quantitative changes can be observed (Fig. 4d). The microscopical and submicroscopical anatomical pattern of the regenerated arm seems, therefore, to be complete. The arm can perform all the functions with all the movements required. Its following growth will not involve any further differentiation but only an increase in length and thickness.

CONCLUSIVE REMARKS

In conclusion in <u>Antedon</u> the earliest regenerative stages seem to be the most crucial for the development and the differentiation of the whole structure. The results seem to unequivocally demonstrate that the coelom and the nervous system are the main "organizers" responsible for the development of all the other structures. These results conversely indicate that the selective local presence of specific aminergic neurotransmitters is an indispensable prerequisite, so that the organizers may exert their inducing action. The control of morphogenesis seems, therefore, to mostly depend on a critical contribution of neurotransmitters, which could play,

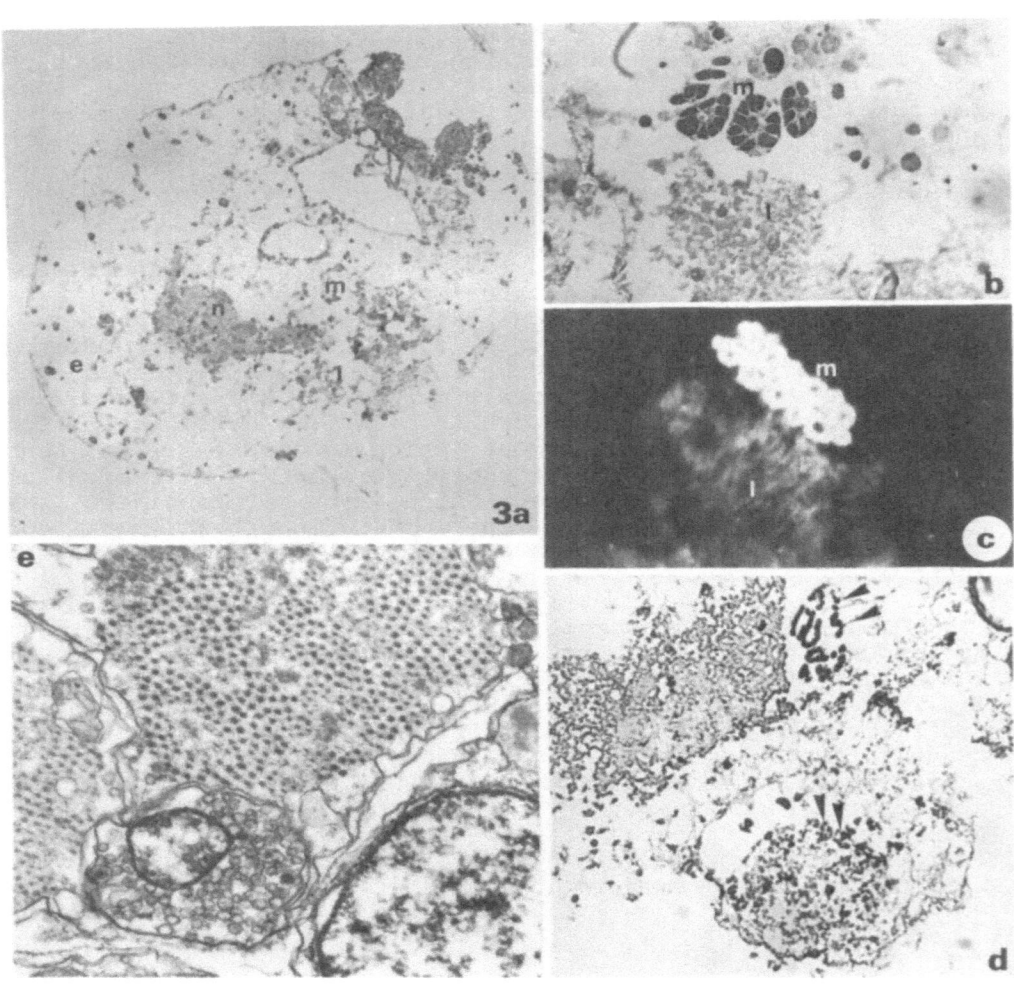

Fig. 3. Different aspects of the <u>nervous system-dependent</u> stage.

a) 12th day of regeneration. LM cross section of a whole regenerating arm. Lateral branches of the nervous trunk innervate the regions where other specialized tissues are developing. n=nervous trunk, m=presumptive muscle, 1=presumptive ligament, e=presumptive endoskeleton,(x220).

b) 13th day of regeneration. The LM detail in cross section shows muscle (m) and ligament (1) differentiating in parallel, (x1000).

c) 15th day of regeneration. Monoamine histofluorescence. Detail of muscle bundle (m) and ligament (1) in cross section. Both these structures are intensely fluorescent (serotonin yellow-fluorescence, x800).

d) 15th day of regeneration. Immunohistochemistry. Detail in cross section showing the distribution of 5-HT immunoreactivity (arrows) visualized using anti-5-HT antibodies. The reaction is positive in some cells of the nervous trunk and in the muscle, whereas is less evident in the ligaments, (x500).

e) 15th day of regeneration. TEM detail in cross section. The muscle fibres are arranged around a nervous ending which contains many different vesicles (x2000).

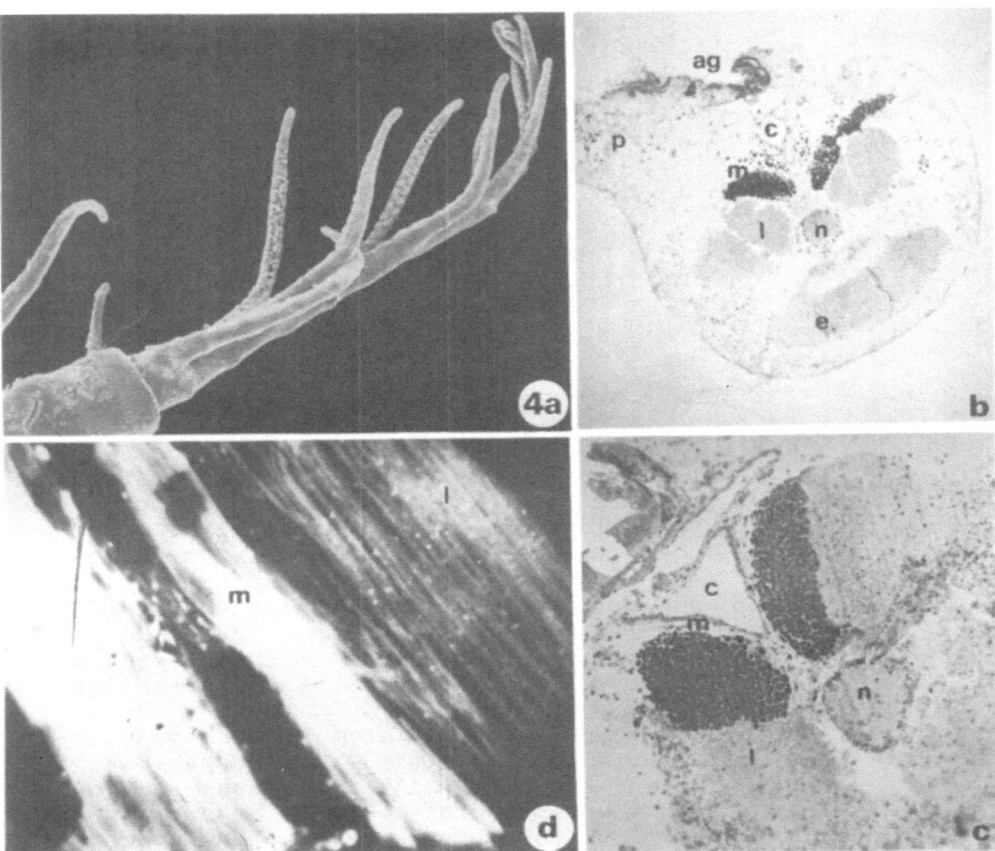

Fig. 4. Different aspects of the skeleton-dependent stage: a) 25th day of regeneration. SEM lateral view showing the new arm perfectly regenerated in its complete functional morphology, (x20); b) 22th day of regeneration. LM cross section showing the detailed anatomy of the regenerating arm. n=nervous trunk, m=muscle, l=ligament, c=coelom, ag=ambulacral groove, p=pinnula, e=endoskeleton, (x80); c) 28th day of regeneration. LM cross section. The internal organized pattern is now completely developed. n=nervous trunk, m=muscle, l=ligament, c=coelom, (x220); d) 28th day of regeneration. Monoamines histofluorescence. Longitudinal section of the arm. An intense fluorescence (serotonin yellow-fluorescence) is localized in the muscle (m) and in the ligament (l) (x500).

before and after the regeneration of the nervous system itself, an important promotional role possibly related to cell movements, as well as to changes in cell shape or cell-to-cell interactions (Franquinet, 1979; Franquinet et Catania, 1979; Huet, 1977; Huet and Franquinet, 1981; Lauder et al., 1988; Lenique and Feral, 1977).

REFERENCES

Candia-Carnevali, M.D. and Saita A., 1985a, Muscle system organization in the echinoderms: II. Microscopic anatomy and functional significance of the muscle-ligament-skeleton system in the arm of the comatulid Antedon mediterranea, J. Morphol., 185:59-74.

Candia-Carnevali, M.D. and Saita, A., 1985b, Muscle system organization in the echinoderms: III. Fine structure of the contractile apparatus of the arm flexor muscle of the comatulid Antedon mediterranea, J. Morphol., 185:75-87.

Cobb, J.L.S., 1969, The distribution of mono-amines in the nervous system of echinoderms, Comp. Biochem. Physiol., 28:967-971.

Cobb, J.L.S., 1986, Neurobiology of the Echinodermata, in: "Nervous System in Invertebrates", M.A. Ali ed., NATO ASI Series A, vol. 14, Plenum Press, New York.

De Biasi, S., Vitellaro-Zuccarello L. and Blum, I., 1984, Histochemical localization of monoamines and cholinesterases in Mytilus pedal ganglion, Histochemistry, 81:561-565.

Franquinet, R., 1979, Rôle de la sérotonine et des catécholamines dans la régénération de la planaire Polycelis tenuis, J. Embryol. exp. Morphol., 51:85-95.

Franquinet, R. and Catania R., 1979, Localization histofluorimetrique et étude microspectrofluorimetrique de la sérotonine et des catécholamines chez une Planaire entière et en cours de régénération, C.R. Acad. Sc. Paris, 289:339-342.

Holland, N.D. and Grimmer, J.C., 1981, Fine structure of syzygial articulations before and after arm autotomy, Zoomorphology, 98:169-183.

Huet, M., 1975, Le rôle du système nerveux au cours de la régénération du bras chez une étoile de mer, Asterina gibbosa Penn. (Echinoderme, Astéride), J. Embryol. exp. Morph., 33:535-552.

Huet, M. and Franquinet, R., 1981, Histofluorescence study and biochemical assay of catecholamines (Dopamine and Noradrenaline) during the course of arm-tip regeneration in the starfish Asterina gibbosa (Echinodermata, Asteroidea), Histochemistry, 72:149-154.

Hyman, L.H., 1955, "The Invertebrates. Echinodermata". Vol. 4, McGraw Hill, New York, Toronto.

Lauder, J.M., Tamir, H. and Sandler, T.W., 1988, Serotonin and morphogenesis. I. Sites of serotonin uptake and binding protein immunoreactivity in the midgestration mouse embryo, Development, 102:709-720.

Lenique, P.M. and Feral, J.P., 1977, Effects of biogenic amines on the regeneration of small pieces of the pedal disc of the sea anemone Metridium senile (Linnaeus), Comp. Biochem. Physiol., 57 C:91-93.

Minckert, A.R., 1905, Regeneration bei Comatuliden, Arch. Naturgesch., 71, (pt. 1):163-244.

Mladenov, P., 1983, Rate of arm regeneration and potential causes of arm loss in the feather star Florometra serratissima (Echinodermata, Crinoidea), J. Canad. Zool., 61:2873-2879.

Perrier, E., 1873, L'anatomie et la régénération des bras de la Comatule, Arch. Zool. Exp. Gen., 2:29-86.

Przibram, H., 1901, Experimentelle Studien über Regeneration, Arch. Entw. Mech. Organismen, 11:321-345.

Reichensperger, H., 1912, Beitrage zur Histologie und zum Verlauf der Regeneration bei Crinoiden, Ztschr. Wiss. Zoll., 101:1-69.

Scholer, J. and Armstrong, W.E., 1982, Aqueous aldehyde (Faglu) histofluorescence for catecholamines in 2 μm sections using polyethylene glycol embedding, Brain Res. Bull., 9:27-31.

Wilkie, I.C., 1984, Variable tensility in echinoderm collagenous tissues: A review, Mar. Behav. Physiol., 11:1-34.

INTERCALARY REGENERATION AND SUPERNUMERARY LIMBS AFTER PROXIMAL OR DISTAL TRANSPOSITION OF THE DEVELOPING LIMB BUD OF THE ANURAN <u>BUFO BUFO</u> IN CONTRALATERAL TRANSPLANTATIONS (*)

Panayotis Kostaridis and Costas Zafeiratos

University of Athens, Zoological Laboratory
Panepistimiopolis, 157 84 Athens, Greece

INTRODUCTION

It is known that developing limbs in Urodeles have the ability for intercalary regeneration. When blastemata of distal origin are grafted proximally without rotation of the transverse axes they interact with the stump so that the deleted skeletal patterns are replaced (Iten and Bryant, 1975; Stocum, 1975). However, when mature hands are grafted to a more proximal level, intercalary regeneration between mature tissues is very poor (Bryant and Iten, 1977, Michael and Hassona, 1982). A similar result is observed in Anurans (Maden, 1981, Michael and Hassona, 1982). Another feature of urodele limb transplantation is the production of supernumerary limbs when distal and proximal tissues are confronted (Iten and Bryant, 1975; Stocum, 1975; Stocum and Melton, 1977; Tank, 1978).

MATERIALS AND METHODS

In this study two types of transplant operations were performed on the developing hindlimb bud of <u>Bufo bufo</u> tadpoles at stage IV (Rossi, 1959). At this stage a complete limb is regenerated after amputation through all levels of the proximo-distal axis of the limb, but if the level of the amputation is very proximal, the regenerating limb is smaller than the normal.

1. Intercalary Regeneration (proximo-distal intercalation)

A total of 14 limb buds were amputated through the distal third and the remaining buds were amputated again in the middle. The second portion of the bud cut off was discarded and the first one was grafted in normal orientation without any rotation of the transverse axes (ipsilateral transplantations).

2. Contralateral Transplantations Inverting the Dorso-ventral Axis

The right limb bud of 15 tadpoles was amputated through the distal third and the left one (of the same animal) through the proximal third. Just after the blastemata were cut off they were cross transplanted inverting the dorsoventral axis. On the right side there was a proximal graft (left) on a distal stump (proximal to distal transplants) and on the left side a distal graft (right) on a proximal stump (distal to proximal transplants). Besides these experimental limbs 20 supplementary limb buds were used as controls. The blastemata when first amputated were immediately placed back at their normal position. The whole limbs were stained with alcian blue and cleared in glycerine.

RESULTS AND DISCUSSION

The results are shown in the Table.

1. Ipsilateral Transplantations

From the table it can be seen that only 2 of the 14 limbs receiving a transplant formed good intercalary regenerates and were completely normal in structure. In all other additional limbs, skeletal elements were missing or they were not well formed.

2. Contralateral Transplantations

Ten out of the 15 amputated animals formed either one or two supernumerary limbs on the right side (proximal to distal transplants) and 11 on the left side (distal to proximal transplants). The missing pieces of the latter experiment of the stump or the graft did not regenerate. When two supernumerary limbs developed one of them always consisted of skeletal elements appropriate to the level of the transplant. The other in most of the cases was appropriate to the level of the stump or it was not well formed (unclear skeletal elements were missing or curved). Most of the supernumerary limbs arose at ventral and dorsal location with respect to the stump although the position of origin was not always clear.

All the above results indicate that the limb bud of _Bufo_ (at stage IV) has a very low or no ability for intercalary regeneration along the proximodistal axis. On the other hand, supernumerary limbs were well formed after contralateral transplantations. This probably means that pattern regulation of cells around the circumference of the limb is independent from interactions along the proximodistal axis.

REFERENCES

Bryant, S.V. and Iten, L.E., 1977, Intercalary and supernumerary regeneration and mature limbs of _Notophthalmus viridescens_, J. Exp. Zool., 202:1-16.
Iten and Bryant, 1975, The interaction between the blastema and stump in the establishment of the anterior-posterior and proximal-distal organization of the limb regenerate, Dev. Biol., 44:119-147.

Table. Development of the Transplanted Limb Buds-Intercalary Regeneration

				Skeletal structure of the operated limbs							Skeletal structure of the supernumerary limbs						
	N	Sup.	Digits	S+Z+A	S+abZ+A	S+A	S+Z+abA	S+abZ+abA	abS+abA	abS+abZ+abA	Z+A	abZ+A	A	Z+abA	abZ+abA	abA	Double sup.limbs
Ipsilateral Transplantations																	
distal to proximal (one side)	14	2	1	2	5	1	2	2	1	1							
		2	2														
		–	3														
		–	4														
		–	5														
		2	polydactyly														
proximal to distal (right side)		5	0								2	2	4	2	4	11**	6
		5	1														
		–	2														
		1	3														
		2	4														
		10	5														
		7	polydactyly														
Contralateral Transplantations (both sides)	15																
distal to proximal (left side)		–	1								–	–	–	–	–	18	4
		1	2														
		3	3														
		5	4														
		6	5														
		3	polydactyly														
		4	regener.*														
Controls	20	2	1														

N=Number of operated animals. Sup.=Supernumerary limbs or digits. S=Stylopodium. Z=Zeugopodium. A=Autopodium. ab=Abnormal (skeletal elements were missing or anomalous). *=In these cases the graft dropped off and a normal limb was formed. **=including spikes.

Maden, M., 1981, Experiments on Anuran limb buds and their significance for principles of vertebrate limb development. _J. Embryol. exp. Morph._, 63:243-265.

Michael, I.M., and Hassona, A.A., 1982, Intercalary and supernumerary regeneration in limbs of stages of the Egyptian Toad, _Bufo regularis_, Reuss, _J. Exp. Zool._, 220:207-218.

Rossi, A., 1959, Tavole cronologiche dello svillupo embryonale e larvale del _Bufo bufo_(L). _Monitore zool. ital._, 66:2-3.

Stocum, D.L., 1975, Regulation after proximal or distal transposition of limb regeneration blastemas and determination of the proximal boundary of the regenerate. _Dev. Biol._, 45:112-136.

Stocum, D.L., and D.S. Melton, 1977, Self organizational capacity of distally transplanted limb regeneration blastemas in larval salamanders. _J. Exp. Zool._, 201:451-462.

Tank, P.W., 1978, The occurrence of supernumerary limbs following blastemal transplantation in the regenerating forelimbs of the axolotl, _Ambystoma mexicanum_, _Dev. Biol._, 62:143-161.

DIGIT REGENERATION IN TRITURUS CRISTATUS

AFTER PATTERN TRANSFORMING INTERVENTIONS

S. Koussoulakos and V. Kiortsis

University of Athens - Zoological Laboratory
Panepistimiopolis 157 84, Athens, Greece

INTRODUCTION

Experimental interventions during limb regeneration in Anurans (Maden 1981) and Urodele amphibians (Wallace, 1981) have been widely exploited for a better understanding of the rules governing the reappearance of the structures removed. It has been suggested and experimentally supported that limb morphogenesis is dictated at every level by a system of positional codings operating along the three conventional axes of the limb, namely the proximodistal (PD), the anteroposterior (AP) and the dorsoventral (DV). These positional codings are thought to be a property of the individual cells and constitute their posititional value, which specifies their fate during development (Wolpert, 1971). The cells of a developing or regenerating urodele amphibian limb become endowed with these codings according to the position they occupy in the growing structure. At each limb level the cells are thought to possess the same PD positional information, whereas there are considerable variations along the transverse axes (Bryant and Gardiner, this volume).

The bulk of the above cited information originates from experiments performed mainly at the stylopod and zeugopod level, leaving digit regeneration out of serious consideration. Digit regeneration in Urodeles was first reported by Bonnet (1777) and pursued further by Tornier (1896). Interest in this field has since fallen until recently, when some significant technical and developmental advantages over stylopod and zeugopod regeneration were described (Smith, 1978; Stock and Bryant, 1981; Alberch and Gale, 1983; Shubin and Alberch 1986). However, reports on growth characteristics of individual digits do not contain much instructive details. Our current knowledge on digit regeneration and their pattern forming abilities, lags far away behind the corresponding data for more proximal limb levels.

The present work was conducted to fill in this gap. In order to achieve our purpose, the following experiments were performed with the hope to answer several questions:

(a) experiment: bilateral amputation of digits I, III, and V in hindlimbs of various-sized animals.
question: is the growth rate of a digit dependent on its initial length (long versus short digits) or on its position (posterior-central-anterior)?

(b) experiment: vitamin A administration.
question: does a digit produce more and/or longer phalanges (proximalization?); does a digit-stump produce a posterized or ventralized regenerate ?

(c) experiment: 180° rotation of digit buds
question: do supernumerary digits appear?

MATERIALS AND METHODS

Animal Care

Adult _Triturus cristatus_ were used throughout. They were kept individually in ordinary tap water at constant temperature (21±1°C) and a 12 h photoperiod. They were fed twice a week on minced beef muscle. Amputations were performed with iridectomy scissors under a dissecting microscope. MS 222 (1:1000) was used to immobilize the animals.

Operations

(1) 21 specimens of both sexes and various sizes were separated in 3 groups (I, III, V) of 7. The animals in each group had their corresponding hindlimb digit bilaterally amputated (i.e. group I, digit I, etc.).

(2) 20 small and 20 big animals had all 3 digits (I, III, V) bilaterally amputated.

(3) 21 animals were amputated as indicated in paragraph (2). They were kept in a solution of vitamin A palmitate (Serva Feinbiochemika, Heidelberg, water-miscible preparation, 15 IU/ml), for 3 (7 animals), 6 (7 animals) and 9 (7 animals) days after amputation. Thereafter they were transfered into tap water.

(4) In 3 groups of 7 animals each, digits I, III, and V were transected, correspondingly. After 30 days the regenerates were reamputated. In each group 6 blastemata (3 left - 3 right) were replaced without rotation, whereas the remaining 8 were rotated and placed on their positions. The animals were kept immobile for 5 days at +2 °C. After that period, animals were returned to their aquaria. Daily inspections were performed to ascertain blastema survival.

Measurements

Camera lucida drawings were made at regular time intervals to monitor digit elongation and morphogenesis. Approx. 400 digits were measured for about 120 days after amputation.

RESULTS

Growth of Regenerates

The growth rates of individual digits in each series displayed considerable variations (from 0.0007 to 0.0017 cm/day for digit I, from 0.0008 to 0.0022 cm/day for digit V, and from 0.0022 to 0.0034 cm/day for digit III). There was no direct correlation between the initial length of a digit and the growth rate of the regenerate. For example, concerning digit I, the mean growth rates for the shorter (0.18 cm) and the longer (0.36 cm) digits were 0.0014 and 0.0013 cm/day respectively. For digit III, the growth rate for the shorter ones (0.44 cm) was 0.0034 cm/day, whereas that for the longer ones (0.85 cm) was again 0.0034 cm/day. The same data for digit V were 0.21 cm - 0.0017 cm/day for the shorter, and 0.48 cm - 0.0016 cm/day for the longer. Table 1 summarizes some representative data concerning the relationship between initial length, elongation and mean growth rates of the individual digits.

Vitamin A Administration

Since the growth of the regenerates was still in progress at the time of writing, we were not able to evaluate the construction of the skeletal elements by Victoria blue staining and clearing of the regenerated parts. However it was obvious that no supernumeraries were induced. The only significant observation was a retardation in the growth rate of the digits by about 30% of the non-treated animals. This retardation was shown to be independent of the time window of drug administration.

180° Rotation

The survival of the transected and replaced blastemata was approximately 40%. Since we have not used any marker, we were not able to detect eventual derotation of the blastemata. No supernumeraries were produced. However, growth rates were considerably affected. Digit I does not elongate at all. The growth rate for digit V remained at 0.0013 cm/day, whereas that for digit III was reduced to 0.0013 cm/day.

Table 1. Correlation Between Initial Digit Length, Elongation and Growth Rates of the Regenerates Among the Various Digits.

Digits	I	III	V
mean initial length (cm)	0.285	0.646	0.335
elongation (cm in 90 days)*	0.103	0.257	0.125
mean growth rate (cm/day)	0.0011	0.0028	0.0013

*It is obvious that the restoration of the initial digit length is a very slow process.

DISCUSSION

It is generally admitted that amputation at every level on a urodele limb results in the production of a perfect replica of the missing structure. However this is not entirely true; if adjacent digits are amputated at the level of the webbing, then a single blastema is formed at the amputation site and a great variety of abnormalities in the regenerated parts is produced (Stock and Bryant, 1981). Moreover, the development of digits is characterized by some histological (e.g. presence of the reticular lamina of the basement membrane under the basal lamina, Salpeter and Singer, 1960) and molecular (e.g. diminution of the reactivity of the monoclonal antibody WE 3, Tassava, 1988) differences when compared to regeneration from other limb levels. Therefore digits make an attractive and promising choice for extending regeneration studies.

In the present work we found that the rate of regeneration does not depend on the initial length of a digit, although the elongation of digit III (usually the longest, 3 phalanges), was faster than that of digits I and V (both 1 phalanx, V>I). Our results coincide with those of Smith (1978) for the longer forearm digit (III). In both works, the net elongation was about 2-2.5 mm in 50 days. In addition, the highest growth rate was observed around the 30th day after amputation.

The effects of vitamin A on the morphogenesis of the regenerated digits are very difficult to evaluate. More or longer phalanges in a digit does not necessarily indicate proximalization. Appearance of a digit I with more phalanges, or a digit III with one phalanx might however indicate posteriorization. Length measurements do not allow us to draw such conclusions at the present time. Nevertheless, the reduction of the growth rate indicates that actually the cells of the regenerating digits are subject to the well established phenomenon of growth retardation after vitamin A administration (Maden, 1983).

Although the number of survived rotated blastemata was not high enough to allow the drawing of confident conclusions, it was obvious that no supernumeraries were produced. We do not yet know whether this is due to lack of positional disparities or to derotation of the blastemata. The retardation of the growth rates could be due to either elimination of some circumferential values or - more probably - to delay in revascularization and reinnervation.

ACKNOWLEDGEMENTS

The authors wish to express their gratitude to the students Panoutsakopoulou V., Mina A., Panagojannopoulou E., Moissi T., Vlassis P. and Dimas K., whose technical assistance helped to generate some of the above data.

REFERENCES

Alberch P. and Gale E., 1983, Size dependence during the development of the amphibian foot. Colchicine-induced digital loss and reduction, J. Embryol. exp. Morphol., 76:177-197.

Bonnet C., 1777, Observations sur la Physique. 1. Mem. Tome X, 388-405, 2. Mem. Tome XIII, 1-18.

Bryant S. and Gardiner D., 1988, Position dependent growth and pattern formation in limb regeneration, (This volume).

Maden M., 1981, Experiments on anuran limb buds and their significance for principles of vertebrate limb development, J. Embryol. exp. Morphol., 63:243-265.

Maden M., 1983, The effect of vitamin A on the regenerating axolotl limb, J. Embryol. exp. Morphol., 77:273-295.

Salpeter M. and Singer M., 1960, Differentiation of the submicroscopic adepidermal membrane during limb regeneration in adult Triturus, including a note on the use of the term "basement membrane", Anat. Rec., 136:27-40.

Shubin N.H. and Alberch P., 1986, A morphogenetic approach to the origin and basic organization of the tetrapod limb, Evol. Biol., 20:319-387.

Smith R.A., 1978, Digit regeneration in the amphibian Triturus cristatus, J. Embryol. exp. Morphol., 44:105-112.

Stock B.G. and Bryant S., 1981, Studies of digit regeneration and their implications for theories of development and evolution of vertebrate limbs, J. Exp. Zool., 216:423-433.

Tassava R., 1988, Limb regeneration in newts: an immunological search for developmentally significant antigens. in: "Regeneration and Development", S. Inoue, ed., Okada, Maebashi, Japan.

Tornier G., 1896/1897. Hyperdaktylie, Regeneration und Vererbung, Arch. Entw. Mech. 3 und 4.

Wallace H., 1981, "Vertebrate limb regeneration", Wiley, Chichester.

Wolpert L., 1971, Positional information and pattern formation, Curr. Top. Dev. Biol., 6:183-224.

ONTOGENESIS AND REGENERATION OF EMBRYONIC HINDLIMB BUD OF
ANURAN AMPHIBIAN (BUFO) AFTER TRANSPLANTATION ON THE TADPOLE
TAIL (*)

S. Zafeiratou, P. Kostaridis, and C. Zafeiratos

University of Athens, Department of Biology
Panepistimiopolis, 157 84 Athens, Greece

SUMMARY

Hindlimb bud from tadpole-donor in the paddle stage was
transplanted as homograft on the tail of tadpole-host in the
cone and five digit stages. The heterochronic development of
the grafted hindlimb bud resulted in differential effects on
their developmental pattern. Smaller or larger limb-segments
than the normal were developed. It was concluded that the limb
size is a parameter independed on the growth and the pattern
formation during the hindlimb development.

INTRODUCTION

There is a lot of information concerning the ontogenesis
and regeneration of the vertebrate limb and the factors which
possibly direct these processes (Faber, 1971; Wallace, 1981;
Muneoka and Bryant, 1984).

As far as the tail and the limbs of anuran Amphibians are
concerned, we know that:

a) The tail of the anuran tadpole appears in the early
larval stages and it is an organ well differentiated,
consisting of muscle cells, notochord, a wide vascular system,
lymphatic-blood, and neural cord. Furthermore, it has
important regenerative abilities and during the climax period
the tail shortens and finally disappears as the animal
acquires the froglet form (Fox, 1973).

b) Hindlimb bud appears as an heterogonic development
going through the stage of cone, paddle and indication of the
digits and at the end of the larval period there is a perfect
leg. We must note that the developing limb has regenerating
abilities which begins to lose quite early when it is small
with poorly formed digits (Marcucci, 1916; Zafeiratos, 1965).

c) The tail and the limbs of the tadpole are self-
organizing systems (Faber, 1965; Weber, 1969; Fabian and
Mervitz, 1973).

d) Anuran limb buds appear to be sensitive to any dose of thyroid hormone starting from zero, but equally degenerating response of the tail tissues does not begin until a considerable concentration of the thyroid hormone is reaching the blood (Champy, 1922; Etkin, 1964).

MATERIALS AND METHODS

a) Hindlimb buds at the stage of cone, paddle and indication of digits from the anuran _Bufo bufo_ and _Bufo viridis_ were amputated near the basis (pelvic area) of the limb (Tscumi, 1957) and were transplanted to the damaged muscles of the tail as autografts (_Bufo bufo_), homografts (_Bufo bufo_) and heterografts (_Bufo bufo_, host and _Bufo viridis_, donor and vice versa).

b) Mesenchyme cells of several cone and paddle limb bud stages were transplanted into the tail-fin and under the tail skin.

Over five hundred animals were used for all the experiments.

RESULTS

1. Mesenchyme cells of the limb bud cone and paddle stages which were transplanted into tail-fin and under tail skin never gave any outgrowth but they were absorbed after a short period of time.

2. The transplanted limb buds on the tadpole tail, continue their development during the whole larval period of the host, even during the climax period when drastic changes on the tail happen and hystolytical activities shrink the tail, thus transporting the grafted limb to the back of the froglet. That limb is generally smaller than the normal of the host and it isn't a functional leg.

3. Hindlimb bud stages cone, paddle and indication of digits which were transplanted as: a) Autografts gave legs little smaller than the normal limb. b) Homografts where the ages of the host and the donor are different, in such a way that we can have young and younger limb buds in old and older hosts and vice versa. The results here are various, depending on the combination of the host and donor ages. Smaller than the normal but complete legs are formed when the host is older than the donor. On the contrary, they are bigger than the normal ones of the host when the grafted limb bud comes from an older donor (Fig. 1).

In Fig. 2 one can see three cases: autografts (A), homografts (H1) and homografts (H2). c) Heterografts when i) host was _Bufo viridis_ and donor was _Bufo bufo_: there was a very small percentage of 3% where the grafts were rejected a few days after transplantation, a percentage of 80% where the host did not develop, but after a long period of time died, and 17% of the grafts developed the legs which had the morphological features of the host (morphology of the limb and distribution of the pigment cells of the skin). ii) Host was _Bufo bufo_

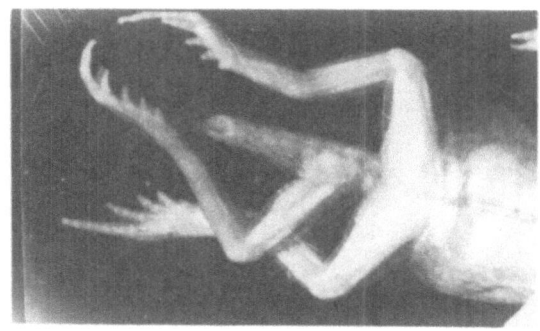

Fig. 1. Homograft: the developed leg on the tail has the femur longer than the normal-control of the host.

and donor <u>Bufo viridis</u>: a percentage of 10% where the grafts were rejected a few days after the transplantation and a percentage of 90% remained and developed the limbs, which indicated morphological similarities to the host limb (morphology of the limb and distribution of the pigment cells on the skin, Fig. 2).

4. Regardless the kind of transplantation we had in some cases the appearance of supernumerary limbs, mirror imaged to the grafted leg, (Fig. 3).

5. Parts of the limb buds from all the stages which were grafted on the tail developed the parts of the limb which were from the level of the amputation towards the distal side of the proximodistal axis.

Fig. 2. Going from left to right: the first leg is the right normal of the host (<u>Bufo viridis</u>), the second is the graft (right leg from donor <u>Bufo bufo</u>) and the third is the left normal leg of the donor <u>Bufo bufo</u>.

Fig. 3. Homograft: the graft limb bud has developed
supernumerary mirror-imaged limb.

DISCUSSION

We have followed the method of transplantation of the
hindlimb buds on the tail, and we wanted to find out which of
the recognized histological limb structures and features
(Zwilling, 1968) could change under these conditions. Thus,
we would estimate the behavior of the morphogenetic features
of the mesoderm cells of the cone and paddle, in the heterochronic supply of the thyroid hormone and the influence of the
new environment of the regenerating tail on the ontogenesis of
the grafted limb.

From the histological analysis of the microscopic
sections in the conjunction areas of the grafted limb with the
tail, we found out that its development starts with chondrogenesis of the pelvis and femur, near the tail notochord
and within the muscles of the tail. No conjunction of the
neural fibres of the limb with that of the tail was found.
This is probably the cause of the lack of function of the leg
graft which was developed on the tail and transformed to the
back of the froglet (Fig. 5).

Although further research is going to take place we can
comment as follows:

1. In autograft transplantations, the developed legs are
always smaller than the normal ones and the control limbs of
the host. According to our opinion these are the results of
the operation, the time which is required of the wound healing
and also the time of the reconnection of the leg with the
tail.

2. The appeared differences in the limb sizes of the
developed legs as homografts (H2) in comparison to the normal
ones of the host and the donor are probably due to the
disturbance of a stoichiometric age-dependent mechanism (Hsu
et al., 1971) which connects the growth of the limb (Tata,
1971; Dournon and Chibon, 1974). For example the older

474

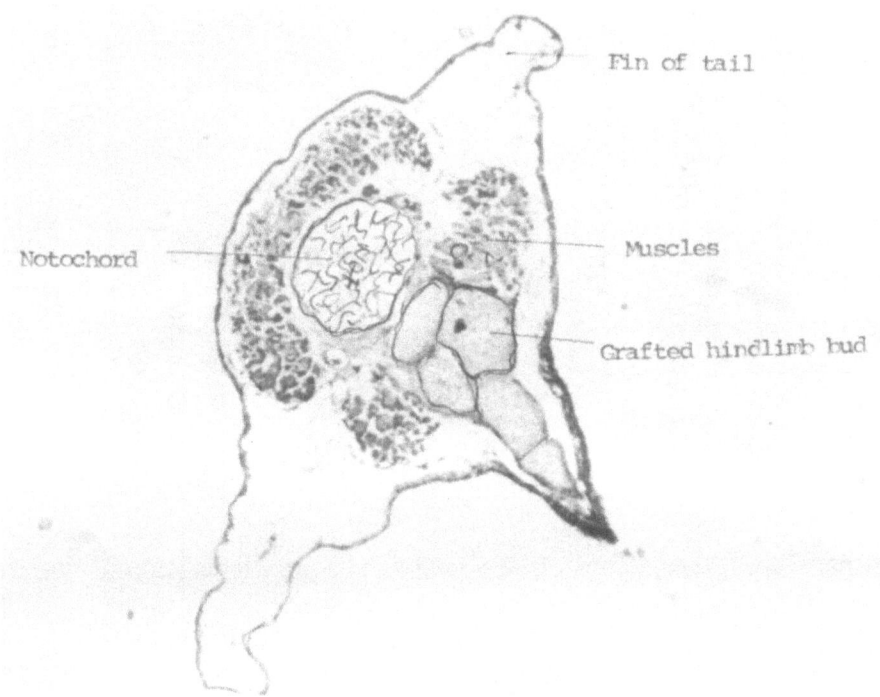

Notochord

Fin of tail

Muscles

Grafted hindlimb bud

Fig. 4. Microscopical section. The grafted limb is developed within the tadpole-tail near the notochrod, into muscles.

tadpoles have increased amounts of thyroid hormone than the younger ones (Etkin, 1964). In our experiments we have very small legs which developed on hosts which at the moment of transplantation were older than the donor. On the contrary, we had bigger legs which developed on younger hosts than the donor. These results support the idea of a feedback mechanism which regulates the cellular activities for the control of the limb growth (Rose, 1960; Rabes, 1969).

3.a. The disability of the limb to continue the development as heterograft (in some cases) is due to the presence of the graft which produces an immune reaction-response by the host's vascular system (Harris, 1941; Edds, 1958).

b. In heterograft transplantations, a large percentage of them failed to develop when the host was _Bufo viridis_, while the opposite happened when the host was _Bufo bufo_. This probably is due to the different time of establishment of incompatibility (Etkin and Harris, 1945).

4. On the heterografted limbs morphogenetic changes are observed: morphogenetic features such as the size of the limb, skeletal pattern and the pattern of distribution of the pigment cells on the skin are in accordance with these of the host limb (either the host is _Bufo bufo_ or _Bufo viridis_). Similar results were obtained on regenerated blastemata on

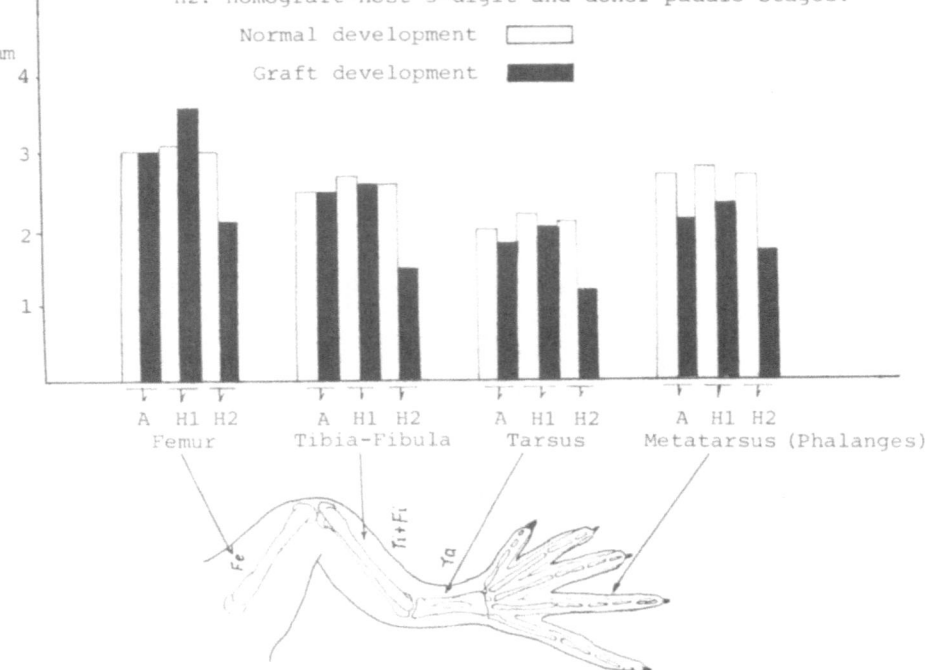

A: Autograft-host and donor in paddle stage.

H1: Homograft-host cone and donor paddle stages.

H2: Homograft-host 5-digit and donor paddle stages.

Normal development ▢

Graft development ▬

Fig. 5. Quantitative changes in segment sizes of the grafted limb, in comparison to the normal one.

Ambystoma from the influence of homologous and heterologous stumps (Lazard, 1959).

Probably these morphogenetic features of the limb, are not established, yet, neither in the mesodermic cells nor in the skin, which are at the stages of cone and paddle, so they undergo quantitative changes under the appropriate conditions.

5. The production of supernumerary limb mirror-imaged to the grafted limb is an important observation because it demonstrates the capability of mesodermic cells (stages cone and paddle) in the amputation area to regenerate a limb far away from the limb field. There is no doubt that an inductive interaction is responsible for this production and should be searched among many causes:

a. In a Polarizing Zone activity in the amputation area.
b. In the activity of the regenerating muscular cells in the area of conjunction with the grafted limb bud.
c. In the presence of the tail notochord near the grafted limb bud.

In all these cases there are tissues with different developmental history and properties which come in close association (Grobstein, 1956; Muneoka and Bryant, 1984).

CONCLUSION

In the regulating factors which control the formation and the size of the hindlimb in anuran amphibian during ontogenesis, thyroid hormone is participating in defined amounts of secretions which are dependent on the age of the tadpole. Changes in the amounts of the thyroid secretions at defined ages of the tadpole can alter the formation and the size of the limb. This means that the differentiation of the limb segments of the tadpoles takes place on defined (from the growth) limb sizes. Still the sizes are not indispensably stable but can be altered during the development of the hindlimb.

REFERENCES

Champy, C., 1922, L'action de l'extrait thyroidien sur la multiplication cellulaire, Arch. Morph. Gen. Exper., 4:1-58.

Dournon, C. and Chibon, F., 1974, Effect of temperature, aging and hormonal conditions (thyroxin) on cell proliferation, in young tadpole and during metamorphosis of toad Bufo bufo, Wilhelm Roux's Archiv. Entw. Mech. Org., 175:27-47

Edds, M.V., 1958, Work conference on Immunology and Development. in: "The Cell", University of Chicago Press.

Etkin, R.M., and Harris, M., 1945, Incompatibility between amphibian host and xenoplastic grafts as related to host age, J. Exp. Zool., 98:35-64.

Etkin, W., 1964, Metamorphosis, in: "Physiology of Amphibian Metamorphosis", J. Moore, ed., New York, Academic Press.

Faber, J., 1965, Autonomous morphogenetic activities of the amphibian regeneration blastema, in: "Regeneration in Animals and Related Problems" (V. Kiortsis, and H.A.L. Trampusch, eds, North Holland, Amsterdam.

Faber, J., 1971, Vertebrate limb ontogeny and limb regeneration: Morphogenetic paralles, in: "Advances in Morphogenesis", Academic Press, New York.

Fabian, B.C. and Mervitz, P.S., 1973, The stimulation of DNA synthesis in regeneration blastemas from the hindlimbs of tadpoles - an organ culture study, in: "Proceedings of the Physiological Society", March, pp. 50-51.

Fox, H., 1973, Degeneration of the tail notochord of Rana temporaria at metamorphic climax. Examination by electron microscopy, Z. Zellforsch. micr. Anat., 138:371-386.

Grobstein, C., 1956, Inductive tissue interaction in development, Advan. Cancer. Res., 4:187-236.

Harris, M., 1941, The establishment of tissue specificity in tadpoles of Hyla regina, J. Exp. Zool., 88:373-397.

Horton, D.J., 1969, Ontogeny of immune response to skin allografts in relation to lymphoid organ development in Amphibian Xenopus laevis Daudin, J. Exp. Zool., 170:449-466.

Hsu, C.V., Yu, N.W. and Liang, H.M., 1971, Age factor in the induced metamorphosis of thyroidectomized tadpoles, _J. Embryol. exp. Morph._, 25:331-338.

Lazard, L., 1959, Influence des greffes homologues et heterologues sur la morphologie des regenerates de membres chez _Amblystoma punctatum_, _C.N. Acad. Sci. Paris_, 249:468-469.

Marcucci, E., 1916, Capacita regenerative degli arti nelle larve de anuri e condizioni che ne determinano la perdita, _Arch. Zool. Ital. Napoli_, 8:89-117.

Muneoka, K., and Bryant, S., 1984, Regeneration and development of vertebrate appendages, _Symp. Zool. Soc. Lond._, 52:117-196.

Rabes, H., 1969, Demonstration and analysis of organ specificity of tissue extracts by radioactive tracer methods, _in_: "Symposia of the International Society for Cell Biology", 7:147-164, Academic Press.

Ranzi, S., 1962, The proteins in embryonic and larval development, _in_: "Advances in Morphogenesis", 2:211-255, Academic Press.

Rose, S.M., 1960, A feedback mechanism of growth control in tadpoles, _Ecology_, 41:188-199.

Tata, J.R., 1971, Hormonal regulation of metamorphosis, _in_: "Symposium of the Society for Experimental Biology", 25:163-181, Davies, D. and Balls, M., eds., Cambridge University Press.

Tscumi, P., 1957, The growth of the hindlimb bud of _Xenopus laevis_ and its dependence upon the epidermis, _J. Anat._, 91:149-173.

Wallace, H., 1981, Vertebrate Limb Regeneration, John Wiley and Sons, Chichester.

Weber, R., 1969, The isolated tadpole tail as model system for studies on the mechanism of hormone dependent tissue involution, _Gen. Comp. Endocr. Sup._, 2:408-416.

Zafeiratos, C., 1965, Influence of the tail notochord on the limb regeneration of Anuran Amphibian, Ph.D., University of Athens.

Zwilling, E., 1968, Morphogenetic phases in development, _Devel. Biol._, Suppl. 2:184-207.

THE EFFECTS OF 4-NITROQUINOLINE-N-OXIDE (4NQO) AT DIFFERENT

STAGES OF TRITURUS LIMB REGENERATION (*)

N.P. Zilakos

University of Athens, Zoological Laboratory
Panepistimiopolis, GR 157 84, Athens, Greece

INTRODUCTION

The concept that an inverse association may exist between the regenerative capacity of an animal and its susceptibility to neoplasia is old (Waddington, 1935; Prehn, 1971) but even in our days it remains popular (Tsonis, 1987). There is a considerable degree of contradiction in this field since various studies have revealed that some carcinogenic substances do not affect, whereas others have shown that carcinogens arrest regeneration (Tsonis, 1983).

MATERIALS AND METHODS

4NQO is a known carcinogen. A quantity of about 10 μm of 4NQO in crystal form was directly administered in the blastema at the following stages of regeneration (Iten and Bryant, 1973) of adult Triturus cristatus:

Stage a: 2-3 days post amputation (dpa). Wound healing stage.
Stage b: 8-12 dpa. In the late dedifferentiation stage.
Stage c: 22-27 dpa. In the late bud stage.

We focused on the following questions:

1. Is 4NQO able to induce any kind of tumors?

2. Does 4NQO affect growth and morphogenesis of the regenerating limb?

3. How is the regeneration of the limb affected by repeated applications (4,10 and 30 dpa) or a high dose (40 μg) of 4NQO?

4. Can the effects of 4NQO be transferred to the re-amputated secondary regenerates?

RESULTS - DISCUSSION

1. No obvious tumor was observed for a period of four months or more after amputation when regenerates were cut off and were prepared for anatomical study of their skeletal elements.

2. (i) At stage b, the regenerative cone emerged dorsally (i.e. from the area at which 4NQO had been implanted), in 20% of the cases and with a great delay (about 55 dpa), in comparison with the control animals that always showed growth along the proximodistal axis.

   (ii) As shown in Table 1, the form of the skeletal elements of the regenerates was particularly affected only in the cases when the carcinogen had been administered in the stages of late dedifferentiation (stage b), and late bud (stage c). A remarkable characteristic is that approx. 50% of the regenerates of stage b, showed abnormalities throughout the regenerates. On the other hand at stage c, approx. 50% of the regenerates showed abnormalities only at the zeugopodium (usually absence of the radius or the ulna). On the contrary at stage a, (wound healing) the cases of abnormal regenerates were rare.

   The above mentioned abnormalities can not be attributed either to the toxic effect of the carcinogen or to the small injury which was caused during the process of the implantation of the carcinogen since it has been noticed on control animals that, toxic or non-toxic salts such as sodium cacodylate sodium chloride, as well as a plain wound of the blastema with a thin needle were not capable of inducing serious skeletal abnormalities.

3. It is possible that a high dose (40 µg 4NQO) shows non-specific toxic effects and in some cases arrests regeneration. Repeated administrations destroy already regenerated structures (see Table 1).

4. Re-amputation of the abnormal regenerates was performed as follows:
   (a) Just proximal to the initial amputation,, led to normal regeneration. This shows the local effect of 4NQO.
   (b) Through the abnormal area, or at the tip, of arrested regenerates. In this case our early results show that abnormal morphogenesis is repeated (Zilakos and Tsonis, unpublished results).

ACKNOWLEDGEMENTS

I would like to thank Prof. V. Kiortsis for encouraging me in this project and Prof. C. Zafiratos for his help. I would also like to express my sincere gratitude to Prof. L. Margaritis for his most valuable help in matters of electron microscopy, during my studies. Finally I would like to thank my good friend P.G. Karadelioglou (W.O. Hellenic Navy), for his help in arranging this text and the general appearance of it.

480

Table 1.   General Morphology of Regenerates

Treatment Morphology of Regenerates	10 μ gr 4NQO at stage a	10 μ gr 4NQO at stage c	40 μ gr 4NQO at stage b	Administr'n of 10 μ gr at 4,10,30 dpa.	Simple puncture	10 μ gr NaCl at stage b	10 μ gr NaCl at stage c	10 μ gr of NaCacody-stage b
Abnormalities in Stylopodium	0	0	0	0	0	0	0	0
Abnormalities in Zeugopodium	1	2	10	1	2	0	0	0
Abnormalities in Autopodium	2	4	2	0	1	3(*)	1(*)	2(*)
Abn. throughout the regenerate	0	11	0	1	0	0	0	0
Arest of the regeneration	0	1	4	3	0	0	0	0
Accessory limbs	0	0	1	0	0	0	0	0
Normally regenerated	11	3	6	22	12	12	6	8
Total number of regen's examined	14	21	23	27	15	15	6	10

(*): These regenerates showed slight abnormalities mainly in fingers.

481

REFERENCES

Iten, L.E., and Bryant, S.V., 1973, Regeneration from differ-
    rent levels of amputation in T. viridescens forelimbs.
    Length, rate and stages, Wilhelm Roux Archiv.,
    173:263-282.
Prehn, R.T., 1971, Immunosurveillance, regeneration and
    oncogenesis, Prog. Exp. Tumor Res., 14:1-24.
Tsonis, P.A., 1983, Effects of carcinogens on regenerating and
    non-regenerating limbs in Amphibia, Ant. Research,
    3:195-202.
Waddington, C.H., 1935, Cancer and the theory of organisers,
    Nature, 135:606-608.

Commentary

by Susan V. Bryant, on

PATTERN FORMATION OF REGENERATES

In the sessions of the conference devoted to pattern formation, a broad variety of approaches, from description of patterns of structures of various types and their reformation during regeneration, through experimental analyses of the pattern forming process, to the design and testing of theoretical models of pattern formation, were presented. It is undoubtedly the case that since the concept of pattern formation spans the gamut from the final complex pattern to its simplest antecedents, that the study of pattern formation will be wide ranging in scope.

Although most of the papers dealt with patterning events associated with vertebrate limb regeneration, a few papers discussed experimental data from invertebrates. Nubler-Jung reported on her interesting studies on the restoration and regeneration of pattern in an insect segment (_Dysdercus intermedins_). She has studied two different features of the segment pattern: the anterior-posterior gradient in pigmentation, and the polarity pattern as demonstrated by the posteriorly pointing cuticular denticles. These two features are products of the underlying, pattern generating mechanisms. Abnormalities in the anterior-posterior gradient created by grafting are repaired either by cell sorting (graft rotation) or when this is topologically impossible, by intercalation of the appropriate intervening pattern elements. When treated with colchicine, the normally uniform posterior orientation of the denticles is disrupted. The disoriented denticles now form small arrays with uniform orientation, which bear a striking similarity to the patterns that form in confluent fibroblast cultures. It was concluded that the polarity pattern repairs itself locally by cell-cell interactions that operate independently of a gradient.

Candia-Carnevali and co-workers presented a description of the different phases of regeneration in the arms of a crinoid echinoderm, _Antedon mediterranea_. Using light microscopy, electron microscopy and immunohistochemistry, they described the progression, through a stage in which the coelom is organized, a stage in which the nervous system is organized and finally a stage in which the skeleton is organized. Both dopamine and serotonin are present during the regenerative process. It will be interesting to look for parallels in the regeneration process of this distant relative of the vertebrates and of vertebrates themselves.

Chandebois presented an interesting, global overview of development and regeneration. Drawing on numerous examples from embryonic development and from her own work on planarians, she emphasized the oversimplification of pattern formation that has occurred in recent years as a result of the widespread use of hypothetical morphogen gradients as explanations. After tracing the origins of gradient concepts to a time when little specific information existed about the cellular activities taking place in development, she presented her view of the emergence of final form. In this view, the final pattern is seen as emerging step by step by means of a progressive series of cellular interactions. Each new developmental event is conditioned by and builds upon the previous ones to generate the final complexity of the differentiated organism or organ.

Intercalary regeneration along the proximal-distal axis of the amphibian limb was the subject of two studies presented at the meeting. Kostaridis and Zafeiratos presented a poster of their work on Bufo bufo. Using animals at stage IV of development, blastemata were transplanted ipsilaterally from a distal to a proximal level, and contralaterally from distal to proximal and from proximal to distal. In most of the contralateral grafts, supernumerary limbs developed as expected, but in the majority of both types of grafts, no intercalation along the proximal-distal axis occurred. According to the authors, normal regeneration can occur from all proximal-distal levels at this stage in development. Hence it would appear that the abilities to regenerate from an amputation plane and to form supernumerary limbs, can be separated from the ability to intercalate along the proximal-distal limb axis.

Anton also presented a number of experiments concerning intercalary regeneration along the proximal-distal axis of limbs. By transplanting the distal part of the limb of Triturus alpestris to a more proximal level, he showed that as limb development proceeds, the ability for intercalary regeneration declines and the degree of abnormality increases. Anton has shown that grafts between differentiated limb parts are unable to interact to form intercalary regenerates. He was further able to demonstrate that if distal limb grafts were damaged by amputation of the digits, then intercalary regeneration once again occurred. It was proposed that the removal of digits stimulates dedifferentiation of the graft, which, by some means as yet not understood, facilitates the necessary interactions for intercalation. Since it has been shown previously by Pescitelli and Stocum that the mesodermal component of the intercalary regenerate is derived from the proximal partner in the interaction, presumably the dedifferentiated state of the graft influences the stump cells to also dedifferentiate and then to intercalate the appropriate missing parts of the pattern. The finding in cross-species grafts between Salamandra salamandra and Ambystoma mexicanum that pigment cells of distal graft origin can be found in the intercalated region could be a reflection of the migratory potential of these epidermally located and controlled cells.

In the paper presented by Papageorgiou and colleagues, experimental investigations into predictions of the polar coordinate model (PCM) and Papageorgiou's extended version of the PCM, known as the hierarchical PCM (HPCM) were presented. According to the HPCM, ipsilateral small angle (90°, 270°) blastema rotations should lead to a low frequency of supernumerary limbs (lower than for 180° rotation). These supernumerary limbs should contain fewer digits than normal limbs, and they should be mainly double-ventral or double-dorsal in character. Contralateral grafts at small angles of rotation, such that the DV axis of the graft coincides with the AP axis of the stump, are predicted to lead to supernumerary limbs of normal structure and stump handedness. Results of experiments of <u>Triturus cristatus</u> by Papageorgiou and colleagues were completely in accord with these predictions. It was concluded that the PCM and HPCM provide a more suitable description of amphibian supernumerary limb formation than do models based on cartesian coordinates. Another interesting facet of HPCM is its ability to account for apparent anatomical discontinuities in certain experimentally generated limbs by proposing that sometimes twisted contours will occur.

Bryant and Gardiner presented recent results on regenerating axolotl limbs in the context of the polar coordinate model, according to which growth and pattern formation are coordinately controlled. Experiments which show that regenerative ability in <u>Xenopus</u> is an intrinsic property of limb cells were reported. Position-dependent growth was used to map the internal positional values of the limb, which surprisingly turned out not to be a simple projection of peripheral information onto the center. Position-dependent growth was also used to show that both the patterning and cell signalling mechanisms involved im limb outgrowth are compatible in urodeles and anurans. The evidence from grafting and cell marking experiments that lead to the conclusion that fibroblasts are the limb cells with positional information and the ability for position-dependent growth was presented. Preliminary studies that demonstrate position-dependent growth <u>in vitro</u> were described. Putting it all together, position-dependent growth was shown to be capable of accounting for the initiation, maintenance and the termination of growth during limb regeneration.

The paper of Koussoulakos and Kiortsis dealt with a little studied aspect of urodele limbs: regeneration of the digits. Their results show that the rate of regeneration from long and short digits is not dependent upon the initial length of the digit. Further, as would be expected if digits are symmetrical in terms of positional values, and contain only a subset of the angular positional values present at more proximal levels, 180° rotation of digit regenerates did not provoke the formation of any supernumerary structures.

Two papers were presented that dealt with the regenerative abilities in the limbs of vertebrates other than urodeles. Michael presented a retrospective survey of his studies on regenerative ability and regenerative decline in <u>Bufo regularis</u> limbs. As in other anurans, regenerative ability is lost in a proximal to distal direction during development. Abnormal limbs formed during the period of decline show a distal to proximal sequence of abnormalities, the least affected limbs showing reductions in the phalanges and the most affected, spike-like outgrowths. Comparison of the regenerative abilities of fore- and hindlimbs showed that in animals of the same stage, a distal forelimb amputation regenerates as well as a proximal hindlimb amputation. This cranial to caudal difference in regenerative abilities may reflect the fact that at early stages of development the forelimb is morphologically more advanced than the hindlimb. Michael also presented some data on mechanical and chemical stimulation of enhanced regeneration in stages that normally do not regenerate. The mechanism by which this trauma is effective in prolonging regenerative ability is not known.

Zafeiratos and colleagues presented a poster of their studies on the behavior of _Bufo_ limb buds grafted to the tail. They found that limb buds continue to develop normally in their new location, and that in cases where older buds were transplanted to younger larvae, the grafted limbs were at all times larger than the host limbs. Supernumerary limbs were sometimes observed, indicating that the host tail site provided the necessary positional cues. Limb buds were also able to regenerate in their new location on the tail, demonstrating again the limb-intrinsic nature of regeneration.

The relationship between tumorigenesis and regeneration was investigated in a study by Zilakos. After treatment of the amputated limbs of _Triturus cristatus_ with a known carcinogen (4-nitroquinoline-N-oxide: 4NQO) tumors did not develop, but some limbs regenerated abnormally. Re-amputation through the affected limb region resulted in a second round of abnormal regenerates, suggesting that the abnormalities generated by 4NQO are stable features of the pattern.

Muneoka presented a paper which reviewed his pioneering attempts, in collaboration with Wanek and Bryant, to study mouse limb development _in situ_. He reported on the _exo utero_ technique for surgery, and on studies which have staged and fate-mapped the limb bud during stages that are accessible for experimental manipulation. In addition, he described experiments that demonstrate the regulative ability of mouse limbs following amputation, or grafting of an anterior wedge of tissue into a posterior location. In the case of amputation, regeneration of the peripheral digits (1 and 5), as well as digit 2 can occur when the rudiment for the digit is transected. After wedge grafting, supernumerary digits develop as a result of positional confrontations. Finally, he presented some preliminary data (in collaboration with Chris Trevino) using retrovirally marked 3T3 cells injected into different positions in the limb bud to act as reporter cells for differences in the growth promoting ability of the different regions. A clear difference in the labeling index of proximally (13.1%) and distally (20.7%) located reporter cells was shown, which parallels but is larger than the difference between the labeling index of proximally (11.8%) and distally (16%) located limb cells. With the combination of an ability to intervene in mouse development, and the innovative use of cell markers it is likely that important new information about limb development and regeneration in the mouse will be forthcoming.

It is clear that the more traditional approaches to the study of pattern formation, that is, grafting and deletion experiments, continue to provide new clues about the underlying processes. In addition, theoretical treatments of pattern formation continue to guide and inform the design of experiments. With the ready applicability of the tools of molecular biology, it is anticipated that in the next decade we will witness a shift in emphasis towards studies of the molecular nature of pattern forming signals and their response systems. Progress in this area will be greatly assisted by the interaction that retinoids have with the pattern forming system; by the development of _in vitro_ assay systems for position-dependent growth; and by the much needed development of position-specific cell markers.

GENERAL DISCUSSION

by Hugh Wallace

The final session of the workshop was deliberately left
open to allow a wider discussion of topics raised in earlier
sessions or even of matters which anyone felt had been
neglected, entirely at the discretion of the participants.
This democratic procedure worked surprisingly well, as the
discussion was subtly propelled from one topic to another by
the two chairpersons (Tassava and Bryant). Their initial
invitation to us to identify the major subjects for review
determined the course of the session. Only a lack of time and
stamina prevented us from exploring further the relationships
of regeneration to the immune system or to cancer, or from any
critical analysis of how best to approach pattern formation
and morphogenesis.

The session started with an explanation by Brockes of how
hypomethylation at specific sites of particular structural
genes could be used to trace the conversion of cells from one
differentiated state to another during regeneration. He
illustrated this by a newt myosin gene (whose characteristics
will be published elsewhere) which indicated a contribution of
myocytes to regenerated cartilage that was much too great to
be explained by contamination. The general applicability of
this technique excited even those of us who considered there
was already overwhelming evidence for cellular transformation
or metaplasia, and stimulated the doubters into a deeper
examination of what we really mean by the term dedifferentiation. It was agreed to be essentially a histological description and some clearly wished to confine it to tissues, while
conceding the logic of extending the term to those cells
released into the blastemal mesenchyme. Anton reminded us
that such cells synthesize nucleic acids and proteins and show
no sign of degeneration. Bryant agreed that dedifferentiation
is easily observed but suggested that its importance has been
exaggerated, as most of the mesenchyme is normally derived
from dermal fibroblasts which were not greatly differentiated
beforehand and which could impose their morphogenetic pattern
on the minority of cells coming from other tissues. Stocum
then pointed out that appearances are deceptive, in that most
dermal fibrocytes have secreted a matrix with collagen fibrils
and consequently are quite as differentiated as any other cell
type. Their escape from the dermis is therefore just as much
dedifferentiation as seen in other tissues. Egar mentioned
her suspicion that some references to fibroblasts were merely
cryptic appeals to reserve cells, but nobody present was prepared to admit the existence of such cells or even of neo-
blasts in planaria. So perhaps that myth is now dead.

Some criticism was voiced of previous attempts to demonstrate metaplasia, particularly regarding the possibilities
that satellite cells might travel from the shoulder to reform
muscle whenever grafted tissue allowed the regeneration of an
X-irradiated arm stump, or that the graft might stimulate some
recovery of locally irradiated cells. Wallace discounted such
explanations, but some doubts evidently remain and will only
be dispelled by the use of more precise cell markers. Tassava

enquired if muscle satellite cells could themselves show metaplasia. Carlson replied he knew of no evidence (which we all
interpreted as meaning no evidence exists) of that, but considered the recent successful cloning of satellite cells in
culture should make such studies feasible for the first time.
Lemoigne amplified on this theme, relating that cultures
derived from muscle tissue displayed aberrant, multinucleate
and polyploid cells initially, but the satellite population
defined by continued proliferation mainly differentiated into
myotubes in older cultures. Tassava also asked if there was
not a contradiction between the extrinsic control of muscle
regeneration described by Carlson and the intrinsic control of
limb blastemata championed by Bryant. The latter saw no contradiction, if myoblasts were democratically subservient to
the majority fibroblast population in terms of pattern formation as outlined previously.

That seemed to bring the discussion back full circle,
until Tassava provocatively suggested that limb regeneration
also depends on extrinsic factors such as growth factors and
the nerve supply. Bryant countered that she regarded them as
permissive factors which allowed cellular proliferation but
did not dictate morphogenesis. That view gained support from
participants who mentioned that partial denervation indicates
an excessive supply is normally present and that supplementing
it with extra ganglia had little effect on growth or pattern,
or even on the frequency of cells entering S phase. The cellular response to external agents might be determined by its
specific receptors, whereas the immediate contact between
adjacent cells seemed of even greater importance to several
participants. This healthy divergence of opinion probably
reflected the various aspects of regeneration on which different speakers had focused their attention. I doubt if we could
ever agree on their relative importance.

Cox rather innocently remarked that it would be simpler
and more logical to consider growth by cellular proliferation
separately from the pattern formation that underlies morphogenesis. This met considerable resistance from those who
considered (but had not proved) these two processes must be
intimately connected, in the belief that positional discrepan-
cies provoke cell division - except in mature tissues as Anton
pointed out. Carlson attempted to elicit some evidence for the
debate by asking how new patterns arise without growth in planarian morphallaxis. Chandebois was not sure that she understood the question, so reminded us that planaria are routinely
starved during regeneration but still exhibit growth. She
thought the number of dedifferentiated cells increased for at
least two weeks near the cut surface of the worm and stated
that at least two populations of undifferentiated cells were
present then. Lemoigne also remembered that cell division
reached a peak as early as ten hours after bisecting a planarian. This exchange provoked two thoughts among a rather bew-
ildered audience. Firstly, measuring planaria must be like
asking how long is a piece of elastic and, secondly, most of
us simply did not know enough about planarian regeneration to
form a judgement concerning either recent or ancient disputes
about it. The miniature structures produced by newt arms
which had been denervated late in regeneration were agreed to
be irrelevant to the issue, as their pattern had surely been
determined earlier. Tassava requested Brockes to comment on

the reaction of his notorious antibody 22/18 to blastemal mesenchyme but not to the apparently equivalent cells in aneurogenic regenerates. Brockes emphasized that the lack of binding to limb bud mesenchyme or the slightly older aneuro- genic mesenchyme of _Pleurodeles_ larvae must indicate that nerves impose a new identity on the mesenchyme of normal limb regenerates, where the reaction is widespread.

The discussion was then redirected to the section of articles on retinoids. Maden elaborated on the evidence that cytoplasmic binding proteins lacked specific DNA-binding sites, whereas the more recently discovered retinoic acid receptor does. His comment that binding proteins could serve as a cytoplasmic buffer or reservoir was supported by Stocum, who referred to a direct binding of retinoic acid by a recep- tor in leukaemia cells which possessed little or no binding protein. Our hopes that retinoic acid might turn out to be one of the elusive morphogens were tempered by the revelation of several complicating factors. Maden explained that cells generally make their own retinoic acid by converting retinol which they gain from the circulation, and so do not need an external supply. Stocum remarked that the available measure- ments of binding proteins do not distinguish between the mole- cules which are bound to retinoic acid and those which are not. Lheureux and Bryant both questioned the assumption that retinoids must exert their effects on regeneration by directly altering gene transcription. They mentioned an apparently direct effect on surface proteoglycans, gap junctions and cell motility. Retinoic acid has also been reported to stimulated macrophages to secrete interleukin 2, and that could apply to other growth factors. Finally, Bryant enquired how the con- centration of a single molecule could affect all three axes of a limb regenerate. Maden suggested rather tentatively that retinoic acid might accentuate any inherent gradient but admitted the problem was less apparent in chick limb buds, where only the antero-posterior axis seemed to be sensitive to retinoic acid treatment. Despite such reservations, the meet- ing expressed enthusiasm for this novel means of analysing morphogenesis. Niazi's persistence over the past decade has produced a notable revolution in our subject and we wish him further success in whatever form of life he adopts on retire- ment.

Turning briefly to the subject of cell division in blastemata, Maden wondered if the postulated existence of transiently quiescent cells would make all cell cycle measure- ments meaningless. Tassava disagreed emphatically, as the transitions between active cycling and quiescence would be relatively infrequent but conceded that such transitions may have produced minor overestimates of the phase durations. Nobody, not even Geraudie, could account for the irregular distribution of these apparently quiescent cells after pro- longed thymidine labelling. A local absence of nerve endings seemed an unsatisfactory explanation, because neurotrophic factors (Singer insisted there must be at least two of them) certainly diffuse fairly readily into cultured blastemata. Brockes then attempted to turn the whole neurotrophic theory on its head by demanding how we could exclude the possibility that denervation provoked Schwann cells to produce an inhibi- tor of cell division, which persists until the Schwann cells redifferentiate around regrowing axons. The novelty of his concept was too great for anyone to recall that Schwann cells themselves divide under these circumstances (but so do root cells of the autumn crocus while it exudes colchicine). Muneoka preferred to consider that the proteolytic enzymes secreted by nerve sprouts prevent any excessive accumulation of matrix and thus allow cell proliferation to continue. Mescher countered this by stating that nerves actually stimulate the production of a novel matrix enriched in glycosaminoglycans. Perhaps Taban was the most receptive to Brocke's heresy, commenting that either stimulatory or inhibitory mechanisms could be accommodated in a general theory of nervous information controlling regenerate growth.

By then, however, our attention had been diverted to the possible functions of matrix components. Singer considered they could provide the basis of positional information. That idea evidently appealed to both Cox and Mescher, although the latter cautioned us about our ignorance of most matrix components. That did not deter us, especially as information and ignorance are the basis of all speculation. The swelling which results from secreted hyaluronidate could result in a gel which facilitates growth, according to Holder. Bryant agreed with Mescher's question that the matrix could at least provide a pathway for the apical migration of mesenchyme cells and possibly provide space for the later expansion of the blastema. Bouilly reiterated his observation of fibroblast growth factor binding to heparin and thought that might apply more generally to other growth factors being trapped in the matrix. Mescher developed this theme by considering the secretion of heparin sulphate and proteoglycans from the wound epithelium should sequester the growth factors and thus stimulate a local cell proliferation. Liversage asked for more details of the nervous control of hyaluronidate secretion. Mescher responded that the hyaluronidate component of the matrix decreases when cells stop dividing after denervation, but then recovers as nerves grow back into the blastema and cell division is resumed.

Before we completely ran out of time and steam, Tassava informed us we had an obligation to consider the future direction of research into regeneration. Kiortsis reinforced, this appeal by quoting and translating an aphorism of Guyénot to the effect that any research seems to proceed smoothly until coming up to a brick wall. What recommendations could we offer about old unsolved problems or emerging new ones? I believe we were all aware that an apparently satisfactory answer to one question will inevitably uncover several new ones, hence the advantage of rhetorical questions. It was already evident that our best prospects lay in the molecular-genetic approach. The labour involved had been touched on during the meeting, but it surely provides a means of characterising cell-types and probably of identifying molecules involved in the cellular interactions which seem to regulate both blastemal growth and morphogenesis. Taban added that the cellular receptors for such molecules were equally important and that we had to find a way of deciding which were likely to be the most rewarding, rather than searching for them indiscriminately. It gradually emerged during this discussion that there was a lack of unanimity over several problems which many of us thought had been settled years ago, but now found our solutions were in conflict or our attempted demonstrations were not universally accepted. That is surely the main strength of a small workshop in fostering uninhibited but amicable debate, while sequestered from normal distractions. There were no obvious Pauline conversions at the meeting, but many of us must have been stimulated into reconsidering our cherished beliefs and revising our research priorities. I confidently expect the initial outcome will be an upsurge of Epistles to the Corinthians and perhaps some later Revelations. One necessary weakness of a small workshop was also apparent, in that the emphasis on vertebrates and particularly their limbs entailed a comparative neglect of invertebrates. I was enthralled by the accounts of crinoid regeneration and the perturbed pattern visible in a regenerated insect exoskeleton, and intrigued by several of the posters. Our failure to relate them to the general discussion surely reflects a shortage of expertise rather than any lack of interest.

Our sympathy was mostly directed to Koussoulakos, whose epic struggle to subdue a recalcitrant slide-projector culminated in wrestling with two of them simultaneously. Unlike Laocoon, he emerged unscathed from that but may be permanently scarred by the multitude of minor troubles we all caused him. Our admiration and gratitude are not enough. The Gods (and Kiortsis assured me that NATO is the God of Peace) should reward conference organizers, or at least provide pensions for the widows and orphans of less resilient organizers than those responsible for the unqualified success of this workshop on regeneration research.

AUTHOR INDEX

INDEX